MARINE MACROECOLOGY

MARINE MACROECOLOGY

JON D. WITMAN AND KAUSTUV ROY

The University of Chicago Press : CHICAGO AND LONDON

JON D. WITMAN is professor in the Department of Ecology and Evolutionary Biology at Brown University. **KAUSTUV ROY** is professor in the Section of Ecology, Behavior, and Evolution at the University of California, San Diego.

The University of Chicago Press, Chicago 60637
The University of Chicago Press, Ltd., London

Printed in the United States of America

17 16 15 14 13 12 11 10 09   1 2 3 4 5

ISBN-13: 978-0-226-90411-5 (cloth)
ISBN-13: 978-0-226-90412-2 (paper)

ISBN-10: 0-226-90411-3 (cloth)
ISBN-10: 0-226-90412-1 (paper)

Library of Congress Cataloging-in-Publication Data

Marine macroecology / Jon D. Witman and Kaustuv Roy.
p. cm.
Includes bibliographical references and index.
ISBN-13: 978-0-226-90411-5 (cloth : alk. paper)
ISBN-13: 978-0-226-90412-2 (pbk. : alk. paper)
ISBN-10: 0-226-90411-3 (cloth : alk. paper)
ISBN-10: 0-226-90412-1 (pbk. : alk. paper) 1. Marine ecology. 2. Macroecology. I. Witman, Jon D. II. Roy, Kaustuv, 1959–
QH541.5.S3M2829 2009
577.7—dc22

2009004545

♾ The paper used in this publication meets the minimum requirements of the American National Standard for Information Sciences—Permanence of Paper for Printed Library Materials, ANSI Z39.48-1992.

CONTENTS

INTRODUCTION

JON D. WITMAN AND KAUSTUV ROY

For over two centuries, biologists have been documenting how patterns of biological diversity vary in space and time, and have been trying to understand the processes that produce such trends. In fact, the observation that species diversity is nonrandomly distributed across the globe, with tropical areas harboring many more species and higher taxa compared to high latitudes, may be the oldest documented pattern in ecology (Hawkins 2001; Turner 2004). Attempts to understand the causes of such spatial gradients in biodiversity have occupied biologists for an equally long time, starting with Alexander von Humboldt and Alfred Russel Wallace. But despite such a long history, studies of large-scale patterns in the distribution and abundance of species had lacked a common focus. As Robert MacArthur pointed out in his book, *Geographical Ecology*, ecologists tended toward a "machinery oriented" approach to the problem, concentrating on understanding contemporary processes that structure communities, while evolutionary biologists, paleontologists, and biogeographers largely emphasized the role of historical processes in shaping present day biodiversity patterns. This difference in focus also led to differences in the spatial and temporal scales of analyses. The vast majority of ecological studies have been carried out on spatial scales of a few to tens of meters and have lasted less than five years (Levin 1992; May 1999), while biogeographers and paleontologists have sought to understand how processes unfolding over thousands to ten of thousands of years have shaped the biodiversity patterns we see today. This problem of scale remains a major obstacle to process-based understanding of large-scale ecological and biodiversity patterns. Since the pioneering work of Arrenhius (1921) and Gleason (1922), showing that species diversity increases as a function of area, ecologists have known that ecological patterns depend on spatial scale of study and have acknowledged that the "scale of resolution chosen by ecologists is perhaps the most important decision in their research program, because it largely pre-determines the questions, the procedures, the observations and the results." (Dayton and Tegner 1984, page 458). Thus it is difficult, if not impossible, to understand the large-scale patterns and processes—so crucial for making generalizations in ecology—by extrapolating from small-scale studies conducted at local sites (Brown 1995).

Macroecology, as defined by James H. Brown and Brian A. Maurer in a 1989 paper in *Science* (and subsequently expanded by Brown in a 1995 book), seeks to bridge this gap in resolution and emphasis between ecological and historical approaches to understanding large-scale ecological patterns. More specifically, the goal is to gain "insights that can come from applying the questions posed by ecologists to the spatial and temporal scales normally studied by biogeographers and macroevolutionists." (Brown and Maurer 1989, page 1149). As originally defined, macroecology is explicitly empirical in nature, largely involving statistical analyses of information on species abundance, diversity, body size and range size (Brown and Maurer 1989, Brown 1995). Subsequently, the scope of macroecology has been expanded to include the development and empirical tests of process-based models that seek to explain large-scale ecological patterns (Brown et al. 2004; Holyoak et al. 2005; Connolly, this volume). More recently, the increasing feasibility of conducting and replicating experiments on large spatial scales has also opened up the possibility of experimental tests of some of these models, an endeavor that can be called experimental macroecology (Witman and Roy; Sanford and Bertness; Connell and Irving, in this volume).

Since its inception, macroecology has largely been a terrestrial endeavor. Brown and Maurer (1989) viewed macroecology as "the division of food and space among species on *continents*" (emphasis added). And the first book devoted to the topic, *Macroecology*, (Brown 1995) defined the field by elegantly outlining the macroecological perspective based on continental scale analyses of terrestrial biota. The second book, *Patterns and Process in Macroecology*, by Gaston and Blackburn (2000) expanded the discussion with detailed treatment of patterns of species richness, abundance, body size, and range size using British birds as an example. Similarly, only two chapters out of twenty-one of a subsequent edited volume on macroecology (Blackburn and Gaston 2003) are on marine organisms. Syntheses of some marine macroecological relationships can also be found in Kritzer and Sale (2006), although the main focus of that volume is on marine metapopulations. A February 2008 search of the primary literature using the ISI Web of Science database using the term "macroecology" as a topic yielded 383 papers, while the term "macroecology and marine" yielded only thirty-five papers. Both of these numbers are serious underestimates—there are clearly more than thirty-five papers on marine macroecology, as evidenced by the citations in this volume—apparently reflecting the fact that many macroecological studies do not use the word *macroecology* in the title or abstract. Nonetheless, our experience with the literature suggests that this order of magnitude difference would hold even if one did a more comprehensive search. On the other hand, it is also clear that interest in quantifying and understanding large-scale diversity and ecological patterns in the sea has been rapidly increasing over the last two decades (Witman and Roy, this volume), and such analyses are becoming feasible as global-scale environmental, biogeographic, and ecological data become increasingly available. The time is thus right for us to ask not only where marine macroecology stands today but also to lay out the prospects and opportunities that lie ahead.

To achieve that goal, we invited a group of scientists whose work has led to new and interesting insights into large-scale ecological patterns in the ocean to not only provide an overview of our current knowledge of marine macroecology, but also to set an agenda for future research in marine macroecology.

Overview of This Volume

Part 1 of this volume starts with an overview of marine diversity patterns by Valentine that summarizes large-scale patterns of species distribution and diversity in the world oceans as well as the environmental and ecological framework for understanding the processes responsible for large-scale diversity gradients. The rest of the chapters in this section are focused on specific habitats and/or taxonomic groups, ranging from the pelagic to the deep sea and from microbes and invertebrates to fishes and algae. Given the tremendous differences in the environmental conditions and habitat structure of, say the deep sea versus the pelagic ecosystems, or the differences in life habits and ecologies of sessile benthic invertebrates versus fish or algae, it is an open question whether macroecological relationships are generalizable across various groups or from the surface to the ocean trenches. Is the slope of the relationship between body size and abundance similar for all the different groups of organisms living in the ocean? What about the distribution of body sizes or geographic range sizes of species living in the shallowest waters or in the plankton versus those in the deep sea? At this point, the data needed to answer many such questions are lacking, but a comparison of the information available for individual groups and habitats, such as those discussed in the individual chapters in this section, should provide some preliminary insights. Chapter 2, by Li, not only provides quantitative information on spatial patterns of diversity, abundance, and distribution of marine plankton, but also conveys how technological innovations are opening up exciting new opportunities for macroecological analyses of microorganisms living in the open ocean. And the data currently available show that while some macroecological relationships, such as the allometric scaling of abundance and body size, may indeed be generalizable from marine plankton to fish—others may not. Chapter 3, by McClain, Rex, and Etter, provides an overview of macroecological relationships in the least-known part of the world oceans—the deep sea. Despite being the largest habitat on the planet, biological diversity in the deep sea remains very poorly known because of the logistical challenges of working there. As McClain, Rex, and Etter point out, even though deep-sea macroecology is still in its infancy, available data have revealed a number of interesting trends, some of which are surprisingly similar to those seen in shallow marine and terrestrial ecosystems. Chapter 4, by Roy and Witman, reviews the spatial patterns of species diversity of shallow marine invertebrates and our current understanding of the underlying processes. Quest for a better understanding of the processes that determine the species richness of a particular region has been one of the most active areas of macroecological research. Spatial patterns of species diversity remain poorly quantified for most marine invertebrate groups, but some of the most direct insights about how history constrains present-day species diversity come from marine invertebrates (e.g., Jablonski, Roy, and Valentine 2006). Chapter 5, by Macpherson, Hastings, and Robertson, provides an overview of our current knowledge of macroecology of marine fishes and discusses mechanisms that potentially drive the observed patterns. Due to the long history of fisheries research, fish represent one of the best-sampled group of marine animals and hold a tremendous potential for testing process-based hypotheses about macroecological patterns. But fish are also a group where many species have declined due to tremendous pressures of industrial fishing. A better understanding of macroecological relationships can potentially lead to better fisheries management strategies (Fisher and Frank 2004; Tittensor, Worm, and Myers, this volume). Finally, chapter 6, by Santelices, Bolton, and Meneses provides a much-needed synthesis of what we currently know about the large-scale distribution and diversity patterns in marine algae. Despite their tremendous ecological importance, marine algae remain seriously understudied, and this chapter should provide a useful starting point for future studies.

The six chapters in Part 2 of this volume discuss various process-based explanations for marine macroecological patterns. Nee and Stone, in chapter 7, discuss Hutchinson's classic "paradox of the plankton"—how so many species of plankton can coexist in what appears to be a homogenous habitat—in the context of both Hubbell's neutral theory as well as many of the recent discoveries about microbial diversity in the open ocean. They suggest that the paradox of the plankton is readily explained by a combination of niche differentiation and the presence of specialist parasites.

Oceanographic processes must play a central role in creating large-scale patterns in the ocean's biota. Despite early work by fisheries biologists linking fish recruitment to oceanographic processes, benthic ecologists didn't fully embrace the key role of oceanographic processes until the severe El Niño of 1982–1983 impacted coral reefs and kelp forests. Work on benthic effects of oceanographic processes such as internal waves continued (Pineda 1991; Witman et al. 1993), and fortunately, for progress in marine macroecology, accelerated in rocky intertidal shores after the discovery of the importance of recruitment variation (Gaines and Roughgarden 1985); how it is driven by upwelling, and how upwelling potentially controls food webs from bottom trophic levels (Menge et al. 1997). Nonetheless, there are a number of lesser-known oceanographic processes capable of producing macroecological patterns, besides upwelling. The goal of chapter 8, by Leichter and Witman, is to outline the oceanographic mechanisms with great potential to create macroecological patterns in marine populations and communities.

Chapter 9, by Gaines and coauthors, provides an overview of the state of knowledge about dispersal, a critically important process influencing macroecological patterns such as species ranges, patterns of abundance, and diversity. They focus on the link between dispersal and species ranges, particularly the size of geographic ranges, the locations of species borders, and how individuals are distributed within a range. Gaines et al. not only suggest that there is little relation between range size and dispersal ability of species, but also that support for the general hypothesis that species attain maximum abundance in the center of their ranges is weak for marine populations. More research on dispersal is clearly needed to settle these issues. Chapter 10, by Clarke, evaluates the role of temperature in regulating marine species diversity. Temperature has featured prominently in discussions about processes that determine patterns of species richness on both land and in the ocean, such as the species-energy hypothesis and the metabolic theory of ecology. But Clarke provides compelling arguments that history rather than temperature may be the primary driver of marine diversity gradients. Chapter 11, by Connolly, represents an attempt to develop macroecological theory relating oceanographic and environmental variables to species distributions. From the very beginning, macroecological analyses have primarily derived inferences about processes from statistical relationships between ecological and environmental variables. But as Connolly points out, in order to better understand the processes driving patterns of species diversity we need to move beyond such correlative approaches and develop predictive, process-based macroecological models that can be tested using observational and experimental approaches. Chapter 12, the final chapter in this section, by Tittensor, Worm, and Myers, focuses on how human exploitation is changing the very nature of marine ecosystems. Industrial fishing affects virtually every part of the world oceans and has negatively impacted the distributions and abundances of many marine species, both vertebrate and invertebrate. Yet as Tittensor, Worm, and Myers point out, we still know very little about how human exploitation affects macroecological relationships.

Part 3 of this volume presents a new approach to understanding the processes driving macroecological patterns in the sea. Experimental macroecology involves testing process-based hypothesis using relatively small-scale experiments replicated over regional to global spatial scales. As chapter 13, by Witman and Roy discusses, such an experimental approach, despite its drawbacks, has the potential to not only complement the statistical analyses of observational data that form the foundation of macroecology, but also resolve some of the long-standing debates about the processes driving macroecological relationships. While descriptive marine population and community studies commenced at large spatial scales, experimental studies began at small local spatial scales but have expanded greatly over the past two decades. However, site replication remains low in experimental marine ecology, with nearly half of experimental studies conducted at only one site. Thus, there is still a long way to go before generalizations about mechanisms governing patterns of community structure are available for large spatial scales in the ocean. Witman and Roy present a conceptual model of macroecology, illustrating how both descriptive and experimental approaches can yield process-based generalizations on broad scales. Chapter 14, by Sanford and Bertness, outlines the development of the comparative experimental approach, as it has been applied, to understand one of the oldest observations in ecology, that of latitudinal variation in species interactions. An excellent example of this approach is Vermeij's initial observations of greater anti-predator adaptations in tropical versus temperate mollusks (Vermeij 1978) which led to experimental tests of tropical-temperate differences in predation pressure and community structure (Bertness, Garrity, and Levins 1981; Menge and Lubchenco 1981). Sanford and Bertness's review of coastal macroecology (rocky intertidal and salt marsh) indicates that the comparative experimental approach has successfully demonstrated geographic variation in predation, and other biotic interactions in response to large-scale gradients in the physical environment such as temperature and coastal upwelling. Latitudinal variation in biotic interactions is undoubtedly more complex than originally thought, and as the authors point out, additional factors such as physiological acclimation, genetic variation, and mesoscale variability in environmental conditions need to be considered when seeking process-based explanations. Sanford and Bertness conclude the chapter with many recommendations for future research. Finally, in chapter 15, Connell and Irving share their insights from experiments designed with spatial scale as a treatment to unravel interactions between canopy-forming kelps, algal understorey, and grazers from Eastern to Western Australia. Connell and Irving advance general hypotheses for the creation of regional patterns and provide evidence for regional patterns being derived from local processes. They conclude with a model of trade-offs of experimental and descriptive approaches to macroecology.

While this book presents information about macroecological patterns involving a variety of marine organisms and habitats, as well as a broad, and sometimes divergent, set of perspectives on the underlying processes, it does not cover all of the important topics. For example, discussion of macroecological relationships in corals and other colonial organisms is lacking, although it has been the focus of recent studies (e.g., Cornell and Karlson 2000; Connolly, Bellwood, and Hughes 2003; Karlson, Cornell, and Hughes 2004; Connolly et al. 2005). This volume also has very little information about marine arthropods, despite the tremendous species richness of this phylum, or many of the other organisms that play crucial roles in the functioning of marine ecosystems. Most of these gaps are representative of the gaps in our current knowledge about marine macroecology, but a few also stem from the constraints of covering such a broad set of topics in one volume.

This book is aimed toward students as well as established researchers in a variety of subdisciplines, ranging from marine ecology and biogeography to paleontology and evolutionary biology. The attraction of macroecology lies in the fact that it strives to bring together insights from different disciplines with the common goal of identifying and understanding large-scale patterns. The marine environment, with a rich history of observational and experimental work in ecology, a wealth of biogeographic information, and an excellent fossil record, provides an ideal system for macroecological analyses. The marine environment is also fundamentally different from terrestrial ones in many important ways. For example, both the magnitude and rhythm of environmental variability differs between land and sea (Steele 1985; Halley 2005) and recruitment dynamics, so critical for marine population biology, has virtually no analog in terrestrial animal ecology (Paine 2005). Yet, as the chapters in this volume show, many of the macroecological relationships seen in the ocean are very similar to those on land. Are these similarities superficial in that they are underlain by fundamentally different processes? Or do they point to a more interesting possibility—that over longer temporal and spatial scales the differences between marine and terrestrial systems become less important and common dynamics emerge (e.g., Halley 2005)? We do not yet know, but the answer to questions such as these are needed in order to answer the bigger question of whether there are general laws in ecology or macroecology (Brown 1995; Lawton 1999; Brown et al. 2004). Much remains to be done in marine macroecology before we can answer these questions. Better data about the distribution and abundances of marine species today, as well as in the historical and geological past, are sorely needed, along with information about how environmental variables change over different temporal and spatial scales. Also needed is a body of predictive, process-based theory of marine macroecology. As the chapters in this volume show, important progress is being made on all of these areas, and we hope that by providing an overview of marine macroecology this book will inspire further research. Macroecology evolved on land, and despite an early invasion into the oceans, so far it has diversified more rapidly on land. It is our hope that this book will provide the impetus needed for proliferation of macroecological analyses involving marine organisms. In fact, this book will be most successful if it inspires enough work to make it obsolete in a few years.

This book grew out of a symposium at the 2005 Ecological Society of America annual meeting in Montreal, Canada. We thank the authors and the external reviewers for providing thoughtful and constructive feedback on individual chapters. Two reviewers (sadly anonymous to us) took the time to read the entire manuscript and provided some very insightful suggestions that greatly improved the volume. Finally, our thanks to Christie Henry and the editorial staff at the University of Chicago Press, who remained patient, encouraging, and supportive throughout the somewhat long gestation period for this book.

REFERENCES

Arrhenius, O. 1921. Species and area. *Journal of Ecology* 9:95–99.

Bertness, M. D., S. D. Garrity, and S. C. Levins. 1981. Predation pressure and gastropod foraging: A tropical-temperate comparison. *Evolution* 35:995–1007.

Blackburn, T. M., and K. J. Gaston. 2003. Macroecology: Concepts and consequences. Oxford: Blackwell Science.

Brown, J. H. 1995. Macroecology. Chicago: University of Chicago Press.

Brown, J. H., J. F. Gillooly, A. P. Allen, V. M. Savage, and G. B. West. 2004. Toward a metabolic theory of ecology. *Ecology* 85:1771–1789.

Brown, J. H., and B. A. Maurer. 1989. Macroecology: The division of food and space among species on continents. *Science* 243:1145–50.

Connolly, S. R., D. R. Bellwood, and T. P. Hughes. 2003. Geographic ranges and species richness gradients: A re-evaluation of coral reef biogeography. *Ecology* 84:2178–90.

Connolly, S. R., T. P. Hughes, D. R. Bellwood, and R. H. Karlson. 2005. Community structure of corals and reef fishes at multiple scales. *Science* 309:1363–65.

Cornell, H. V., and R. H. Karlson. 2000. Coral species richness: Ecological versus biogeographic influences. *Coral Reefs* 19:37-49.

Dayton, P. K., and M. J. Tegner. 1984. The importance of scale in community ecology: A kelp forest example with terrestrial analogs, Pages 457-481 In *A new ecology: Novel approaches to interactive systems, ed.* P. W. Price, C. N. Slobodchikoff, and W. S. Gaud, 457–81. New York: Wiley.

Fisher, J. A. D., and K. T. Frank. 2004. Abundance-distribution relationships and conservation of exploited marine fishes. *Marine Ecology Progress Series* 279:201–13.

Gaines, S. D., and J. Roughgarden. 1985. Larval settlement rate: A leading determinant of structure in an ecological community of the marine intertidal zone. *Proceedings of the National Academy of Sciences, USA* 82:3707–11.

Gaston, K. J., and T. M. Blackburn. 2000. *Pattern and process in macroecology.* Oxford: Blackwell Science.

Gleason, H. A. 1922. On the relation between species and area. *Ecology* 3:158–62.

Halley, J. M. 2005. Comparing aquatic and terrestrial variability: At what scale do ecologists communicate? *Marine Ecology Progress Series* 304:274–80.

Hawkins, B. A. 2001. Ecology's oldest pattern? *Trends in Ecology & Evolution* 16:470.

Jablonski, D., K. Roy, and J. W. Valentine. 2006. Out of the tropics: Evolutionary dynamics of the latitudinal diversity gradient. *Science* 314:102-106.

Holyoak, M., M. A. Leibold and R. D. Holt. 2005. *Metacommunities: Spatial dynamics and ecological communities.* Chicago: University of Chicago Press.

Karlson, R. H., H. V. Cornell, and T. P. Hughes. 2004. Coral communities are regionally enriched along an oceanic biodiversity gradient. *Nature* 429:867–70.

Kritzer, J. P., and P. F. Sale 2006. *Marine metapopulations.* Amsterdam: Elsevier.

Lawton, J. H. 1999. Are there general laws in ecology? *Oikos* 84:177–92.

Levin, S. A. 1992. The problem of pattern and scale in ecology. *Ecology* 73:1943–67.

May, R. 1999. Unanswered questions in ecology. *Philosophical Transactions of the Royal Society of London B* 354:1951–59.

Menge, B. A., B. A. Daley, P. A. Wheeler, E. Dahloff, E. Sanford, and P. T. Strub. 1997. Benthic-pelagic links and rocky intertidal communities: bottom-up effects on top-down control? *Proceedings of the National Academy of Sciences, USA* 94:14530–35.

Menge, B. A., and J. Lubchenco. 1981. Community organization in temperate and tropical rocky intertidal habitats: Prey refuges in relation to consumer pressure gradients. *Ecological Monographs* 51:429–50.

Paine, R. T. 2005. Cross environment talk in ecology: Fact or fantasy. *Marine Ecology Progress Series* 304:280–83.

Pineda, J. 1991. Predictable upwelling and the shoreward transport of planktonic larvae by internal tidal bores. *Science* 253:548–51.

Steele, J. H. 1985. A comparison of terrestrial and marine ecological systems. *Nature* 313:355–58.

Turner, J. R. G. 2004. Explaining the global biodiversity gradient: Energy, area, history and natural selection. *Basic and Applied Ecology* 5:435–48.

Vermeij, G. J. 1978. *Biogeography and adaptation: Patterns of marine life.* Cambridge: Harvard University Press.

Witman, J. D., J. J. Leichter, S. J. Genovese, and D. A. Brooks. 1993. Pulsed phytoplankton supply to the rocky subtidal zone: Influence of internal waves. *Proceedings of the National Academy of Sciences, USA* 90:1686–90.

PART I
MACROECOLOGICAL PATTERNS IN THE SEA

CHAPTER ONE
OVERVIEW OF MARINE BIODIVERSITY

JAMES W. VALENTINE

Major Environmental Framework of the Oceans

The purpose of this chapter is to review the environmental framework in the sea and its relation to major biodiversity patterns, and briefly to review some of the leading hypotheses proposed to explain those relationships. The major biotic patterns are global or regional, and it is plausible to relate them to global or regional patterns of ecological parameters in the marine environment. Most hypotheses of major biodiversity patterns involve temperature, environmental heterogeneity, trophic resources, and biotic interactions. Here I deal with such factors as they are distributed in larger environmental partitions, at levels from oceanic to subprovincial, as a framework for subsequent chapters, in which correlates of local and regional diversities for distinctive faunal elements and ecological settings are examined in much more detail by workers with expertise in those systems.

Major Trends in Marine Biodiversity

Biodiversity is sometimes treated as composed of two aspects: richness, the number of individual taxa or their attributes that are under study; and disparity, some measure of distinctiveness among the taxa or their attributes (population size or morphological distance, for example). Here the unqualified term *diversity* will refer simply to richness, and thus diversities based on metrics that involve aspects of disparity are not necessarily comparable with the patterns described here. Knowledge of marine biodiversities is bedeviled by sampling problems, which vary among taxa, and by the enormity of the effort required to obtain accurate taxonomic censuses in such a heterogeneous, dynamic, and largely remote system. Even the best-sampled marine taxa are quite incompletely known, and understudied taxa may contain more species than well-known groups (see Poore and Wilson 1993). Yet at oceanic and global scales, biodistributional patterns of such major marine taxa of the shallow sea as fish, mollusks, arthropods, and echinoderms are similar, as was recognized by the mid-nineteenth century (see Ekman 1953; Briggs 1974). While biodiversity patterns established by those taxa are constantly being refined, their large-scale features seem unlikely to be altered dramatically by new data. Many of the patterns cited here are of mollusks, simply because they are the group I know least poorly; fortunately they are very diverse and their trends are likely to be representative of most of the marine benthic fauna.

Figure 1.1 is an attempt to represent the diversity pattern of benthic marine organisms at shelf depths, based on published summaries of biodiversity for a number of important invertebrate taxa of the shallow sea, chiefly bivalve and prosobranch mollusks, bryozoans, and scleractinian corals. These groups are all rather diverse, and have the advantage of having good fossil records, so that their histories may be used to help interpret their present patterns. Figure 1.1 is not underpinned by actual species counts over the world ocean. In some cases major taxa have different patterns, even at the scale of major oceanic compartments. For example, Taylor (1997) has pointed out that bivalves (chiefly suspension feeders) are most diverse in the relatively eutrophic compartments of continental coastal environments and upwelling regions, while prosobranchs (chiefly grazers and carnivores) are most diverse in communities in the more oligotrophic oceanic compartments. If the diversity of either of these molluscan groups would be considered alone it would produce a biased picture of actual trends. Diversity data are usually presented for a single taxonomic group in a given region, and there are enough understudied groups and regions that accurate, quantitative data on marine biodiversity patterns cannot be assembled as yet. The somewhat subjective shelf diversity pattern of figure 1.1 will certainly be subject to modification as new data emerge.

The most pervasive trend in marine biodiversity is the latitudinal diversity gradient (LDG), found in the open ocean, on continental shelves, and even in the deep sea, with high diversities in the tropics grading to low diver-

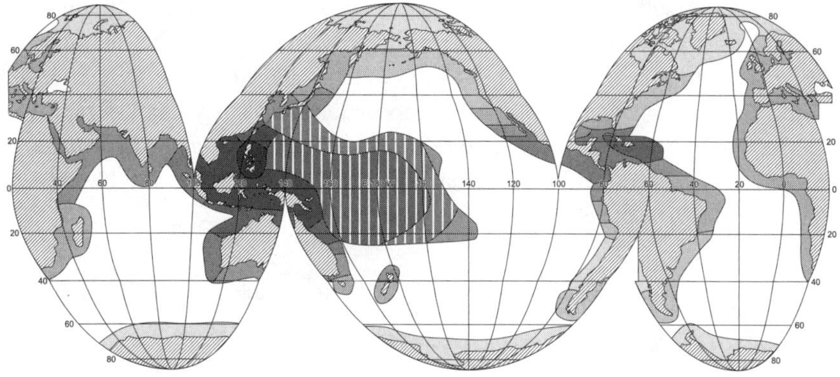

FIGURE 1.1 An hypothesis of shallow-sea diversity trends, based on skeletonized invertebrates. Darker regions have higher diversities. As data are spotty and it is necessary to use different groups at different localities or even in different regions, the map indicates diversity trends in a general way, but correspondence between absolute levels of diversity in disjunct regions may not always be very precise. In the central Pacific the shallow-water environments are found only on scattered island systems of very small areas, but diversities are mapped as if shelves extended over the whole region; shading there is marked by vertical stripes to indicate it is neither oceanic nor deep-sea. The widths of the shelves are exaggerated for visual effect, and are not to scale.

sities in high latitudes. A fundamental question about the LDG was posed by Stebbins (1974): do the tropics have high diversities because speciation is high there—are they a cradle of new lineages—or because species have accumulated there over geological time—are they a museum of old lineages? There are also striking longitudinal variations in marine biodiversity among oceans, between coasts within oceans, and even within oceanic provinces, as with reef coral and fish diversity in the Indo-Pacific (Connoly, Bellwood, and Hughes 2003).

Aside from the broad biodiversity trends, there are some regional anomalies. The clearest of the broad trends are, for the most part, associated with the more north-south trending coasts, which makes sense, since the main environmental gradients are strongly latitudinal. However, where there are chiefly east-west features, such as continental coastlines in the Arabian region, or broad east-west reentrants such as the south Caribbean Sea, or island chains such as the Aleutians, or where two north-south coastal faunas coalesce as at capes, diversities commonly show unusual or at least theoretically unexpected levels, that do not always appear in the broad diversity classes of figure 1.1. Nevertheless, these exceptions deserve special attention.

Deep-sea benthic diversities, while poorly known, are clearly fairly high, slope faunas being comparable to or exceeding diversities in the shallow sea

(Levin et al. 2001 and references therein). There is evidence of a latitudinal gradient or at least trend in deep-sea diversity among mollusks, although it may not be present in all basins (e.g., Stuart and Rex 1994; Rex, Etter, and Stuart 1997). Finally, pelagic diversity patterns (as of planktonic foraminifera, a group with a good fossil record (Stehli, Douglas, and Kafescioglu 1972; Bé 1977) also show a latitudinal diversity trend (and see Angel 1997 for a general review of pelagic diversity patterns).

Environmental Heterogeneity

Ecospace

The oceanic environment displays a rich tapestry of physical conditions—features such as light, temperature, substrate conditions, water motions, and so forth, that can be parameterized and used to describe a multidimensional space, positions in which correspond to possible combinations of those factors. Organisms have evolved so as to exploit, defend against, and otherwise interact with each other, and when such interactions are also employed as additional dimensions of that hyperspace, possible combinations of the many conditions of life are modeled. The portion of this hyperspace that is occupied by an individual or by a specified group of organisms may be termed the *ecospace* of those forms (Valentine 1973), which subsumes the niche of species and the adaptive zone of higher taxa.

Biotic units inhabiting the global marine ecospace are usually regarded as forming a hierarchy, with levels that include individuals, populations, communities, regional biotas or provinces, and oceanic-scale biotas; communities and the geographically based units are sometimes subdivided into more levels. The ecological units vary greatly in size and importance and sometimes require arbitrary decisions as to their positions, as is true within all natural hierarchical structures. The organisms inhabiting this hierarchy are, of course, the living marine representatives of the tree of life, evolved along myriad branching pathways stretching back billions of years. When studying ecological interactions involving individuals and species, it seems appropriate to consider the conditions in the ambient environment, with due regard for their variability and very recent histories. However, when studying higher ecological and biogeographical levels, there is always a chance (and in some cases a certainty) that the present situation reflects in part historical factors in play during long-vanished environmental conditions that may date back many millions of years, and that affected the lineages now represented by living species.

Marine macroecology deals in part with patterns of ecospace occupation and partitioning, usually by comparative methods among species or species associations. The levels and distributions of taxonomic richness, range sizes, population sizes, and functional features such as reproduction and feeding types, vary by geography and habitat, speaking to regulation by evolutionary and/or ecological processes.

Spatial Heterogeneity

It is clear from generations of study that environmental heterogeneity tends to enhance diversity, other things being equal. This relation seems to hold across all scales. At the smallest scale, experiments and manipulations, chiefly in terrestrial settings, indicate that increasing local spatial heterogeneity within habitats permits increased local species diversity. These studies range from classic tests of competitive exclusion within habitats in which beetle species are permitted to coexist only in more heterogeneous conditions (e.g., Crombie 1945), to a recent experiment with twig-dwelling ants in which ant diversity increased when twig spatial heterogeneity was raised (without regard to which twig species were present; Armbrecht, Perfecto, and Vanderrmeer 2004). Observations in the marine environment are consistent with such experiments; for example, benthic diversity in the deep sea has been correlated with the diversity of particle sizes in the substrate (Etter and Grassle 1992; see also Rex, Etter, and Suart 1997, Levin et al. 2001). Spatial environmental heterogeneity also rises as the number of habitat types increases, supporting higher levels of biodiversity. Coasts with varied habitats support more species than do environmentally monotonous coasts (say, rocky shorelines with bays and lagoons versus those without such features), as also demonstrated by experiments with sunken ships and artificial reefs that enhance local species diversity. And at still larger scales, barriers to the spread of species between regions, provinces, or ocean basins produce environmental heterogeneity on a biogeographic scale that enhances species diversity in the world ocean. Such within-community, between-community, and regional aspects of ecospace heterogeneity represent steps that tie local ecological processes into a macroecological system.

A common observation in diversity studies, particularly in terrestrial settings, is that higher diversities are associated with larger geographic areas, and it is sometimes concluded that area per se is responsible (see Rosenzweig 1995 and references therein). However, other workers hypothesize that it is actually an increase in the heterogeneity of habitats found in the larger regions that may account for the relationship (see MacArthur and Wilson 1967). Certainly in the sea the diversity/area relation does not hold regionally. For example, in the northeastern Pacific, tropical shelves are relatively narrow, while subarctic and arctic shelves are exceptionally broad and have about five times the area of tropical shelves, yet prosobranch diversities at both community and provincial levels are much higher in the tropics (about three times higher when binned by provinces; Roy et al. 1998); indeed, no relation between area and diversity is found for mollusks on either the North Pacific or the North Atlantic shelves (for bivalve diversities see Roy, Jablonski, and Valentine 2000). The same general pattern is true for many invertebrate groups, and for the shelf fauna as a whole.

Temporal Heterogeneity

Fluctuations in environmental conditions can also be represented by ecospace dimensions. For macroecology, the dominant temporal fluctuation is clearly seasonality, which is associated with fluctuations in insolation, temperature, salinity, strength of currents, intensity of upwelling, and in biotic factors such as productivity, spawning, prey abundance, and so on. Although the gross effects of seasonality grade with latitude, regional and local hydrographic conditions introduce irregularities. A difficulty in interpreting the effects of seasonality is that mean temperature and its correlates, as well as various aspects of productivity, all of which clearly have major ecological effects, also vary with latitude and hydrographic conditions (see the following). Thus, the effects of temperature and of productivity per se are difficult to disentangle from those of seasonality. Long-term environmental changes such as glacial cycles, or even longer-term climatic trends, can also affect the taxonomic composition of a fauna, with macroecological consequences.

Diversity-Dependent Factors

Two classes of ecologically significant factors, those that are diversity dependent and those that are diversity independent (Valentine 1972, 1973), can be distinguished in part by their relations to evolution. If a factor does not limit the number of species that evolutionary processes could produce, it is diversity independent. For example, in the marine realm, temperature per se is hypothetically a diversity-independent factor, for evolution can produce species that are adapted to any level across the entire range of ordinary marine temperatures. There does not seem to be any reason that cold waters should not be able to support as many species as warm waters, other things being equal, as is certainly indicated by the faunas of the continental slopes (see the following). Spatial habitat heterogeneity, on the other hand, is a diversity-dependent factor; in theory habitats may be fully occupied, setting a diversity limit (in the absence of extinction; Walker and Valentine 1984). An important caveat to these cases is that under certain conditions almost any diversity-independent factor may produce ecological diversity effects if there is only a limited pool of available species that can tolerate certain conditions. For example, when temperature variations contribute to spatial heterogeneity, either locally (e g., upwelling patches) or regionally (e.g., climate zones), they do affect diversity by providing additional thermal habitat types, but within those habitats temperature is diversity independent. The principle is that diversity dependence involves factors that can be used up, so that evolution cannot partition them among species beyond a certain limit, while diversity independence involves factors that are not used up and to which evolution may produce an adaptive response irrespective of and independent of diversity.

Trophic Resources

Trophic resources function as a classic density-dependent factor and presumably function in diversity dependence as well. While sunlight and nutrients underlie the primary productivity that forms the base of the trophic pyramid, their relation to diversity is not straightforward. On an oceanic scale, primary productivity can now be inferred from satellite data (e.g., Sathyendranath et al. 1995 and refs. therein). Longhurst (1998) has synthesized and summarized these data, binning them within geographic compartments ("provinces") that represent distinctive oceanographic entities, such as water masses and current systems. In table 1.1, Longhurst's productivity data are summarized by level and range for his compartments. Although the productivity data within Longhurst's compartments are averages over broad regions that naturally obscure many important local conditions, they are useful at the regional level and can provide a basis for a general discussion of relations between productivity and diversity. For each compartment, a measure of annual primary productivity (summed from average monthly productivity) and a measure of the range of productivity (the difference between the highest and lowest monthly average) is given. Annual variabililty in marine primary productivity from Longhurst's data is mapped in figure 1.2 by shading for each 10gC/m^2/month, from 0–10 (blank) to over 100 (black).

One might expect that regions of high productivity would support more diverse faunas, but this is clearly not the case, at least not on the scale of these major oceanographic compartments. If anything, there is a tendency for diversity to be lower in regions of higher productivity, though the relation is clearly variable. For oceanic compartments, productivity is highest in high latitudes and in regions of convergence where nutrients are in relatively large supply, at least seasonally, in the euphotic zone, a pattern that does not correlate with diversity trends. Coastal compartment productivities, while generally high in high latitudes also, are recorded as being highest in regions

TABLE 1.1. Average yearly productivity levels and ranges for Longhurst's (1998) oceanic compartments (labeled in fig. 1.2); the data are in g C per meter^2 per month. "Primary Productivity" is the average monthly figures summed over a year; "Range" is the difference between months with highest and lowest productivities.

Compartment	Primary Productivity	Range
1. Pacific Subarctic gyres	201	33
2. North Pacific transition	173	21
3. Kuroshio Current	197	33
4. North Pacific tropical gyres	58	1
5. North Pacific equatorial countercurrent	105	2
6. Pacific Equatorial divergence	113	3
7. Western Pacific warm pool	81	2
8. Archipelagic deep basins*	99	3
9. South Pacific subtropical gyre	89	8
10. South subtropical convergence**	138	16
11. North Atlantic drift	252	45
12. Gulf Stream	180	25
13. North Atlantic subtropical gyres	95	11
14. North Atlantic tropical gyre	108	2
15. Eastern tropical Atlantic	158	10
16. Western tropical Atlantic	129	4
17. South Atlantic gyre	77	4
18. Indian monsoon gyres	104	7
19. Indian south subtropical gyre	70	3
20. Subantarctic water ring	120	18
21. Antarctic	123	24
22. Boreal polar	397	81
23. N. Pacific epicontinental sea	368	49
24. Alaska downwelling coastal	661	110
25. California Current	395	31
26. Central American coastal	343	21
27. Humboldt Current coastal***	273	25
28. Austral polar**	219	58
29. China Sea	630	42
30. E India coastal	359	36
31. W India coastal	374	53
32. Red Sea	619	56
33. NW Arabian upwelling	461	87
34. Sunda-Arafuru shelves*	329	12
35. E Australia coastal	237	23
36. New Zealand coastal	310	49
37. Australia-Indonesia coastal	199	13
38. E. Africa coastal	193	13
39. Atlantic Arctic	366	75
40. Atlantic Subarctic	302	64
41. Mediterranean, Black Seas	222	19
42. Eastern (Canary) coastal	706	60
43. Guinea Current coastal	495	39
44. Benguela Current coastal	320	32

TABLE 1.1. Continued

Compartment	Primary Productivity	Range
45. NW Atlantic shelves	544	62
46. Caribbean	192	8
47. Guianas coastal	694	51
48. Brazil Current coastal	311	13
49. SW Atlantic shelves	468	42
50. Tasman Sea	164	20

Notes: *The Sunda-Arafuru shelves and Archipelagic deep basins provinces intertwine and are mapped together.
**Note province present both in Atlantic and Pacific.
***ENSO effects are not included in these figures.

where special conditions prevail, such as southern Alaska (low diversity) and in the Red and Arabian Seas (high diversity). (As chlorophyll measurements are difficult in coastal regions such figures should be regarded as provisional so far as shelf waters are concerned, especially for inshore areas, which may be poorly represented by compartmental averages.)

Another aspect of productivity, seasonal variability (measured as the difference between highest and lowest monthly averages, which form unimodal curves in Longhurst's data), appears to fit the diversity patterns much more closely: lower productivity ranges correlate with higher diversities (table 1.1 and fig. 1.2). This is most apparent in the oceanic compartments, where low-latitude seasonal ranges are generally less than 10gC/m^2/month, while high-latitudes range into the 30s and 40s, higher in the north. Coastal compartments also follow this trend, especially along north-south coasts; local exceptions chiefly involve east-west features, such as the Alaska downwelling compartment (no. 24) and the Northwest Arabian upwelling compartment (no. 33), which exhibit anomalously high seasonalities. The high Alaskan productivity is well-documented (e.g., Glover, Wroblewski, and McClain 1994) but although the general hydrographic framework is known (Royer 1998) the conditions that permit the high nutrient concentrations required for the seasonal productivity peak are not understood. The Northwest Arabian region is subjected to a strong seasonal monsoon that alters the direction of the major currents, an effect that is particularly strong from Somalia to the western coast of India, and is associated with intense seasonal upwelling in a number of regions, especially off Oman (Shetye and Gouveia 1998); these events are reflected in the seasonality of productivity. Additionally, discharge of nutrient-laden waters from rivers can create important hot spots of productivity, some of considerable extent (as within the Guianas

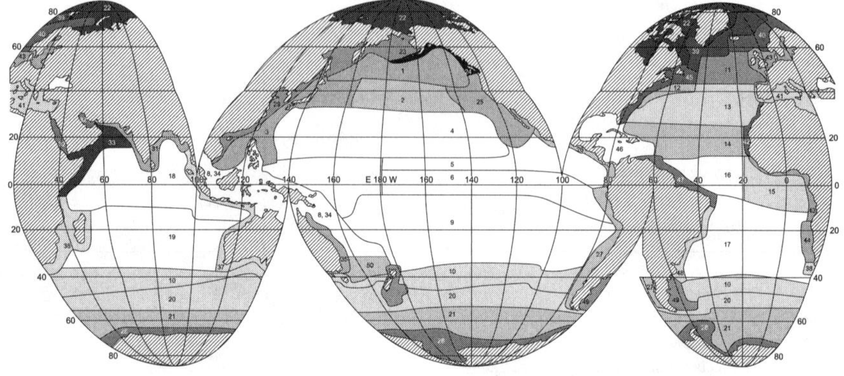

FIGURE 1.2. Average annual variability in productivity of the oceans binned in Longhurst's oceanographic compartments. Scaled in intervals of 10gC/m^2/month, from 0–10 (blank) to over 100 (darkest). Data from Longhurst 1998 (see table 1.1).

coastal compartment (no. 47; Geyer et al. 1996), and the levels of discharge are usually highly seasonal.

On the shelves, production from phytoplankton is supplemented by benthic production and by the regeneration of nutrients on the sea floor. The processes of regeneration, at least, should have the effect of damping seasonality for many species, by extending the time when nutrient supplies, and trophic supplies at other levels, are available. In the deep sea, where primary productivity is limited to chemautotrophy localized to vents and seeps, regeneration is nearly the entire trophic story, and trophic supplies depend on the delivery of organic materials to the sea floor, which can be seasonal but is certainly highly patchy and includes relatively unpredictable components (see the following and McClain, Rex and Etter, this volume).

Thus a clear negative trend between seasonality and diversity is best displayed among north-south trending series of compartments, either oceanic or coastal (figs. 1.1 and 1.2; Valentine 1983). And even in the deep sea, an LDG may be related to seasonal fluctuations in trophic supplies descending from increasingly higher and narrower peaks of productivity to poleward in the euphotic zone. Lack of correlation between benthic shelf diversity and seasonality in the coastal compartments is usually found within east-west trending features, but not in any simple relationship.

Habitat Heterogeneity

The notion that more heterogeneous habitats encourage high diversity is consistent with observations in tropical forests and, in the sea, on coral reefs. The tropical ecospace associated with forest plant associations and marine reef structures is clearly broken into intricate mosaics of habitats from the standpoint of their inhabitants, which form rich communities and associations of communities that presumably would not be present otherwise. As for reefs, many clades have radiated to exploit specialized habitats and food sources found there, representing a sort of evolutionary feedback at several levels that greatly enhances diversity. Examples include the studies of Kohn (1997 and references therein) and Taylor (1997 and references therein) for predatory gastropods.

In nonreefal settings, Williams and Reid (2004) studied diversity and speciation of the gastropod *Echinolittorina* ($N = 59$ evolutionary significant units) on tropical rocky shores, and found that the highest diversities resulted from specializations related to habitat heterogeneity along oceanic/continental gradients. Valdovinos, Navarette, and Marquet (2002) suggest that the relatively high benthic diversity on the southwestern South American shelf is partly owing to high habitat heterogeneity, perhaps associated with the glacial submarine topography there.

The situation in the deep sea has been thought to provide a counterexample to hypotheses of any necessarily strong relation between high habitat heterogeneity and high species diversity. The nature of niche differentiation within a rich fauna in what seems to be a relatively homogeneous realm has been a major puzzle. But the deep-sea species may not perceive their environment as particularly homogeneous; patchiness in substrate conditions and in the delivery of trophic resources, many of which are localized and ephemeral, coupled with a high incidence of trophic specialization, may produce microhabitats that form the basis of niche heterogeneity (Grassle and Maciolek 1992; Grassle 1994; Rice and Lambshead 1994). The deep-sea fauna changes most rapidly with depth, compared with change along depth contours (e.g., Sanders and Hessler 1969), and diversity appears to decrease from the shallower bathyal to the deeper abyssal depths (see Rex, Etter, and Stuart 1997, McClain, Rex, and Etter, this volume). Paralleling the diversity decrease with depth is a seeming decrease in habitat heterogeneity with depth (Etter and Caswell 1994 and references therein).

Provinciality

Provinciality is a case of environmental heterogeneity at a level higher than that of habitat. Provinciality could be expressed by some metric of species turnover with distance or area—turnover as defined by the geographic range margins or end-points of species (or other taxa). Bioprovinces are regions with distinctive biotas, the criterion of difference varying among workers; they usually represent rather broad regions of relatively low geographic turnover separated by narrow borders or regions that show relatively high turnover. For the shelf benthos, the earliest large-scale descriptions of bioprovinces were based on climate zones (Dana 1853), with the boundaries defined by isocrymes. As the geographic ranges of species became better known, it became apparent that conjunct bioprovincial boundaries do not necessarily fall at the borders of climate zones, but may occur at hydrographic convergences or other discontinuities that separate water types that are distributed in ocean currents. If shallow-sea benthic bioprovinces were based on climates per se, they would presumably be broadly intergrading latitudinally; it must be the hydrographic localization of range-limiting factors (e.g., Gaines et al. this volume), or their former presence under past regimes, that are chiefly responsible for the boundaries that we find. Shelf bioprovincial boundaries evidently extend from shore right across the shelf to the thermocline, below which a bathyal biodistribution system is found (e.g., Jablonski and Valentine 1981). Nevertheless, a common assumption is that temperature contrasts set limits to many species' ranges, although details of the limiting factors—whether due to reproductive or adult temperature limits, to exclusion by biotic factors, to habitat failure, or even to simple blockage of larval transport by opposing currents (Gaylord and Gaines 2000, Gaines et al. this volume)—are uncertain for many species at any particular boundary.

In the euphotic zone of the open sea, conjunct oceanographic compartments as defined by Longhurst (1958) differ in regimes of temperature and productivity. Although their status as bioprovinces remains to be thoroughly documented, many of the compartments are biotically distinctive, although many pelagic forms have extensive ranges (e.g., Brinton 1962, Bé 1977, Bé and Gilmer 1977; see Li this volume). In the deep sea, many of the very abundant forms have extremely broad distributions, and while most species are rare and these are more restricted, data are still sparse and no comprehensive description of provincial biotas has been attempted. Clear boundaries between assemblages within similar depth zones are best known when the assemblages are confined within basins that are compared below the basinal sill depths.

A Quasi-diversity-dependent Factor: Species Pools

In island biogeographic theory, the pool of species from which immigrants can be recruited is one of the important determinants of island species diversity (MacArthur and Wilson 1967), and the effects of regional species diversity on local diversity has been demonstrated for epifaunal marine communities in general by Witman, Etter, and Smith (2004). Enrichment of a marine biota by immigration from a regional species pool has been reported for reef corals (Karlson, Cornell, and Hughes 2004) and modeled for terrestrial plants in non-island situations (Loreau and Mouquet 1999). A similar effect may operate on the scale of marine provinces and subprovinces. For an oceanic example, planktonic foraminiferal diversity within the Transition Zone between subtropical and subarctic waters is nearly as high as in subtropical waters, yet there is only a single endemic species (Bé 1977). This situation must be due at least in part to the openness of boundaries between the three-dimensional oceanic zones. Many subtropical and tropical species may be recruited to the Transition Zone, and although they are generally rare there (and may or may not reproduce) their presence inflates Transition diversity, which presumably would be very much lower if the reservoir of species in the subtropics were not present.

As for possible examples in the molluscan shelf benthos, northeastern Pacific diversity within the latitudinal band from 50°–60° rises above the general latitudinal trend, interrupting the otherwise monotonic gradient (Roy et al. 1998 for prosobranchs, Roy, Jablonski and Valentine 2000 for bivalves). This region coincides with the islands of the Aleutian arc, whose shelves form stepping-stones to the Commander Islands and the mainland Asian shelf. Although there are few, if any, molluscan species in the Aleutians that do not also range to the North American mainland, it is a plausible hypothesis that the high diversity of this segment is partly owing to past recruitment of species from both the Asian and American faunas, which together form a large species pool for those latitudes. Additionally, the varied habitats that stretch for two thousand miles or so along those islands must greatly increase the capacity of that 10° latitudinal segment to accommodate species once they are there.

Molluscan diversity within the most southern segment of the eastern South American shelf, from near Chiloé (42° S) and southward, is higher than shelf diversity to northward (Valdovinos, Navarette, and Marquet 2002), thus reversing the latitudinal gradient there. This region shares many forms with the southern portion of the Argentine shelf, together comprising what is often termed the Magellanic fauna, which inhabits the shelves of both the southern Humboldt and the Falklands hydrographic compartments (see Briggs 1974 and references therein). This fauna occupies over twice the latitudinal spread found on the Chilean coast, and presumably the varied and heterogeneous shelf environments can thus harbor a larger species pool than would only a single coastline. Another possible reason for the lower diversity along the more equatorward Humboldt Current coastal compartment is the periodic occurrence of strong El Niño–Southern Oscillation (ENSO) events, which are not represented in the seasonal productivity figures in table 1.1

but which create periods of intense upwelling and high productivities and thus would greatly increase the range of productivity if somehow treated in the calculation of seasonal fluctuations. Furthermore, an important oxygen minimum zone is present, usually at shelf depths, which may have an impact on diversity accommodation. This region suffered an important extinction during the late Neogene and perhaps those unusual environmental factors have retarded faunal recovery.

Some shelf segments appear to show conditions that can be interpreted as producing low diversity situations elsewhere, yet their diversities are not obviously depressed. An important case is in the western Indian Ocean, in the NW Arabian upwelling compartment, where the famous monsoonal conditions create very high seasonality, yet where the diverse tropical Indo-Pacific fauna is well represented (as off Oman; see Oliver 1992, for Bivalvia). Another case is furnished by the fauna of the shelf along northern Brazil and the Guianas, in the Guianas coastal compartment, where seasonal turbidity from river discharge, especially from the Amazon, is very high (Geyer et al. 1996), but does not appear to result in notably lowered diversity (see Altena 1971, although diversity on this continental shelf is evidently somewhat lower than diversities around the Caribbean islands). In these and similar cases, it can be hypothesized that the regional diversities draw upon a large species pool, and that this pool permits regions that otherwise might show low diversities to maintain a higher diversity, perhaps owing to range extensions associated with metapopulation dynamics—a sort of metadiversity (see discussions in Maurer and Nott 1998, Stuart and Rex 1994, and Witman, Etter, and Smith 2004). To the extent that this is true, the standing diversity, filling ecospace above a normal level (see the following), would presumably depress speciation and thus act as a quasi diversity-dependent factor.

Diversity-Independent Factors

Temperature

As diversity varies in fairly consistent ways among climatic zones, and as marine species have demonstrable temperature limits that affect their distributions, some workers have hypothesized that temperature is a major component of diversity regulation (e.g., Stehli, Douglas, and Kafescioglu 1972). In this case, as high temperatures generally correlate with high diversities in the shallow sea, the latitudinal gradient would be responding to the ocean temperature pattern. On the other hand, the presence of a given number of cold-adapted species would not seem to preclude the presence of still others simply because of the temperature; temperature per se must be considered a diversity-independent factor (Valentine 1973). Correlations between temperature and diversity would thus be owing to a shared correlation with some diversity-dependent factor(s), perhaps including the seasonality of insolation with its effects on both temperature and productivity. Certainly the seasonality of ocean temperature itself is not involved; the regions where temperatures vary the least seasonally include the highest as well as the lowest latitudes, with the greatest temperature seasonality in middle latitudes, thus displaying no special relation to diversity patterns (Valentine 1972).

However, to the extent that temperature patterns increase the heterogeneity of the environment they favor increased diversity—ecological diversity may be affected by diversity-independent factors. In the shallow sea, water temperatures result from factors that contribute to the conditioning of a given water mass or type, which include not only the latitudinal trend in insolation but upwelling of cooler water, lateral mixing of different water types, and the distribution of the resulting waters in currents. The species within any given region must be thermally adapted to the ambient temperature regime, hence the distinctive faunas of different climate zones or regions. Local distributions may also mirror local temperature patterns. For example, relatively cool-water species appear in upwelling patches, where diversities may be lowered locally (Dawson 1951), while relatively warm-water species sequester in shallow, protected embayments (Hubbs 1948). Regional diversities may certainly be affected when either or both warm- and/or cool-water patches are present.

Perhaps the strongest empirical evidence that diversity and temperature per se are not closely related evolutionarily is provided by the fauna of the deep sea, which has local and regional diversities that compare well with those of the shelf and may have even higher faunal diversities at bathyal depths (Levin et al. 2001). Yet the water temperatures are as low as those of the shallow-sea provinces that have the lowest diversities. Within the deep sea, however, a strong correlation has been demonstrated between temperature fluctuations associated with changing Quaternary climates, and benthic foraminiferal species diversities over the last 130,000 years (Hunt, Cronin, and Roy 2005). The pattern was reasonably interpreted as resulting from a poleward shift in the boundaries of more warmth-adapted species during warming of bottom waters, species that were presumably drawn from a higher-diversity biota in lower latitudes, and the equatorward restriction of those species during cooling trends. Thus, adaptive responses to the (low) deep-sea temperature ranges set by evolution may be evoked in ecological responses to fluctuations in those temperatures.

That temperature might impose diversity limits through its effects on metabolism has been proposed by Allen, Brown, and Gillooly (2002) and Brown et al. (2004), and the predictions of those formulations have been examined for marine shelf bivalves with mixed results (Roy, Jablonski, and Valentine 2004) and for deep-sea foraminifera with more promising results (Hunt, Cronin, and Roy 2005). However, the mechanism that might create diversity-dependence from metabolic functions has not been determined. As Brown et al. note, higher temperatures may correlate with higher mutation rates, but while this might lead to higher rates of change within lineages (and to pseudospeciation), it is not clear how this would lead to higher standing diversities, which are evidently ecologically and biogeographically regulated, at least in part. It appears that relations between thermally regulated metabolic rates and marine diversities are still uncertain, despite the obvious importance of metabolism for other macroecological factors.

Adaptive Strategies and Diversity

The patterns of diversity and of environmental parameters, discussed earlier, suggest that evolution (including macroevolution) tends to generate characteristic adaptive strategies in different environments, which underlie the macroecological trends associated with diversity. Elements or relevant consequences of these strategies may include fecundity, dispersal ability, genetic divergence, geographic range, and speciation and extinction rate. It is hypothesized that such elements interact with diversity-dependent factors to produce the major aspects of global diversity patterns.

Fecundity and Dispersal

Among most marine poikilotherms, fecundity is chiefly a function of body size and of developmental mode: forms with larger body sizes and with planktotrophic development produce more eggs (review for invertebrates in Jablonski and Lutz 1983). A study of body size frequencies of Bivalvia across latitude showed no significant variation among provinces from the tropics to the Arctic (Roy, Jablonski, and Martien 2000). Indeed, body size per se seems to play a surprisingly small role in marine invertebrate macroecological patterns on the shelves (Jablonski 1996) despite the metabolic implications. Invertebrate developmental modes, on the other hand, have been known to vary latitudinally since the classic work of Thorson (1950), with planktotrophy most frequent in low latitudes and direct development most frequent in high latitudes. To be sure, some groups are locked into direct development globally (as are anomalodesmatan and protobranch bivalves), while others display frequent planktotrophy at high latitudes (as do echinoderms; Pearse 1994), but Thorson's rule has been generally confirmed (Jablonski and Lutz 1983; Laptikhovsky 2006). As larval planktotrophy is high among deep-sea forms such as gastropods, it is unlikely that temperature per se is responsible for Thorson's pattern.

The effect of developmental type on dispersal, at least as indicated by geographic range, appears to depend on the character of the geographic framework. On the continental shelves of North America, the average latitudinal geographic range of species of prosobranch gastropods (N = 3,916) found within 5° bins is greatest in low latitudes, where planktotrophs are most frequent, but the trends are not monotonic and in fact latitudinal range lengths rise poleward (with increasing frequency of nonplanktotrophy) on temperate and Arctic Atlantic shelves (Roy et al. 1998), suggesting that ranges are being influenced more by patterns of local environmental discontinuities than by dispersal abilities. On the other hand, a study of the largely planktotrophic, carnivorous gastropod family Cypraeidae in the Indo-West Pacific (Meyer 2003) found that among evolutionary significant units (groups at species and subspecies levels of genetic divergences, N = 210), those that have the longest planktotrophic lives have the broadest geographic ranges. A group of cypraeid species (the Erroneinae), that can be inferred from larval protoconch sizes to have relatively short larval lives, have relatively narrow ranges. In much of the Indo-Pacific province, where shallow-water habitats are widely separated around scattered island systems, dispersal ability is clearly an important factor in range size.

As might be expected, the degree of genetic divergence between distant populations is also related to dispersal ability. For example, in Meyer's (2003) cypraeid study it was found that genetic divergences were higher between those conspecific evolutionary significant units with shorter larval lives. In some cases, nominal subspecies are more divergent than many species. Many of these genetically disparate units have not been recognized taxonomically, so that recognizing genetically divergent units raises the diversity by 38 percent over the formal taxonomic units. As the newly recognized divergent populations tend to inhabit separate islands and to have relatively restricted geographic ranges, local diversities are not much affected. Along geographically contiguous shelves also, contrasts have been found between species with planktonic larvae that tend to be relatively well-mixed genetically among populations, and species that are direct-developing forms, which tend to have greater genetic divergences among populations (Kittiwattanawong 1999, Bohanak 1999).

Geographic Range

Speciation rate has been thought to correlate with geographic range, sometimes positively, sometimes negatively. The geographically more restricted, lower-dispersal species of cypraeids, for example, appear to be speciating more rapidly than high-dispersal forms (Meyer 2003). That this relation is not simply a consequence of the geographic island framework of the Indo-Pacific is indicated by study of fossil ranges and speciation rates along a broad region of contiguous shelves. Gastropods of the late Cretaceous (Campanian-Maestrichtian) of the Gulf and Atlantic Coastal Plains of North America also show a positive correlation between shorter geographic ranges and speciation rates per species (Jablonski and Roy 2003). However, in the two shelf systems where they have been studied, molluscan species' geographic ranges are not universally shorter in low latitudes, where speciation rates appear to be significantly higher than in the extratropics (Roy et al. 1998), indicating that other factors are also involved.

Marine Diversity Gradients through Time

An advantage of working with groups that have durable skeletons is that their diversity patterns may be traced through time in the fossil record, and the macroevolutionary dynamics that produce diversity and mediate its changes can be inferred from patterns of ages and rates of origination and extinction of their taxa. The fossil record is spotty enough at the species level that most studies have been at the generic or familial level to improve accuracy. Pioneering studies by Stehli and Wells (1971) with corals and Stehli, Douglas, and Kafescioglu (1972) with planktonic foraminifera revealed a pattern that may prove to be fairly general. The average age of both the foraminiferal and coral genera studied is least in regions of their greatest diversity, in the tropics, and greatest in regions of their least diversity, at higher latitudes for foraminifera or at the tropical margins for reef corals. Flessa and Jablonski (1996) found that the median age of a large sampling of Recent bivalve genera is also younger in the tropics than in higher latitudes. Crame (2000), working from a database of regional bivalve species diversity patterns, confirmed and extended this pattern (and see Clarke and Crame 2003). Furthermore, even families of marine mollusks show a muted though clear diversity gradient today (Campbell and Valentine 1977), and Crame concluded that the present bivalve gradient arose in large part because of a tropically centered radiation of heteroconch families that began in the Mesozoic. More recently, Jablonski, Roy, and Valentine (2006) determined the earliest appearances of genera and subgenera (hereafter, genera) of marine bivalves during the last eleven million years, and showed that bivalve genera found outside the tropics largely originated in the tropics and then expanded their ranges into higher latitudes, where they tend to outnumber extratropical endemics by about 3/1. Those genera have retained their tropical presence, however, so for bivalves the tropical regions are diverse partly because they contain both older and younger lineages, and are both cradles *and* museums. Younger lineages have not spread as far as older ones, and thus there is a latitudinal gradient in the average age of genera, from younger in the tropics to progressively older in the extratropics (with the oldest average ages in polar latitudes; Jablonski, Roy, and Valentine 2006, Valentine et al. 2008). Clearly, migration is a significant contributor to diversity patterns.

It may be that examining diversity history at different taxonomic scales may eventually provide important clues to large-scale macroecological dynamics. Jablonski (1993) has shown that significant evolutionary novelties such as recognized at the taxonomic level of orders appear preferentially in the tropics. While diversity trends in families also suggest preferential tropical originations (see the preceding), *N* is low, the trends appear over geological periods, and the dynamics involved at lower levels are largely obscure. At the generic level, the history of lineages becomes clearer, for the younger genera are being preferentially generated in the tropics, implying that the poleward paucity in diversity results from a lack of origination of the sort of novel morphologies recognized by taxonomists as genera. At the species level there is as yet not enough hard data on longevities in invertebrate groups to be certain whether there is a latitudinal age differential between tropics and extratropics. The ages of planktonic foraminiferal species reported by Stehli. Douglas, and Kafescioglu (1972) show no important latitudinal differences, but *N* is very low. Recent studies on a bipolar planktonic foraminiferal morphospecies indicate relatively recent (Quaternary) genetic diversification that suggests incipient or actual speciation(s), producing young high-latitude lineages (Darling et al. 2004). On the other hand, Buzas, Collins, and Culver (2002) found that when measured by alpha diversities, the LDG for benthic foraminiferal species has steepened (nearly doubled) through the Neogene, owing chiefly to a greater rate of species increase in lower latitudes. This pattern does not necessarily indicate the relative ages of low- and high-latitude species pools, however.

Adaptive Strategies and Diversity Regulation

If the trends seen in the data previously reviewed hold up, interpretations of diversity dynamics across latitudes are greatly constrained. One interpretation has been that the tropics are a source of new taxa and higher latitudes are sinks, with novelties spreading from low latitudes but with their species numbers progressively declining at higher latitudes through increasing extinctions. But it is hard to see how extinctions can be molding latitudinal diversities, given the age data on genera, families, and higher taxa mentioned previously. Certainly regional extinctions alter the shape of the latitudinal gradient, but whether such events permanently perturb the dynamics of the gradient is not clear. In the late Neogene of the northeastern Pacific, the generic extinction rate is higher in the temperate zone than the tropics but lower in the polar zone (above 60°) than in the temperate zone (Valentine et al. 2008), yet the diversity and age gradients do not reflect this nonmonotonic trend.

These data suggest a progressive restriction on the originations of new lineages with latitude, at least extratropically. Any such restrictions presumably arise from the macroecological effects of diversity-dependent factors. If a resource can be used up, selection might be able to produce strategies that mitigate that limitation. It would seem from the foregoing sections that diversity is highest in the more stable environmental regimes—the tropics on the shelves, in the euphotic zone of the oligotrophic oceanic water masses, and in the deep sea in general—and lowest in high latitudes and commonly in upwelling regions. Just how do species respond to stability-instability gradients in putative diversity-dependent factors? Assuming that seasonality is a leading cause of instability on a regional scale, it can be postulated that adaptation to increasing seasonality of resources might be accomplished by lowering resource demand, by excluding competitors, by increasing the breadth of available resources, and by protecting against temporarily inclement conditions.

Lowering of resource demand would be more likely to be a successful strategy in environments in which resource supply is reliable. Although competitive preclusion should be favored at all levels of stability, a relatively narrow supply of resources could be tolerated in trophically stable conditions (i.e., populations could persist on a specialized diet, for the prey which they are efficient at utilizing would be predictably present) and a zone of preclusion might be quite narrow, even though species might be able to feed somewhat opportunistically on a wider diet. Reducing body size and metabolism would also lower resource demands. In highly seasonal or otherwise trophically unstable environments, however, increasing dietary breadth would be favored where seasonality is so great that any given resource often tends to fail for a season or more, as is surely most common in high latitudes (and perhaps in upwelling regions). Inclement periods can be ridden out by drawing on energy stores obtained during high peaks of seasonal productivity, and/or by feeding at low levels of the trophic pyramid where resources are most abundant. Benthic invertebrates appear to exemplify these strategies; for example, in high latitudes, carnivorous forms are more commonly scavengers or generalized feeders, while in low latitudes a larger percentage are specialized predators.

As for factors associated with spatial heterogeneity, lowering resource demand should involve specializing in a narrow range of habitat parameters appropriate to the specialized diet, which again would be most favored in the presence of abundant competition and tolerated in the more stable regimes, and thus most common in low-latitude forms. Broad habitat tolerances would clearly benefit species in seasonal regimes, permitting larger populations and thus providing some buffer against local extinction. Also, in situations in which local resource failures occur, populations living in most environments would have the best chance of being represented in any habitat that happened to be favored. And habitat diversity should provide access to a greater variety of dietary items.

These adaptive scenarios are by no means new, having been entertained at least in part by many workers involved in large-scale diversity problems. The theoretically optimal strategy for a given environmental regime involves a number of components, and it is probably the case that species in a given regime commonly do not share all of the theoretically appropriate components, and often inhabit somewhat different regions of ecospace. It would not be surprising if species sharing some but not all adaptive components were found to have different, though perhaps overlapping, distributional patterns. Furthermore, in a regime in which generalists tend to be favored, there must (perhaps in principle) be opportunities created for specialists, and vice versa. While the proportion of specialists in high-latitude communities is evidently significantly lower than in low latitudes, specialists, like generalists, probably exist right across the spectrum of environmental conditions.

Summary: Hypothesis of Marine Diversity Determination

Marine patterns seem to suggest that the processes that determine the diversity of the species pool in any given climatic zone or major environmental compartment are related to environmental stability, and operate through the sorts of adaptive strategies that are most successful in a given environmental regime, leading to the following hypotheses. If resources are plotted as a pie diagram, then environments in the tropics permit species to persist on smaller slices of the pie than those in high latitudes, and local populations can become highly specialized. Selection for local specializations produces genetic divergences among species populations, often leading to speciation. The morphological specializations that accompany these divergences are commonly pronounced and lead to the recognition of subgenera and genera, and to the eventual evolution of families as the successful lineages spread, or even to higher taxa if their novel morphological features become the basis for body plan variations. In higher latitudes, though, specialization of species' populations commonly leads to the extinction of specialized demes, and species tend to display less genetic diversity, although within populations there may be a rather high degree of phenotypic plasticity, presumably related to generalist strategies. As a result, specialized groups from low latitudes are poor invaders of higher-latitude environments, and high-latitude faunas tend to contain a higher proportion of old lineages than do low-latitude faunas. In this scenario, the role of extinction is largely to provide openings for the successful invasion of species, some of which will introduce genera into regions from which they had been absent.

Within environmental compartments characterized by a given species pool there is a possibility that regional or local diversities are enriched by metapopulations that would become locally extinct if not reinforced by continued introduction of propagules, even though some might reproduce locally and sustain themselves even if subsequently isolated, until particularly inclement periods when their slice of the pie becomes too slim. Whether metapopulations are present or not, local diversities are sorted further by habitat heterogeneities and by ecological responses to local modifications of the regional environmental norms. Evidently, any explanation of large-scale taxonomic diversity patterns must involve multiple levels of effects set out in seriated and partly nested hypotheses.

ACKNOWLEDGMENTS

Thanks are due David Jablonski, Kaustuv Roy, and Zack Krug for many enlightening discussions, and Michael Rex and Gene Hunt for most helpful manuscript reviews. Figures were prepared by David K. Smith, Museum of Paleontology, University of California, Berkeley. These ideas have grown from research sponsored by the National Science Foundation, by NASA, and by a Faculty Research Grant from the University of California, Berkeley. University of California Museum of Paleontology publication number 1883.

REFERENCES

Allen, A. P., J. H. Brown, and J. F. Gillooly. 2002. Global biodiversity, biochemical kinetics, and the energetic-equivalence rule. *Science* 297:1545–48.

Altena, C. O. van R. 1971. The marine mollusca of Suriname (Dutch Guiana) Holocene and Recent. Part II. Bivalvia and Scaphopoda. *Zoologica Verhandelingen* No. 119:1– 99.

Angel, M. V. 1997. Pelagic biodiversity. In *Marine Biodiversity, Patterns and Processes,* ed. R. F. G. Ormond, J. D. Gage, and M. V. Angel, 35–68. Cambridge: Cambridge University Press.

Armbrecht, I., I. Pefecto, and J. Vandermeer. 2004. Enigmatic biodiversity correlations: ant diversity responds to diverse resources. *Science* 304:284–86.

Bé, A. W. H. 1977. An ecological, zoogeographic and taxonomic review of Recent planktonic foraminifera. In *Oceanic Micropalaeontology,* Vol. 1. ed. A. T. S. Ramsay, 1–10. London: Academic Press.

Bé, A. W. H., and R. W. Gilmer. 1977. A zoogeographic and taxonomic review of euthecosomatous Pteropoda. In *Oceanic Micropalaeontology,* Vol. 1, ed. A. T. S. Ramsay, 733–808. London: Academic Press.

Bohanak, A. 1999. Dispersal, gene flow, and population structure. *Quarterly Review of Biology* 74:21–45.

Briggs, J. C. 1974. *Marine zoogeography.* New York: McGraw-Hill.

Brinton, E. 1962. The distribution of Pacific euphausiids. *Bulletin of the Scripps Institution of Oceanography* 8:51–270.

Brown, J. H., J. F. Gillooly, A. P. Allen, V. M. Savage, and G. B. West. 2004. Toward a metabolic theory of ecology. *Ecology* 85:1771–89.

Buzas, M. A., L. S. Collins, and S. J. Culver. 2002. Latitudinal difference in biodiversity caused by higher tropical rate of increase. *Proceedings of the National Academy of Sciences USA* 99:7841–43.

Campbell, C. A., and J. W. Valentine. 1977. Comparability of modern and ancient marine faunal provinces. *Paleobiology* 3:49–57.

Clarke, A., and J. A. Crame. 2003. The importance of historical processes in global patterns of diversity. In *Macroecology, concepts and consequences,* ed. T. M. Blackburn and K. J. Gaston, 130–51. Oxford: Blackwell.

Connoly, S. R., D. R. Bellwood, and T. P. Hughes. 2003. Indo-Pacific biodiversity of coral reefs: Deviations from a mid-domain model. *Ecology* 84:2178–90.

Crame, J. A. 2000. Evolution of taxonomic diversity gradients in the marine realm: Evidence from the composition of Recent bivalve faunas. *Paleobiology* 26:188-214.

Crombie, A. D. 1945. On competition between different species of grammivorous insects. *Proceedings of the Royal Society, London B* 132:362–95.

Dana, J. D. 1853. On an isothermal oceanic chart, illustrating the geographical distribution of marine animals. *American Journal of Science,* Ser. 2, 16:153–67, 314–27.

Darling, K. F., M. Kucera, C. J. Pudsey, and C. M. Wade. 2004. Molecular evidence links cryptic diversification in polar planktonic protists to Quaternary climate dynamics. *Proceedings of the National Academy of Sciences USA* 101:7657–62.

Dawson, Y. E. 1951. A further study of upwelling and associated vegetation along Pacific Baja California, Mexico. *Journal of Marine Research* 10:39–58.

Ekman, S. 1953. *Zoogeography of the sea.* London: Sidgwick and Jackson.

Etter, R. J., and H. Caswell. 1994. The advantages of dispersal in a patchy environment: Effects of disturbance in a cellular automaton model. In *Reproduction, larval biology, and recruitment of the deep-sea benthos,* ed. C. M. Young and K. J. Eckelbarger, 284–305. New York: Columbia University Press.

Etter, R. J., and J. F. Grassle. 1992. Patterns of species diversity in the deep sea as a function of sediment particle size diversity. *Nature* 360:576–78.

Flessa, K. W., and D. Jablonski. 1995. Biogeography of Recent marine bivalve molluscs and its implications for paleobiogeography and the geography of extinctions: a progress report. *Historical Biology* 10:25–47.

———. 1996. The geography of evolutionary turnover: A global analysis of extant bivalves. In *Evolutionary paleobiology*, ed. D. Jablonski, D. H. Erwin, and J. H. Lipps, 376–97. Chicago: University of Chicago Press.

Gaylord, B., and S. D. Gaines. 2000. Temperature or transport? Range limits in marine species mediated solely by flow. *American Naturalist* 155:769–89.

Geyer, W. R., R. C. Beardsley, S. J. Lentz, J. Candela, R. Limeburner, W. E. Johns, B. M. Castro, and I. D. Soares. 1996. Physical oceanography of the Amazon shelf. *Continental Shelf Research* 16:575–616.

Glover, D. M., J. S. Wroblewski, and C. R. McClain. 1994. Dynamics of the transition zone in coastal zone color scanner-sensed ocean color in the North Pacific during oceanographic spring. *Journal of Geophysical Research* 99:7501–11.

Grassle, J. F. 1994. Ecological patterns in the deep-sea benthos: How are they related to reproduction, larval biology, and recruitment? In *Reproduction, larval biology, and recruitment of the deep-sea benthos*, ed. C. M. Young and K. J. Eckelbarger, 306–14. New York: Columbia University Press.

Grassle, J. F., and N. J. Maciolek. 1992. Deep-sea species richness: regional and local diversity estimates from quantitative bottom samples. *American Naturalist* 139:313–41.

Hubbs, C. L. 1948. Changes in the fish fauna of western North America correlated with changes in ocean temperature. *Journal of Marine Research* 7:459–82.

Hunt, G., T. M. Cronin, and K. Roy. 2005. Species-energy relationship in the deep sea: A test using the Quaternary fossil record. *Ecology Letters* 8:739–47.

Jablonski, D. 1993. The tropics as a source of evolutionary novelty through geological time. *Nature* 364:142–44.

———. 1996. Body size and macroevolution. In *Evolutionary paleobiology*, ed. D. Jablonski, D. H. Erwin, and J. H. Lipps, 256–89. Chicago: University of Chicago Press.

Jablonski, D., and R. A. Lutz. 1983. Larval ecology of marine benthic invertebrates: Paleobiological implications. *Biological Review* 58:21-89.

Jablonski, D., and K. Roy. 2003. Geographical range and speciation in fossil and living molluscs. *Proceedings of the Royal Society of London B* 401–6.

Jablonski, D., K. Roy, and J. W. Valentine. 2006. Out of the tropics: Evolutionary dynamics of the latitudinal diversity gradient. *Science* 314:102–6.

Jablonski, D., and J. W. Valentine. 1981. Onshore-offshore gradients in Recent Eastern Pacific shelf fauna and their paleobiogeographic significance. In *Evolution today: Proceedings of the 2nd international congress of systematic and evolutionary biology*, ed. G. G. E. Scudder and J. L. Reveal, 441–53. Pittsburgh: Carnegie-Mellon University.

Karlson, R. H., H. V. Cornell, and T. P. Hughes. 2004. Coral communities are regionally enriched along an oceanic biodiversity gradient. *Nature* 429:867–70.

Kittiwattanawong, K. 1999. The relation of reproductive modes to population differentiation in marine bivalves and gastropods. *Phuket Marine Biological Center Special Publication* 19:129–38.

Kohn, A. J. 1997. Why are coral reef communities so diverse? In *Marine biodiversity, patterns and processes*, ed. R. F. G. Ormond, J. D. Gage, and M. V. Angel, 201–15. Cambridge: Cambridge University Press.

Krug, A. Z., D. J. Jablonski, and J. W. Valentine, J. W. 2007. Contrarian clade confirms the ubiquity of spatial origination patterns in the production of latitudinal diversity gradients. *Proceedings of the National Academy of Sciences* 104:18129–18134.

Laptikhovsky, V. 2006. Latitudinal and bathymetric trends in egg size variation: A new look at Thorson's and Rass's rules. *Marine Ecology* 27:7–14.

Levin, L. A., R. J. Etter, M. A. Rex, A. J. Gooday, C. R. Smith, J. Pineda, C. T. Stuart, R. R. Hessler, and D. Pawson. 2001. Environmental influences on regional deep-sea species diversity. *Annual Review of Ecology and Systematics* 32:51–93.

Longhurst, A. 1998. *Ecological geography of the sea.* San Diego: Academic Press.

Loreau, M., and N. Mouquet. 1999. Immigration and the maintenance of local species diversity. *American Naturalist* 154:427–40.

MacArthur, R. H., and E. O. Wilson. 1967. *The theory of island biogeography.* Princeton, NJ: Princeton University Press.

Maurer, B. A., and M. P. Nott. 1998. Geographic range fragmentation and the evolution of biological diversity. In *Biodiversity dynamics,* ed. M. L. McKinney and J. A. Drake, 31–50. New York: Columbia University Press.

Meyer, C. P. 2003. Molecular systematics of cowries (Gastropoda: Cypraeidae) and diversification patterns in the tropics. *Biological Journal of the Linnean Society* 79:401–59.

Oliver, P. G. 1992. *Bivalved seashells of the Red Sea.* Verlag Christa Hemmen, Wiesbaden, and National Museum of Wales, Cardiff.

Pearse, J. S. 1994. Cold-water echinoderms break "Thorson's Rule." In *Reproduction, larval biology, and recruitment of the deep-sea benthos,* ed. C. M. Young and K. J. Eckelbarger, 26–39. New York: Columbia University Press.

Poore, G. C. B., and G. D. F. Wilson. 1993. Marine species richness. *Nature* 361:597–98.

Rex, M. A., R. J. Etter, and C. T. Stuart. 1997. Large-scale patterns of species diversity in the deep-sea benthos. In *Marine biodiversity, patterns and processes,* ed. R. F. G. Ormond, J. D. Gage, and M. V. Angel, 94–121. Cambridge: Cambridge University Press.

Rice, A. L., and P. J. D. Lambshead. 1994. Patch dynamics in the deep-sea benthos: The role of heterogeneous supply of organic matter. In *Aquatic ecology: Scale, pattern and process,* ed. P. S. Giller, A. G. Hildrew, and D. G. Raffaelli, 469–97. Oxford: Blackwell Scientific.

Rosenzweig, M. L. 1995. Species diversity in space and time. Cambridge: Cambridge University Press.

Roy, K., D. Jablonski, and K. K. Martien. 2000. Invariant size-frequency distributions along a latitudinal gradient in marine bivalves. *Proceedings of the National Academy of Sciences USA* 970:13150–55.

Roy, K., D. Jablonski, and J. W. Valentine. 2000. Dissecting latitudinal diversity gradients: Functional groups and clades of marine bivalves. *Proceedings of the Royal Society of London B* 267:293–99.

———. 2004. Beyond species richness: Biogeographic patterns and biodiversity dynamics using other metrics of diversity. In *Frontiers of biogeography: New directions in the geography of nature,* ed. M. V. Lomolino and L. R. Heaney, 151–70. Sunderland, MA: Sinauer.

Roy, K., D. Jablonski, J. W. Valentine, and G. Rosenberg. 1998. Marine latitudinal diversity gradients: Tests of causal hypotheses. *Proceedings of the National Academy of Sciences USA* 95:3699–3702.

Royer, T. C. 1988. Coastal processes in the northern North Pacific. In *The sea: The global coastal ocean,* ed. A. R. Robinson and K. H. Brink, 395–414. New York: Wiley.

Sanders, H. L., and R. R. Hessler. 1969. Ecology of the deep-sea benthos. *Science* 163:1419–24.

Sathyendranath, S., A. Longhurst, C. M. Caverhill, and T. Platt. 1995. Regionally and seasonally differentiated primary production in the North Atlantic. *Deep-Sea Research* 42:1773–1802.

Shetye, S. R., and A. D. Gouveia. 1988. Coastal circulation in the North Indian Ocean. In *The sea: The global coastal ocean,* ed. A. R. Robinson and K. H. Brink, 523–56. New York: Wiley.

Stebbins, G. L. 1974. *Flowering plants: evolution above the species level.* Cambridge: Harvard University Press.

Stehli, F. G., and J. W. Wells. 1971. Diversity and age patterns in hermatypic corals. *Systematic Zoology* 20:115–26.

Stehli, F. G., R. G. Douglas, and I. A. Kafescioglu. 1972. Models for the evolution of planktonic foraminifera. In *Models in paleobiology,* ed. T. J. M. Schopf, 116–28. San Francisco: Freeman, Cooper.

Stuart, C. T., and M. A. Rex. 1994. The relationship between developmental patterns and species diversity in deep-sea prosobranch snails. In *Reproduction, larval biology, and recruitment of the deep-sea benthos,* ed. C. M. Young and K. J. Eckelbarger, 118–36. New York: Columbia University Press.

Sverdrup, H. U., M. W. Johnson, and R. H. Fleming. 1942. *The oceans.* New York: Prentice-Hall.

Taylor, J. D. 1997. Diversity and structure of tropical Indo-Pacific benthic communities: Relation to regimes of nutrient input. In *Marine biodiversity,* ed. R. F. G. Ormond, J. D. Gage, and M. V. Angel, 178–200. New York: Cambridge University Press.

Thorson, G. 1950. Reproductive and larval ecology of marine bottom invertebrates. *Biological Reviews* 25:1–45.

Valdovinos, C., S. A. Navarette, and P. A. Marquet. 2002. Mollusk species diversity in the southeastern Pacific: Why are there more species towards the pole? *Ecography* 26:139–44.

Valentine, J. W. 1972. Conceptual models of ecosystem evolution. In *Models in paleobiology,* ed. T. J. M. Schopf, 192–215. San Francisco: Freeman, Cooper.

———. 1973. *Evolutionary paleoccology of the marine biosphere.* Englewood Cliffs, NJ: Prentice Hall.

———. 1983. Seasonality: Effects in marine benthic communities. In *Biotic interactions in recent and fossil benthic communities,* ed. M. J. S. Tevesz and P. L. McCall, 121–56. New York: Plenum.

Valentine, J. W., D. Jabonski, A. Z. Krug, and K. Roy, K. 2008. Incumbency, diversity, and latitudinal gradients. *Paleobiology* 34:169-178.

Walker, T. D., and J. W. Valentine. 1984. Equilibrium models of evolutionary species diversity and the number of empty niches. *American Naturalist* 124:887–99.

Williams, S. T., and D. G. Reid. 2004. Speciation and diversity on tropical rocky shores: A global phylogeny of snails of the genus *Echinolittorina. Evolution* 58:2227–51.

Witman, J. D., R. J. Etter, and F. Smith. 2004. The relationship between regional and local species diversity in marine benthic communities: A global perspective. *Proceedings of the National Academy of Sciences USA* 101:15664–69.

CHAPTER TWO
PLANKTON POPULATIONS AND COMMUNITIES

WILLIAM K. W. LI

The most general problem of marine biology is to understand the distribution and abundance of life in the sea. The approach to this problem must be primarily statistical through the development of significant relationships between large quantities of observations on biological and physical events, occurring often in widely scattered places.

Alfred C. Redfield (1960)

Introduction

More than 70 percent of the Earth's surface is covered by the oceans, and more than 99 percent of the Earth's biosphere volume is contained in seawater. In the ocean, the pelagic realm is the largest habitat, and here, plankton account for the greatest number of organisms and also for the majority of biomass. Plankton, which are organisms that are passively advected in water currents, constitute a tremendous biological diversity that not only spans all three domains of the living world—*Archaea*, *Bacteria*, and *Eukarya*—but also includes viruses that are agents of their mortality. It is in marine plankton that we find the smallest free-living cell (*Pelagibacter ubique*), the smallest photoautotrophic cell (*Prochlorococcus marinus*), and the smallest eukaryote (*Ostreococcus tauri*). Globally, the number of plankton organisms is estimated to be more than several-fold 10^{28}, a hyper-astronomically large number that almost certainly harbors much undiscovered biodiversity.

For a long time, ecological research in biological oceanography has concentrated on phytoplankton, often the diatoms and dinoflagellates, and on metazooplankton, often the crustaceans. However, the paradigm change brought about by microbial oceanography (Kirchman 2000; Karl 2002) now requires a more inclusive view of all the biological components. Most of the smallest forms do not have names assigned from the binomial nomenclature system. Molecular analysis indicates their phylogenetic affinities but not necessarily their in situ metabolic performance. Furthermore, the existence of many other forms is indicated only by their genes in an "environmental genome." Even with so much yet unknown, it seems important that all these elements of the pelagic component need to be considered if comprehensive solutions are sought to maintain biodiversity, to sustain the harvest of fishery resources, and to mitigate climate change. Earth's real biodiversity, it has been said (Nee 2004), is invisible to the naked eye, whether we like it or not.

The study of the abundance, distribution, and diversity of plankton has been a focus of biological oceanography for much of the twentieth century. The questions have often been framed in the context of biogeography: what species are present at a particular place, how many individuals are there of each species, what are defining attributes of characteristic assemblages, and where do they occur or not occur? This approach has been fruitful, especially for the microplankton and mesoplankton, which have had a long history of study because they are quite easily observed under a light microscope. However, even for these groups, difficulties remain in accurate taxonomic identification, in sparseness of sampling locations, in confirming records of species absence, and perhaps in the concept of species itself (Wood and Leatham 1992). Even organisms such as copepods with clear morphological characteristics may be cryptically speciated (Goetze 2003).

In the case of the smallest microbial plankton, the difficulties are more daunting because a taxonomy based on morphological features is unsuitable. The answer to what constitutes a bacterioplankton species is generally based on similarities in genome sequences; but this is not necessarily straightforward because, for example, sequence similarities do not always correlate well with DNA-DNA cross-hybridizations. In addition, prokaryotes share genes through recombination; in some cases the exchange of genetic information is so promiscuous that the degree of linkage equilibrium approaches that of a sexual population (Papke et al. 2004). It is therefore difficult to define the ecologically relevant biological unit for these smallest of organisms. Due to their hyper-astronomical abundances, these organisms might be ubiquitously dispersed and have low rates of local extinction, leading to an expectation that they can be found wherever their required habitats are realized (Fenchel and Finlay 2004). Yet, evidence based on fine molecular resolution indicates that physical isolation or independent adaptation are important to microbial diversification (Papke and Ward 2004). For microbial plankton, the principles of fundamental evolutionary mechanism (Doolittle and Bapteste 2007), biogeographic distribution (Hughes-Martiny et al. 2006), and ecological relationships (Prosser et al. 2007) are still under much consideration.

Whether plankton communities are assessed by microscopes or by microarrays, present research largely emphasizes the distinctions between different kinds of organisms and ecosystems, and on the extensive spatial and temporal variation within ecosystems. Ecological systems are viewed as highly idiosyncratic, being contingent on the organisms present, and the particular circumstances of the environment in space and time. Lawton (1999) has argued that this contingency is tractable at two levels of ecological organization. In relatively simple cases when the number of interacting populations is small, the contingency may be manageable. However, at the intermediate scales of community ecology, the overwhelming number of case histories complicates the contingency to an unmanageable form. Interestingly, the contingency becomes manageable once again in large sets of species, over large scales of space and time, in the form of statistical patterns when local details are subsumed; in other words, macroecology. This therefore is the aim of the present chapter: to find widespread, repeatable patterns emerging from a large collection of plankton observations, in spite of the numerous contingent processes that underlie the collection. The illustrative examples in this chapter are restricted to the smallest microbial plankton, for this is work with which I am most familiar; however, the philosophical underpinning, the statistical approaches and the general conclusions may prove to be more widely applicable. It is sobering to realize that more than half a century has passed since Redfield articulated this holistic view as the primary approach to the most general problem of marine biology, yet we still need to be reminded once in a while (Parsons 2003).

Macroecology and Comparative Analysis

Macroecology is a way of studying relationships between organisms and their environment that involves characterizing and explaining statistical patterns of abundance, distribution, and diversity (Brown 1995). Macroecology is concerned with patterns at large spatial scales, but it is not simply a subdiscipline of biogeography, because its primary emphasis is the understanding of nature from observations of a large number of system entities, no matter the scale (Blackburn and Gaston 2002). In biological oceanography, macroecology can be seen as an extension of the comparative analysis of ecosystems. Comparative ecology describes nature by specifying the possible states of ecological properties and by searching for patterns and trends from the variance (Duarte 1991). It is a well-established method of investigation in aquatic microbial ecology—for example, in constraining the ranges of abundances and activities of particular trophic groups (Gasol and Duarte 2000). To predict the pelagic characteristics of a given area of the ocean, it has been proposed to integrate ecological considerations into the regional physical oceanography, arriving at an ecological geography of the sea (Longhurst 1998). This establishes a rational basis upon which the ocean can be partitioned into distinguishable regions permitting comparisons.

The philosophy of inductive reasoning is similar in comparative analysis and in macroecology. At times, these empirical approaches may be used to extract the probable from among the possible, and therefore are methods for prediction. At other times, they may be used to distinguish among alternate hypotheses concerning the relationship between variables, and therefore are methods for inference. Comparative analysis recognizes differences between ecological systems and seeks similarities between them; macroecology emphasizes universal patterns emerging from the statistical phenomenology of a large number of observations.

The linkage between the two approaches can be illustrated with an example showing the widely studied relationship between the abundance of heterotrophic bacterioplankton and the biomass of phytoplankton (Li, Head, and Harrison 2004). The ecological coupling between these two trophic groups is of significant interest because the immediate fate of a great deal of primary production is determined by whether it is assimilated or respired by bacteria. In the open ocean, heterotrophic bacteria depend on labile dissolved organic substrates originating from the autochthonous phytoplankton, directly or indirectly. These substrates appear as exudates from healthy phytoplankters, as cytosolic components liberated by viral lysis or autolysis, and as egesta of grazers that have consumed the phytoplankton. To greater or lesser extents influenced by allochthonous inputs of utilizable substrates, different pelagic systems display a positive covariation of bacterial and phytoplankton biomass indicating the strength of resource control (Gasol and Duarte 2000). Comparative analysis of different oceanic provinces based on a single statistical model, namely a linear regression of the logarithmic-

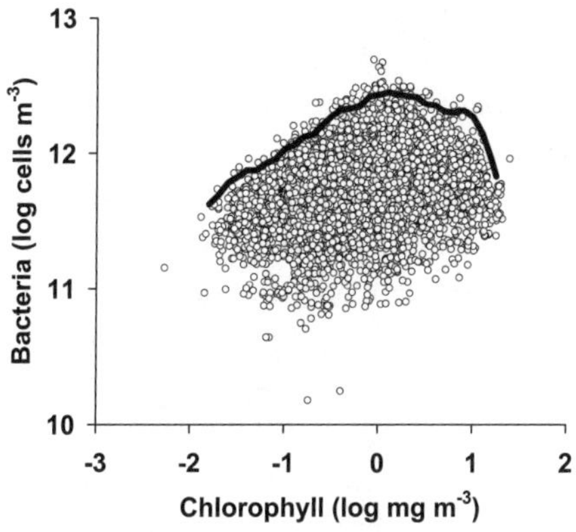

FIGURE 2.1 Consolidating the data relating heterotrophic bacterial abundance to chlorophyll concentration from ten biogeochemical provinces reveals a curvilinear upper boundary that is not strongly evident in separate data sets. The curved line is the 99 percentile of consolidated bacterial distributions within successive binned chlorophyll intervals of 0.1 logarithmic unit. Data sources and description in Li et al. (2004).

transformed variables, indicates that there are differences in the slopes and intercepts (Li, Head, and Harrison 2004). Each province may be examined in detail for regional study, but when all the data are consolidated, a new perspective emerges (see fig. 2.1). Bacterioplankton abundance in the ocean is no longer adequately described by a straight-line model because the upper constraint is defined by a relationship that undergoes a change in the sign of the slope. Fundamentally, the coupling between these trophic groups must be different at low and high levels of phytoplankton, perhaps a change from bottom-up resource control to top-down mortality control. The macroecological approach seeks to identify these emergent perspectives.

Plankton Biodiversity and Biogeography

The total number of living species on earth with a Latin binomial name is about 1,500,000 (May 1988). Of these, it is estimated that 185,000 species are aquatic (Fenchel 1993), but this is greatly uncertain, since the marine subset alone might be as high as 300,000 (Snelgrove 2001). In the marine plankton, there are perhaps 3,000 species of metazooplankton (Longhurst 2001); 3,900 species of free-living protozooplankton (Finlay 2001); and 3,900 species of phytoplankton (Sournia, Chrétiennot, and Ricard 1991). Recognized species of algae, both macro and micro forms, and both freshwater and marine, number about 10,000 to 40,000 (Norton, Melkonian, and Andersen 1996; Falkowski and Raven 1997), of which 5,000 to 10,000 may be diatoms and 2,000 may be dinoflagellates. These estimates are based on the Linnean classification system and rely largely on the concept of morphospecies.

For the smallest forms of plankton, biological diversity must be assessed by the methods of cell or molecular biology, because traditional criteria of morphology and physiology cannot adequately resolve their differences. Since only a very small number of these microbes can be isolated into culture, sensitive methods have been developed to directly interrogate bio-optical properties or genomic sequences in situ. In some cases, characteristic combinations of pigment fluorescence and cellular light scatter permit recognition at the genus level, most notably of the cyanobacteria *Synechococcus* and *Prochlorococcus*. In one case, a unique fluorescence transient detected in the infrared region is indicative of cells of unusual metabolic capabilities—namely aerobic, anoxygenic photoheterotrophy implied by the presence of bacteriochlorophyll *a* in the α-Proteobacterium genus *Erythrobacter* (Kolber et al. 2000). In another case, the gene encoding the photopigment rhodopsin can be found in DNA belonging to an uncultivated γ-Proteobacterium prevalent in the bacterioplankton, thus indicating a new type of light-driven energy-generating mechanism in the sea (Béjà et al. 2000). The native fluorescence properties of many common organisms have still not been well studied. For example, the discovery of ultraviolet-excited blue autofluorescence of the neurotoxin producing diatom *Pseudo-nitzschia multiseries* may prove to be a practical marker for rapid ocean monitoring of this and perhaps related species (Orellana, Petersen, and van den Engh 2004). A full resolution of all cellular pigments can be made by chromatographic separation, but the resulting taxonomic identification is generally at the rank of class, which is relatively high. In all cases, phylogenetic relationships are best revealed by comparative sequencing of homologous genes, most notably those for ribosomal RNA. Thus, the concept of morphospecies is replaced by the concept of ribotype or phylotype. The sequencing of genes from the amplified DNA of unidentified plankton filtered from seawater has revealed not only new lineages in the form of clades in existing phyla, but even the existence of members from another domain, the *Archaea*. These and other methods have led to a multitude of reports on the hidden diversity of the sea (Karl 2002), with justified various claims that the organisms were unsuspected, unexpected, previously unkown, hitherto unrecognized, and novel. Molecular biology has changed the manner in which plankton ecology is prosecuted, most particularly microbial plankton ecology. Molecular methods must reconcile the new and old ways of organism classification (Caron, Countway, and Brown 2004), but in some respects, microbial oceanography is already well into the brave new world (Kirchman and Pedrós-Alió 2007).

The large-scale geographic distribution of plankton organisms is known for only a relatively small number of species. Phytogeographies have long existed for net phytoplankton such as diatoms and large dinoflagellates (Guillard and Kilham 1977; Semina 1997). A renewed sense of urgency to map the distribution for some from this group lies in an apparent global increase in harmful algal blooms. It is estimated that about 300 algal species can achieve abundances sufficient to discolor the water, and that about forty species are toxigenic (Hallegraeff 1993). In a small number of cases, distinctive optical characteristics allow particular species (e.g., *Trichosdesmium*) to be detected by remote sensing, allowing them to be mapped on a synoptic basis at large scale. Zoogeographies refer mainly to metazooplankton such as copepods and euphausiids (McGowan and Walker 1993), but global distributions of heterotrophic protists are also becoming available (Massana et al. 2006). Maps of mesoplankton species tend to be concordant at large spatial scale and reflect the zonal distribution of oceanographic properties (Longhurst 2001). These properties, such as temperature, water-column stratification, and primary production are the likely factors regulating metazooplankton diversity (Gibbons 1997; Beaugrand et al. 2002; Woodd-Walker, Ward, and Clarke 2002).

Although few microbial plankters have been mapped on a global scale, some newly described bacterial distributions are highly resolved and show concordance to oceanographic properties. The photoautotrophic cyanobacterium *Prochlorococcus* has a basis of range, or center of distribution between 40°N and 40°S (Partensky, Hess, and Vaulot 1999), with different ecotypes of this species partitioning the available niches along environmental gradients, most importantly temperature (Johnson et al. 2006). On the other hand, a new phylotype of α-Proteobacterium affiliating with the *Roseobacter* clade occurs in temperate and polar waters, but apparently not in tropical and subtropical regions (Selje, Simon, and Brinkhoff 2004). The other major cyanobacterium (*Synechococcus*) and the dominant bacterioplankton clade SAR11 (*Pelagibacter ubique*) are both ubiquitous (Partensky, Blanchot, and Vaulot 1999; Morris et al. 2002). The large-scale distribution of *Synechococcus* in the southern hemisphere at about 30°S is remarkably similar in the Pacific, Atlantic, and Indian Oceans (Bouman et al. 2006). In each ocean, there is a U-shaped pattern indicating low abundance in the gyre center and high abundance at the gyre edges, roughly concordant with the concentrations of phytoplankton in general (see fig. 2.2). This gives a strong impression that general features of ocean circulation are dominant factors in plankton distribution at this scale. In the pelagic ocean, the dynamic mixing of the hydrographic regime maintains a large-scale pattern of plankton biodiversity that is related to the main circulation patterns of the major ocean basins (Angel 1993).

Physical barriers to dispersion are permeable to some species and formidable to others, meaning that at global scales, patterns of pelagic gene

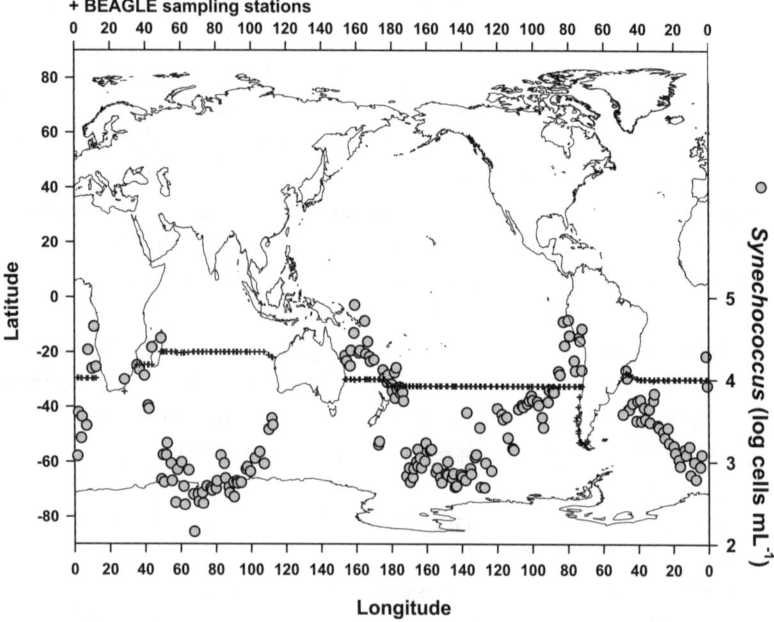

Fig. 2.2 *Synechococcus* in surface waters collected between 20° and 30°S from August 2003 to February 2004 during the Blue Earth Global Expedition has a U-shaped pattern of abundance in each major ocean basin, giving a strong impression that general features of ocean circulation are dominant factors in its large-scale distribution.

flow may be species specific (Goetze 2003). The test for ubiquity, or a cosmopolitan distribution, would be most persuasive for a bipolar species because of the long dispersal distances and the transition through temperature extremes. Conceivably, most so-called bipolar species really have a much wider cold-water distribution by progressive submergence equatorward (Longhurst 2001), or perhaps by using corridors of cool water provided by upwelling, as occurs in boundary currents (Darling et al. 2000). Indeed, supposedly characteristic Antarctic diatoms, such as *Thalassiosira antarctica*, also occur in the Northern hemisphere, giving rise to the suggestion that diatom survival under conditions of the deep ocean may be longer than suspected (Guillard and Kilham 1977). Some specimens of planktonic foraminifers from both poles have completely identical sequences in the small subunit ribosomal DNA, implying that gene flow has occurred between the bipolar populations (Darling et al. 2000). At the same time, this and other studies, such as on the diatom *Pseudo-nitzschia delicatissima* in Naples (Orsini et al. 2004) emphatically show that several distinct molecular lineages can be present in a single morphospecies.

In addition to natural mechanisms of dispersal, there is also global translocation of plankton via human activities. This complicates the study of their ecology and evolution. The inoculation of waters with nonindigenous species by the worldwide exchange of ship ballast water, aquaculture products, and perhaps even discarded laboratory cultures is an effective way to eliminate many natural barriers that might have previously existed.

In principle, ubiquitous dispersal leads to a relatively low number of species, and a high ratio of local-to-global diversity (Finlay and Fenchel 1999). It has been postulated that because asexual microorganisms are astronomically abundant, they have extremely high rates of dispersal and low rates of local extinction. If every microbe can be everywhere, then every habitat contains a pool of species on which the environment selects. If it is possible to explain the distribution of microbes strictly by considering the habitat, then the search for macroecological patterns is presumably freed of historical contingencies. Concurrent information on environmental similarity and community similarity is required to distinguish the contributions of environmental and historical effects on biogeography (Hughes-Martiny et al. 2006), but alternative hypotheses in this context have rarely been tested systematically for marine plankton. In part, the influence of environment versus history depends on what is perceived to be the relevant ecological unit. For example, eukaryote morphotypes may provide evidence favoring a view of cosmopolitan distribution (Fenchel and Finlay 2004), but prokaryote ribotypes may require a more nuanced view (Dolan 2005).

What then is the number of bacterioplankton species in the ocean? Calculations based on a lognormal species-abundance curve, and a limiting condition that the rarest species has only a single individual, indicate at most 163 species in one milliliter of seawater, and at most 2 million species in the entire ocean (Curtis, Sloan, and Scannell 2002). High-throughput DNA shotgun sequencing on the bacterioplankton of the Sargasso Sea gives an estimate of at least 1,824 genomes, and perhaps up to 47,733 genomes (Venter et al. 2004). Considering the species richness versus body size relationship (Fenchel 1993), these estimates suggest a relatively low richness for such small organisms.

It is likely, however, that much of marine microbial diversity resides in an underexplored rare biosphere (Sogin et al. 2006) whose members form a seed bank of global biodiversity (Pedrós-Alió 2006). It is now evident that microbial diversity will be revealed, not in the mere richness of species, but in the complexity of biological organization. Most marine bacterial species appear to be assemblages of numerous strains that differ greatly in gene content, and presumably in ecological niches. For example, co-occurring populations of *Prochlorococcus* (deemed to be closely related on the basis of

16S rRNA sequences) can have very different physiologies and significantly different whole genome sequences. These co-isolates exhibit a clear ecotypic differentiation that can be explained in part by their gene differences (Moore, Rocap, and Chisholm 1998; Rocap et al. 2003). Multiple ecotypes broaden the environmental niches over which a species can thrive, extending the geographic range. The metagenomes of bacterioplankton characterized on a transect from the northwest Atlantic to the eastern tropical Pacific confirm extensive microdiversity at the subtype level (Rusch et al. 2007). Collectively, the ecotypes constitute a virtual pangenome many times larger than any real genome, rendering quixotic any quest for a uniform bacterial species concept (Cullen et al. 2007).

The deciphering of genetic codes of pure culture strains is an important approach in plankton studies. The sequencing of whole genomes has already been completed, or is in progress for an important selection of marine microbes totalling fourteen *Archaea*, forty-seven *Bacteria,* and thirteen *Eukarya* (Moran and Armbrust 2007). These include the α-Proteobacteria *Pelagibacter ubique* and *Silicibacter pomeroyi*, several strains of the picocyanobacteria *Prochlorococcus* and *Synechococcus*, the filamentous cyanobacterium *Trichodesmium erythraeum*, the coccolithophorid *Emiliania huxleyi*, the diatoms *Thalassiosira pseudonana* and *Pseudo-nitzschia multiseries*, as well as the prasinophytes *Ostreococcus tauri* and *Micromonas pusilla*. An important recent advance is the amplification of the genome from a single cell (Zhang et al. 2006). This development circumvents the need to cultivate microbes before DNA sequencing, raising the possibility of genomic studies on single cells directly isolated from the ocean. Indeed, fluorescence-activated cell sorting can be combined with gene amplification to generate sufficient quantities of genomic DNA from single cells for specific reactions, allowing phylogeny and metabolism to be matched on an individual cell basis (Stepanauskas and Sieracki 2007).

These developments bring us to an interesting point in the progress of plankton ecology. More than a quarter century ago, the paradigm of the marine microbial loop was born on the recognition that an exceedingly large percentage of bacterioplankton in the ocean was unaccounted for by laboratory culture collections. Today, successes in the isolation and cultivation of bacterioplankton using innovative protocols and technologies (Rappé et al. 2002; Zengler et al. 2002) herald a resurgence of the experimental approach in autecology. We have therefore almost come full circle and must recall the admonition of Redfield (1960) not to indulge in the laboratory method of biology to the exclusion of observations in nature. The integration of these approaches in a framework of genomic, biochemical, and environmental data appears to point the way ahead (Azam and Worden 2004), but the concept of a molecular continuum from organisms to the environment is hugely complex. To reduce the complexity, we consider a macroecology based on the size of the organism. In this, we have also come in a circle, because the concept of an ecosystem in terms of a size spectrum (Elton 1927), the formulation of this concept as a theory of the pelagic community (Platt and Denman 1977, 1978), and the supporting empirical evidence from the ocean (Quiñones, Platt, and Rodríuez 2003) have all been well established as deliberate attempts to simplify ecological complexity by aggregating taxonomic diversity.

Plankton Size

Body size, or cell size in the case of unicellular organisms, is a fundamental organizing principle in plankton ecology. Beginning from the early days of oceanography, the selective retention of plankton in nets of different mesh sizes has provided a first-order basis for a size-based ecology. In modern research, there is an array of electronic instruments that resolve the size of marine particles to a much higher degree. However, marine plankton range over six decades in linear dimension, and there is no single instrument capable of spanning this spectrum. Complete community size spectra can be constructed by merging data acquired by different analytical methods such as microscopy, conductometry, cytometry, gravimetry, acoustics, and optics. The generally accepted size nomenclature is based on the International System of Units prefixes of the approximate live weight in grams for successive logarithmic intervals of the organism width or length: thus, femtoplankton (0.02–0.2 μm), picoplankton (0.2–2 μm), nanoplankton (2–20 μm), and microplankton (20–200 μm). The largest class is traditionally named mesoplankton (200–20,000 μm).

A compelling property of plankton size spectra is the small variance of the scaling factor connecting abundance (N) to organism size (M) across different trophic levels. It is commonly observed that $N \propto M^b$ where b has a narrow range of about -1 ± 0.02 (Quiñones. Platt, and Rodríguez 2003 and references therein). As a corollary, this leads to long-standing evidence indicating only slight or no size variance of total standing plankton biomass $B = NM \propto M^{b+1}$. In the case where b is exactly -1 (and therefore $b + 1 = 0$), a general explanation can be proposed that combines a within-trophic level scaling of metabolism according to $M^{-3/4}$, together with canonical assumptions of a 10% efficiency of energy transfer and a 10,000 fold difference in average size between trophic levels (Brown and Gillooly 2003). This general explanation provides the macroecological basis for examining the transfer of production from primary producers to successive levels of consumers (Parsons 2003). There are, however, notable exceptions to the canonical scheme that presumably lead to statistical departures from the major trend. In the plankton, the assumption of a predator being very much larger than its prey is sometimes overly simplified, better suited to simple linear food chains than the complexity of the microbial ocean. By far, the most important consumers of picoplankton are the nanoplankton. The latter are generally at most ten-fold greater in linear dimension, and therefore only 1,000-fold greater in size. In some cases, the predator-prey pyramid is even inverted. Myzocytosis, a feeding mode employed by some dinoflagellates in which the cell contents of the prey are sucked out by the dinoflagellate peduncle is a remarkable example of the small consuming the large. An extreme, but local example may be fish mortality due to micropredation by the dinoflagellate *Pfiesteria* (Vogelbein et al. 2002). In reality, animals eating plants (or vice versa) is only one of many types of trophic interactions. Mixotrophy and symbiosis are not uncommon amongst protistan taxa and can alter the efficiencies by which biomass is transferred through the food web (Caron 2000). Some of these complexities can be incorporated into a more general food web model extending from metabolic theory to predict different abundance scalings using other values of the energy transfer efficiency and predator-prey size ratio (Brown et al. 2004).

Instead of a trophic model for the size spectrum, Platt and Denman (1977, 1978) adopted a different approach by modeling a continuous flow of energy from small to large organisms. Allowing for uncertainties in the underlying coefficients, an approximate range for b from -0.82 to -1.23 can be admitted (Quiñones, Platt, and Rodríguez 2003). This gives support to $B \propto M^{-0.22}$, which is in good agreement with actual observations.

Whatever the numerical value of b, and whatever the underlying mechanism, it is evident that plankton size serves importantly to illustrate the emergence of regularity from complex natural systems. Regardless of the slope, community size spectra appear continuous, without strong peaks or domes that would indicate dominance of particular-sized organisms. At the low end of the size spectrum, the community is assembled from the femtoplankton, which consist mainly, if not solely, of viruses. Occurring at densities exceeding 10^7 mL^{-1}, or about ten times that of picoplankton, viruses constitute the smallest and most numerous biological entities in the ocean. Many of the newly discovered free-living organisms belong to the picoplankton and nanoplankton: *Archaea*, chemoheterotrophic and photohetrotrophic bacteria, photoautotrophic cyanobacteria, new stramenopiles (flagellated protists with hollow mastigonemes)—both photosynthetic and nonphotosynthetic, and others. For the most part, existing theories of the size continuum are based on observations that do not fully account for these microbial entities. Reassuringly, recently constructed microbial size spectra appear to conform in shape and slope to spectra of larger plankton. These new spectra are assembled from individual size distributions of heterotrophic bacteria, *Prochlorococcus*, *Synechococcus*, eukaryotic picophytoplankton, and nanophytoplankton. The individual distributions are strongly peaked, but partial overlap in size from one to another results in smoothed ensembles. Where one group (e.g., *Prochlorococcus*) might be missing, there might be more of another (e.g., heterotrophic bacterioplankton) or a broadening of the distribution of another (e.g., *Synechococcus*) to fill the gap (Cavender-Bares, Rinaldo, and Chisholm 2001).

All of this leads to speculation that plankton ecosystems tend to self-organize into states of scale invariance (Rinaldo et al. 2002). Holistic descriptions of self-organization in complex marine ecosystems are not new. Earlier such work tended to be abstract and prescriptive, but they fully anticipated the technological developments for the necessary data to implement a theory of macrobiology (Ulanowicz and Platt 1985). There is now a substantial body of empirical observations to revisit this theme. A closer examination will therefore be made of the statistical distribution of selected microbes after a diversion into how oceanographic research has led to collection of data suitable for macroecology.

Plankton Data

Macroecology is an endeavor that demands extensive observation of nature. Since the late nineteenth century, when *HMS Challenger* circumnavigated the globe and recorded thousands of new marine species, observations of plankton abundance, distribution, and diversity have relied on collection of samples at sea, followed by identification and quantification of the organisms, usually sometime later upon return to the laboratory. Although the capabilities of both the sampling platforms and the analytical methods have vastly improved, the time-consuming and labor-intensive nature of the work have not changed. Sampling designs are always a compromise between spatial and temporal considerations. An extensive spatial grid of stations can only be revisited once in a while, whereas frequent return to a station can only be made at a small number of locations. A long time is needed to build up a large data set, yet neither analytical methodologies nor scientific paradigms remain immutable over the years. Meta-analysis of disparate studies is often a suitable way to undertake a synthesis; sometimes it is the only way. However, it lacks the persuasiveness of an internally consistent data set that can arise from a purposefully designed exercise.

One such exercise was the Joint Global Ocean Flux Study (JGOFS), an international effort with participation from more than 20 nations and a full decade of ocean observations. The primary goal of JGOFS was to determine and understand on a global scale the processes controlling the time-varying fluxes of carbon and associated biogenic elements in the ocean. To this end, many observations were made on plankton stocks and processes in diverse provinces and in sustained time-series fashion at a small number of important oceanic locations. Another remarkable undertaking is the Continuous Plankton Recorder (CPR) Survey. Using a plankton sampling instrument designed to be towed from merchant ships on their normal sailings, this survey monitors the near-surface plankton on a network of routes, largely in the North Atlantic, but also in the North Pacific and elsewhere. With a history of more than seventy years, this survey is an important repository of information concerning annual and decadal changes in marine phytoplankton and zooplankton. Yet another program of sustained opportunistic large-scale sampling is the Atlantic Meridional Transect (AMT) program. Twice a year, a vessel sails between the United Kingdom and the Falkland Islands: southbound in September and northbound in April. First begun in 1995, this ongoing program samples plankton across several major oceanic provinces. These three oceanographic programs lie alongside hundreds of other multinational, national, regional, and local projects that include observations of plankton abundance and diversity. They vary from opportunistic sampling on primarily nonscientific expeditions to integrated ecosystem-based monitoring programs in support of responsible fisheries management. The need for a worldwide electronic atlas for mapping marine biological data in multi-dimensions has been recognized by the community of researchers.

Concomitant with greater sample availabilty has been a diversification of the methods used to analyze the organisms. These have already been mentioned briefly in connection with size spectra and environmental genomics. One method in particular has become an indispensable facet in modern oceanographic surveys. Fluorescence-activated cell analysis, commonly called flow cytometry, is a technique for rapid, quantitative resolution of fluorescence and size-related characteristics of single cells. Unicellular plankton microbes suspended in their natural seawater mileu are hydrodynamically focused in a laminar stream and detected, one by one in single file, as they flow past a point of laser excitation. Since the electronic response time is very short, tens of thousands of cells can be counted in a few minutes. Because different species have overlapping fluorescence-size signatures, the taxonomic resolution afforded by flow cytometry is low. In routine applications, analysis is made on both the chlorophyll-bearing cells (based on detection of pigment autofluorescence) and the heterotrophic picoplankters (based on detection of fluorescence from a nucleic acid-staining fluorochrome). Without molecular refinements such as immunological or hybridizational targeting, the routine plankton profile is typically limited to the prokaryotic picocyanobacteria *Prochlorococcus* and *Synechococcus*, the eukaryotic phytoplankters partitioned into pico- and nano-size classes, and the total heterotrophic bacterioplankton. Conventionally, unicellular plankton are examined by microscopy, a method that reveals much taxonomic detail at the expense of low sample throughput. Flow cytometry sacrifices taxonomic detail in favor of a greatly enhanced speed by which cells can be counted. With this capability, it is feasible to map—in both space and time—some properties of selected marine microbes at a resolution limited only by the number of water samples recovered in hydrographic surveys. Recent technological advances point the way to a high throughput of samples examined at high taxonomic resolution. For example, picoplankton identified by tyramide signal amplification following whole-cell fluorescence in situ hybridization with rRNA-targeted oligonucleotide probes can be analyzed by computer-assisted automated digital microscopy (Pernthaler, Pernthaler, and Amann 2003) or by flow cytometry (Biegala et al. 2003).

Yet, there remains an inescapable truth. No map of a plankton variable derived from shipboard surveys can capture all the variance at relevant space and time scales. Autonomous moored or free-floating sensors deployed by the thousands in an integrated global observation system is a fascinating vision for biologists. Already, there are specialized submersible flow cytometers and devices for in situ detection of specific organisms by molecular probes. However, it would be a herculean task to scale-up these biological devices to the order currently attainable by hydrographic devices. Notwithstanding the fact that marine plankton will long be undersampled, oceanographers are in the enviable state of already knowing the high-resolution distribution of one key plankton variable everywhere in the ocean every few days. Earth-orbiting satellites carry spectroradiometers designed to sense water-leaving radiances, from which can be derived the photosynthetic biomass expressed as the concentration of chlorophyll *a* pigment. The synoptic mapping of phytoplankton distribution at kilometer-scale resolution for the entire world ocean is a revolutionary achievement. In conjunction with similarly scaled maps of other remotely sensed variables, notably sea surface temperature, the domain for phytoplankton macroecology can be defined at the largest scale.

A global chlorophyll field at 4-km resolution obtained by the MODIS

(Moderate Resolution Imaging Spectroradiometer) Aqua system is binned to 8,640 columns for longitude and 4,320 rows for latitude, giving more than thirty-seven million pixels. Even after applying a mask to eliminate the land, an extraordinarily large number of ocean datapoints remains. In figure 2.3, the relationship between chlorophyll and sea surface temperature is shown, using the 2003 annual composite of MODIS data resampled to a lower resolution grid of 255 × 255 pixels. The global distribution of chlorophyll is unimodal at about 0.2 mg m^{-3}, and temperature is trimodal at about 1°, 6°, and 27°C. Throughout most of the temperature range, there is a first-order decrease of chlorophyll with temperature at a modest rate of about 0.02 log units per degree. High chlorophyll values (1–10 mg m^{-3}) in the interval 26°–28°C contraindicating this general trend appear to originate from the Arafura and Sunda shelves of the coastal Pacific Ocean. Many of the points exceeding 31°C are artifacts of the data-resampling procedure. At any given temperature, there is a 1.5 order of magnitude difference in chlorophyll between the 5 and 95 percentiles. These quantiles (or some others, say 1 and 99 percentiles) outline a polygon within which the steady-state ocean mainly operates.

Superimposed on these estimates of chlorophyll derived from remotely sensed ocean color are direct measurements of chlorophyll filtered from seawater collected by Niskin bottles (fig. 2.3). A dataset in the Bedford Institute of Oceanography (BIO) has been constructed from measurements made on thirty cruises in the North Atlantic Ocean during the period 1989 to 2003, largely but not entirely in an area bounded approximately by 38°N, 61°N, 42°W, and 67°W (Li and Harrison 2001; Li 2002; Li, Head, and Harrison 2004). Out of 11,487 paired observations of chlorophyll and temperature, the subset of 3,376 data pairs in the upper 20 m of the water column was extracted, representing about 1,080 separate hydrocasts. This depth stratum approximates the sensing penetration of the satellite radiometer, which corresponds roughly to 22% of the euphotic depth. The sea measurements from this particular part of the world largely validate the global satellite estimates, in particular the first-order decrease of chlorophyll with temperature and the range of variation at any given temperature. Within this one all-inclusive domain lies essential information for a partition of the pelagic ocean into four biomes and fifty-one biogeochemical provinces (Longhurst 1998).

This is not to say that combinations of these two variables outside the limiting quantile boundaries are nonpermissable in nature. Indeed, in eutrophic coastal embayments within the temperate zone, the large annual excursion of temperature is the backdrop against which blooms of chlorophyll biomass recur in spring and autumn. Thus high chlorophyll levels can occur in both cold and warm periods. We might imagine that different local eco-

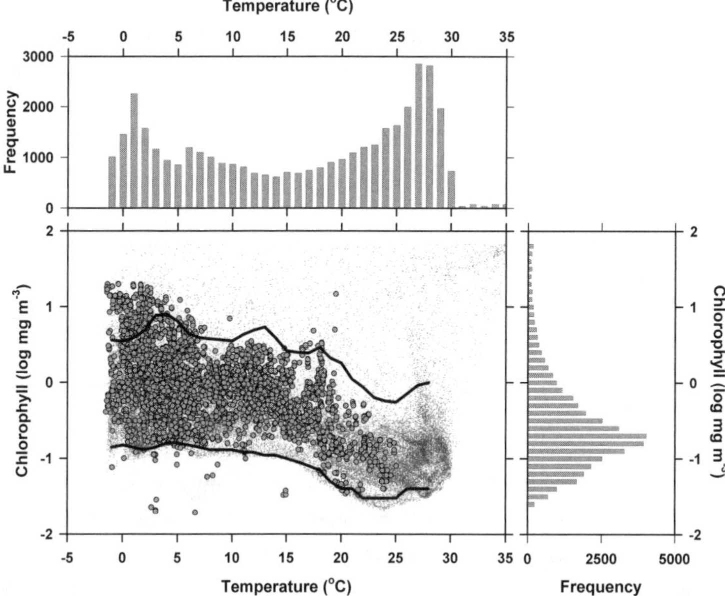

FIGURE 2.3 The co-distribution of chlorophyll and temperature indicates the limits of variability of the steady state ocean. The global data set (open small circles) is the 2003 annual composite of 4-km resolution data obtained by the Moderate Resolution Imaging Spectroradiometer Aqua system (http://oceancolor.gsfc.nasa.gov/) resampled to 65,025 points. The 95 and 5 percentile lines are constructed from data binned into 1°C temperature intervals. The frequency distribution of resampled data is indicated in the top panel for the partition by temperature, and in the right panel for the partition by chlorophyll. Data above 31°C are largely artifacts from data resampling. The North Atlantic data set (filled large circles) of the surface layer (≤20 m) is from shipboard measurements of the Bedford Institute of Oceanography (Li, Head, and Harrison 2004).

systems have seasonal responses that impart different "footprints" in the coordinate plane. These footprints would be variously shaped and possibly extend widely across the plane. However, a measure of its central tendency would likely be located within the boundaries defined by the global annual composite. It is in the large-scale regional geographic context that we now search for ecological patterns and constraints in microbial plankters (Li 2002; Li, Head, and Harrison 2004).

Phytoplankton Abundance

Exploratory Data Analysis

The aim in this section is to examine the statistical phenomenology of certain relations between variables. The starting point is to recognize that pelagic ecosystems are fundamentally structured by physical processes. Phytoplankton ecology is largely an expression of the turbulence and nutrient conditions presented by ocean physics (Cullen et al. 2002). The different possible combinations of various states (from a high to a low degree) of these conditions inform of the adaptive traits of the phytoplankton resident therein. Such arguments lead to the view that the physical oceanography in each region is the primary determinant of ecosystem structure (Longhurst 1998). The phytoplankton species are structured appropriately, and these characteristic flora are in turn associated with characteristic fauna, and presumably microflora as well, although these other assemblages may show less linkage to the physical environment. Bottom-up control is mediated by the delivery of inorganic nutrients to the phytoplankton in the euphotic zone, whereas top-down pressures are subsidiary modulators. It is no coincidence then that temperature is such an important indicator of plankton structure (Bouman et al. 2003): high temperatures manifest the physics generally associated with conditions of low turbulence and low nutrients; conversely, low temperatures manifest those of high turbulence and high nutrients.

Underlying the chlorophyll-temperature relation (fig. 2.3) are the patterns we seek for separate components of the phytoplankton. We use the picophytoplankton as an illustrative example, keeping in mind that this size class is itself an aggregation of different taxa. The BIO North Atlantic data set contains about 10,000 measurements of picophytoplankton from the illuminated depths of the water column. At times when we examine these data, we focus on the surface layer of 20 m, giving a subset of about 2,000 measurements, which is still sufficient for a robust analysis. By this restriction, we neglect the deep layers of relatively high chlorophyll prevalent in warm oligotrophic waters, choosing instead to more closely match our analysis to the optical layer sensed by satellites. Exploratory analysis begins with an examination of bivariate plots of picophytoplankton cell density (hereafter called "abundance" according to common usage in marine microbial ecology) versus the three variables known to be important: temperature, nitrate concentration, and chlorophyll biomass (see fig. 2.4). The data appear as clouds of points that visually conform to existing knowledge of these cells (Agawin, Duarte, and Agustí 2000). Thus, surface layer picophytoplankton are generally more abundant in subtropical oceans than in the polar seas (fig. 2.4, panel A); generally more abundant in nutrient-poor waters (fig. 2.4, panel B); and generally less abundant in chlorophyll-rich waters (fig. 2.4, panel C). The contribution of picophytoplankton to the standing stock of photosynthetic biomass, indicated here as the ratio of cell abundance to chlorophyll, increases with temperature (fig. 2.4, panel D). Although these impressions

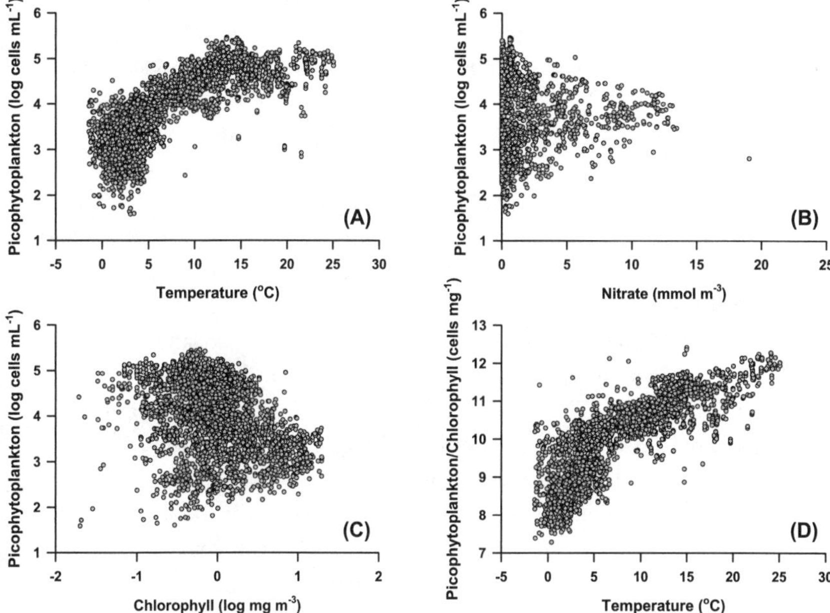

FIGURE 2.4 The abundance of North Atlantic picophytoplankton in the surface layer (≤20 m) is related to temperature (A), ambient nitrate concentration (B), and chlorophyll concentration (C), but the underlying relationships are confounded because the predictor variables are correlated. The contribution of picophytoplankton to chlorophyll biomass increases with temperature (D).

are validated by simple statistical correlations between abundance and the predictor variables taken one at a time, they do not reveal the underlying relationships because the variables are correlated themselves, and because the various taxonomic components have different responses.

Standard Statistical Analysis

The underlying relationships may be studied in several ways. Some, such as path analysis, are effective in making distinctions between alternate structural models specifying the linkages between variables. The interrelationships between phosphorus, bacterioplankton, and phytoplankton have been analyzed in this way (Currie 1990). For predictive purposes, we turn instead to multiple regression, in which the effect of each independent variable is assessed under the condition that all others are mathematically held constant. A substantial percentage (66 percent) of the variance in picophytoplankton abundance can be explained by just three variables: temperature, nitrate, and chlorophyll, with temperature being the dominant factor. This result is highly encouraging, since global climatological fields exist for all three variables. Including two other widely measured variables (silicate concentration, bacterial abundance) and a simple descriptor (day of year) increases the explained variance to 82 percent. There is a high degree of correlation among these predictor variables but their incremental contribution to a robust empirical description cannot be dismissed.

Beyond the predictive utility of this result, such an analysis reveals hidden biological insight. In particular, the partial regression coefficients for both nitrate and chlorophyll are positive, even though their simple correlation coefficients are negative. This disparity is due to correlations between the independent variables themselves. Thus, holding temperature and chlorophyll (or nitrate) constant, the influence of nitrate (or chlorophyll) on picophytoplankton is positive. This predicts some interesting experimental outcomes. For example, at a given temperature and a given standing stock of chlorophyll, adding more nitrate would increase picophytoplankton abundance. Whether this could be demonstrated is uncertain because the strength of the standard partial regression coefficient for nitrate is weak, serving to increase the explained variance in abundance by a mere 2 percent above the 63 percent achieved by temperature alone.

Nevertheless, because the standard partial effect of nitrate is statistically significant, we can delve further by partitioning the picophytoplankton into its component taxa: *Synechococcus*, *Prochlorococcus*, and picoeukaryotic algae. Bivariate scatterplots show that the dependence of abundance on temperature is strong for *Synechococcus*, being positive below 14°C and negative above 14°C (fig. 2.5, panel A). It is also strongly positive for *Prochlorococcus* (fig. 2.5, panel B), but weak overall for picoeukaryotes (fig. 2.5, panel C). *Prochlorococcus* has rarely been found below 10°C. However, above this temperature, its inverse relation to *Synechococcus* (and picoeukaryotes to some extent) is the basis for relatively constant total picophytoplankton abundance at higher temperatures (fig. 2.4, panel A). This is a striking example of niche complementarity previously noted (Chisholm 1992). Compellingly, there is a distinct difference in the effect of nitrate on these taxa: the partial regression coefficient for nitrate is positive for both *Synechococcus* and picoeukaryotes, but not significant for *Prochlorococcus*. This is a very different conclusion than one derived from simple correlation, which indicates negative nitrate correlations for both *Prochlorococcus* and *Synechococcus*. Assuringly, there appears to be a biological basis for the lack of response by *Prochlorococcus* to nitrate: they do not contain the genes required for nitrate uptake and reduction (Moore et al. 2002). Conversely, the responsiveness of *Synechococcus* to nitrate addition is well known (Glover et al. 1988).

An oceanographic transect that proceeds along a line of latitude provides

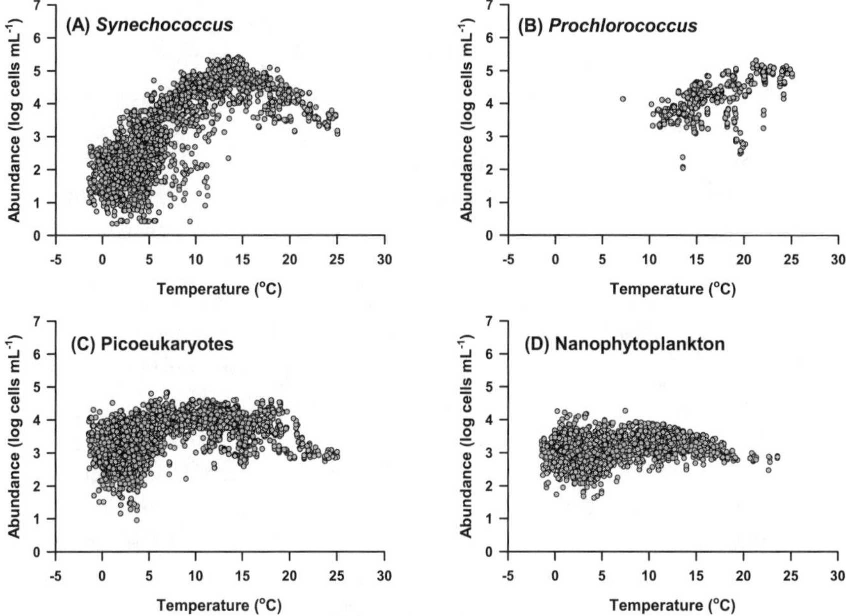

FIGURE 2.5 The abundance of North Atlantic phytoplankton in the surface layer (≤20 m) is differently related to temperature for *Synechococcus* (A), *Prochlorococcus* (B), picoeukaryotic algae (C), and nanophytoplankton (D).

plankton data for which temperature is held to small changes in the face of large changes in other environmental variables, such as nutrient concentrations. In effect, the observation regime serves to minimize the effect of a particular variable. The large change in abundance of *Synechococcus* in the face of relatively constant temperatures between 20° and 30°S (fig. 2.2) attests to the importance of other factors in shaping the global scale distribution of these cells.

Partitioned Statistical Analysis

At this point, it is evident that a large data set has more to offer beyond greater statistical confidence given by the many degrees of freedom. The possibility exists to perform the same statistical evaluation on subsets of the data that have been stratified according to criteria chosen in the context of a particular query. For example, we may wish to enquire about factors influencing picophytoplankton in the extreme limits set by temperature, near the carrying capacity and near the minimum threshold. We may also ask about the factors that are important in frequently observed cases near the median abundance. When the data are unstratified and submitted to statistical analysis as a whole, we obtain insight about the ensemble average. Given a value of x (or $x_1 \ldots x_n$), we seek the single value of y around which the observations cluster. The corpus of plankton ecology is largely derived from such standard analysis. On the other hand, it is increasingly recognized that interesting questions may be posed concerning the entire range that is occupied by values of y. In other words, the internal structure and boundaries of bivariate point clouds are ecologically significant (Duarte 1991), and these can be extracted by various statistical methods (Kaiser, Speckman, and Jones 1994; Thomson et al. 1996; Scharf, Juanes, and Sutherland 1998).

We take an intuitive approach based on the idea that the quantiles of a statistical distribution provide a robust partitioning or slicing of the data (Koenker and Hallock 2001). Returning to the picophytoplankton-temperature example (fig. 2.4, panel A), it is possible to construct the frequency histograms for cell abundance in successive 1° temperature intervals. These would display, for each temperature, the central location, dispersion, and range of the abundance measurement as well as the volume of data. We are now in a position to select slices of the data for standard statistical analysis. At each 1° temperature interval, we consider the upper extreme values (≥90 percentile), the lower extreme values (≤10 percentile), and the median values (45–55 percentile) for multiple regression analysis. Each subset contains only 10 percent of the approximately 2,000 measurements, but this still allows for a sufficiently large degree of freedom.

As before, the results for the effect of nitrate are intriguing. The partial regression coefficient for nitrate changes from positive at low abundances, to nonsignificant at median abundances, to negative at high abundances. Thus, at any given temperature, although picophytoplankton can occur at varying abundances, the ones found near the lower limit are those that increase with nitrate additions; the ones found near frequently observed levels are those that appear unresponsive to nitrate; and the ones found near the upper limit apparently decrease with nitrate. This tri-phasic pattern recalls the typical population response to a resource gradient: limitation, saturation, and inhibition. However, the situation is complicated by the fact that it is a community response, and that the parameters must change across the entire temperature range. The limits of abundance move up with temperature, and so if the community at these limits retain their characteristics (e.g., for nitrate response), they must adapt as a unit. For one thing, *Prochlorococcus* are absent at low temperatures; therefore the assemblage of *Synechococcus* and picoeukaryotes under these conditions would impart community characteristics equivalent to the expression when all three taxa are present at high temperatures.

There is a multitude of ways to partition multidimensional data, the so-called "curse of dimensionality." In our example, we selected the quantile slices based on temperature bins. Other questions will require other views of the same data. Inspection of the picophytoplankton abundance versus chlorophyll scatterplot (fig. 2.4, panel C) indicates that as biomass exceeds 0.5 mg m^{-3}, abundance converges toward a value where the falling upper limit meets the rising lower limit. In this case, selection of quantile slices based on chlorophyll bins would be the appropriate choice for study. Partitioning of data by chlorophyll concentration can be particularly informative because it may be akin to sampling ecosystems of different trophic states. In considering how bacterioplankton production is related to temperature, a partitioned analysis effectively removes the statistical distortion imposed by the fact that there are few high-resource situations under warm water conditions (López-Urrutia and Morán 2007). It is not far-fetched to imagine that macroecologists may eventually require the sophisticated electronic visualization techniques currently used in other fields to undertake virtual-reality exploration of high-dimensional data.

Phytoplankton Biomass and Size

As a rule, there are more small organisms than large ones. Underlying this apparently simple notion is a wealth of ideas to be explored (Brown et al. 2004). In phytoplankton macroecology, we approach the body size versus abundance relationship from the perspective of cross-community scaling (White et al. 2007), in which the average size of an individual in an assemblage is related to the total number of individuals in that assemblage. Each data pair is from a spatially or temporally distinct community. Thus in a unit volume of seawater, let N be the number of phytoplankton cells, M be the average cell carbon mass of the taxonomically diverse assemblage, then $B = NM$ is the phytoplankton carbon biomass. All phytoplankton contain the light-harvesting pigment chlorophyll *a* (or its divinyl form); therefore it is often used as a measure of photosynthetic biomass. Thus chlorophyll concentration $C = B\theta^{-1} = NM\theta^{-1}$, where θ is the carbon-to-chlorophyll ratio. A bivariate plot of B (calculated from flow cytometric measurements of N and M) versus C (measured from pigments collected on a filter) using the BIO data set (fig. 2.6, panel A) indicates a best-fit value of just slightly over 30 for θ, representative of many nutrient-sufficient conditions. Importantly, the slope of the power relationship is not significantly different from 1, indicating a direct proportionality between the two measures of standing stock. Closer inspection reveals, however, that at high chlorophyll concentrations,

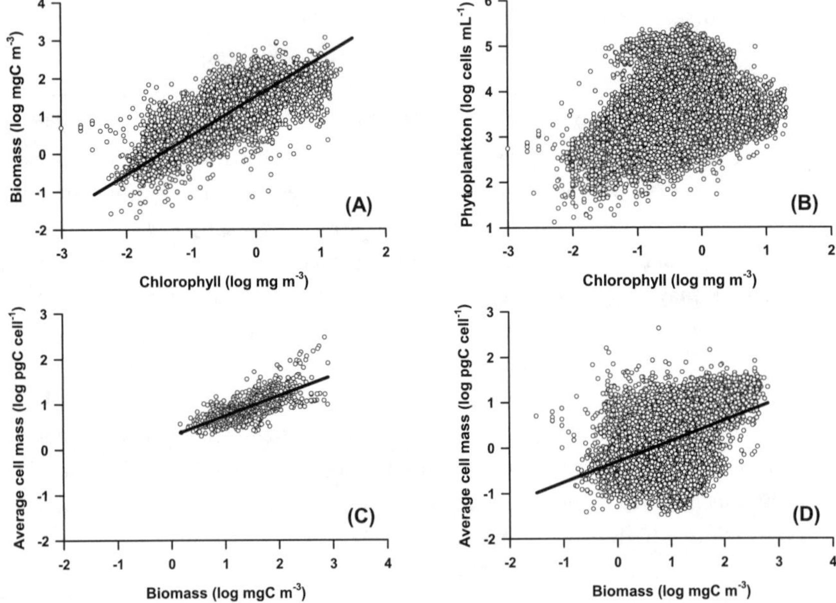

FIGURE 2.6 The biomass of North Atlantic phytoplankton estimated by flow cytometric analysis of single cells is in direct proportion ($\log Y = \log 31.4 + 1.03 \log X$) to the concentration of chlorophyll measured from filtered seston (A), but the relationship between phytoplankton abundance and chlorophyll (B) is a filled triangular polygon indicative of different cell quotas for chlorophyll. Average cell mass increases with biomass, clearly evident in (C) the data of Irigoien, Huisman, and Harris (2004), which does not include picoplankton ($\log Y = 0.30 + 0.44 \log X$), but the BIO dataset (D) which includes picoplankton shows a much larger domain and a similar slope ($\log Y = -0.30 + 0.45 \log X$).

flow cytometry systematically underestimates biomass—undoubtedly because the larger cells are undetected.

A bivariate plot of N versus C informs us of θM^{-1}, or its inverse, which is the chlorophyll content per cell. The macroecological distribution of N versus C (fig. 2.6, panel B) is essentially a filled polygon whose boundaries form a triangular shape. The montonically rising lower boundary indicates cells of high chlorophyll content. The upper boundary has both a rising edge and a falling edge: the former indicates cells of low chlorophyll content, while the latter indicates a change to higher chlorophyll content. The transition from positive to negative slope in the upper boundary occurs at approximately 0.5 mg Chl m^{-3}, this being the carrying capacity for picophytoplankton in the ocean (Chisholm 1992). At this point, the apex of the triangular polygon, the difference is greatest between maximum and minimum recorded values of abundance. In other words, there are a great many ways to assemble a phytoplankton community of half milligram chlorophyll standing stock: a large number of small cells, a small number of big cells, or any combination of them.

An equivalent analysis, but one that provides a different perspective, is to plot *M* versus *B*, or equivalently versus $C\theta$. This informs us of N^{-1}, or its inverse, which is cell abundance. The macroecological distribution of *M* versus $C\theta$ for the BIO data set (fig. 2.6, panel D) is somewhat a mirror image of *N* versus *C* (fig. 2.6, panel B), namely an inverted triangle with a nadir of *M* at a value of *B* approximately equal to 15 mg C m^{-3} (= 0.5 mg Chl $m^{-3} \times$ 30 mg C mg Chl^{-1}). As a whole, there is a positive correlation between *M* and *B*, meaning that to a first order, where phytoplankton biomass is high (e.g., spring blooms), the average cell size is large (e.g., diatoms). A similar plot of *M* versus *B* by Irigoien, Huisman, and Harris (2004) using a data set that did not include picophytoplankton overlaps the BIO data set near the upper boundary region (fig. 2.6, panel C), but clearly does not fill the entire domain. At high *B*, there is in fact a rather equitable distribution of phytoplankters, by numbers, in various size classes (Li 2002). In this situation, picophytoplankton remain present, but the assemblage *M* mainly reflects the contribution of approximately equal numbers of larger phytoplankton. The predictable pattern by which phytoplankton of different size classes change in relation to standing stock leads to the next section, addressing patterns of diversity in relation to environmental constraints.

Phytoplankton Diversity

Temperature, it seems, explains a great deal of statistical variance in the picophytoplankton. At the organism level, physiological tolerances prescribe the temperature range within which a species can be found. Thus, for *Prochlorococcus*, laboratory culture experiments indicate the lower limit to be near 10°C (Moore, Goericke, and Chisholm 1995), matching well the distribution of these cells in nature (fig. 2.5, panel B). Within the physiological range, temperature controls biochemical reaction rates according to activation energy theory. Metabolic control at the biochemical level translates to predictable temperature dependencies at virtually all higher levels of biological organization (Allen, Brown, and Gillooly 2002; Brown et al. 2004). Moreover, differences in activation of photosynthesis and respiration predict a metabolic imbalance in the ocean, tending toward greater net heterotrophy under warming climate scenarios (López-Urrutia et al. 2006).

Temperature, however, is also important in its hydrodynamic control of water-column structure. Whether phytoplankton receive sufficient light to photosynthesize is dependent on whether they are retained for a sufficiently long time in the upper sunlit zone. Conversely, whether phytoplankton receive sufficient nutrients is dependent on whether they are mixed deeply enough to enjoy the benefits of being near the nutricline. In other words, photosynthesis in the ocean is significantly dependent on physical controls. Temperature sets a maximum to their physiological capabilities in the form of dark reaction enzymes, but when phytoplankton spend their lives being mixed up and down the water column under sub-saturating light intensities, they are generally not performing at their maximum. In the pelagic ocean, plankton diversity is evidently related to temperature, both directly and indirectly. For instance, even a prediction by temperature that explains as much as 90 percent of geographic variation in planktic foraminiferal diversity must be interpreted in terms of upper-ocean thermal structure (Rutherford, D'Hont, and Prell 1999). Conversely, the temperature field may bear little resemblance to what is clearly a structured spatial distribution of plankton cells (fig. 2.2).

In the North Atlantic, the smallest phytoplankters (*Prochlorococcus* and *Synechococcus*) are strongly influenced by processes indicated by temperature (fig. 2.5, panels A, B). Picoeukaryotic algae, although in the same nominal size class as the cyanobacteria, are larger and show less dependence on temperature (fig. 2.5, panel C). Nanophytoplankton, larger still, are completely uncorrelated with temperature (fig. 2.5, panel D). Together, this photoautotrophic microbial assemblage creates a pattern of diversity that displays a dependence on temperature and size, but is also a likely response to upper-ocean physics. This becomes evident when we forgo the mixed classifications (cyanobacteria, picoeukaryotes, nanophytoplankton) in favor of a measure of diversity expressing the notions of richness and evenness for an assemblage that is characterized by the ataxonomic criteria of flow cytometry. By the same token that diversity indices can be calculated for phytoplankton differentiated according to photosynthetic pigments or Coulter volume (Parsons 1969), we can calculate an ataxonomic diversity index for phytoplankton that is differentiated on the combined basis of red fluorescence intensity (chlorophyll per cell) and light-scatter intensity (cell size). The so-called "cytometric diversity" is the Hill number of order 1, which is the exponential of the Shannon-Wiener index, applied to flow cytometric data (Li 1997). Although cogent criticisms have been raised against diversity metrics borrowed from information theory, these indices remain widespread in ecology because they (i.e., the entire family of Hill numbers) convey the concept that diversity is related to entropy, and are demonstrably useful measures of ecological diversity in bacterial communities (Hill et al. 2003).

Cytometric diversity attains its highest values when richness is counterbalanced by evenness (fig. 2.7). At low phytoplankton abundance, richness is low but evenness is high. Evidently, no species is particularly dominant when all are represented by sparse numbers. Conversely, at high abundance, richness is high but evenness is low. For example, *Prochlorococcus marinus*, represented by various ecotypes, is exceedingly abundant in oligotrophic central gyres. *Prochlorococcus* is clearly the dominant phenetic and genetic cluster in waters that are otherwise considered species rich. At an intermediate abundance of about 2,000 cells mL^{-1}, a balance of moderate richness and evenness yields high cytometric diversity (fig. 2.7, panel C; Li 2002).

From what is already known about the temperature dependence of picophytoplankton and nanophytoplankton (fig. 2.5), the phenomenology of cytometric diversity may be easily surmised. Above 7°C, total abundance is almost always greater than 2,000 cells mL^{-1} (98 percent of cases). Below this temperature, total abundance is almost equally distributed below (48 percent) and above (52 percent) this level. It is therefore no surprise that cyto-

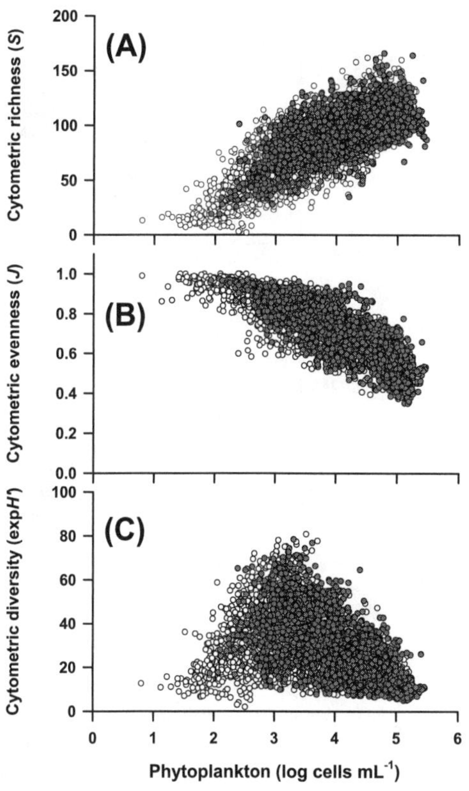

FIGURE 2.7 Cytometric diversity (C) attains its highest values when cytometric richness (A) is counterbalanced by cytometric evenness (B). At low phytoplankton abundance, richness is low but evenness is high. Conversely, at high abundance, richness is high but evenness is low. At an intermediate abundance of about 2,000 cells mL^{-1}, a balance of moderate richness and evenness yields high cytometric diversity. North Atlantic samples in the surface layer (≤20 m) are indicated by filled symbols; deeper samples in the euphotic zone are indicated by open symbols.

metric diversity is generally low at higher temperatures, but varies widely at lower temperatures (fig. 2.8, panel A). Additionally, since picophytoplankton make up an increasing fraction of the total abundance with temperature (fig. 2.4, panel D), it is also no surprise that the average cell mass of the assemblage (which decreases with a larger fraction of picophytoplankton) has the opposite effect on cytometric diversity as temperature (fig. 2.8, panel B). These phenomenologies concur with general perceptions that in picophytoplankton-dominated warm waters, there is little uncertainty that a random sample will likely reveal only a few major taxa (e.g., *Prochlorococcus*) even though many minor ones also exist. Conversely, in picophytoplankton-poor cold waters, there is greater uncertainty because the fewer taxa are more evenly matched.

In the community of microbial phytoplankters, there are indeed many more observations of low cytometric richness in subarctic waters than subtropical waters (fig. 2.8, panels C, D), echoing the often-stated pattern of a decrease in macrofaunal richness from equator to poles (Brown 1995). It

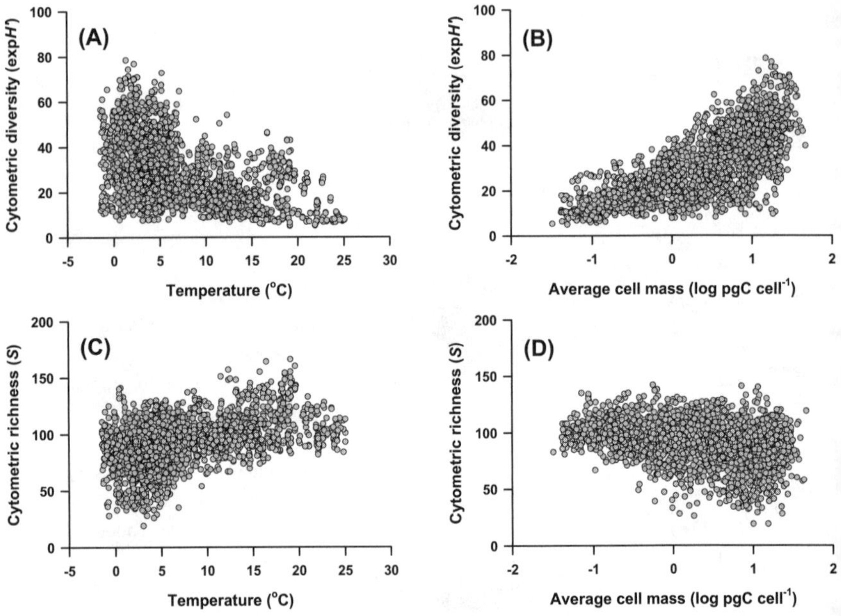

FIGURE 2.8 Cytometric diversity of North Atlantic phytoplankton in the surface layer (≤20 m) is low at high temperatures (A), in spite of high cytometric richness (C) because of the low cytometric evenness of highly abundant picophytoplankton populations. Cytometric diversity is low in phytoplankton assemblages of low average cell mass (B), in spite of high cytometric richness (D) because of the same reason.

is equally true, however, that maximum cytometric richness changes little across geographic latitudes (fig. 2.8, panels C, D). In other words, conditions exist in both subarctic and subtropical waters leading to the same community characteristic. A latitudinal decline in diversity from the tropics to the poles is strongly dependent on body mass, with the largest organisms showing the strongest decline (Hillebrand 2004a, 2004b). It appears that microorganisms such as diatoms do not exhibit such a latitudinal decline, perhaps because they have a high transportability and their astronomical abundances greatly favor the chance of long dispersal (Hillebrand and Azovsky 2001). Consequently, the diversity of unicellular organisms increases only slowly with the size of the geographic area, meaning that the ratio of local to global diversity is high (Azovsky 2000). The ecological processes underlying the strong phenomenologies indicated by temperature and cell size (fig. 2.8) are to be sought at the local scale in which the plankton perceive their environment.

The spatial distribution of plankton often matches ocean circulation patterns, but water masses encompass a wide range of physical and chemical properties, so that these do not provide a good indication of causation. However, the observation that patterns of seasonality and biological productivity can match the physical patterns suggests a probable linking mechanism (Angel 1993). A clue to what these processes might be for phytoplankton is provided by examining the relationship between their diversity and biomass, the latter being a proxy for primary productivity. The upper constraint on this relationship appears to show that diversity peaks at an intermediate level of biomass (Irigoien, Huisman, and Harris 2004). At low phytoplankton biomass or productivity, diversity is conjectured to be low because there is strong competition for limited nutrients. At high phytoplankton biomass or productivity, diversity is conjectured to be low because of selective grazing and self-shading of light. At intermediate phytoplankton biomass or productivity, these pressures are weakened from both sides, allowing diversity to attain high values (Irigoien, Huisman, and Harris 2004).

This hypothesis of multiple resource competition and grazing is complicated by the fact that phytoplankton are passively carried through opposing vertical gradients of light and nutrient by the motion of water. Alternatively, the entire phytoplankton community through the full depth of the photic zone might be viewed as a single unit. In this way, the biologically integrated response to vertical changes in light, nutrients, and grazing pressure can be expressed in a weighted index. In the North Atlantic Ocean, the weighted cytometric diversity of phytoplankton under a square meter of sea surface is highest when the water column is moderately mixed (Li 2002). Where vertical stratification of the water column is strong, conditions are favorable for prolonged biological interactions and nutrient competition among co-occurring organisms. Conversely, where stratification is weak, conditions favor those who take advantage of reduced light income but benefit from increased inorganic resources. Measurements in extremely well-mixed conditions will always be rare because those are the very conditions (e.g., winter deep convection of the Labrador Sea) in which oceanographic sampling is hindered by harsh weather. Strong and weak stratification are the end points of a continuum between which many possible combinations of adaptive responses can be expressed (Reynolds 2001; Cullen et al. 2002). At an intermediate level of stratification, the milder restraints imposed by both biological and physical factors permit a greater range for cytometric diversity. As in other (nonplanktonic) ecosystems, it seems that intermediate disturbance may be a possible explanation for high diversity. There are many other plausible explanations that merit examination (Brown et al. 2004, Morin and Fox 2004).

Concluding Remarks

Patterns in ecology are not easily discerned, and even less easily explained. Some, such as the allometric scaling of abundance to body size, are clearly evident in the marine plankton. Others are not so obvious, but seem to emerge as the boundary constraints when a large number of possibilities are considered. Shifts in the relative importance of bottom-up and top-down control, multiple resource competition, niche complementarity, and intermediate disturbance are examples indicating that biological processes in the pelagos share many characteristics with those elsewhere. Plankton communities, unlike others, are peculiarly shaped by the turbulent hydrodynamic environment, but the causal mechanisms underlying patterns of abundance, distribution, and diversity are general. It is useful to view the plankton, especially the microbial forms, as members at the low end of the space and time ranges in the ocean. In this way, it may be possible to assimilate plankton patterns into the commonalities of life. We are reminded by the chapter's opening quote that our philosophical approach is not new, just our data.

ACKNOWLEDGMENTS

This research was supported by the Strategic Science Fund of the Canadian Department of Fisheries and Oceans and by the Program of Energy Research and Development of Natural Resources Canada. I thank all sea-going staff of the Biological Oceanography Section at the Bedford Institute of Oceanography, and the officers and crews of CCGS *Hudson* for expedition work in the North Atlantic. I am grateful to James Gillooly and Glen Harrison for their comments on the manuscript, but all deficiencies and errors remain my own.

REFERENCES

Agawin, N. S. R., C. M. Duarte, and S. Agustí. 2000. Nutrient and temperature control of the contribution of picoplankton to phytoplankton biomass and production. *Limnology and Oceanography* 45:591–600.

Allen, A. P., J. H. Brown, and J. F. Gillooly. 2002. Global biodiversity, biochemical kinetics, and the energetic-equivalence rule. *Science* 297:1545–48.

Angel, M. V. 1993. Biodiversity of the pelagic ocean. *Conservation Biology* 7:760–72.

Azam, F., and A. Z. Worden. 2004. Microbes, molecules, and marine ecosystems. *Science* 303:1622–24.

Azovsky, A. I. 2000. Concept of scale in marine ecology: Linking the words or the worlds? *Web Ecology* 1:28–34. Online serial at http://www.oikos.ekol.lu.se.

Beaugrand, G., F. Ibañez, J. A. Lindley, and P. C. Reid. 2002. Diversity of calanoid copepods in the North Atlantic and adjacent seas: Species associations and biogeography. *Marine Ecology Progress Series* 232:179–95.

Béjà O., L. Aravind, E. V. Koonin, M. T. Suzuki, A. Hadd, L. P. Nguyen, S. B. Jovanovich, et al. 2000. Bacterial rhodopsin: Evidence for a new type of phototrophy in the sea. *Science* 289:1902–1906.

Biegala I. C., F. Not, D. Vaulot, and N. Simon. 2003. Quantitative assessment of picoeukaryotes in the natural environment by using taxon-specific oligonucleotide probes in association with Tyramide Signal Amplification—Fluorescent In Situ Hybridization and flow cytometry. *Applied and Environmental Microbiology* 69:5519–29.

Blackburn, T. M., and K. J. Gaston. 2002. Scale in macroecology. *Global Ecology & Biogeography* 11:185–89.

Bouman, H. A., T. Platt, S. Sathyendranath, W. K. W. Li, V. Stuart, C. Fuentes-Yaco, H. Maass, E. P. W. Horne, O. Ulloa, V. Lutz, and M. Kyewalyanga. 2003. Temperature as an indicator of optical properties and community structure of marine phytoplankton: Implications for remote sensing. *Marine Ecology Progress Series* 258:19-30.

Bouman, H. A., O. Ulloa, D. J. Scanlan, K. Zwirglmaier, W. K. W. Li, T. Platt, V. Stuart, et al. 2006. Oceanographic basis of the global surface distribution of *Prochlorococcus* ecotypes. *Science* 312:918–21.

Brown, J. H. 1995. *Macroecology*. Chicago: University of Chicago Press.

Brown, J. H., and J. F. Gillooly. 2003. Ecological food webs: High-quality data facilitate theoretical unification. *Proceedings of the National Academy of Sciences* 100:1467–68.

Brown, J. H., J. F. Gillooly, A. P. Allen, V. M. Savage, and G. B. West. 2004. Towards a metabolic theory of ecology. *Ecology* 85:1771–89.

Caron, D. A. 2000. Symbiosis and mixotrophy among pelagic microorganisms. In *Microbial ecology of the oceans*. ed. D. Kirchman, 495–523. New York: Wiley-Liss.

Caron, D. A., P. D. Countway, and M. V. Brown. 2004. The growing contributions of molecular biology and immunology to protistan ecology: Molecular signatures as ecological tools. *Journal of Eukaryotic Microbiology* 51:38–48.

Cavender-Bares, K. K., A. Rinaldo, and S. W. Chisholm. 2001. Microbial size spectra from natural and nutrient enriched ecosystems. *Limnology and Oceanography* 46:778–89.

Chisholm, S. W. 1992. Phytoplankton size. In *Primary productivity and biogeochemical cycles in the sea*, ed. P. G. Falkowski and A. D. Woodhead, 213–37. New York: Plenum.

Cullen, J. J., W. F. Doolittle, S. A. Levin, and W. K. W. Li. 2007. Patterns and prediction in microbial oceanography. *Oceanography* 20 (2):34–46.

Cullen, J. J., P. J. S. Franks, D. M. Karl, and A. Longhurst. 2002. Physical influences on marine ecosystem dynamics. In *The sea, vol. 12*, ed. A. R. Robinson, J. J. McCarthy, and B. J. Rothschild, 297–336. New York: Wiley.

Currie, D. J. 1990. Large-scale variability and interactions among phytoplankton, bacterioplankton, and phosphorus. *Limnology and Oceanography* 35:1437–55.

Curtis, T. P., W. T. Sloan, and J. W. Scannell. 2002. Estimating prokaryotic diversity and its limits. *Proceedings of the National Academy of Sciences* 99:10494–99.

Darling, K. F., C. M. Wade, I. A. Stewart, D. Kroon, R. Dingle, and A. J. L. Brown. 2000. Molecular evidence for genetic mixing of Arctic and Antarctic subpolar populations of planktonic foraminifers. *Nature* 405:43–47.

Dolan, J. R. (coordinator) 2005. Biogeography of aquatic microbes. *Aquatic Microbial Ecology* 41:39–102.

Doolittle, W. F., and E. Bapteste. 2007. Pattern pluralism and the Tree of Life hypothesis. *Proceedings of the National Academy of Sciences* 104:2043–49.

Duarte, C. M. 1991. Variance and the description of nature. In *Comparative analyses of ecosystems: Patterns, mechanisms and theories*, ed. J. Cole, G. Lovett, and S. Findlay, 301–18. New York: Springer-Verlag.

Elton, C. 1927. *Animal ecology*. New York: Macmillan.

Falkowski, P. G., and J. A. Raven. 1997. *Aquatic photosynthesis*. Oxford: Blackwell Science.

Fenchel, T. 1993. There are more small than large species? *Oikos* 68:375–78.

Fenchel, T., and B. J. Finlay. 2004. The ubiquity of small species: Patterns of local and global diversity. *Bioscience* 54:777–84.

Finlay, B. J. 2001. Protozoa. In *Encyclopedia of biodiversity*, ed. S. A. Levin, 4:901–15. San Diego: Academic Press.

Finlay, B. J., and T. Fenchel. 1999. Divergent perspectives on protist species richness. *Protist* 150:229–33.

Gasol, J. M., and C. M. Duarte. 2000. Comparative analyses in aquatic microbial ecology: How far do they go? *FEMS Microbiology Ecology* 31:99–106.

Gibbons, M. J. 1997. Pelagic biogeography of the South Atlantic Ocean. *Marine Biology* 129:757–68.

Glover, H. E., B. B. Prezelin, L. Campbell, M. Wyman, and C. Garside. 1988. A nitrate-dependent *Synechococcus* bloom in surface Sargasso Sea water. *Nature* 331:161–63.

Goetze, E. 2003. Cryptic speciation on the high seas; Global phylogenetics of the copepod family Eucalanidae. *Proceedings of the Royal Society of London B* 270:2321–31.

Guillard, R. R. L., and P. Kilham. 1977. The ecology of marine planktonic diatoms. In *The biology of diatoms*, ed. D. Werner, 372–469. Berkeley, CA: University of California Press.

Hallegraeff, G. M. 1993. A review of harmful algal blooms and their apparent global increase. *Phycologia* 32:79–99.

Hill, T. C. J., K. A. Walsh, J. A. Harris, and B. F. Moffett. 2003. Using ecological diversity measures with bacterial communities. *FEMS Microbiology Ecology* 43:1–11.

Hillebrand, H. 2004a. On the generality of the latitudinal diversity gradient. *American Naturalist* 163:192–211.

Hillebrand, H. 2004b. Strength, slope and variability of marine latitudinal gradients. *Marine Ecology Progress Series* 273:251–67.

Hillebrand, H., and A. I. Azovsky. 2001. Body size determines the strength of the latitudinal diversity gradient. *Ecography* 24:251–56.

Hughes Martiny, J. B., B. J. M. Bohannan, J. H. Brown, R. K. Colwell, J. A. Fuhrman, J. L. Green, M. C. Horner-Devine, et al. 2006. Microbial biogeography: Putting microorganisms on the map. *Nature Reviews Microbiology* 4:102–12.

Irigoien, X., J. Huisman, and R. P. Harris. 2004. Global biodiversity patterns of marine phytoplankton and zooplankton. *Nature* 429:863–67.

Johnson, Z. I., E. R. Zinser, A. Coe, N. P. McNulty, E. M. S. Woodward, and S. W. Chisholm. 2006. Niche partitioning among *Prochlorococcus* ecotypes along ocean-scale environmental gradients. *Science* 311:1737–40.

Kaiser, M. S., P. L. Speckman, and J. R. Jones. 1994. Statistical models for limiting nutrient relations in inland waters. *Journal of the American Statistical Association* 89:410–23.

Karl, D. M. 2002. Hidden in a sea of microbes. *Nature* 415:590–91.

Kirchman, D. L., ed. 2000. *Microbial ecology of the oceans*. New York: Wiley-Liss.

Kirchman, D. L., and C. Pedrós-Alió. 2007. Predictions for the future of microbial oceanography. *Oceanography* 20 (2): 166–71.

Koenker, R., and K. F. Hallock. 2001. Quantile regression. *Journal of Economic Perspectives* 15:143–57.

Kolber, Z. S., C. L. Van Dover, R. A. Niederman, and P. G. Falkowski. 2000. Bacterial photosynthesis in surface waters of the open ocean. *Nature* 407:177–79.

Lawton, J. H. 1999. Are there general laws in ecology? *Oikos* 84:177–92.

Li, W. K. W. 1997. Cytometric diversity in marine ultraphytoplankton. *Limnology and Oceanography* 42:874–80.

———. 2002. Macroecological patterns of phytoplankton in the northwestern North Atlantic Ocean. *Nature* 419:154–57.

Li, W. K. W., and W. G. Harrison. 2001. Chlorophyll, bacteria and picophytoplankton in ecological provinces of the North Atlantic. *Deep-Sea Research II* 48:2271–93.

Li, W. K. W., E. J. H. Head, and W. G. Harrison. 2004. Macroecological limits of heterotrophic bacterial abundance in the ocean. *Deep-Sea Research I* 51:1529–40.

Longhurst, A. 1998. *Ecological geography of the sea*. San Diego: Academic Press.

———. 2001. Pelagic biogeography. In *Encyclopedia of ocean sciences*, ed. J. H. Steele, S. A. Thorpe, and K. K. Turekian, 2114–22. San Diego: Academic Press.

López-Urrutia, Á., E. San Martin, R. P. Harris, and X. Irigoien. 2006. Scaling the metabolic balance of the oceans. *Proceedings of the National Academy of Sciences* 103:8739–44.

López-Urrutia, Á., and X. A. G. Morán. 2007. Resource limitation of bacterial production distorts the temperature dependence of oceanic carbon cycling. *Ecology* 88:817–22.

Massana, R., R. Terrado, I. Forn, C. Lovejoy, and C. Pedrós-Alió. 2006. Distribution and abundance of uncultured heterotrophic flagellates in the world oceans. *Environmental Microbiology* 8:1515-1522.

May, R. M. 1988. How many species are there on earth? *Science* 241:1441–49.

McGowan, J. A., and P. W. Walker. 1993. Pelagic diversity patterns. In *Species diversity in ecological communities*, ed. R. E. Ricklefs and D. Schluter, 203–14. Chicago: University of Chicago Press.

Moore, L. M., R. Goericke, and S. W. Chisholm. 1995. Comparative physiology of *Synechococcus* and *Prochlorococcus*: Influence of light and temperature on growth, pigments, fluorescence and absorptive properties. *Marine Ecology Progress Series* 116:259–75.

Moore, L. M., A. F. Post, G. Rocap, and S. W. Chisholm. 2002. Utilization of different nitrogen sources by the marine cyanobacteria *Prochlorococcus* and *Synechococcus*. *Limnology and Oceanography* 47:989–96.

Moore, L. M., G. Rocap, and S. W. Chisholm. 1998. Physiology and molecular phylogeny of ocexisting *Prochlorococcus* ecotypes. *Nature* 393:464–67.

Moran, M. A., and E. V. Armbrust. 2007. Genomes of sea microbes. *Oceanography* 20 (2): 47–55.

Morin, P. J., and J. W. Fox. 2004. Diversity in the deep blue sea. *Nature* 429:813–14.

Morris, R. M., M. S. Rappé, S. A. Connon, K. L. Vergin, W. A. Siebold, C. A. Carlson, and S. J. Giovannoni. 2002. SAR11 clade dominates ocean surface bacterioplankton communities. *Nature* 420:806–10.

Nee, S. 2004. More than meets the eye. *Nature* 429:804–5.

Norton, T. A., M. Melkonian, and R. A. Andersen. 1996. Algal biodiversity. *Phycologica* 35:308–26.

Orellana, M. V., T. W. Petersen, and G. van den Engh. 2004. UV-excited blue autofluorescence of *Pseudo-nitzschia multiseries* (Bacillariophyceae). *Journal of Phycology* 40:705–10.

Orsini, L., G. Procaccini, D. Sarno, and M. Montresor. 2004. Multiple rDNA ITS-types within the diatom *Pseudo-nitzschia delicatissima* (Bacillariophyceae) and their relative abundances across a spring bloom in the Gulf of Naples. *Marine Ecology Progress Series* 271:87–98.

Papke, R. T., J. E. Koenig, F. Rodríguez-Valera, and W. F. Doolittle. 2004. Frequent recombination in a saltern population of *Halorubrum*. *Science* 306:1928–29.

Papke, R. T., and D. M. Ward. 2004. The importance of physical isolation to microbial diversification. *FEMS Microbiology Ecology* 48:293–303.

Parsons, T. R. 1969. The use of particle size spectra in determining the structure of a plankton community. *Journal of the Oceanographic Society of Japan* 25:172–81.

———. 2003. Macroecological studies of the oceans. *Oceanography in Japan* 12:370–74.

Partensky, F., W. R. Hess, and D. Vaulot. 1999a. *Prochlorococcus*, a marine photosynthetic prokaryote of global significance. *Microbiology and Molecular Biology Reviews* 63:106–27.

Partensky, F., J. Blanchot, and D. Vaulot. 1999b. Differential distribution and ecology of *Prochlorococcus* and *Synechococcus* in oceanic waters: A review. *Bulletin de l'Institut océanographique, Monaco, numero spécial* 19:457–75.

Pedrós-Alió, C. 2006. Marine microbial diversity: Can it be determined? *Trends in Microbiology* 14:257–63.

Pernthaler, J., A. Pernthaler, and R. Amann. 2003. Automated enumeration of groups of marine picoplankton after fluorescence in situ hybridization. *Applied and Environmental Microbiology* 69:2631–37.

Platt, T., and K. Denman. 1977. Organization in the pelagic ecosystem. *Helgolander Wissenschaftliche Meeresuntersuchungen* 30:575–81.

———. 1978. The structure of pelagic marine ecosystems. *Rapports et Procès-Verbaux des Réunions du Conseil International pour l'Exploration de la Mer* 173:60–65.

Prosser, J. I., B. J. M. Bohannan, T. P. Curtis, R. J. Ellis, M. K. Firestone, R. P. Freckleton, J. L. Green, et al. 2007. The role of ecological theory in microbial ecology. *Nature Reviews Microbiology* 5:384–92.

Quiñones, R. A., T. Platt, and J. Rodríguez. 2003. Patterns of biomass-size spectra from oligotrophic waters of the Northwest Atlantic. *Progress in Oceanography* 57:405–27.

Rappé, M. S., S. A. Connon, K. L. Vergin, and S. J. Giovannoni. 2002. Cultivation of the ubiquitous SAR11 marine bacterioplankton clade. *Nature* 41:630–33.

Redfield, A. C. 1960. The inadequacy of experiment in marine biology. In *Perspectives in marine biology,* ed. A. A. Buzzati-Traverso, 17–26. Berkeley, CA: University of California Press.

Reynolds, C. S. 2001. Status and role of plankton. In *Encyclopedia of biodiversity,* ed. S. A. Levin, 4:569–99. San Diego: Academic Press.

Rinaldo, A., A. Maritan, K. K. Cavender-Bares, and S. W. Chisholm. 2002. Cross-scale ecological dynamics and microbial size spectra in marine ecosystems. *Proceedings of the Royal Society London B* 269:2051–59.

Rocap, G., F. W. Larimer, J. Lamerdin, S. Malfatti, P. Chain, N. A. Ahlgren, A. Arellano, et al. 2003. Genome divergence in two *Prochlorococcus* ecotypes reflects oceanic niche differentiation. *Nature* 424:1042–47.

Rusch, D. B., A. L. Halpern, G. Sutton, K. B. Heidelberg, S. Williamson, S. Yooseph, D. Wu, et al. 2007. The *Sorcerer II* Global Ocean Sampling Expedition: Northwest Atlantic through eastern tropical Pacific. *PloS Biology* 5:398–431.

Rutherford, S., S. D'Hondt, and W. Prell. 1999. Environmental controls on the geographic distribution of zooplankton diversity. *Nature* 400:749–53.

Scharf, F. S., F. Juanes, and M. Sutherland. 1998. Inferring ecological relationships from the edges of scatter diagrams: Comparison of regression techniques. *Ecology* 79:448–60.

Selje, N., M. Simon, and T. Brinkhoff. 2004. A newly discovered *Roseobacter* cluster in temperate and polar oceans. *Nature* 427:445–48.

Semina, H. J. 1997. An outline of the geographical distribution of oceanic phytoplankton. *Advances in Marine Biology* 32:527–63.

Snelgrove, P. V. R. 2001. Diversity of marine species. In *Encyclopedia of ocean sciences,* ed. J. H. Steele, S. A. Thorpe, and K. K. Turekian, 748–57. San Diego: Academic Press.

Sogin, M. L., H. G. Morrison, J. A. Huber, D. M. Welch, S. M. Huse, P. R. Neal, J. M. Arrieta, and G. J. Herndl. 2006. Microbial diversity in the deep sea and the underexplored "rare biosphere." *Proceedings of the National Academy of Sciences* 103:12115–20.

Sournia, A., M.-J. Chrétiennot-Dinet, and M. Ricard. 1991. Marine phytoplankton: How many species in the world ocean? *Journal of Plankton Research* 13:1093–99.

Stepanauskas, R., and M. E. Sieracki. 2007. Matching phylogeny and metabolism in the uncultured marine bacteria, one cell at a time. *Proceedings of the National Academy of Sciences* 104:9052–57.

Thomson, J. D., G. Weiblen, B. A. Thomson, S. Alfaro, and P. Legendre. 1996. Untangling multiple factors in spatial distributions: Lilies, gophers, and rocks. *Ecology* 77:1698–1715.

Ulanowicz, R. E., and T. Platt, eds. 1985. Ecosystem theory for biological oceanography. *Canadian Bulletin of Fisheries and Aquatic Sciences* 213, Ottawa.

Venter, J.C., K. Remington, J. F. Heidelberg, A. L. Halpern, D. Rusch, J. A. Eisen, D. Wu, I. Paulsen, et al. 2004. Environmental genome shotgun sequencing of the Sargasso Sea. *Science* 304:66–74.

Vogelbein, W. K., V. J. Lovko, J. D. Shields, K. S. Reece, P. L. Mason, L. W. Haas, and C. C. Walker. 2002. *Pfiesteria shumwayae* kills fish by micropredation not exotoxin secretion. *Nature* 418:967–70.

White, E. P., S. K. M. Ernest, A. J. Kerkoff, and B. J. Enquist. 2007. Relationships between body size and abundance in ecology. *Trends in Ecology and Evolution* 22:323–30.

Wood, A. M., and T. Leatham. 1992. The species concept in phytoplankton ecology. *Journal of Phycology* 28:723–29.

Woodd-Walker, R. S., P. Ward, and A. Clarke. 2002. Large-scale patterns in diversity and community structure of surface water copepods from the Atlantic Ocean. *Marine Ecology Progress Series* 236:189–203.

Zengler, K., G. Toledo, M. Rappé, J. Elkins, E. J. Mathur, J. M. Short, and M. Keller. 2002. Cultivating the uncultured. *Proceedings of the National Academy of Sciences* 99:15681–86.

Zhang, K., A. C. Martiny, N. B. Reppas, K. W. Barry, J. Malek, S. W. Chisholm, and G. M. Church. 2006. Sequencing genomes from single cells by polymerase cloning. *Nature Biotechnology* 24:680–86.

CHAPTER THREE
PATTERNS IN DEEP-SEA MACROECOLOGY

CRAIG R. MCCLAIN, MICHAEL A. REX,
AND RON J. ETTER

Introduction

In *Macroecology*, Brown (1995, 10) defined this emerging field as "a way of studying relationships between organisms and their environment that involves characterizing and explaining statistical patterns of abundance, distribution, and diversity." It is a relatively new and fundamentally different approach to ecology that centers on large-scale phenomena. Macroecology grew partly out of the recognition that the results of short-term, small-scale experiments could not be extrapolated readily to larger scales (see Roughgarden, Gaines, and Possingham 1988, and Wares and Cunningham 2001 for examples in the marine realm). It also explores the intriguing and often puzzling relationships between community structure and the ecogeographic properties of species that had simply never been seriously examined. Established as both an unabashedly empirical and inductive discipline, the scaling relationships that emerged in macroecology have contributed significantly to the development of a synthetic metabolic theory of ecology, a conceptual framework that has the potential to unify ecology across multiple levels of organization (Brown et al. 2004).

Most macroecological analyses have focused on terrestrial systems because large databases on geographic ranges of species, body size, abundance, diversity, and relevant environmental variables were already available for regional and often global spatial scales. Much less is known about marine macroecology, a shortcoming that this volume is intended to remedy. Least well known is the deep sea. Ecological investigation in the deep sea began only forty years ago (Sanders, Hessler, and Hampson 1965). Its vast size and extreme environment make exploration technically difficult and expensive. Macroecology would seem to provide a useful and practical perspective for understanding the structure and function of deep-sea ecosystems. Small-scale manipulative experiments conducted on annual time scales in the deep sea have provided important insights into the causes of local species coexistence (Snelgrove and Smith 2002). However, it is now clear that community structure varies on local, regional, and global spatial scales, and on temporal scales ranging from annual to cycles of orbital forcing (Stuart, Rex, and Etter 2003). Local diversity is affected by oceanographic processes that operate on very large scales in both surface and benthic environments (Levin et al. 2001). With present technology, it seems unlikely that experiments could be deployed on geographic and temporal scales sufficiently large to capture the full range of factors that regulate deep-sea benthic diversity. A comparative approach like macroecology seems promising both to integrate our present understanding and to shape an agenda for future research. Ultimately, experiments will be necessary to test inferences from macroecological studies, but large-scale comparative studies will help identify and limit the range of hypotheses that need to be tested experimentally (Brown 1995, Menge et al. 2003).

In this chapter, we present the basic macroecological features of the deep-sea fauna, including geographic variation in standing stock, species diversity, species ranges, and body size. We also explore the relationships among body size, diversity, and abundance, and between body size and metabolic rate. Whenever possible, we compare these trends to those found in other ecosystems. We concentrate on the deep North Atlantic Ocean, which is by far the most intensively sampled region of the World Ocean. We emphasize communities that inhabit the soft-sediment habitats covering most of the seafloor. An excellent summary of biogeographic patterns in deep-sea chemosynthetic habitats is provided by Van Dover (2000). Many of our case studies involve mollusks, because their taxonomy and biogeography are relatively well known, owing to the extensive published work of Philippe Bouchet, Anders Warén, John Allen, Howard Sanders, and their colleagues. While the data are limited and geographically restricted, they begin to provide a macroecological context for studying deep-sea assemblages.

Standing Stock

The pattern of standing stock with depth is the most well-established feature of community structure in the deep-sea benthos. The biomass and abundance of the macrofauna in the western North Atlantic are shown in figure 3.1. Both decrease exponentially by two to three orders of magnitude from the continental shelf to the abyssal plain. Abyssal macrobenthic standing stock is extremely low (10s–100s individuals m^{-2} and <1 g m^{-2}). Bathymetric decreases in standing stock also occur in bacteria (Deming and Yager 1992; Aller, Aller, and Green 2002), the meiofauna (Soltwedel 2000), invertebrate megafauna (Lampitt, Billett, and Rice 1986) and demersal fishes (Haedrich and Rowe 1977).

The primary source of food for the benthos is sinking phytodetritus augmented by the sporadic occurrence of sinking plant and animal remains. Food availability decreases with increasing distance from productive coastal waters and terrestrial runoff, and because of remineralization during descent through a progressively deeper water column. Thus, the decline in standing stock is driven by the decrease in the rate of organic carbon input with increasing depth and distance from land. The entire process of surface-benthic coupling is complicated and incompletely understood. It involves the biotic and abiotic factors affecting surface production, the export of organic material to the deep ocean, horizontal dispersal by currents in the water column, transformation during sinking, and redistribution within the benthic community. Despite this complexity, deep-sea macrobenthic standing stock can be accurately predicted by estimates of organic carbon flux at depths determined from satellite imagery of overhead surface production and empirical models of downward flux (Johnson et al. 2005).

Rowe (1983) first showed that the exponential decrease in standing stock is a global phenomenon. Subsequently, it has become clearer that the pattern is modulated in a predictable way by unusual circumstances of food availability at depth. For example, elevated standing stock is caused by organic carbon loading associated with proximity to oxygen minimum zones (Levin and Gage 1998), lateral advection and deposition (Blake and Hilbig 1994), upwelling systems (Sanders 1969), exposure of reactive sediments or deposition of sediments by strong bottom currents (Aller 1997), or where topography concentrates food in canyons (Vetter and Dayton 1998) or trenches (Gambi, Vanreusel, and Danovaro 2003). By contrast, depressed standing stock is found in oligotrophic ocean basins such as the Arctic Sea (Kröncke et al. 2000) and the Mediterranean Sea (Tselepides et al. 2000). Overall, how-

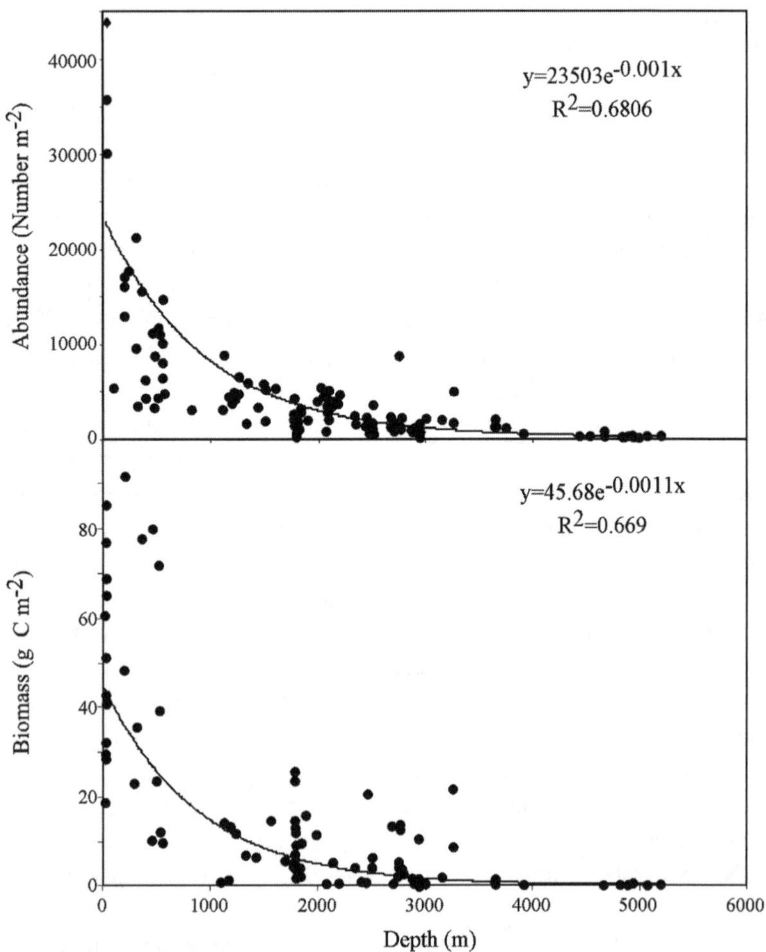

FIGURE 3.1 Standing stock of macrobenthos with depth in the deep western North Atlantic. Data are from Sanders, Hessler, and Hampson (1965); Rowe, Polloni, and Horner (1974); Smith (1978); Rowe, Polloni, and Haedrich (1982); Maciolek et al. (1987b).

ever, as we show in our discussion of body size with a new global analysis of abundance in the macrofauna and meiofauna, there is a strong and clear tendency for standing stock to decrease with depth and distance from land when geography is statistically controlled. Benthic standing stock appears to be the best available correlate of food availability in the deep sea (Smith et al. 1997), and arguably represents the single most significant environmental gradient affecting geographic patterns of biodiversity and evolutionary potential of the deep-sea benthos.

Patterns of Species Diversity

The quantitative study of community structure in the deep-sea benthos began with Hessler and Sanders' (1967) momentous discovery that species diversity is surprisingly high. It had been assumed for a century before that the deep-sea fauna was depauperate, and prior to then that the great depths were essentially sterile. Sanders (1968), in his influential comparative study of marine benthic diversity, showed that bathyal diversity exceeded coastal diversity in the temperate zone and approached shallow-water tropical diversity. The development and deployment of more effective sampling gear (Hessler and Jumars 1974) has indicated that diversity is probably even higher than Sanders estimated (Grassle and Maciolek 1992). Given the brief period of exploration, the enormous size of the environment, and the difficulty of sampling, it is not surprising that our knowledge of ecology and biogeography in the deep sea remains far behind that for terrestrial and coastal systems. All the same, Hessler and Sanders' discovery has inspired remarkably rapid progress, and a picture of diversity in time and space is beginning to take shape. In this section we summarize patterns of diversity on local, regional, and global scales.

On relatively small scales, the number of species coexisting in the deep sea is surprisingly high, exceeding 300 macrofaunal species m^{-2} at bathyal depths in the western North Atlantic (Etter and Mullineaux 2001; Levin et al. 2001). Despite recent controversy (Gray 1994, 2002, Gray et al. 1997), diversity is considerably higher than in nearby shallow-water communities. While true that some deep-sea communities (e.g., western North Atlantic) are less diverse than some shallow-water communities elsewhere in the world (e.g., Australia), it is not clear what this means or how it will help us to identify the ecological and evolutionary forces that regulate diversity. When comparisons are controlled for spatial scale, geography, taxonomy, sampling methods, and habitat, the number of species coexisting at small scales in the deep sea is considerably higher than in shallow-water communities (Etter and Mullineaux 2001, Levin et al. 2001).

The greater diversity in what appears to be a more homogeneous environment has long perplexed marine ecologists (Sanders 1968; Gage 1996; Gray 2002) and remains a major theoretical challenge. Numerous hypotheses have been proposed, including competition, facilitation, predation, disturbance, productivity, environmental heterogeneity, and patch dynamics (reviewed in Etter and Mullineaux 2001, Levin et al. 2001, and Snelgrove and Smith 2002). Existing experimental and comparative evidence suggest that no single factor is responsible. Diversity within local deep-sea assemblages is apt to reflect a complex dynamical process that integrates a number of interdependent forces operating at different space and time scales and changing in relative importance along various environmental gradients.

Much of our understanding of the patterns and potential causes of deep-sea biodiversity comes from regional-scale sampling studies, particularly along bathymetric gradients (Levin et al. 2001). Sanders (1968) showed that local species diversity of bivalve mollusks and polychaete worms increased from the continental shelf to lower bathyal depths in the western North Atlantic. When the analysis was extended to abyssal depths and more taxa, the overall diversity-depth trend appeared to be unimodal; diversity increased to a peak in the mid- to lower-bathyal zone, and then decreased in the abyss (Rex 1973, 1981, 1983). These early studies relied on estimating diversity from qualitative samples by normalizing the number of species to a common number of individuals with rarefaction, a numerical method to resample the relative abundance distribution devised by Sanders (1968) and formalized statistically by Hurlbert (1971). Recent intensive quantitative sampling with precision box corers confirmed Sanders' finding that diversity increases with depth below the continental shelf (Etter and Mullineaux 2001; Levin et al. 2001), and suggest that peak diversity of the whole macrofaunal assemblage occurred at around 1,000–1,500 m on the continental slope (Etter and Grassle 1992). The depth of maximum diversity may depend on the taxa considered, species' ranges and dispersion patterns, and the spatial scales covered by different sampling gears (Stuart, Rex, and Etter 2003). However, the general unimodel shape of diversity-depth patterns seems to be typical in the western North Atlantic. No other region of the World Ocean has been sampled so intensively. Polychaetes, the most abundant and diverse macrofaunal taxon, show unimodal diversity-depth patterns in the eastern North Atlantic (Paterson and Lambshead 1995) and in the eastern equatorial Atlantic (Cosson-Sarradin et al. 1998). But limited data on other taxa and geographic regions suggest that unimodal patterns may not be universal; and moreover, that the causes of known unimodal patterns may vary geographically (Rex, Etter, and Stuart 1997; Flach and deBruin 1999; Gage et al. 2000; Stuart, Rex, and Etter 2003).

Studies of bathymetric diversity trends have centered on variation in alpha (sample) diversity within basins. Here, we present a different approach, based on species ranges, and extend the analysis to much larger spatial scales that include eastern and western corridors of the North Atlantic. Figure 3.2 shows depth ranges of all protobranch bivalve mollusks collected by the Woods Hole Oceanographic Institution's Benthic Sampling Program (Sanders 1977) from sampling transects in the North American and West European Basins

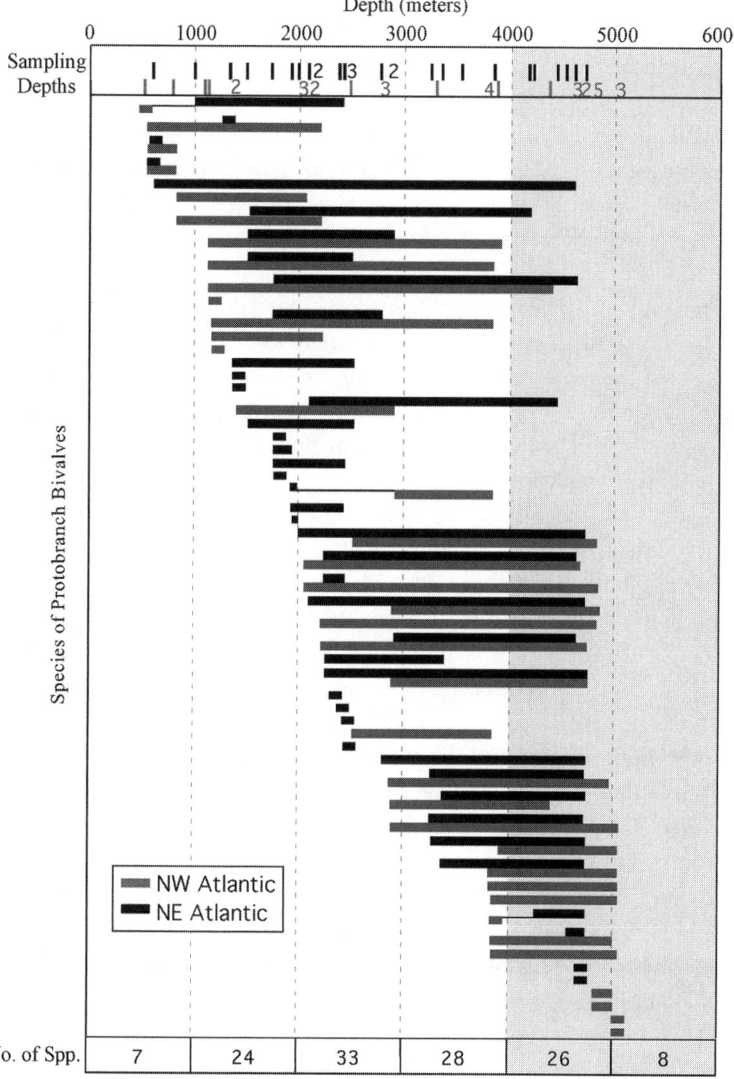

FIGURE 3.2 Depth ranges of protobranch bivalves from the eastern and western North Atlantic. Data are from Allen and Sanders (1996). Subspecies are combined. Where ranges do not overlap between the eastern and western North Atlantic, the distributions are connected with a thin line. Depth locations of samples on which the ranges are based are given at the top of the figure (ticks represent individual samples, and numbers represent multiple samples that are located close together). The number of coexisting species in 1,000 m depth intervals is indicated at the bottom. The species and their depth ranges (western and eastern North Atlantic respectively) are given in Appendix A.

(Allen and Sanders 1996). All of the material was collected with epibenthic sleds (Hessler and Sanders 1967). The depth distribution of samples in both basins is indicated at the top of figure 3.2. While sampling is not perfectly equable between the eastern and western North Atlantic, both transects extend from upper bathyal to abyssal depths, represent all major seafloor physiographic features, and include a similar number of samples (thirty-five west, twenty-eight east). Species accumulation curves suggest that the faunas are reasonably well characterized, and that the eastern fauna is somewhat more diverse over much of the depth range (Allen and Sanders 1996).

The unimodal diversity-depth pattern discussed previously for alpha diversity of individual samples is also apparent when the number of coexisting species ranges is summed over 1,000 m depth intervals (fig. 3.2, bottom panel). Diversity peaks in the 2,000–3,000 m region and is depressed at upper bathyal and abyssal depths. Eastern and western faunas show similar diversity-depth patterns.

Figure 3.2 reveals a high rate of faunal turnover, or β-diversity, along the depth gradient. High rates of zonation are also common in the megafauna (Haedrich, Rowe, and Polloni 1980; Hecker 1990; Howell, Billett and Tyler 2002), macrofauna as a whole (Rowe, Polloni, and Haedrich 1982; Blake and Grassle 1994; Gage et al. 2000) and meiofauna (Coull 1972). A surprisingly high proportion of protobranch species (twenty-four out of fifty-six, or 43 percent) are shared between the eastern and western North Atlantic (fig. 3.2). Even more remarkable, twenty-one of the shared species (88 percent) have depth ranges that overlap between basins. Only three of these species (connected by thin lines, figure 3.2) have disjunct depth ranges, and even so, occur in the same basic physiographic feature and are separated by <1,000 m. As Sanders and Hessler (1969) conjectured, based on fewer data, some basic features of faunal zones within basins appear to extend as bands around the North Atlantic, at least in protobranchs. This large-scale faunal redundancy has important implications for projecting global biodiversity in the deep-sea benthos, and suggests that diversity might be lower than the 10,000,000 species projected by Grassle and Maciolek (1992).

Allen and Sanders (1996) showed that deeper-dwelling protobranch species tend to be more cosmopolitan, as is borne out in figure 3.2 for the North Atlantic. The proportion of species that occur in both eastern and western basins increases from 44 percent at upper bathyal depths (500–2,000 m) to 54 percent at lower bathyal depths (2,000–4,000 m) to 60 percent in the abyss (>4,000 m). Rex et al. (2005) pointed out that within-basin abyssal endemism in mollusks appeared to be low. On a between-basin scale it appears to be even lower. For example, *Ledella aberrata* would appear to be an abyssal endemic in the eastern North Atlantic (see Appendix A), but has a lower bathyal distribution in the western North Atlantic. *Malletia polita* is another case of an apparent eastern North Atlantic abyssal endemic that occurs at bathyal and abyssal depths in the western North Atlantic (Allen and Sanders 1973; Allen, Sanders, and Hannah 1995). Two other apparent western North Atlantic abyssal endemics, *Silicula macalisteri* and *Yoldiella similiris* are known from upper bathyal sites in western South Atlantic (Allen and Sanders 1973; Allen, Sanders, and Hannah 1995). Of the other abyssal endemic species that are described, only one *Ledella galathea,* is known exclusively from abyssal depths (the West European Basin and off West Africa; Allen and Hannah 1989). Three undescribed unique species (*Ledella* sp., *Spinula* sp., and *Tindariopsis* sp.) are also potential candidates for true abyssal endemics (J. Allen, personal communication). A conspicuous feature of figure 3.2 is that the vast majority of species with abyssal distributions (79 percent; 86 percent if we exclude *S. mcalisteri* and *Y. similiris* as discussed previously) are range extensions of bathyal species. This has important implications for the causes of diversity discussed in the following.

Variation in species diversity and composition also occurs on oceanwide interbasin scales in the Atlantic (Allen and Sanders 1996; Wilson 1998). There is some indication of latitudinal gradients of diversity in the deep-sea fauna, though this is based on much less sampling than in terrestrial and coastal systems (Roy, Jablonski, and Valentine 1994, 1998; Hawkins, Porter, and Dinîz-Filho 2003; Hillebrand 2004a, 2004b, Witman, Etter, and Smith 2004). Rex et al. (1993; Rex, Stuart, and Coyne 2000) found poleward decreases in the diversity for gastropods, bivalves, and isopods in the North Atlantic and Norwegian Sea. The South Atlantic is more poorly sampled. It shows strong regional variation in diversity, and a weak latitudinal signal for mollusks, but not isopods (Rex et al. 1993). Gage et al. (2004) found poleward declines in the diversity of deep-sea cumaceans for the entire Atlantic (a significant parabolic regression with a peak at tropical latitudes). The eastern corridor of the North Atlantic showed a significant latitudinal gradient, but the western corridor did not, suggesting strong interbasin differences and mixed evidence for a simple consistent hemisphere-wide latitudinal gradient. Among the meiofauna, deep-sea foraminiferans show latitudinal gradients in the North and South Atlantic (Culver and Buzas 2000). Nematodes do not show a clear monotonic poleward decline (c.f. Lambshead et al. 2000 and Rex, Stuart, and Etter 2001), but may show peak diversity at midlatitudes in the North Atlantic (Mokievsky and Azovsky 2002).

The causes of geographic patterns of diversity on local, regional, and global scales have been reviewed several times recently (Rex, Etter, and

Stuart 1997; Levin et al. 2001; Etter and Mullineaux 2001; Snelgrove and Smith 2002; Stuart, Rex, and Etter 2003; Rex et al. 2005), and will only be summarized briefly here, since our chapter is concerned primarily with patterns. Unimodal bathymetric gradients of diversity have been attributed to mid-domain effects based on boundary constraints imposed by the coast and seafloor (Pineda 1993), but recent analyses show that diversity trends depart significantly from the predictions of mid-domain models (Pineda and Caswell 1998; McClain and Etter 2005). Just as in other marine environments, a wide variety of biotic and abiotic factors that act on different scales of time and space appear to affect deep-sea diversity. Apart from catastrophic events such as burial by submarine landslides (Rothwell, Thomson, and Kahler 1998) and ash from volcanic eruptions (Hess and Kuhnt 1996) or global anoxic events (Kennett and Stott 1991), much of the variation observed in deep-sea diversity appears to be related, directly or indirectly, to productivity in the form of carbon flux to the benthos from surface production. Within the deep sea, the relationship between diversity and productivity is unimodal as it is frequently, but not universally, in coastal and terrestrial environments (Rosenzweig 1995; Mittelbach et al. 2001). The most accurate indication available of average carbon flux over large spatiotemporal scales is the standing stock of the benthos (Smith et al. 1997). The unimodal diversity-depth pattern evident in figure 3.2 coincides with the monotonic decline in standing stock with depth shown in figure 3.1. Rex (1983) suggested that depressed diversity at upper bathyal depths, where standing stock is high, might be due to accelerated rates of local competitive displacement driven by pulsed carbon loading from high seasonal productivity in coastal waters (Rex 1981). All of the circumstances mentioned earlier where heavy carbon loading associated with upwelling, topographic focusing of sinking organic material, sediment erosion, and deposition result in high standing stock also show depressed diversity irrespective of depth. Even oxygen-minimum zones on continental margins that limit diversity through severe physiological constraints are ultimately caused by unusually high rates of overhead production and downward carbon flux (Levin and Gage 1998).

Rex et al. (2005) recently proposed that continental margins and abyssal plains of the North Atlantic may constitute a source-sink system for many species. As can be seen in figure 3.2, the vast majority of abyssal protobranchs in both eastern and western basins represent range extensions of bathyal species. These abyssal range extensions are very sparsely occupied because of the low density of all abyssal macroinvertebrates (fig. 3.1). Abyssal protobranch densities have been estimated to be on the order of one to three individuals m^{-2} for the few commonest species, one to five individuals 100 m^{-2} for most species, and two individuals 1,000 m^{-2} for the rarest species (Rex et al. 2005). Adult protobranchs are minute organisms with low mobility, low gramete production, and separate sexes. Their larvae are lecithotrophic and disperse demersally, potentially over considerable distances in the frigid deep-sea environment. These conditions suggest that abyssal populations of many species are sinks that experience chronic local extinction as an Allee Effect and are maintained by immigration from more abundant bathyal source populations through larval dispersal. In this view of deep-sea community ecology, bathyal diversity may be regulated by essentially the same equilibrial and nonequilibrial mechanisms that govern community structure in coastal and terrestrial systems where population densities are relatively high (Bertness, Gaines, and Hay 2001)—though the relative importance and operation of these mechanisms in the deep-sea remain far from clear. Much of the abyssal macrofauna may exist as a mass effect from bathyal populations. While source-sink dynamics may help explain abyssal molluscan diversity, at least for rarer species, its potential relevance to other elements of the abyssal fauna has not been examined.

At very large interbasin scales, both differences in regional ecology and the evolutionary-historical development of faunas may be important. Depressed diversity at high latitudes in some contemporary taxa may be caused in part by high-pulsed organic carbon loading resulting from high and seasonal surface production (Campbell and Aarup 1992). However, isopods show a gradient of decreasing diversity from the South to the North Atlantic that Wilson (1998) has attributed to a relatively new wave of invasion from shallow water in the Southern Hemisphere that augments the diversity of an earlier deep-sea in situ radiation. Stuart and Rex (1994) demonstrated that local diversity was a function of regional diversity in deep-sea gastropods, suggesting that the size of the species pool, presumably originating from regional-scale adaptive radiation, influences local diversity. The gradual historical formation of latitudinal gradients in deep-sea foraminifera during the Cenozoic can be traced in deep seabed cores (Thomas and Gooday 1996).

Body Size

Body size is related to a variety of life-history, physiological, and ecological traits (Peters 1983; Brown 1995; Gillooly et al. 2001, Brown et al. 2004, Savage et al. 2004), and thus may provide a link between processes at the individual level and higher levels of organization such as species diversity. The most immediate impression when looking at deep-sea samples is the extraordinarily small size of most species. As early as 1880, Mosely commented on the pervasive dwarfism of deep-sea animals. More recently, Thiel (1975, 593) referred to the deep sea as a "small organism habitat." Gage (1977) showed that the mean weight per individual in coastal waters of Loch Creran in Scotland was 0.039 g, compared to 0.002 g in the Rockall Trough (2,875 m), a full order of magnitude difference. Similar methods have been used to show meiofaunal miniaturization in the deep sea (Shirayama 1984; Pfannkuche 1985; Tietjen 1992; Vincx et al. 1994). Paradoxically, some deep-sea arthropods (isopods, amphipods, pycnogonids, ostracods) are much larger than shallow-water representatives (Gage and Tyler 1991).

Examples of the dramatic difference in body size of snails between the continental shelf and deep sea are shown in figures 3.3 and 3.4. Estimates of size in the deep-sea assemblages collected south of New England (McClain, Rex, and Jabbour 2005) are compared to those on Georges Bank, the adjacent continental shelf (Maciolek and Grassle 1987). The deep-sea fauna shows a smaller average size (deep sea: 8.2mm, and Georges Bank: 21mm). Large deep-sea gastropods are known; for example, *Troschelia berniciensis* reaches 10 cm in the bathyal eastern Atlantic (Olabarria and Thurston 2003), and *Guivillea alabastrina*, the largest deep-sea gastropod known, reaches 16.8 cm in the Southern Hemisphere (Knudsen 1973). But such large individuals and species appear to be very rare in the deep sea. In figure 3.4, we compare the sizes of eastern North Pacific continental shelf gastropods belonging to three families (Roy 2002) to those same families in the deep sea. Body size (geometric mean of length and width of the shell) is significantly

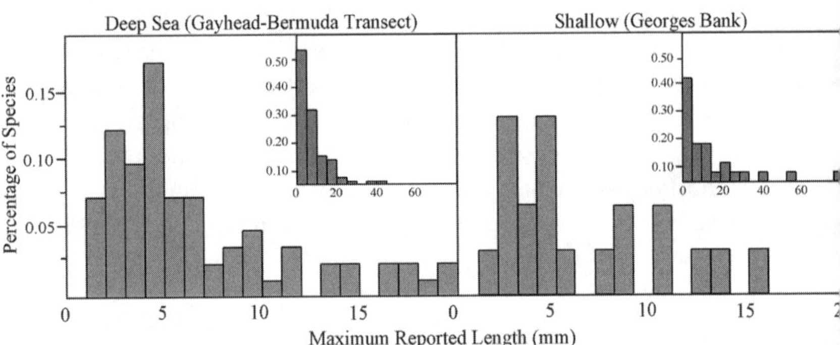

FIGURE 3.3 A comparison of body sizes between coastal and deep-sea benthic gastropods. Deep-sea gastropods were collected south of New England in the western North Atlantic (see McClain, Rex, and Jabbour 2005). Coastal gastropods represent the fauna of Georges Bank (Maciolek and Grassle 1987). Inset histograms reflect the percentage of species in 5mm length bins between 0–80mm. Main histograms reflect the percentage of species in 1mm length bins between 0–20mm.

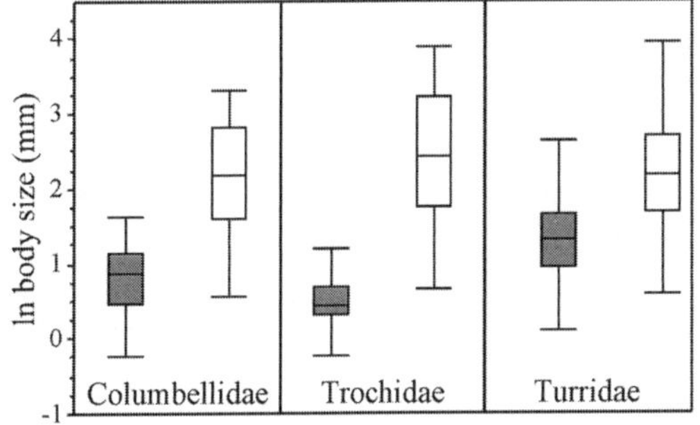

FIGURE 3.4 A comparison of size in deep-sea gastropods (McClain, Rex, and Jabbour 2005) from the western North Atlantic (solid bars) to continental shelf gastropods (open bars) from the Pacific (Roy 2002) for three separate families.

greater on the Pacific shelf (both t-test and Median Test, p-values < 0.0001) in all cases. As a striking example of the size difference between coastal and deep-sea gastropods, a back-of-the-envelope calculation based on data from McClain, Rex, and Jabbour (2005) shows that all of the deep-sea snails collected from the western North Atlantic by the Woods Hole Oceanographic Institution's Benthic Sampling Program (forty-four samples, 20,561 individuals) would fit comfortably into a single large shell of the common New England knobbed whelk *Busycon carica.*

The average size of deep-sea organisms continues to decrease with depth below the shelf-slope transition. Thiel (1979) was the first to demonstrate this trend by regressing abundance of the smaller meiofauna and larger macrofauna against depth. An ANCOVA showed that the meiofaunal regression had a higher intercept and lower slope than did the macrofauna, therefore the meiofauna comprise a larger proportion of the total assemblage as depth increases (average size decreases with depth). The analysis was based on meiofaunal densities off Portugal and East Africa, and macrofaunal densities (from Rowe 1971) off New England, Brazil, the Gulf of Mexico, and Peru. Here we repeat Thiel's analysis using a much larger dataset representing most major ocean basins (sixty-five studies and 705 observations for the meiofauna, sixty-one studies and 912 observations for the macrofauna). To remove regional influences, we regressed the partial residuals of abundance (effects of longitude and latitude removed) against depth (fig. 3.5). The resulting abundance-depth relationships are highly significant, with the meiofauna regression showing a higher elevation, as expected. An ANCOVA (see figure caption) shows that the macrofauna has a significantly steeper slope confirming, on a global basis, Thiel's conclusion that average metazoan size decreases with depth.

A closer look at size-depth patterns within individual taxa reveals a very mixed picture; size can increase, decrease, or show no pattern with depth

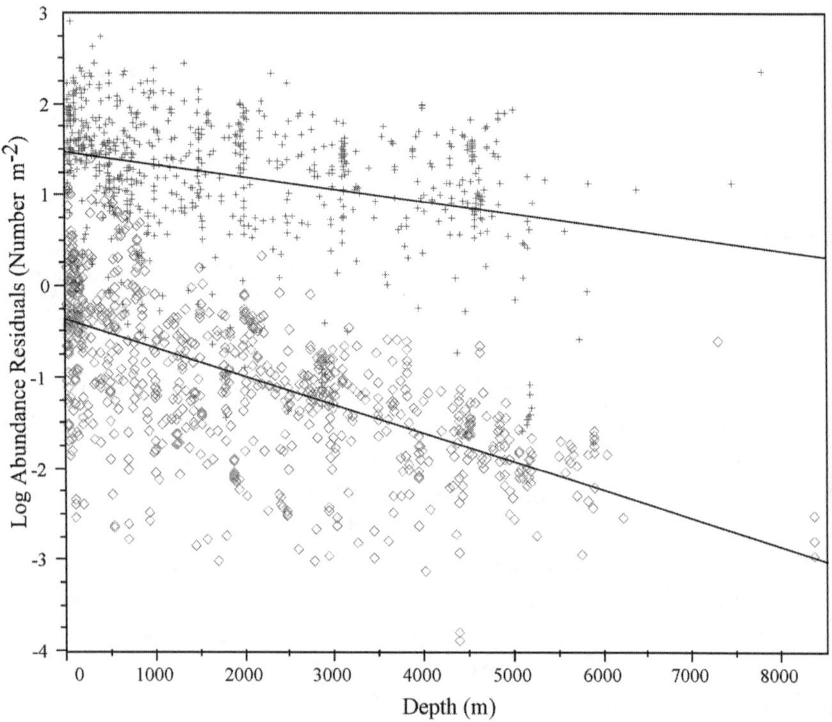

FIGURE 3.5 Regressions of the partial residuals of abundance (with the effects of longitude and latitude removed) against depth for deep-sea meiofauna and macrofauna. Meiofaunal abundance is higher and decreases with depth less rapidly than does macrofaunal abundance. This indicates that the average size of organisms decreases with depth. An ANCOVA shows that the slopes are significantly different ($F = 99.934$, d.f. = 1,1513, $P < 0.0001$). Regression equations are: Meiofauna $Y = 5.761 - 0.000135\ X$, $R^2 = 0.109$, $N = 705$, $F = 85.931$, $P < 0.001$; Macrofauna $Y = 3.588 - 0.000312X$, $R^2 = 0.452$, $N = 912$, $F = 667.913$, $P < 0.001$. Data references are given in Appendix B.

(reviewed in Rex and Etter 1998, Soetaert, Muthumbi, and Heip 2002). Part of this variation among taxa probably relates to methodological differences in the way size was measured and statistically analyzed, and some of it may reflect differences in the overall biological properties of the taxa and regional ecological differences. It is also important to remember that geographic variation in size might merely be a phenotypic plastic response. The degree to which size-depth trends represent actual adaptations to the environment is best studied at the level of individual species. If adaptive, bathymetric trends at higher taxonomic levels, and within and between functional groups, must result from clinal effects within species or depth-correlated replacement of species that differ in size.

Bergmann's rule states that body size increases toward higher latitudes.

Although the explanations for this trend are contentious, there is some support for increased size toward the poles for homeotherms (Brown and Lee 1969; Brodie 1975; Forsman 1991; Scharples, Fa, and Bell 1996). For ectotherms, Bergmann's rule applies for some insect taxa (e.g., Cushmann, Lawron, and Manly 1993; Hawkins and Lawton 1995; Arnett and Gotelli 1999) but not others (Hawkins 1995; Hawkins and Lawton 1995). Few studies have been conducted for marine invertebrates. Roy and Martien (2001) found no relationship between size and latitude for eastern coastal Pacific bivalves. For deep-sea faunas, only two studies have investigated latitudinal-size relationships. Latitude appears to be only a weak predictor of gastropod body size and is often subordinate to depth in multiple regression analyses (McClain and Rex 2001, Olabarria and Thurston 2003).

Many hypotheses have been proposed to explain spatial gradients of size in other systems, including temperature (e.g., Atkinson and Sibly 1997), predation (e.g., Blumeshine, Lodge, and Hodgson 2000), energy input (Blackburn and Gaston 1996), oxygen availability (Chapelle and Peck 1999) and sediment diversity (Schwinghamer 1985). It is unlikely that temperature, relatively invariant throughout much of the deep sea, plays a significant role in determining sizes of deep-sea organisms (McClain and Rex 2001), but this relationship has not been examined statistically. Schwinghamer (1985) proposed that the tendency toward smaller organisms in the deep sea is related to sediment diversity as a reflection of greater habitat diversity. However, the relationship between body size and sediment heterogeneity has not been borne out in coastal and shelf benthic habitats (Duplisea and Drgas 1999; Parry et al. 1999; Leaper et al. 2001). Sediment-organism interactions do appear to be important in the deep sea (Etter and Grassle 1992), but they have not been related to body size. Chapelle and Peck (1999) demonstrated that maximum potential size is limited by oxygen availability in benthic amphipod crustaceans from coastal and freshwater environments. Larger size at more oxygenated sites is also found in deep-sea gastropods (McClain and Rex 2001). Spicer and Gaston (1999) suggested that oxygen content of water should not affect size in aquatic environments, and that these relationships are a spurious consequence of temperature gradients. However, it is clear that body size and oxygen availability are related independent of temperature for both shallow and deep-water organisms (McClain and Rex 2001; Peck and Chapelle 2003). Body size can also be regulated by the effects of oxygen on development, cell size, and cell number (Frazier, Woods, and Harrison 2001; Peck and Chapelle 2003).

Although all of these factors, and others, may account for some of the variation in body size, the most important determinant is likely to be carbon flux to the benthos (Thiel 1975; Rex and Etter 1998). Support comes from the inference that benthic standing stock decreases exponentially with depth (fig. 3.1), and that standing stock is the best proxy available for carbon flux to the seabed (Rowe 1983; Smith et al. 1997). According to the optimality theory of body size (Sebens 1982, 1987), optimal size should decrease with depth as rates of food intake decrease and the costs of foraging increase (Rex and Etter 1998). In general, this prediction is supported by the decrease in average metazoan size with depth (fig. 3.5). However, as mentioned earlier, size-depth clines within species show considerable variation. This may be because other selective advantages of large size (metabolic efficiency, escape from predation, and the ability to exploit more food resources) displace populations away from optimal size in a taxon-specific way. For gastropods there is a shift from positive to negative size-depth clines within species with increasing depth (McClain, Rex, and Jabbour 2005). There is also an indication of an increase in size across bathyal depths and a decrease in the abyss for gastropod assemblages as a whole and in demersal fishes (fig. 3.6). In the upper- to mid-bathyal zone, where population densities are relatively high, other advantages of larger size may offset selection for optimal size based on maximizing the energy available for reproduction. However, at the extremely low densities found at lower bathyal and abyssal depths, the relative rates of energy intake, and cost may finally enforce smaller size. Alternatively, abyssal snails may be smaller either because severe energy constraints favor small-bodied species as the energy demands become too great for large organisms to maintain reproductively viable population sizes (Thiel 1975), or because individuals of many populations experience retarded growth in an unfavorable sink environment (Rex et al. 2005), or some combination of these phenomena.

Some differences in the shape of size-depth trends also may be due to changes with depth in the fundamental causative agents. Soetaert, Muthumbi, and Heip (2002) showed that the average size of nematodes decreases with depth in the eastern North Atlantic. However, this was a complex response to both food availability and the biogeochemical properties of sediments that limit vertical distribution. Nematodes that occupy the topmost oxygenated layer of sediment scarcely change in size with depth. The overall miniaturization with depth is attributable to larger nematodes being able to utilize deeper more anoxic layers at bathyal depths, but not at abyssal depths.

A major objective of macroecology is to explore how body size relates to the structure of ecological communities (Lawton 1990; Brown 1995; Allen, Brown, and Gilloly 2002; Brown et al. 2004). In particular, this has centered on the relationship between body size, species richness, and abundance in

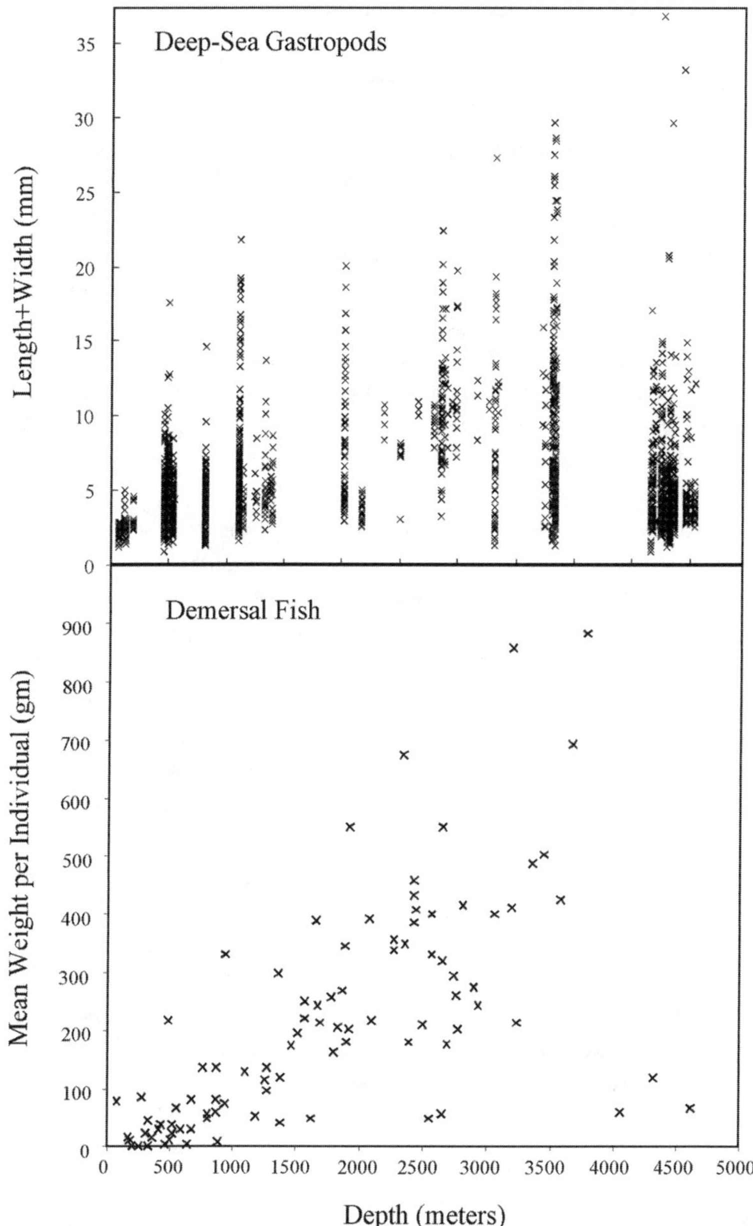

FIGURE 3.6 Maximum Size (length + width) for deep-sea gastropods, each point represents an individual(McClain et al. 2005) and mean weight per individual for demersal fishes (from Polloni et al. 1979) across a depth gradient South of New England, points represent a mean size per trawl.

terrestrial and marine systems, with little research devoted to these trends in the deep sea (but see McClain 2004, and Rex and Etter 1998). The statistical relationships among these variables have important implications for estimating biodiversity and could suggest how resource availability helps determine community structure. Siemann, Tilman, and Haarstad (1996, 1999), using data from a comprehensive sampling study of grassland insects, showed that diversity and abundance had fairly symmetrical unimodal relationships to body size, as had been frequently observed in other studies. When the relationships were reviewed in three dimensions they described a parabola with highest diversity at intermediate abundant size classes. The projection of this parabola on the diversity-abundance plane revealed that the number of species (S) scaled to the number of individuals within size classes (I) according to the expression $S = I^{0.5}$, independent of body size. A study of diversity, size and abundance in rocky intertidal mollusks showed a similar set of relationships (Fa and Fa 2002).

McClain (2004) examined these relationships for deep-sea gastropods (fig. 3.7). Again, the relationships of diversity and abundance to size are unimodal (fig. 3.7, panels A, B), although the diversity-body size relationship is right skewed. This contrasts with the log-normal trends in terrestrial arthropods (Siemann, Tilman, and Haarstad 1996) and coastal mollusks (Roy, Jablonski, and Valentine 2000; Fa & Fa 2002), but agrees with the prevalent right-skewed pattern observed in a variety of other organisms (Brown 1995). The skewed distribution also, of course, affects the shape of the curve in the three dimensional representation (fig. 3.7, panel D), and the distribution of size classes about the diversity abundance regression (fig. 3.7, panel C). Nonetheless, diversity scales to abundance in a way that is similar to terrestrial insects and coastal mollusks. That studies of three such different communities produce such a similar set of relationships suggests a common set of underlying causes. While the causal links remain obscure they are likely linked to scale, energy availability, available niche space, size-biased speciation and extinction, and constraints on minimum size (Brown 1995; Kozlowski and Gawelczyk 2002; McClain 2004).

Metabolic Rates and the Metabolic Theory of Ecology

Metabolism underlies all biological rates because it is the sum of expenditures toward growth, maintenance, and reproduction. The metabolic theory of ecology (MTE) is a potentially unifying framework that seeks to link the factors controlling metabolic rates to higher-order macroecological patterns and processes at population, community, and ecosystem levels (Gillooly

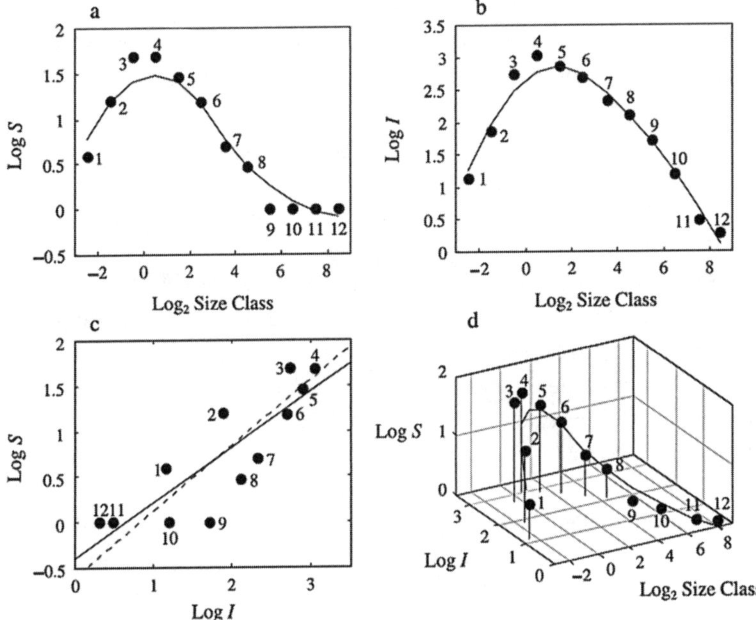

FIGURE 3.7 Panel A: Relationship between log-species richness (*S*) per $\log_2$ size class (mm^3). B. Relationship between log abundance (*I*) per $\log_2$ size class. C. Relationship between log species richness (*S*) and log abundance (*I*) per $\log_2$ size class. The lines represent fitted regressions (OLS: $S = -0.406I^{0.616}$, $R^2 = 0.719$, $p=0.0005$; RMA: $S = -0.613I^{0.726}$, $R^2 = 0.848$, $p = 0.0500$). D. Relationships between log of species richness (*S*), log of number of individuals (*I*), and $\log_2$ size class. All plots are from McClain (2004). Lines represents a kernel smoothing curve fitted to the data and numbers refer to the $\log_2$ size classes.

et al. 2001; Brown et al. 2004). The metabolic theory of ecology has generated substantial interest and criticism (see, e.g., special issue of *Functional Ecology* vol. 18, 2004; and *Ecology*, vol. 85, 2004). It predicts that metabolic rates of all organisms are primarily determined by individual body mass and temperature. Individual mass-specific metabolism is expected to scale with body size at a –1/4 power (West, Brown, and Enquist 1997, 1999a, 1999b) and scale exponentially with temperature according to Boltzmann/Van't Hoff-Arrhenius equation, with an activation energy between 0.60–0.70 eV (Gillooly et al. 2001). The value of MTE in deep-sea ecology, if it can be verified, would be to help explain large-scale macroecological patterns in an environment where both experimentation and precision sampling programs are logistically difficult. Because of the recent development of MTE, little research has been conducted to establish its generality, or its applicability to deep-sea ecology. Metabolic rates of deep-sea organisms are temperature and size dependent, but selective pressures depending on habitat and life styles may lead to some additional variation (Childress et al. 1980; Childress 1995).

Conclusions and a Future Agenda for Research in Deep-Sea Macroecology

The study of macroecological patterns in the deep sea has changed our perception of this remote, enormous, and complex ecosystem and helped to identify many of the ecological and evolutionary processes that might be important in regulating these communities. Several broad-scale patterns have emerged that provide the basic framework for exploring deep-sea ecosystems. Biomass, density, and body size all decrease with depth, presumably reflecting the exponential decline in carbon flux from surface production. Species diversity (or richness) typically peaks at intermediate depths and might also be related to energy flux, but not in a simple monotonic way. The consensus so far is that several interdependent processes that change in magnitude and vary in relative importance across various environmental gradients regulate diversity (Levin et al. 2001). One of the most intriguing hypotheses recently advanced to explain depressed abyssal diversity integrates a number of macroecological, life history, and natural history patterns to suggest that the low diversity at abyssal depths occurs because the abyssal zone acts as a sink habitat (sensu Pulliam 1988), relying on more productive bathyal regions as a source of larvae to sustain populations (Rex et al. 2005). If correct, this theory would explain lower diversity on the abyss.

As in other ecosystems, there appears to be a link between abundance, body size, and diversity that changes with spatial scale and energy availability. The relationships appear to be general but remain largely untested. Based on MTE (Brown et al. 2004), metabolic rates, governed by body size and temperature, and energy availability interact to regulate the basic time scales of biological processes, which in turn control population dynamics, rates of biotic interactions and the structure of communities. How applicable MTE is to life in the deep ocean is uncertain.

At global scales, species diversity locally and regionally declines poleward, paralleling similar patterns in shallow-water and terrestrial environments. Because the patterns are manifested at both local and regional scales, the gradient may reflect the evolutionary buildup of the regional species pools and the way in which these pools respond to ecological processes.

Although we have made significant progress, deep-sea macroecology is in its infancy and much remains to be done. The most pressing need is for more samples, collected over a broader geographic range. Fundamental to any macroecological study is the distribution of individual species, which is poorly documented in the deep sea. New sampling programs should be based on existing knowledge and undertaken in a coordinated fashion at strategic locations to test specific hypotheses (e.g., abyssal source-sink), determine the generality of existing patterns, and expand our knowledge to other taxa, basins, and oceans. Most macroecological studies are based on a limited set of taxa, raising the possibility that observed patterns are not reflective of entire communities. This is especially true for broad-scale studies of species diversity because the patterns for a single taxon (or size category, e.g., macrofauna) may differ from the whole community due to tradeoffs among major taxonomic groups within local assemblages (Wilson 1998).

Most deep-sea research has focused on documenting geographic and bathymetric patterns and inferring mechanisms that might shape these patterns. The next phase of research needs to include an experimental component to test the role of these putative processes in shaping macroecological patterns and establish how those roles vary on large scales. It should also incorporate an evolutionary perspective because macroecological patterns may ultimately reflect evolutionary processes (Brown 1995; Gaston and Blackburn 2000; Brown et al. 2004). For example, recent population genetic (Cardillo 1999; Martin and McKay 2004; Williams and Reid 2004; Xiang et al. 2004) and paleontological (Jablonski 1993; Buzas, Collins, and Culver 2002) evidence suggests that differences in evolutionary rates might be a key factor in generating latitudinal gradients in diversity. Similarly, bathymetric gradients in diversity may reflect differences in the potential for evolution at different depths (Etter and Rex 1990; Rex et al. 2005; Etter et al. 2005). Recent advances in molecular genetics (Chase et al. 1998) make it possible to identify where and how evolution has unfolded in the deep sea, which should allow us to test the role of historical processes. To ultimately understand the origin and maintenance of macroecological patterns, we will need to quantify and critically test the influence of processes operating at a variety of different spatial and temporal scales.

APPENDIX A

Yoldiella frigida (503–808 m, 609 m), *Yoldiella lucida* (503–808 m, 609 m), *Ennucula bushae* (530 m, 1,006–2,430 m), *Ennucula granulosa* (530–2,178 m, 1,336 m), *Ledella pustulosa* (—, 609–4,632 m), *Yoldiella curta* (808–2,178 m, 1,500–4,190 m), *Ennucula similis* (808–2,044 m, —), *Neilonella salicensis* (1,102–3,834 m, 1,500–2,494 m), *Deminucula atacellana* (1,102–3,900 m, 1,500–2,890 m), *Yoldia inconspicua* (1,102–4,400 m, 1,739–4,632 m), *Ledella solidula* (1,144 m, —), *Yoldiella enata* (1,144 m, —), *Phaseolus* sp. (1,144–2,196 m, —), *Malletia johnsoni* (1,144–3,834 m,

1,739–2,777 m), *Ledella similis* (—, 1,336 m), *Ledella acuminata* (—, 1,336–2,503 m), *Nuculana commutata* (—, 1,336 m), *Yoldiella obesa* (1,383–2,886 m, 2,091–4,466 m), *Yoldiella lata* (—, 1,500–2,503 m), *Microgloma turnerae* (—, 1,739–1,922 m), *Tindaria hessleri* (—, 1,739 m), *Bathyspinula filatovae* (—, 1,739 m), *Portlandia lenticulata* (—, 1,739–2,430 m), *Yoldiella insculpta* (—, 1,922–2,430 m), *Bathyspinula subexisa* (—, 1,993 m) *Yoldiella fibula* (2,496–4,825 m, 1,993–4,734 m), *Ledella sublevis* (2,022–4,680 m, 2,209–4,632 m), *Pristogloma nitens* (2,022–4,853 m, 2,209–2,430 m), *Pristogloma alba* (2,178–4,833 m, —), *Brevinucula verrilli* (2,196–4,749 m, 2,890–4,632 m), *Bathyspinula hilleri* (—, 2,209–3,358 m), *Ledella bushae* (—, 2,379 m), *Yoldiella veletta* (—, 2,430 m), *Portlandica minuta* (—, 2,494 m), *Lametia abyssorum* (2,496–3,834 m, —), *Portlandica fora* (—, 2,503 m), *Yoldiella biscayensis* (—, 2,777–4,734 m), *Neilonella whoii* (2,862–4,970 m, 3,250–4,734 m), *Yoldiella jeffreysi* (2,862–4,862 m, 2,091–4,734 m), *Yoldiella ella* (2,862–4,400 m, 3,358–4,734 m), *Malletia cuneata* (2,864–4,749 m, 2,209–4,734 m), *Malletia abyssorum* (2,864–5,042 m, 3,250–4,734 m), *Silicula fragilis* (2,886–3,834 m, 1,922 m), *Ledella ultima* (3,806–5,042 m, 3,358–4,734 m), *Tindaria callistiformis* (3,806–5,042 m, —), *Silicula filatovae* (3,828–5,042 m, —), *Ledella aberrata* (3,834 m, 4,240–4,734 m), *Yoldiella americana* (3,834–5,042 m, —), *Malletia polita* (3,834–5,007 m, 4,550–4,734 m), *Yoldiella subcircularis* (3,900–5,042 m, 3,250–4,734 m), *Ledella* sp. (4,800–5,000 m, —), *Spinula* sp.(4,800–5,000 m, —), *Silicula mcalisteri* (5,000 m, —), *Yoldiella similiris* (5,000 m, —), *Tindariopsis* sp. (—, 4,734 m), *Ledella galathea* (—, 4,734 m).

APPENDIX B

Meiofaunal data from Aller et al. 2002 (NW Atlantic), Alongi and Pichon 1988 (SW Pacific), Alongi 1992 (SW Pacific), Ansari, Paurlekar, and Jagtap 1980 (Arabian Sea), Coull et al. 1977 (NW Atlantic), Danovaro et al. 1995 (Mediterranean), Danovaro et al. 2000 (Mediterranean), Danovaro, Gambi, and Della Croce 2002 (SE Pacific), DeBovée, Guidi, and Soyer 1990 (Mediterranean), Dinet and Vivier 1977 (NE Atlantic), Dinet 1973 (SE Atlantic), Dinet 1974 in Soltwedel 2000 (Norwegian Sea), Dinet 1976 (Mediterranean), Dinet in Vincx et al. 1994 (Tropical E Atlantic), Duineveld et al. 1997 (Tropical W Indian), Escobar et al. 1997 (Gulf of Mexico), Fabiano and Danovaro 1999 (Antarctic), Ferrero in Vincx et al. 1994 (NE Atlantic), Flach, Muthumbi, and Heip 2002 (NE Atlantic), Gage 1977 (NE Atlantic), Gage 1979 (NE Atlantic), Galéron et al. 2000 (Tropical E Atlantic), Galéron et al. 2001 (NE Atlantic), Gooday in Vincx et al. 1994 (NE Atlantic), Heip et al. 2001 (NE Atlantic), Herman and Dahms 1992 (Antarctic), Kamenskaya and Galtsova 1996 (NE Atlantic), Khripounoff, Desbruyéres, and Chardy 1980 (Tropical W Atlantic), Kröncke et al. 2000 (Arctic), Lambshead and Ferrero in Vincx et al. 1994 (NE Atlantic), Levin and Thomas 1989 (Tropical Central Pacific), Levin, Huggett, and Wishner 1991 (Tropical E Pacific), Parekular et al. 1983 in Soltwedel 2000 (Antarctic), Pequegnat, Gallaway, and Pequenat 1990 (Gulf of Mexico), Pfannkuche and Thiel 1987 (Arctic), Pfannkuche, Theeg, and Thiel 1983 (NE Atlantic), Pfannkuche et al. 1990 in Vincx et al. 1994 (NE Atlantic), Pfannkuche 1985 (NE Atlantic), Rachor 1975 (NE Atlantic), Relexans et al. 1996 (Tropical E Atlantic), Romano and Dinet 1981 in Soltwedel 2000 (Arabian Sea), Rutgers van der Loeff and Lavaleye 1986 (NE Atlantic), Shirayama and Kojima 1994 (NW Pacific), Shirayama 1983 (Tropical W and NW Pacific), Sibuet et al. 1984 (Tropical W Atlantic), Sibuet et al. 1989 (NE, Tropical, SE Atlantic), Snider, Burnett, and Hessler 1984 (Central N Pacific), Soetaert, Heip, and Vincx 1991 (Mediterranean), Soltwedel and Thiel 1995 (Tropical E Atlantic), Soltwedel, Mokievsky, and Schewe 2000 (Arctic), Soltwedel 1997 (Tropical E Atlantic), Sommer and Pfannkuche 2000 (Arabian Sea), Tahey et al. 1994 (Mediterranean), Thiel 1966, 1975 (Tropical W. Indian), Thiel 1975 (Norwegian Sea, NE Atlantic), Thiel 1979 (Red Sea), Thiel 1982 (NE Atlantic), Thistle, Yingst, and Fauchald 1985 (NW Atlantic), Tietjen 1971 (NW Atlantic), Vanaverbeke et al. 1997 (NE Atlantic), Vanhove et al. 1995 (Antarctic), Vanreusel and Vincx in Vincx et al. 1994 (NE Atlantic), Vanreusel et al. 1992 (NE Atlantic), Vanreusel et al. 1995 (NE Atlantic), Vivier 1978 (Mediterranean), Wigley and McIntyre 1964 (NW Atlantic).

Macrofauna data from Aller, Aller, and Green 2002 (NW Atlantic), Alongi 1992 (SW Pacific), Blake and Grassle 1994 (NW Atlantic), Blake and Hilbig 1994 (NW Atlantic), Carey and Ruff 1974 in Rowe 1983 (Arctic), Carey, Jr. 1981 (NE Pacific), Clough et al. 1997 in Rowe 1983 (Arctic), Cosson, Sibuet, and Galeron 1997 (Tropical E Atlantic), Dahl et al. 1976 (Norwegian Sea), Daule, Herman, and Heip 1998 (North Sea), Desbruyères, Bervas, and Kripounoff 1980 (NE Atlantic), Duineveld et al. 2000 (Mediterranean), Spiess et al. 1987 (Tropical E Pacific), Flach and Heip 1996 (NE Atlantic), Flach, Muthumbi, and Heip 2002 (NE Atlantic), Frankenberg and Menzies 1968 (Tropical E Pacific), Gage 1977 (NE Atlantic), Gage 1979 (NE Atlantic), Galéron et al. 2000 (Tropical E Atlantic), Galéron et al. 2001 (NE Atlantic), Grassle and Morse-Porteous 1987 (NW Atlantic), Grassle 1977 (NW Atlantic), Griggs, Carey, and Kulm 1969 (NE Pacific), Hecker and Paul 1979 (Tropical E Pacific), Hessler and Jumars 1974 (Central N Pacific), Houston and Haedrich 1984 (NW Atlantic), Hyland et al. 1991 (NE Pacific), Jazdzewski et al. 1986 (Antarctic), Jumars and Hessler 1976 (Central N Pacific), Kripounoff, Desbruyéres, and Chardy 1980 (Tropical W Atlantic), Kröncke 1998 (Arctic), Kröncke, Türkay, and Fiege 2003 (Mediterranean), Kröncke et al. 2000 (Arctic), Laubier and Sibuet 1979 (NE Atlantic), Levin and Thomas 1989 (Tropical Central Pacific), Levin, Huggett, and Wishner 1991 (Tropical E Pacific), Levin et al. 2000 (Arabian Sea), Maciolek and Grassle 1987 (NW Atlantic), Maciolek et al. 1987a, 1987b (NW Atlantic), Nichols and Rowe 1977 (Tropical E Atlantic), Pfannkuche. Theeg, and Thiel 1983 (NE Atlantic), Rhoads et al. 1985 (NW Pacific), Richardson et al. 1985 (Caribbean), Richardson et al. 1995 (Tropical W Atlantic), Romero-Wetzel and Gerlach 1991 (Norwegian Sea), Rowe and Menzel 1971 (Gulf of Mexico), Rowe 1971 (Tropical E Pacific), Rowe et al. 1982 (NW Atlantic), Rowe et al. 1975 (NW Atlantic), Rowe et al. 1974 (NW Atlantic, Gulf of Mexico), Sanders 1969 (SE Atlantic), Sanders et al. 1965 (NW Atlantic), Schaff et al. 1992 (NW Atlantic), Shirayama 1983 (Tropical W and NW Pacific), Sibuet et al.1984 (Tropical W Atlantic), Sibuet et al. 1989 (NE, Tropical, SE Atlantic), Smith 1978 (NW Atlantic), Smith 1987 (Central N Pacific), Tselepides and Eleftheriou 1992 (Mediterranean), Tselepides et al. 2000 (Mediterranean), Witte 2000 (Arabian Sea).

ACKNOWLEDGMENTS

We thank Gene Hunt, Brad Seibel, and James Childress for critically reading the manuscript. Ingrid Kröncke kindly provided unpublished data on macrofaunal abundance in the Mediterranean Sea, and Thomas Soltwedel generously provided the database for his review paper on meiofaunal abundance. Kaustuv Roy graciously provided data on the body size of coastal gastropods used in figure 3.4. Jeremy Morris, Nicholas Johnson, and Jennifer Crouse helped to compile the data used in figure 3.5. This research is supported by NSF grant OCE-0135949 to MAR.

REFERENCES

Allen, A. P., J. H. Brown, and J. F. Gillooly. 2002. Global biodiversity, biochemical kinetics, and the energetic-equivalence rule. *Science* 297:1545–48.

Allen, J. A., and F. J. Hannah. 1989. Studies on the deep-sea Protobranchia: The subfamily Ledellinae (Nuculanidae). *Bulletin of the British Museum of Natural History (Zoology)* 55:123–71.

Allen, J. A., and H. L. Sanders. 1973. Studies on deep-sea Protobranchia (Bivalvia): The families Siliculidae and Lametilidae. *Bulletin of the Museum of Comparative Zoology* 145:263–309.

Allen, J. A., and H. L. Sanders. 1996. The zoogeography, diversity and origin of the deep-sea protobranch bivalves of the Atlantic: The epilogue. *Progress in Oceanography* 38:95–153.

Allen, J. A., H. L. Sanders, and F. J. Hannah. 1995. Studies on the deep-sea Protobranchia (Bivalvia): The Subfamily Yoldiellinae. *Bulletin of the Natural History Museum of London (Zoology)* 61:11–90.

Aller, J. Y. 1997. Benthic community response to temporal and spatial gradients in physical disturbance within a deep-sea western boundary region. *Deep-Sea Research I* 44:39–69.

Aller, J. Y., R. C. Aller, and M. A. Green. 2002. Benthic faunal assemblages and carbon supply along the continental shelf/shelf break-slope off Cape Hatteras, North Carolina. *Deep-Sea Research II* 49:4599–4625.

Alongi, D. M. 1992. Bathymetric patterns of deep-sea benthic communities from bathyal to abyssal depths in the western South Pacific (Solomon and Coral Seas). *Deep-Sea Research* 39:549–65.

Alongi, D. M., and M. Pichon. 1988. Bathyal meiobenthos of the western Coral Sea: Distribution and abundance in relation to microbial standing stocks and environmental factors. *Deep-Sea Research* 35:491–503.

Ansari, Z. A., A. H. Parulekar, and T. G. Jagtap. 1980. Distribution of sub-littoral meiobenthos off Goa coast. *India Hydrobiologia* 74:209–14.

Arnett, A. E., and N. J. Gotelli. 1999. Bergmann's rule in the ant lion *Myrmeleon immaculatus* DeGeer (Neuroptera: Myrmeleontidae): Geographic variation in body size and heterozygosity. *Journal of Biogeography* 26:275–83.

Atkinson, D., and R. M. Sibly. 1997. Why are organisms usually bigger in colder environments? Making sense of a life history puzzle. *Trends in Ecology and Evolution* 12:235–39.

Bergquist, D., F. M. Williams, and C. R. Fisher. 2000. Longevity record for deep-sea invertebrate. *Nature* 403:499–500.

Bertness, M. D., S. D. Gaines, and M. E. Hay, eds. 2001. *Marine community ecology.* Sunderland, MA: Sinauer.

Blackburn, T. M., and K. J. Gaston. 1996. Spatial patterns in body sizes of bird species in the New World. *Oikos* 77:436–46.

Blake, J. A., and J. F. Grassle. 1994. Benthic community structure on the U.S. South Atlantic slope off the Carolinas: Spatial heterogeneity in a current-dominated system. *Deep-Sea Research II* 41:835–74.

Blake, J. A., and B. Hilbig. 1994. Dense infaunal assemblages on the continental slope off Cape Hatteras, North Carolina. *Deep-Sea Research II* 41:875–99.

Blumenshine, S. C., D. M. Lodge, and J. R. Hodgson. 2000. Gradient of fish predation alters body size distributions of lake benthos. *Ecology* 81:374–86.

Brodie, P. F. 1975. Cetacean energetics, an overview of intraspecific size varaition. *Ecology* 56:152–61.

Brown, J. H. 1995. *Macroecology.* Chicago: University of Chicago Press.

Brown, J. H., and A. Lee. 1969. Bergmann's rule and climatic adapatation in woodrates (*Neatoma*). *Evolution* 23:329–38.

Brown, J. H., J. F. Gillooly, A. P. Allen, V. M. Savage, and G. B. West. 2004. Toward a metabolic theory of ecology. *Ecology* 85:1771–89.

Buzas, M. A., L. A. Collins, and S. J. Culver. 2002. Latitudinal differences in biodiversity caused by higher tropical rate of increase. *Proceedings of the National Academy of Sciences.* 99:7841–43.

Campbell, J. W., and T. Aarup. 1992. New production in the North Atlantic derived from seasonal patterns of surface chlorophyll. *Deep-Sea Research* 39:1669–94.

Cardillo, M. 1999. Latitude and rates of diversification in birds and butterflies. *Proceedings of the Royal Academy, B* 266:1221–25.

Carey Jr., A. G. 1981. A comparison of benthic infaunal abundance on two abyssal plains in the northeast Pacific Ocean. *Deep-Sea Research* 28:467–79.

Chapelle, G., and L. S. Peck. 1999. Polar gigantism dictated by oxygen availability. *Nature* 399:114–15.

Chase, M. R., R. J. Etter, M. A. Rex, and J. M. Quattro. 1998. Bathymetric patterns of genetic variation in a deep-sea protobranch bivalve, *Deminucula atacellana. Marine Biology* 131:301–8.

Childress, J. J. 1995. Are there physiological and biochemical adaptations of metabolism in deep-sea animals? *Trends in Ecology and Evolution* 10:30–36.

Childress, J. J., S. M. Taylor, G. M. Cailliet, and M. H. Price. 1980. Patterns of growth, energy utilization and reproduction in the some meso- and bathypelagic fishes off Southern California. *Marine Biology* 61:27–40.

Clough, L. M., W. G. Ambrose, Jr., J. K. Cochran, C. Barnes, P. E. Renaud, and R. C. Aller. 1997. Infaunal density, biomass and bioturbation in the sediments of the Arctic Ocean. *Deep-Sea Research II* 44:1683–1704.

Cosson, N., M. Sibuet, and J. Galeron. 1997. Community structure and spatial heterogeneity of the deep-sea macrofauna at three contrasting stations in the tropical northeast Atlantic. *Deep-Sea Research I* 44:247–69.

Cosson-Sarradin, N., M. Sibuet, G. L. J. Paterson, and A. Vangriesheim. 1998. Polychaete diversity at tropical Atlantic sites: Environmental effects. *Marine Ecology Progress Series* 165:173–85.

Coull, B. C. 1972. Species diversity and faunal affinities of meiobenthic Copepoda in the deep sea. *Marine Biology* 14:48–51.

Coull, B. C., R. L. Ellison, J. W. Fleeger, R. P. Higgins, W. D. Hope, W. D. Hummon, R. M. Rieger, et al. 1977. Quantitative estimates of the meiofauna from the deep sea off North Carolina, USA. *Marine Biology* 39:233–40.

Culver, S. J., and M. A. Buzas. 2000. Global latitudinal species diversity gradient in deep-sea foraminifera. *Deep-Sea Research I* 47:259–75.

Cushman, J. H., J. H. Lawton, and B. F. J. Manly. 1993. Latitudinal patterns in the European ant assemblages: Variation in species richness and body size. *Oecologia* 83:316–24.

Dahl, E., L. Laubier, M. Sibuet, and J.-O. Strömberg. 1976. Some quantitative results on benthic communities of the deep Norwegian Sea. *Astarte* 9:61–79.

Danovaro, R., N. Della Croce, A. Eleftheriou, M. Fabiano, N. Papadopoulou, C. Smith, and A. Tselepides. 1995. Meiofauna of the deep Eastern Mediterranean Sea: Distribution and abundance in relation to bacterial biomass, organic matter composition and other environmental factors. *Progress in Oceanography* 36:329–41.

Danovaro, R., D. Marrale, A. Dell'Anno, N. Della Croce, A. Tselepides, and M. Fabiano. 2000. Bacterial response to seasonal changes in labile organic matter composition on the continental shelf and bathyal sediments of the Cretan Sea. *Progress in Oceanography* 46:345–66.

Danovaro, R., C. Gambi, and N. Della Croce. 2002. Meiofauna hotspot in the Atacama Trench, eastern South Pacific Ocean. *Deep-Sea Research I* 49:843–57.

Dauwe, B., P. M. J. Herman, and C. H. R. Heip. 1998. Community structure and bioturbation potential of macrofauna at four North Sea stations with contrasting food supply. *Marine Ecology Progress Series* 173:67–83.

DeBovée, F., L. D. Guidi, and J. Soyer. 1990. Quantitative distribution of deep-sea meiobenthos in the Northwestern Mediterranean (Gulf of Lions). *Continental Shelf Research* 10:1123–45.

Deming, J. W., and P. L. Yager. 1992. Natural bacterial assemblages in deep-sea sediments: Toward a global view. In *Deep-sea food chains and the global carbon cycle*, ed. G. T. Rowe and V. Pariente, 11–27. The Netherlands: Kluwer, Dordrecht.

Desbruyères, D., J. Y. Bervas, and A. Khripounoff. 1980. Un cas de colonisation rapide d'un sédiment profond. *Oceanologica Acta* 3:285–91.

Dinet, A. 1973. Distribution quantitative du méiobenthos profond dans la région de la dorsale de Walvis (Sud-Ouest Africain). *Marine Biology* 20:20–26.

———. 1976. Études quantitative du méiobenthos dans le sector nord de la Mer Égée. *Acta Adriatica* 18:83–88.

Dinet, A., and M. H. Vivier. 1977. Le méiobenthos abyssal du golfe de Gascogne I. Considérations sur les données quantitatives. *Cahiers de Biologie Marine* 18:85–97.

Duineveld, G. C. A., P. A. W. J. de Wilde, E. M. Berghuis, A. Kok, T. Tahey, and J. Kromkamp. 1997. Benthic respiration and standing stock on two contrasting continental margins in the western Indian Ocean: The Yemen-Somali upwelling region and the margin off Kenya. *Deep-Sea Research II* 44:1293–1317.

Duineveld, G. C. A., A. Tselepides, R. Witbaard, R. P. M. Bak, E. M. Berghuis, G. Nieuwland, J. van der Weele, and A. Kok. 2000. Benthic-pelagic coupling in the oligotrophic Cretan Sea. *Progress in Oceanography* 46:457–80.

Duplisea, D. E., and A. Drgas. 1999. Sensitivity of a benthic, metazoan, biomass size spectrum to differences in sediment granulometry. *Marine Ecology Progress Series* 177:73–81.

Escobar, E., M. López, L. A. Soto, and M. Signoret. 1997. Density and biomass of the meiofauna of the upper continental slope in two regions of the Gulf of Mexico. *Ciencias Marinas* 23:463–89.

Etter, R. J., and J. F. Grassle. 1992. Patterns of species diversity in the deep sea as a function of sediment particle size diversity. *Nature* 360:576–78.

Etter, R. J., and L. S. Mullineaux. 2001. Deep-sea communities. In *Marine community ecology*, ed. M. D. Bertness, S. D. Gaines, and M. E. Hay, 367–93. Sunderland, MA: Sinauer.

Etter, R. J., and M. A. Rex. 1990. Population differentiation decreases with depth in deep-sea gastropods. *Deep-Sea Research* 37:1251–61.

Etter, R. J., M. A. Rex, M. Chase, and J. Quattro. 2005. Population differentiation decreases with depth in deep-sea bivalves. *Evolution* 59, 1479–91.

Fa, D. A., and J. E. Fa. 2002. Species diversity, abundance and body size in rocky-shore Mollusca: A twist in Siemann, Tilman and Haarstad's parabola? *Journal of Molluscan Studies* 68:95–100.

Fabiano, M., and R. Danovaro. 1999. Meiofauna distribution and mesoscale variability in two sites of the Ross Sea (Antarctica) with contrasting food supply. *Polar Biology* 22:115–23.

Flach, E., and W. deBruin. 1999. Diversity patterns in macrobenthos across a continental slope in the NE Atlantic. *Journal of Sea Research* 42:303–23.

Flach, E., and C. Heip. 1996. Vertical distribution of macrozoobenthos within the sediment on the continental slope of the Goban Spur area (NE Atlantic). *Marine Ecology Progress Series* 141:55–66.

Flach, E., A. Muthumbi, and C. Heip. 2002. Meiofauna and macrofauna community structure in relation to sediment composition at the Iberian margin compared to the Goban Spur (NE Atlantic). *Progress in Oceanography* 52:433–57.

Forsman, A. 1991. Variation in sexual size dimorphism and maximum body size among adder populations: Effect of prey size. *Journal of Animal Ecology* 60:253–67.

Frankenberg, D., and R. J. Menzies. 1968. Some quantitative analyses of deep-sea benthos off Peru. *Deep-Sea Research* 15:623–26.

Frazier, M. R., H. A. Woods, and J. F. Harrison. 2001. Interactive effects of rearing temperature and oxygen on the development of Drosophila melanogaster. *Physiological and Biochemical Zoology* 74:641–50.

Gage, J. D. 1977. Structure of the abyssal macrobenthic community in the Rockall Trough. *European Symposium on Marine Biology* 11:247–60.

———. 1979. Macrobenthic community structure in the Rockall Trough. *Ambio Special Report* 6:43–46.

———. 1996. Why are there so many species in deep-sea sediments? *Journal of Experimental Marine Biology and Ecology* 200:257–86.

Gage, J. D., P. A. Lamont, K. Kroeger, G. L. J. Paterson, and J. L. G. Vecino. 2000. Patterns in deep-sea macrobenthos at the continental margin: Standing crop, diversity and faunal change on the continental slope off Scotland. *Hydrobiologia* 440:261–72.

Gage, J. D., P. J. D. Lambshead, J. D. D. Bishop, C. T. Stuart, and N. S. Jones. 2004. Large-scale biodiversity pattern of Cumacea (Peracarida: Crustacea) in the deep Atlantic. *Marine Ecology Progress Series* 277:181–96.

Gage, J. D., and P. A. Tyler. 1991. *Deep-sea biology: A natural history of organisms at the deep-sea floor.* Cambridge: Cambridge University Press.

Galéron, J., M. Sibuet, M.-L. Mahaut, and A. Dinet. 2000. Variation in structure and biomass of the benthic communities at three contrasting sites in the tropical Northeast Atlantic. *Marine Ecology Progress Series* 197:121–37.

Galéron, J., M. Sibuet, A. Vanreusel, K. Mackenzie, A. J. Gooday, A. Dinet, and G. A. Wolff. 2001. Temporal patterns among meiofauna and macrofauna taxa related to changes in sediment geochemistry at an abyssal NE Atlantic site. *Progress in Oceanography* 50:303–24.

Gambi, C., A. Vanreusel, and R. Danovaro. 2003. Biodiversity of nematode assemblages from deep-sea sediments of the Atacama Slope and Trench (South Pacific Ocean). *Deep-Sea Research I* 50:103–17.

Gaston, K. G., and T. M. Blackburn. 2000. *Pattern and process in macroecology.* Oxford: Blackwell Science.

Gillooly, J. F., J. H. Brown, G. B. West, V. M. Savage, and E. L. Charnov. 2001. Effects of size and temperature on metabolic rate. *Science* 293:2248–51.

Grassle, J. F. 1977. Slow recolonisation of deep-sea sediment. *Nature* 265:618–19.

Grassle, J. F., and N. J. Maciolek. 1992. Deep-sea species richness: Regional and local diversity estimates from quantitative bottom samples. *American Naturalist* 193:313–41.

Grassle, J. F., and L. S. Morse-Porteous. 1987. Macrofaunal colonization of disturbed deep-sea environments and the structure of deep-sea benthic communities. *Deep-Sea Research* 34:1911–50.

Gray, J. S. 1994. Is deep-sea species diversity really so high? Species diversity of the Norwegian continental shelf. *Marine Ecology Progress Series* 112:205–9.

———. 2002. Species richness of marine soft sediments. *Marine Ecology Progress Series* 244:285–97.

Gray, J. S., G. C. B. Poore, K. I. Ugland, R. S. Wilson, F. Olsgard, and O. Johannessen. 1997. Coastal and deep-sea benthic diversities compared. *Marine Ecology Progress Series* 159:97–103.

Griggs, G. B., A. G. Carey, Jr., and L. D. Kulm. 1969. Deep-sea sedimentation and sediment-fauna interaction in Cascadia Channel and on Cascadia Abyssal Plain. *Deep-Sea Research* 16:157–70.

Haedrich, R. L., and G. T. Rowe. 1977. Megafaunal biomass in the deep sea. *Nature* 269:141–42.

Haedrich, R. L., G. T. Rowe, and P. T. Polloni. 1980. The megabenthic fauna in the deep sea south of New England, USA. *Marine Biology* 57:165–79.

Hawkins, B. A. 1995. Latitudinal body-size gradients for the bees of the eastern United States. *Ecological Entomology* 20:195–98.

Hawkins, B. A., and J. Lawton. 1995. Latitudinal gradients in butterfly body sizes: Is there a general pattern? *Oecologia* 102:31–36.

Hawkins, B. A., E. E. Porter, and J. A. F. Diniz-Filho. 2003. Productivity and history as predictors of the latitudinal diversity gradient in terrestrial birds. *Ecology* 84:1608–23.

Hecker, B. 1990. Variation in megafaunal assemblages on the continental margin south of New England. *Deep-Sea Research* 37:37–57.

Hecker, B., and A. Z. Paul. 1979. Abyssal community structure of the benthic infauna of the eastern Equatorial Pacific: Domes sites A, B, and C. In *Marine geology and oceanography of the Pacific manganese nodule province,* ed. J. L. Bischoff and D. Z. Piper, 287–308. New York: Plenum.

Heip, C. H. R., G. Duineveld, E. Flach, G. Graf, W. Helder, P. M. J. Herman, M. Lavaleye, et al. 2001. The role of the benthic biota in sedimentary metabolism and sediment-water exchange processes in the Goban Spur area (NE Atlantic). *Deep-Sea Research II* 48:3223–43.

Hermann, R. L., and H. -U. Dahms. 1992. Meiofauna communities along a depth transect off Halley Bay (Weddell Sea-Antarctica). *Polar Biology* 12:313–20.

Hess, S., and W. Kuhnt. 1996. Deep-sea benthic foraminiferal recolonization of the 1991 Mt. Pinatubo ash layer in the South China Sea. *Marine Micropaleontology* 28:171–97.

Hessler, R. R., and P. A. Jumars. 1974. Abyssal community analysis from replicate box cores in the central North Pacific. *Deep-Sea Research* 21:185–209.

Hessler, R. R., and H. L. Sanders. 1967. Faunal diversity in the deep-sea. *Deep-Sea Research* 14:65–78.

Hillebrand, H. 2004a. On the generality of the latitudinal diversity gradient. *American Naturalist* 163:192–211.

Hillebrand, H. 2004b. Strength, slope and variability of marine latitudinal gradients. *Marine Ecology Progress Series* 273:251–67.

Houston, K. A., and R. L. Haedrich. 1984. Abundance and biomass of macrobenthos in the vicinity of Carson Submarine Canyon, northwest Atlantic Ocean. *Marine Biology* 82:301–5.

Howell, K. L., D. S. M. Billett, and P. A. Tyler. 2002. Depth-related distribution and abundance of seastars (Echinodermata: Asteroidea) in the Porcupine Seabight and Porcupine Abyssal Plain, N.E. Atlantic. *Deep-Sea Research I* 49:1901–20.

Hurlbert, S. H. 1971. The nonconcept of species diversity: A critique and alternative parameters. *Ecology* 52:577–86.

Hyland, J., E. Baptiste, J. Campbell, J. Kennedy, R. Kropp, and S. Williams. 1991. Macroinfaunal communities of the Santa Maria Basin on the California outer continental shelf and slope. *Marine Ecology Progress Series* 78:147–61.

Jablonski, D. 1993 The tropics as a source of evolutionary novelty through geological time. *Nature* 364:142–44.

Jazdzewski, K., W. Jurasz, W. Kittel, E. Presler, P. Presler, and J. Sicinski. 1986. Abundance and biomass estimates of the benthic fauna in Admiralty Bay, King George Island, South Shetland Islands. *Polar Biology* 6:5–16.

Johnson, N. A., J. W. Campbell, T. S. Moore, M. A. Rex, C. R. McClain, and M. D. Dowell. 2005. Using SeaWiFS satellite imagery to predict standing stock of the deep-sea macrobenthos. Unpublished manuscript.

Jumars, P. A., and R. R. Hessler. 1976. Hadal community structure: Implications from the Aleutian Trench. *Journal of Marine Research* 34:547–60.

Kamenskaya, O. E., and V. V. Galtsova. 1996. Meiobenthos of the Canary Upwelling. *Oceanology* 36:380–84.

Kennett, J. P., and L. D. Stott. 1991. Abrupt deep-sea warming, palaeoceanographic changes and benthic extinctions at the end of the Palaeocene. *Nature* 353:225–29.

Knudsen, J. 1973. *Guivillea alabastrina* (Watson, 1882), an abyssal volutid (Gastropoda: Mollusca). *Galathea Report* 12:127–31.

Kozlowski, J., and A. T. Gawelczyk. 2002. Why are species' body size distributions usually skewed to the right? *Functional Ecology* 16:419–32.

Kripounoff, A., D. Desbruyères, and P. Chardy. 1980. Les peuplements benthiques de la faille Vema: données quantitiatives et bilan d'énergie en milieu abyssal. *Oceanologica Acta* 3:187–98.

Kröncke, I. 1998. Macrofauna communities in the Amundsen Basin, at the Morris Jesup Rise and at the Yermak Plateau (Eurasian Arctic Ocean). *Polar Biology* 19:383–92.

Kröncke, I., M. Türkay, and D. Fiege. 2003. Macrofauna communities in the eastern Mediterranean deep-sea. *Marine Ecology* 24:193–216.

Kröncke, I., A. Vanreusel, M. Vincx, J. Wollenburg, A. Mackensen, G. Liebezeit, and B. Behrends. 2000. Different benthic size-compartments and their relationship to sediment chemistry in the deep Eurasian Arctic Ocean. *Marine Ecology Progress Series* 199:31–41.

Lambshead, P. J. D., J. Tietjen, T. Ferrero, and P. Jensen. 2000. Latitudinal diversity gradients in the deep sea with special reference to North Atlantic nematodes. *Marine Ecology Progress Series* 194:159–67.

Lampitt, R. S., D. S. M. Billett, and A. L. Rice. 1986. Biomass of the invertebrate megabenthos from 500 to 4100 m in the northeast Atlantic Ocean. *Marine Biology* 93:69–81.

Laubier, L., and M. Sibuet. 1979. Ecology of the benthic communities of the deep North East Atlantic. *Ambio Special Report* 6:37–42.

Lawton, J. H. 1990. Species richness and population-dynamics of animal assemblages-patterns in body size-abundance space. *Philosophical Transactions of the Royal Society of London, B* 330:283–91.

Leaper, R., D. Raffaelli, C. Emes, and B. Manly. 2001. Contraints on body-size distributions: An experimental test of the habitat architecture hypothesis. *Journal of Animal Ecology* 70:248–59.

Levin, L. A., R. J. Etter, M. A. Rex, A. J. Gooday, C. R. Smith, J. Pineda, C. T. Stuart, et al. 2001. Environmental influences on regional deep-sea species diversity. *Annual Review of Ecology and Systematics* 32:51–93.

Levin, L. A., and J. D. Gage. 1998. Relationships between oxygen, organic matter and the diversity of bathyal macrofauna. *Deep-Sea Research II* 45:129–63.

Levin, L. A., C. L. Huggett, and K. F. Wishner. 1991. Control of deep-sea benthic community structure by oxygen and organic-matter gradients in the eastern Pacific Ocean. *Journal of Marine Research* 49:763–800.

Levin, L. A., J. D. Gage, C. Martin, and P. A. Lamont. 2000. Macrobenthic community structure within and beneath the oxygen minimum zone, NW Arabian Sea. *Deep-Sea Research II* 47:189–226.

Levin, L. A., and C. L. Thomas. 1989. The influence of hydrodynamic regime on infaunal assemblages inhabiting carbonate sediments on central Pacific seamounts. *Deep-Sea Research* 36:1897–1915.

Maciolek, N. J., J. F. Grassle, B. Hecker, B. Brown, J. A. Blake, P. D. Boehm, R. Petrecca, S. Duffy, E. Baptiste, and R. E. Ruff. 1987. Study of biological processes on the U.S. North Atlantic slope and rise. Final Report Prepared for U.S. Department of the Interior, Minerals Management Service, Washington, D.C., 362 pp and appendices A–L.

Maciolek, N., J. F. Grassle, B. Hecker, P. D. Boehm, B. Brown, B. Dade, W. G. Steinhauer, E. Baptiste, R. E. Ruff, and R. Petrecca. 1987. Study of biological processes on the U.S. Mid-Atlantic slope and rise. Final Report Prepared for the U.S. Department of the Interior, Minerals Management Service, Washington, D.C., 310 pp. and appendices A–M.

Maciolek, N. J., and J. F. Grassle. 1987. Variability of the benthic fauna, II: The seasonal variation, 1981–1982. In *Georges Bank*, ed. R. H. Backus, 303–309. Cambridge, MA: MIT Press.

Martin, P. R., and J. K. McKay. 2004. Latitudinal variation in genetic divergence of populations and the potential for future speciation. *Evolution* 58:938–43.

McClain, C. R. 2004. Connecting species richness, abundance and body size in deep-sea gastropods. *Global Ecology and Biogeography* 13:327–34.

McClain, C. R., and R. J. Etter. 2005. Mid-domain models as predictors of species diversity patterns: A case study using bathymetric diversity gradients in the Deep Sea. *Oikos* 109:555–66.

McClain, C. R., and M. A. Rex. 2001. The relationship between dissolved oxygen concentration and maximum size in deep-sea turrid gastropods: an application of quantile regression. *Marine Biology* 139:681–85.

McClain, C. R., M. A. Rex, and R. Jabbour. 2005. Deconstructing bathymetric body size patterns in deep-sea gastropods. *Marine Ecology Progress Series* 297:181–87.

Menge, B., J. Lubchenco, M. E. S. Bracken, F. Chan, M. M. Foley, T. L. Freidenburg, S. D. Gaines, et al. 2003. Coastal oceanography sets the pace of rocky intertidal community dynamics. *Proceedings of the National Academy of Sciences* 100:12229–34.

Mittelbach, G. G., C. F. Steiner, S. M. Scheiner, K. L. Gross, H. L. Reynolds, R. B. Waide, M. R. Willig, S. I. Dodson, and L. Gough. 2001. What is the observed relationship between species richness and productivity? *Ecology* 82:2381–96.

Mokievsky, V., and A. Azovsky. 2002. Re-evaluation of species diversity patterns of free-living marine nematodes. *Marine Ecology Progress Series* 238:101–8.

Nichols, J., and G. T. Rowe. 1977. Infaunal macrobenthos off Cap Blanc, Spanish Sahara. *Journal of Marine Research* 35:525–36.

Olabarria, C., and M. H. Thurston. 2003. Latitudinal and bathymetric trends in body size of the deep-sea gastropod *Troschelia berniciensis* (King). *Marine Biology* 143:723–30.

Parry, D. M., M. A. Kendall, A. A. Rowden, and S. Widdicombe. 1999. Species body size distribution patterns of marine benthic macrofauna assemblages from contrasting sediment types. *Journal of Marine Biology Association of the United Kingdom* 79:793–801.

Paterson, G. L. J., and P. J. D. Lambshead. 1995. Bathymetric patterns of polychaete diversity in the Rockall Trough, northeast Atlantic. *Deep-Sea Research I* 42:1199–1214.

Peck, L. S., and G. Chapelle. 2003. Reduced oxygen at high altitude limits maximum size. *Proceedings of the Royal Society of London B* 270:S166–S167.

Pequegnat, W. E., B. J. Gallaway, and L. H. Pequegnat. 1990. Aspects of the ecology of the deep-water fauna of the Gulf of Mexico. *American Zoologist* 30:45–64.

Peters, R. H. 1983. *The ecological implications of body size.* Cambridge: Cambridge University Press.

Pfannkuche, O. 1985. The deep-sea meiofauna of the Porcupine Seabight and abyssal plain (NE Atlantic): Population structure, distribution, standing stocks. *Oceanologica Acta* 8:343–53.

Pfannkuche, O., R. Theeg, and H. Thiel. 1983. Benthos activity, abundance and biomass under an area of low upwelling off Morocco, Northwest Africa. *Meteor Forschungsergebnisse, Reihe D Biologie* 36:85–96.

Pfannkuche, O., and H. Thiel. 1987. Meiobenthic stocks and benthic activity on the NE-Svalbard Shelf and in the Nansen Basin. *Polar Biology* 7:253–66.

Pineda, J. 1993. Boundary effects on the vertical ranges of deep-sea benthic species. *Deep-Sea Research I* 40:2179–92.

Pineda, J., and H. Caswell. 1998. Bathymetric species-diversity patterns and boundary constraints on vertical range distribution. *Deep-Sea Research II* 45:83–101.

Polloni, P., R. Haedrich, G. Rowe, and C. H. Clifford. 1979. The size-depth relationship in deep ocean animals. *Internationale Revue der gasamten Hydrobiologie* 64:39–46.

Pulliam, H. R. 1988. Sources, sinks, and population regulation. *American Naturalist* 132:652–61.

Rachor, E. 1975. Quantitative Untersuchungen über das Meiobenthos der nordostatlantischen Tiefsee. *Meteor Forschungsergebnisse, Reihe D. Biology* 21:1–10.

Relexans, J.-C., J. Deming, A. Dinet, J.-F. Gaillard, and M. Sibuet. 1996. Sedimentary organic matter and micro-meiobenthos with relation to trophic conditions in the tropical northeast Atlantic. *Deep-Sea Research I* 43:1343–68.

Rex, M. A. 1973. Deep-sea species diversity: Decreased gastropod diversity at abyssal depths. *Science* 181:1051–53.

———. 1981. Community structure in the deep-sea benthos. *Annual Review of Ecology and Systematics* 12:331–53.

———. 1983. Geographic patterns of species diversity in the deep-sea benthos. In *The sea,* vol. 8, ed. G. T. Rowe, 453–72. New York: Wiley.

Rex, M. A., and R. J. Etter. 1998. Bathymetric patterns of body size: Implications for deep-sea biodiversity. *Deep-Sea Research II* 45:103–27.

Rex, M. A., R. J. Etter, and C. T. Stuart. 1997. Large-scale patterns of species diversity in the deep-sea benthos. In *Marine biodiversity: Patterns and processes,* ed. R. F. G. Ormond, J. D. Gage, and M. V. Angel, 94–121. Cambridge: Cambridge University Press.

Rex, M. A., C. R. McClain, N. A. Johnson, R. J. Etter, J. A. Allen, P. Bouchet, and A. Warén. 2005. A source-sink hypothesis for abyssal biodiversity. *American Naturalist* 165:163–78.

Rex, M. A., C. T. Stuart, and G. Coyne. 2000. Latitudinal gradients of species richness in the deep-sea benthos of the North Atlantic. *Proceedings of the National Academy of Sciences, USA* 97:4082–85.

Rex, M. A., C. T. Stuart, and R. J. Etter. 2001. Do deep-sea nematodes show a positive latitudinal gradient of species diversity? The potential role of depth. *Marine Ecology Progress Series* 210:297–98.

Rex, M. A., C. T. Stuart, R. R. Hessler, J. A. Allen, H. L. Sanders, and G. D. F. Wilson. 1993. Global-scale latitudinal patterns of species diversity in the deep-sea benthos. *Nature* 365:636–39.

Rhoads, D. C., D. F. Boesch, T. Zhican, X. Fengshan, H. Liqiang, and K. J. Nilsen. 1985. Macrobenthos and sedimentary facies on the Changjiang delta platform and adjacent continental shelf, East China Sea. *Continental Shelf Research* 4:189–213.

Richardson, M. D., K. B. Briggs, F. A. Bowles, and J. H. Tietjen. 1995. A depauperate benthic assemblage from the nutrient-poor sediments of the Puerto Rico Trench. *Deep-Sea Research I* 42:351–64.

Richardson, M. D., K. B. Briggs, and D. K. Young. 1985. Effects of biological activity by abyssal benthic macroinvertebrates on a sedimentary structure in the Venezuela Basin. *Marine Geology* 68:243–67.

Romero-Wetzel, M. B., and S. A. Gerlach. 1991. Abundance, biomass, size-distribution and bioturbation potential of deep-sea macrozoobenthos on the Voring Plateau (1200–1500 m, Norwegian Sea). *Meeresforschung* 33:247–65.

Rosenzweg, M. L. 1995. *Species diversity in space and time.* Cambridge: Cambridge University Press.

Rothwell, R. G., J. Thomson, and G. Kahler. 1998. Low-sea-level emplacement of a very large Late Pleistocene 'megaturbidite' in the western Mediterranean Sea. *Nature* 392:377–80.

Roughgarden, J., S. Gaines, and H. Possingham. 1988. Recruitment dynamics in complex life cycles. *Science* 241:1460–66.

Rowe, G. T. 1971. Observations on bottom currents and epibenthic populations in Hatteras Submarine Canyon. *Deep-Sea Research* 18:569–81.

———. 1983. Biomass and production of the deep-sea macrobenthos. In *The sea,* vol. 8, ed. G. T. Rowe, 97–122. New York: Wiley.

Rowe, G. T., and D. W. Menzel. 1971. Quantitative benthic samples from the deep Gulf of Mexico with some comments on the measurement of deep-sea biomass. *Bulletin of Marine Science* 21:556–66.

Rowe, G. T., P. T. Polloni, and R. L. Haedrich. 1975. Quantitative biological assessment of the benthic fauna in deep basins of the Gulf of Maine. *Journal of Fisheries Research Board of Canada* 32:1805–12.

———. 1982. The deep-sea macrobenthos on the continental margin of the northwest Atlantic Ocean. *Deep-Sea Research* 29:257–78.

Rowe, G. T., P. T. Polloni, and S. G. Horner. 1974. Benthic biomass estimates from the northwestern Atlantic Ocean and the northern Gulf of Mexico. *Deep-Sea Research* 21:641–50.

Roy, K. 2002. Bathymetry and body size in marine gastropods: A shallow water perspective. *Marine Ecology Progress Series* 237:134–49.

Roy, K., D. Jablonski, and J. W. Valentine. 2000. Dissecting latitudinal diversity gradients: Functional groups and clades of marine bivalves. *Proceedings of the Royal Society of London B* 267:293–99.

———. 1994. Eastern Pacific molluscan provinces and latitudinal diversity gradient: No evidence for "Rapoport's rule." *Proceedings of the National Academy of Sciences, USA* 91:8871–74.

Roy, K., D. Jablonski, J. W. Valentine, and G. Rosenberg. 1998. Marine latitudinal diversity gradients: Tests of causal hypotheses. *Proceedings of the National Academy of Sciences, USA* 95:3699–3702.

Roy, K., and K. K. Martien. 2001. Latitudinal distribution of body size in north-eastern Pacific marine bivalves. *Journal of Biogeography* 28:485-493.

Rutgers van der Loeff, M. M., and M. S. S. Lavaleye. 1986. Sediments, fauna, and the dispersal of radionuclides at the N. E. Atlantic dumpsite for low-level radioactive waste. Report of the Dutch DORA program. Netherlands Institute for Sea Research, 1–134.

Sanders, H. L. 1968. Marine benthic diversity: A comparative study. *American Naturalist* 102:243-282.

———. 1969. Benthic marine diversity and the stability-time hypothesis. *Brookhaven Symposia in Biology* 22:71–81.

———. 1977. Evolutionary ecology and the deep-sea benthos. In *The changing scenes in natural sciences 1776–1976*, ed. C. E. Goulden, 223–43. Philadelphia: Philadelphia Academy of Natural Sciences Special Publication.

Sanders, H. L., and R. R. Hessler. 1969. Ecology of the deep-sea benthos. *Science* 163:1419–1424.

Sanders, H. L., R. R. Hessler, and G. R. Hampson. 1965. An introduction to the study of deep-sea benthic faunal assemblages along the Gay Head-Bermuda transect. *Deep-Sea Research* 12:845–67.

Savage, V. M., J. F. Gillooly, J. H. Brown, G. B. West, and E. L. Charnov. 2004. Effects of body size and temperature on population growth. *American Naturalist* 163:429–41.

Schaff, T., L. Levin, N. Blair, D. DeMaster, R. Pope, and S. Boehme. 1992. Spatial heterogeneity of benthos on the Carolina continental slope: large (100 km)-scale variation. *Marine Ecology Progress Series* 88:143–60.

Scharples, C. M., J. E. Fa, and D. J. Bell. 1996. Geographic variation in size in the European rabbit *Oryctolagus cuninculus* (Lagomorpha: Leporidae) in western Europe and North Africa. *Zoological Journal of the Linnean Society* 117:141–58.

Schwinghamer, P. 1985. Observation on size-structure and pelagic coupling of some shelf and abyssal benthic communities. *European Symposium on Marine Biology* 19:347–59.

Sebens, K. P. 1982. The limits to indeterminate growth: An optimal size model applied to passive suspension feeders. *Ecology* 63:209–22.

———. 1987. The ecology of indeterminate growth in animals. *Annual Review of Ecology and Systematics* 18:371–407.

Shirayama, Y. 1983. Size structure of deep-sea meio- and macrobenthos in the Western Pacific. *Internationale Revue der gasamten Hydrobiologie* 68:799–810.

———. 1984. The abundance of deep-sea meiobenthos in the western Pacific in relation to environmental factors. *Oceanologica Acta* 7:113–21.

Shirayama, Y., and S. Kojima. 1994. Abundance of deep-sea meiobenthos off Sanriku. Northeastern Japan. *Japanese Journal of Oceanography* 50:109–17.

Sibuet, M., C. E. Lambert, R. Chesselet, and L. Laubier. 1989. Density of the major size groups of benthic fauna and trophic input in deep basins of the Atlantic Ocean. *Journal of Marine Research* 47:851–67.

Sibuet, M., C. Monniot, D. Desbruyères, A. Dinet, A. Khripounoff, G. Rowe, and M. Segonzac. 1984. Peuplements benthiques et caractéristiques trophiques du milieu dans la plaine abyssale de Demerara. *Oceanologica Acta* 7:345–58.

Siemann, E., D. Tilman, and J. Haarstad. 1996. Insect species diversity, abundance and body size relationships. *Nature* 380:704–6.

Siemann, E., D. Tilman, and J. Haarstad. 1999. Abundance, diversity, and body size: Patterns from a grassland arthropod community. *Journal of Animal Ecology* 68:824–35.

Smith, C. R., W. Berelson, D. J. DeMaster, F. C. Dobbs, D. Hammond, D. J. Hoover, R. H. Pope, and M. Stephens. 1997. Latitudinal variations in benthic processes in the abyssal equatorial Pacific: Control by biogenic particle flux. *Deep-Sea Research II* 44:2295–2317.

Smith, Jr., K. L. 1978. Benthic community respiration in the N.W. Atlantic Ocean: In situ measurements from 40 to 5200 m. *Marine Biology* 47:337–47.

———. 1987. Food energy supply and demand: A discrepancy between particulate organic carbon flux and sediment community oxygen consumption in the deep ocean. *Limnology and Oceanography* 32:201–20.

Snelgrove, P. V. R., and C. R. Smith. 2002. A riot of species in an environmental calm: The paradox of the species-rich deep-sea floor. *Oceanography and Marine Biology: An Annual Review* 40:311–42.

Snider, L. J., B. R. Burnett, and R. R. Hessler. 1984. The composition and distribution of meiofauna and nanobiota in a central North Pacific deep-sea area. *Deep-Sea Research* 31:1225–49.

Soetaert, K., C. Heip, and M. Vincx. 1991. The meiobenthos along a Mediterranean deep-sea transect off Calvi (Corsica) and in an adjacent canyon P.S.Z.N.I. *Marine Ecology* 12:227–42.

Soetaert, K., A. Muthumbi, and C. Heip. 2002. Size and shape of ocean margin nematodes: Morphological diversity and depth-related patterns. *Marine Ecology Progress Series* 242:179–93.

Soltwedel, T. 1997. Meiobenthos distribution pattern in the tropical East Altantic: Indication for fractionated sedimentation of organic matter to the sea floor? *Marine Biology* 129:747–56.

———. 2000. Metazoan meiobenthos along continental margins: A review. *Progress in Oceanography* 46:59–84.

Soltwedel, T., V. Mokievsky, and I. Schewe. 2000. Benthic activity and biomass on the Yermak Plateau and in adjacent deep-sea regions northwest of Svalbard. *Deep-Sea Research I* 47:1761–85.

Soltwedel, T., and H. Thiel. 1995. Biogenic sediment compounds in relation to marine meiofaunal abundances. *Internationale Revue der gasamten Hydrobiologie* 80:297–311.

Sommer, S., and O. Pfannkuche. 2000. Metazoan meiofauna of the deep Arabian Sea: Standing stocks, size spectra and regional variability in relation to monsoon induced enhanced sedimentation regimes of particulate organic matter. *Deep-Sea Research II* 47:2957–77.

Spicer, J. I., and K. J. Gaston. 1999. Amphipod gigantism dictated by oxygen availability? *Ecology Letters* 2:397–401.

Spiess, F. N., R. Hessler, G. Wilson, and M. Weydert. 1987. Environmental effects of deep sea dredging. Contract Number 83-SAC-00659.

Stuart, C. T., and M. A. Rex. 1994. The relationship between development pattern and species diversity in deep-sea prosobranch snails. In *Reproduction, larval biology and recruitment in the deep-sea benthos,* ed. C. M. Young and K. J. Eckelbarger, 119–36. New York: Columbia University Press.

Stuart, C. T., M. A. Rex, and R. J. Etter. 2003. Large-scale spatial and temporal patterns of deep-sea benthic species diversity. In *Ecosystems of the world 28, Ecosystems of the deep oceans,* ed. P. A. Tyler, 297–313. Amsterdam: Elsevier.

Tahey, T. M., G. C. A. Duineveld, E. M. Berghuis, and W. Helder. 1994. Relation between sediment-water fluxes of oxygen and silicate and faunal abundance at continental shelf, slope and deep-water stations in the northwest Mediterranean. *Marine Ecology Progress Series* 104:119–30.

Thiel, H. 1966. Quantitative Untersuchungen über die Meiofauna des Tiefseebondens. Veröffentlichungen des Instituts für Meeresforschung Bremerhaven Supplement II:131–148.

———. 1975. The size structure of the deep-sea benthos. *Internationale Revue der gasamten Hydrobiologie* 60:575–606.

———. 1979. First quantitative data on the deep Red Sea benthos. *Marine Ecology Progress Series* 1:347–350.

———. 1982. Zoobenthos of the CINECA area and other upwelling regions. *Rapports et Procès-Verbaux des Réunions du Conseil International pour l'Exploration de la Mer* 180:323–34.

Thistle, D., J. Y. Yingst, and K. Fauchald. 1985. A deep-sea benthic community exposed to strong near-bottom currents on the Scotian Rise (Western Atlantic). *Marine Geology* 66:91–112.

Thomas, E., and A. J. Gooday. 1996. Cenozoic deep-sea benthic foraminifers: tracers for changes in oceanic productivity? *Geology* 24:355–58.

Tietjen, J. H. 1971. Ecology and distribution of deep sea meiobenthos off N. Carolina. *Deep-Sea Research* 18:941–57.

———. 1992. Abundance and biomass of metazoan meiobenthos in the deep sea. In *Deep-sea food chains and the global carbon cycle*, ed. G. T. Rowe and V. Pariente, 45–62. Dordrecht: Kluwer.

Tselepides, A., and A. Eleftheriou. 1992. South Aegean (Eastern Mediterranean) continental slope benthos: Macroinfaunal-environmental relationships. In *Deep-sea food chains and the global carbon cycle*, ed. G. T. Rowe and V. Pariente, 139–56. Dordrecht: Kluwer.

Tselepides, A., K.-N. Papadopoulou, D. Podaras, W. Plaiti, and D. Koutsoubas. 2000. Macrobenthic community structure over the continental margin of Crete (South Aegean Sea, NE Mediterranean). *Progress in Oceanography* 46:401–28.

Vanaverbeke, J., K. Soetaert, C. Heip, and A. Vanreusel. 1997. The metazoan meiobenthos along the continental slope of the Goban Spur (NE Atlantic). *Journal of Sea Research* 38:93–107.

Van Dover, C. L. 2000. *The ecology of deep-sea hydrothermal vents.* Princeton, NJ: Princeton University Press.

Vanhove, S., J. Wittoeck, G. Desmet, B. Van den Berghe, R. L. Herman, R. P. M. Bak, G. Nieuwland, et al. 1995. Deep-sea meiofauna communities in Antarctica: Structural analysis and relation with the environment. *Marine Ecology Progress Series* 127:65–76.

Vanreusel, A., M. Vincx, D. Schram, and D. Van Gansbeke. 1995. On the vertical distribution of the metazoan meiofauna in shelf break and upper slope habitats of the NE Atlantic. *Internationale Revue der gasamten Hydrobiologie* 80:313–26.

Vanreusel, A., M. Vincx, D. Van Gansbeke, and W. Gijselinck. 1992. Structural analysis of the meiobenthos communities of the shelf break area in two stations of the Gulf of Biscay (N.E. Atlantic). *Belgian Journal of Zoology* 122:185–202.

Vetter, E. W., and P. K. Dayton. 1998. Macrofaunal communities within and adjacent to a detritus-rich submarine canyon system. *Deep-Sea Research II* 45:25–54.

Vincx, M., B. J. Bett, A. Dinet, T. Ferrero, A. J. Gooday, P. J. D. Lambshead, O. Pfannkuche, T. Soltwedel, and A. Vanreusel. 1994. Meiobenthos of the deep Northeast Atlantic. *Advances in Marine Biology* 30:1–88.

Vivier, M. H. 1978. Influence d'un déversement industriel profond sur la nématofauna (Canyon de Cassidaigne, Mediterranée). *Tethys* 8:307–21.

Wares, J. P., and C. W. Cunningham. 2001. Phylogeography and historical ecology of the North Atlantic intertidal. *Evolution* 55:2455–69.

West, G. B., J. H. Brown, and B. J. Enquist. 1997. A general model for the origin of allometric scaling laws in biology. *Science* 276:122–26.

———. 1999a. The fourth dimension of life: Fractal geometry and allometric scaling of organisms. *Science* 284:1677–79.

———. 1999b. A general model for the structure and allometry of plant vascular systems. *Nature* 400:664–67.

Wigley, R. L., and A. D. McIntyre. 1964. Some quantitative comparisons of offshore meiobenthos and macrobenthos south of Martha's Vineyard. *Limnology and Oceanography* 9:485–93.

Williams, S. T., and D. G. Reid. 2004. Speciation and diversity on tropical rocky shores: A global phylogeny of snails of the genus *Echinolittorina*. *Evolution* 58:2227–51.

Wilson, G. D. F. 1998. Historical influences of deep-sea isopod diversity in the Atlantic Ocean. *Deep-Sea Research II* 45:279–301.

Witman, J. D., R. J. Etter, and F. Smith. 2004. The relationship between regional and local species diversity in marine benthic communities: A global perspective. *Proceedings of the National Academy of Sciences USA* 101:15664–69.

Witte, U. 2000. Vertical distribution of metazoan macrofauna within the sediment at four sites with contrasting food supply in the deep Arabian Sea. *Deep-Sea Research II* 47:2979–97.

Xiang, Q., W. H. Zhang, R. E. Ricklefs, H. Qian, Z. D. Chen, J. Wen, and J. H. Li. 2004. Regional differences in rates of plant speciation and molecular evolution: A comparison between eastern Asia and eastern North America. *Evolution* 58:2175–84.

CHAPTER FOUR

SPATIAL PATTERNS OF SPECIES DIVERSITY IN THE SHALLOW MARINE INVERTEBRATES: PATTERNS, PROCESSES, AND PROSPECTS

KAUSTUV ROY AND JON D. WITMAN

Introduction

Like on land, species diversity in the ocean changes along both latitude and longitude as well as with depth. How species richness varies along these gradients has been the focus of investigations for well over a century. Despite this long tradition, such patterns are well documented for only a few invertebrate groups and only for certain regions of the world ocean, largely outside the tropics. It is an unfortunate fact that for the vast majority of marine invertebrates living on the continental shelves of the world oceans we know little about species-level distributional patterns. Similarly, for most marine habitats, especially those in the most diverse tropical areas, we are still far from having adequate inventories of local or regional species pools for even the better-known groups of invertebrates, such as mollusks (Bouchet et al. 2002). Despite these serious gaps in our knowledge, some generalizations about how species richness varies on the continental shelves have emerged, and as discussed below, these trends are generally well supported by data from different groups of invertebrates.

Latitudinal Gradients: Patterns

The latitudinal diversity gradient (LDG), with high richness of species and higher taxa in the tropics and declining toward the poles, is considered to be one of the fundamental patterns of biological diversity on the planet (Willig, Kaufman, and Stevens 2003, Hillebrand 2004a; Hillebrand 2004b). The presence of a latitudinal gradient in taxonomic richness is well established in groups like marine mollusks, especially in the northern hemisphere (Fischer 1960; Roy, Jablonski, and Valentine 1994; Roy et al. 1998), but the trend is less well documented for many other benthic invertebrates. This had led to the obvious question whether a tropical-polar cline in richness is a general pattern in the oceans, especially given the fact that several groups of benthic marine invertebrates show relatively high species richness in the higher latitudes of the southern ocean (Clarke 1992; Gray 2002; Valdovinos, Navarrete, and Marquet 2003). Based on work done over the last decade or so, we now know that the latitudinal cline in richness holds not just for well-studied invertebrate groups like mollusks (Roy, Jablonski, and Valentine 1994, 2000; Roy et al. 1998) but is also present in other groups ranging from crustaceans (Astorga et al. 2003), bryozoans (Clarke and Lidgard 2000), epifaunal invertebrates (Witman, Etter, and Smith 2004) and cephalopods (Macpherson 2002) to benthic foraminifera (Buzas, Collins, and Culver 2002), gammaridean amphipods (Dauvin and Bellan-Santini 2004) and sabellid polychaetes (Giangrande and Licciano 2004). While there are groups that do not appear to conform to this trend (e.g., shallow-water amphipods; Myers 1996), the generality of the latitudinal diversity gradient in the oceans has been demonstrated through a metaanalysis of published studies, which also revealed that the strength of the marine gradients was similar to those on land (Hillebrand 2004b).

While a decline in taxonomic richness from the tropics to the poles is a general biodiversity pattern on continental shelves of the world oceans, it is also becoming clear that the nature of this relationship varies considerably from group to group and from one part of the oceans to another. For example, while both brachyuran and anomuran crabs show a clear latitudinal trend in richness along the east and west coasts of south America, for both groups the slope of the relationship is much steeper along the southwestern Atlantic shelves (Astorga et al. 2003). Similarly, the latitudinal distribution of species richness in other groups such as cephalopods, stomatopods, and decapod crustaceans, also differs substantially between eastern and western Atlantic shelves (Macpherson 2002). In contrast, marine gastropods in the northern hemisphere showed remarkably similar latitudinal patterns of species richness along northeastern Pacific and northwestern Atlantic coasts despite considerable differences in coastal oceanography and faunal compositions (Roy et al. 1998). There are also many regional anomalies in species richness on the continental shelves. Upwelling areas along the west African coast are characterized by anomalously low richness for their latitudinal positions and a similar anomaly is present along the western Atlantic coast near the mouth of the Amazon river, presumably due to substrate and/or salinity effects (Macpherson 2002; Roy et al. 1998). On the other hand, areas like the Mediterranean can rival some parts of the tropics in terms of species richness (Macpherson 2002). Another general pattern that seems well supported is that, on average, the latitudinal diversity gradient is steeper in the northern hemisphere compared to the southern hemisphere (Gray 2002; Valdovinos, Navarrete, and Marquet 2003). There is emerging evidence that these differences in the slopes of the gradient are influenced not only by differences in present-day oceanographic conditions (Valdovinos, Navarrete, and Marquet 2003), but also by biotic factors such as larval modes and dispersal abilities (Astorga et al. 2003) and by historical processes such as past speciation, extinction, and dispersal dynamics (e.g., Jablonski, Roy, and Valentine 2006; Goldberg et al. 2005). But the relative contributions of historical versus contemporary processes in determining the strength of the latitudinal diversity gradient remains poorly explored.

Our discussion of latitudinal richness patterns have so far ignored the issue of spatial scale, but it is clear that spatial patterns of richness in any system are scale dependent (Willis and Whittaker 2002). Most of the studies cited earlier quantified latitudinal patterns of richness using the range through assumption—individual species are assumed to be present in all the latitudinal bins between their northern and southern range limits. Such an assumption is a practical necessity because information on range limits is available for many species but inventories of marine benthic communities at a local scale are scarce outside of a few well-studied areas. The range through assumption provides a maximum estimate of richness at a particular place and in most cases the actual richness, as revealed by ecological surveys, tend to be lower. This raises the question whether a latitudinal cline in taxonomic richness exists at the scale of local communities or whether the pattern is only manifested at the regional or higher level (Clarke 1992; Clarke and Lidgard 2000). Clearly the processes producing a latitudinal gradient in within-habitat or α- diversity are likely to be different from those that generate latitudinal differences in between-habitat (β) or regional (γ) richness. Within-habitat patterns should largely reflect ecological processes while evolutionary and other historical effects are most likely to be important at larger spatial scales (Ricklefs 1987; Ricklefs and Schluter 1993). In fact, some of the divergent views about the nature of the latitudinal deployment of species richness in the oceans stem from a confusion about spatial scale. This is perhaps best illustrated by studies of infaunal species living in soft-bottom habitats. In an influential paper, Thorson (1957) argued that for such species, there was little difference in richness between tropical and arctic areas. This hypothesis was challenged by Sanders (1968) but subsequent shallow-water studies that investigated within-habitat diversity of the infauna generally supported Thorson's hypothesis (Kendall and Aschan 1993; Richardson and Hedgpeth 1977; Warwick and Ruswahyuni 1987). While these studies do suggest a lack of latitudinal difference in within-habitat richness of infaunal communities, they were based on only a few tropical samples, which makes it difficult to know whether the pattern is robust to incomplete sampling. On the other hand, using data for range limits of over 700 species of infaunal bivalves and a range through assumption, Roy, Jablonski, and Valentine (2000) found a strong latitudinal gradient in richness of along the northeastern Pacific shelf. Thus, taken at face value, it appears that the soft-bottom infauna exhibit a latitudinal gradient in richness only at regional scales rather than at the level of individual communities, but at present sampling of local assemblages in the tropics is too incomplete to draw strong conclusions.

For epifaunal taxa, Coates (1998) found little difference in the richness of mollusks and barnacles between a tropical rocky intertidal assemblage from Australia and a temperate one from New Zealand. In a much larger-scale study of shallow-water bryozoans along the North Atlantic continental shelves, Clarke and Lidgard (2000) found evidence for a latitudinal diversity gradient at the level of regional faunas but not at the level of local assemblages. However, they also noted substantial geographical patchiness in sampling intensity, which can influence the within-habitat trend. At present, the only global study that has used sample-standardized estimates of richness to investigate how local and regional diversities of epifaunal communities vary along latitude is by Witman, Etter, and Smith (2004). They sampled the α diversity of multiphyletic assemblages of epifaunal invertebrates (sponges, corals, sea anemones, octocorals, tube-dwelling polychaetes, bivalves, bryozoans, brachiopods, ascidians, etc.) attached to shallow subtidal rock walls at nearly fifty local sites from 62° South to 63° North latitude. While latitude predicted more variation in regional species richness than local richness in this data set, both local and regional species richness displayed hump-shaped patterns with latitude, peaking at low latitudes and decreasing toward the high latitudes (fig 4.1). The latitudinal diversity gradient was evident at the scale of local sites because local species richness was linearly related to regional species richness (regional richness explained 73–76 percent of variation in local richness, Witman, Etter, and Smith 2004). Recent studies have indicated that the size of the regional species pool is generally

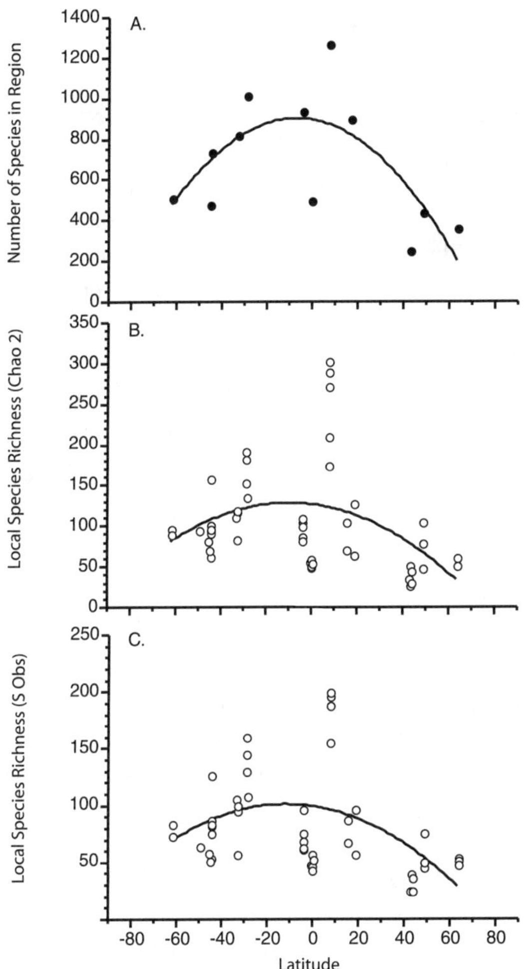

FIGURE 4.1 Latitudinal patterns in the species richness of epifaunal invertebrates on subtidal rock walls sampled by standardized sampling at 10–15 m depth. Lines represent significant, best fits to second-order polynomial equations. A. Regional species richness. $r^2 = 0.393$, $p = .043$. B. Local species richness based on the Chao 2 estimate $r^2 = 0.136$, $p = .013$. C. Local species richness as *S* observed. $r^2 = 0.155$, $p = .007$. (From Witman, Etter, and Smith 2004.)

an important determinant of the local richness of marine communities, suggesting that the regional pool effect must be considered in any explanations of large-scale patterns of marine diversity. For example, the number of species in the regional species pool explained 94–97 percent of variation in the local species richness of reef corals across the Pacific (Karlson, Cornell, and Hughes 2004). A positive linear relationship between regional and local species richness has also been documented in intertidal invertebrate communities (Rivadeneira, Fernandez, and Naverrete 2002; Russell, Wood, Allison, and Menge 2006).

Latitudinal Gradients: Processes

Our knowledge about latitudinal trends in species richness on the continental shelves has increased substantially over the last decade or so, both at the regional and local scales. But even though many hypotheses have been proposed to explain the pattern (see Pianka 1966; Rohde 1999; Mittlebach et al. 2007), we still know relatively little about the ecological and evolutionary processes that determine latitudinal deployment of species richness in the sea. This problem is not unique to marine systems—indeed, the processes underlying the latitudinal diversity gradient on land also remain a subject of much debate (Currie et al. 2004). The challenge here is not only to establish the relative contributions of historical processes (i.e., latitudinal differences in speciation and/or extinction rates as well as changes in geographic range limits of taxa over time) versus the role of contemporary environment in determining the strength and shape of this gradient, but also to identify the environmental variables that most influence spatial patterns of species richness.

Species-energy Hypothesis

Climatic and oceanographic variables have long been thought to play a major role in determining the geographic distributions of species and higher taxa and therefore influence the spatial patterns of taxonomic richness (Valentine, this volume). Among the ecological hypotheses proposed to explain global variation in species richness, the species-energy hypothesis has emerged as a leading contender (Hawkins et al. 2003; Willig, Kaufman, and Stevens 2003). According to this hypothesis, species richness of a region is a function of the total or average amount of energy available, and a positive correlation between some measure of energy availability (temperature and/or productivity are the variables commonly used) and patterns of species (or higher taxon) richness is generally taken to indicate that energy plays an important role in shaping spatial patterns of taxonomic richness (e.g., Currie 1991; Hawkins et al. 2003; Currie et al. 2004). Such correlations have been documented for a number of terrestrial groups (see Currie et al. 2004; Hawkins et al. 2003 for review). Far fewer studies have examined species-energy relationships in benthic marine organisms. When sea-surface temperature (SST) is used as a proxy for energy availability, a positive correlation between energy and richness exists for northern hemisphere gastropods and bivalves (Roy, Jablonski, and Valentine 2000; Roy et al. 1998) and in coral reefs (Fraser and Currie 1996) but the relationship does not hold for southern hemisphere mollusks (Valdovinos, Navarrete, and Marquet 2003). Witman et al. (2008)

used remotely sensed chlorophyll a concentration as a proxy for productivity to examine the effect of a species-energy factor (productivity) on the species richness of epifauna and infauna across regional–continental spatial scales in temperate–arctic benthic communities. The evidence for a productivity effect was strongest in the Canadian arctic benthos (polychaetes, mollusks) where chl a concentration explained 34 percent of variation in species richness at local spatial scales (<20 km distance) and 32 percent of variation in richness at landscape spatial scales (>20 and <200 km distance). As observed in terrestrial communities, hump-shaped relationships between productivity and diversity occurred in temperate–arctic marine communities. Covarying effects of salinity influenced overall patterns of species richness, acknowledging the important effect of environmental stress on diversity in coastal communities. However the productivity effect remained, even when salinity effects were excluded from the analyses.

One of the main weaknesses of the species-energy hypothesis is that the processes underlying any observed correlations between energy and richness remain uncertain (Currie et al. 2004; Clarke this volume). The lack of a process-based explanation precludes specific predictions about the nature of the expected relationship between energy and richness. For example, the spatial scale over which this relationship is expected to hold is unspecified, as is the form (linear, nonlinear) and slope of the relationship. The mechanism generally proposed to explain the correlation between energy and richness is the "more individuals" hypothesis where the increased species richness in areas with higher energy stems from an increased number of individuals present there. But why an increase in the number of individuals would necessarily lead to an increase in species richness is not clear (Gaston 2000). More importantly, available empirical evidence, although virtually all of it from terrestrial organisms, provide little support for this idea (Currie et al. 2004). Thus, whether correlations between present-day climatic variables and species richness are sufficient to demonstrate that current environmental factors are the main determinants of species richness of a region remains a subject of debate (Francis and Currie 1998, 2003; Quian and Ricklefs 2004; Ricklefs 2004). Alternatively, historical processes such as past speciation and extinction could be the primary determinants of spatial patterns of species richness today and the correlation with current climatic variables may not necessarily reflect causality (Ricklefs 2004).

Geographic Ranges and Environmental Tolerances

Spatial patterns of species richness is a function of (a) the sizes of the geographic ranges of individual species and (b) how those ranges are distributed across the landscape. Both of these parameters are likely to be influenced by the environmental conditions experienced by individual species. Stevens (1989) argued that for most organisms, average latitudinal range of species increases with latitude, a trend he called Rapoport's Rule, and used this empirical observation to propose a new explanation for why latitudinal diversity gradients exist; low-latitude species have narrower climatic tolerances compared to their high latitude counterparts, thereby allowing more species to coexist in the tropics. In other words, tropical species are adapted to an environment that shows very little seasonal environmental variation, thereby allowing them to subdivide the landscape into a mosaic of distinct microhabitats, which in turn allows the coexistence of more species through a spillover or "mass effect" (Stevens 1989). High-latitude species, on the other hand, have evolved larger environmental tolerances in response to large seasonal fluctuations in temperature and other environmental parameters, and thus are not subject to the spillover effect that can inflate regional species richness (Stevens 1989). However, subsequent analyses of how sizes of geographic ranges vary with latitude have found little support for Rapoport's Rule in general (Gaston, Blackburn, and Spicer 1998), and for marine organisms in particular (Rohde and Heap 1996; Rohde, Heap, and Heap 1993; Roy, Jablonski, and Valentine 1994), thereby casting doubt on the role of environmental tolerances of species in generating the latitudinal diversity gradient. An important difference between shallow marine and terrestrial realms is that unlike on land, seasonal variation in temperature does not increase monotonically with latitude (Helmuth et al. 2003); along many coastlines seasonal changes in temperature reaches a maximum in mid-latitudes rather than in polar regions (Parmesan et al. 2005; Clarke this volume). In addition, there seems to be little relationship between species richness and seasonality of climate, whether in the ocean (Parmesan et al. 2005) or on land (Kerr 1999), which is inconsistent with the idea that climatic stability promotes specialization, which in turn leads to higher species richness (Stevens 1989).

But does the failure to find empirical support for Rapoport's rule mean that environmental tolerances of species do not play a role in generating the tropical-to-polar gradient in diversity? Or, perhaps more importantly, does it necessarily reject the actual process invoked by Stevens (1989) to explain this gradient? The use of latitudinal range as a surrogate for environmental tolerances not only makes the important assumption that range limits of species are determined by their climatic tolerances (Parmesan et al. 2005), but also requires a tight positive correlation between range size and climatic tolerance. Factors that determine species range limits remain poorly understood, and it is unlikely that for marine species climate is the only variable of interest here (Case et al. 2005; Gaines et al., this volume).

In general, how environmental tolerances of marine invertebrate species relate to the size and limits of their geographic ranges remain poorly known, largely because such analyses require information about the physiological tolerances of individual species, data currently not available for the vast majority of marine organisms (but see Compton, Rijkenberg, Drent and Piersma 2007; Stillman and Somero 2000; Tomanek and Somero 1999). There is, however, a clear need for such analyses, since understanding the role of ambient environment in determining the distributional limits of marine species is not only important for understanding the causes of the latitudinal diversity gradient but also for predicting how these species will respond to climate change.

Historical Processes

Despite the basic premise that macroecological patterns are likely to reflect both historical and contemporary processes (Brown 1995), attempts to understand why diversity patterns change from one region to another have largely focused on ecological mechanisms, even though paleontological, phylogenetic, and biogeographic data strongly suggest that present-day diversity gradients have a strong historical component (Ricklefs 1987, 2004, 2006; Ricklefs and Schluter 1993; Wiens and Donoghue 2004). In order to understand the historical underpinnings of large-scale diversity patterns seen today, we need to better understand the interactions between three basic parameters—origination, extinction, and changes in geographic distributions of taxa over time (Roy and Goldberg 2007). More specifically, we not only need information on how origination and extinction rates vary from one place to another (e.g., from low to high latitudes) but also data on post-origination changes in the geographic distributions of taxa. All three of these parameters remain poorly quantified for most groups, marine or terrestrial, but perhaps not surprisingly, marine invertebrate groups with a good fossil record account for many of the existing studies that have examined how patterns of origination and extinction change from one region to another (e.g., Allen et al. 2006; Buzas, Collins, and Culver 2002; Clarke and Crame 1997; Crame 2000, 2002; Flessa and Jablonski 1996; Goldberg et al. 2005; Jablonski 1993; Jablonski, Roy, and Valentine 2006; Stanley, Addicott, and Chinzei 1980; Stehli, Douglas, and Newell 1969; Wei and Kennett 1983). However, the majority of these studies have focused on net diversification rates (speciation-extinction) rather than estimating how origination and extinction rates change along latitude and/or longitude (see Jablonski, Roy, and

Valentine 2006; Roy and Goldberg 2007). Even for groups with a relatively complete fossil record, separately estimating origination and extinction rates has proven to be difficult because of incomplete sampling of the tropical fossil record (Jablonski, Roy, and Valentine 2006). For groups with either a poor or no fossil record, molecular phylogenies can be used to estimate origination and extinction rates (Nee et al. 1994) but for the vast majority of marine invertebrates such information is currently unavailable. In addition, in order to estimate extinction rates using molecular phylogenies, one has to assume that extinction is stochastically constant over time, a problematic assumption given that past extinctions often tend to show phylogenetic or geographic selectivity (McKinney 1997; Paradis 2004; Todd et al. 2002). Thus it is not surprising that the focus of existing studies has largely been on quantifying how net diversification rates, a parameter that can be estimated more easily from paleontological data as well as molecular phylogenies, change along spatial and environmental gradients. Available evidence for marine invertebrates, primarily from mollusks and foraminifera, suggests that net diversification rates may indeed be higher in warmer regions (Buzas, Collins, and Culver 2002; Crame 2000; Crame 2002; Jablonski 1993; Stehli, Douglas, and Newell 1969; Stehli and Wells 1971) although this is not supported by some analyses (Clarke and Crame 1997; Stanley, Wetmore, and Kenett 1988; Wei and Kennett 1983). At present it is difficult to directly compare the results of many of these analyses, since a variety of methods have been used to look at how diversification rates vary along latitudinal or environmental gradients—from age distributions of living taxa (Flessa and Jablonski 1996; Goldberg et al. 2005) to survivorship curves (Stanley, Wetmore, and Kennett 1988) and Lyellian percentages (Stanley, Addicott, and Chinzei 1980) to places of origin of individual lineages (Jablonski 1993; Jablonski, Roy, and Valentine 2006)—and also since the focal taxonomic levels differ between studies.

A more fundamental issue here is that virtually all existing attempts to estimate how diversification rates change with latitude have ignored a critical parameter, how geographic distributions of taxa change after origination and over time (Goldberg et al. 2005; Jablonski, Roy, and Valentine 2006; Roy and Goldberg 2007). This not only implies that distributions of taxa are static over time (i.e., present-day distributions reflect those in the past), an assumption inconsistent with empirical evidence that range limits of many marine taxa can shift substantially in response to environmental change (Roy, Jablonski, and Valentine 2001; Roy and Pandolfi 2005), but also presents methodological problems for estimating how origination or diversification rates vary with latitude (Roy and Goldberg 2007). The failure to separate the effect of post-origination changes in geographic range limits is largely due to the difficulty of estimating the place of origin of a given taxon, but the problem can be addressed using theoretical models that explicitly account for shifts in geographic distributions (Goldberg et al. 2005; Roy and Goldberg 2007), reconstructions of ancestral ranges using phylogenetic information (Ree et al. 2005) or by using high-resolution paleontological data (Jablonski 1993; Jablonski, Roy, and Valentine 2006). Applying a neutral biogeographic model of origination, extinction, and dispersal to age distributions of marine bivalve genera derived from the fossil record, Goldberg et al. (2005) showed that polar regions of the world ocean are enriched in older taxa relative to lower latitudes and represent macroevolutionary sinks. This result suggests that for marine bivalves, differential origination in the tropics during the Cenozoic followed by geographic range expansions have played a major role in shaping the latitudinal diversity gradient seen today on the continental shelves.

More direct evidence for the hypothesis that changes in geographical distributions over evolutionary time plays an important role in shaping the present-day latitudinal diversity gradient comes from analyses of first appearances of bivalve genera in the fossil record. For bivalve genera originating during the last eleven million years (i.e., late Miocene to Recent), originations in the tropics significantly outnumber those outside (by over a factor of two; Jablonski, Roy, and Valentine 2006). More importantly, over 80 percent of the genera that first originated in the tropics have extended their geographic ranges outside the tropics (Jablonski, Roy, and Valentine 2006). The results of Jablonski (1993), Goldberg et al. (2005) and Jablonski, Roy, and Valentine (2006) all contradict the traditional view of tropics either as a cradle of diversity with high origination rates or a museum of diversity with low extinction rates (sensu Stebbins 1974). Instead, the tropics are both a cradle, since taxa preferentially originate there, and a museum, since they persist there over evolutionary time. This dynamic of tropical first occurrences followed by expansion into higher-latitude regions, termed "Out of the tropics" model (fig. 4.2; Jablonski, Roy, and Valentine 2006), suggests that in contrast to existing studies that pose evolutionary and ecological processes as alternative explanations of the latitudinal diversity gradient, the pattern results from the interaction between two distinct sets of processes—those that lead to higher originations in lower latitudes and those that set range limits of taxa. In other words, ecological and evolutionary processes work in concert to shape the gradient and progress on both fronts is needed to better understand the nature of this fundamental biodiversity pattern.

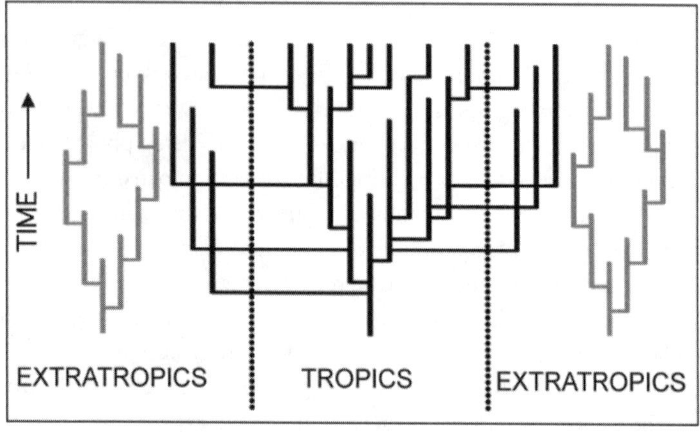

FIGURE 4.2 Hypothetical depiction of the Out of the Tropics (OTT) model of Jablonski, Roy, and Valentine 2006. Geographic distributions of taxa are shown by the horizontal lines connecting sister taxa. Lineages in black all originated in the tropics but some have also expanded to extratropical regions. Lineages in gray represent those endemic to the extratropics. See text for details. (Modified from Jablonski, Roy, and Valentine 2006).

Other Hypotheses

In addition to the general hypotheses discussed previously there are some processes specific to marine organisms that are likely to influence the shape of the latitudinal diversity gradient in the sea. The most important in this regard is the role of oceanography. Ocean currents play a major role in the dispersal of marine organisms, and clusters of species range limits that mark marine provincial boundaries tend to coincide with contacts between major water masses or other changes in oceanographic regime (Hayden and Dolan 1976; Roy, Jablonski, and Valentine 1994; Valentine 1966; Gaines et al., this volume). Thus the distributions of these physical boundaries can exert a strong influence on the shape of the latitudinal diversity gradient (Roy, Jablonski, and Valentine 1994; Valentine 1966). More generally, ocean currents have the potential to constrain the geographic distributions of individual species, although the nature and strength of these constraints depend on species-specific life histories (Gaylord and Gaines 2000; Gaines et al., this volume). Thus, even though oceanographic forcing cannot explain why there are more species in the tropics compared to temperate and polar environments, patterns of flow are likely to be an important determinant of the shape and slope of the latitudinal diversity gradient.

Longitudinal Gradients

Although latitudinal diversity gradients have received much more attention, richness of species and higher taxa also vary along longitude. In general, longitudinal richness gradients remain poorly studied in the ocean, just as they are on land. However, in the oceans one particular example of this pattern has been the focus of a large number of studies and a subject of considerable debate. Within the tropical Indo-West Pacific (IWP) region species richness is unusually high within a relatively small area often called the East Indies Triangle (a triangular area defined by the Philippines, New Guinea, and the Malay Peninsula) and declines rapidly away from this triangle (Briggs 2004). For instance, there is nearly a 60 percent decline in the local species richness of corals from Indonesia to the Society Islands (Karlson, Cornell, and Hughes 2004). While the existence of this trend has been known for well over a century (e.g., Forbes 1856), the factors controlling this gradient are still poorly understood. There is some consensus that this longitudinal gradient has a strong historical component, but the underlying processes remain highly debated (Barber and Bellwood 2005; Briggs 2004; Connolly, Bellwood, and Hughes 2003; McMillan and Palumbi 1995; Mora et al. 2003; Rosen 1988; Vermeij 1987). In particular, some view the East Indies Triangle as a "center of origin" and argue that the richness gradient results from dispersal of species away from this center (Briggs 2004; Ekman 1953; Mora et al. 2003) while others have argued that the high diversity of the East Indies results from either accumulation there of species that evolved in other regions or from "overlap" of distributions with the surrounding regions contributing to the high diversity (Barber and Bellwood 2005; Briggs 2004; Palumbi 1997; Pandolfi 1992). In general, the patchiness of the tropical fossil record, and the lack of information about the evolutionary histories of most invertebrate species living in the region have hindered efforts to understand the origin of the Indo-Pacific diversity (Barber and Bellwood 2005; McMillan and Palumbi 1995) and no clear consensus exists as to which of these competing models best explains the diversity trend. For example, for reef fishes, one of the best-studied groups, some studies favor the center-of-origin hypothesis (Briggs 1999; Mora et al. 2003) while others suggest that speciation in the peripheral areas is more important (Connolly, Bellwood, and Hughes 2003; Hughes, Bellwood, and Connolly 2002; Pandolfi 1992). In addition, Karlson, Cornell, and Hughes (2004) suggest that factors contributing to the size of the regional species pool have a highly significant influence on the decline in local species richness of corals eastward across the Pacific.

The competing models about the origin of the diversity gradient in the Indo-Pacific make different predictions about phylogenetic relationships and biogeographic distributions of species and hence should be testable. However, such tests are hampered by a lack of information about phylogenetic relationships of species as well as robust estimates of the ages of lineages present in different areas (Barber and Bellwood 2005). Because of this, some studies have attempted to test the competing models, using distributional data for reef fishes and the simplifying assumption that endemism is a reflection of evolutionary age with recently derived species being geographically restricted (Mora et al. 2003). However, such an assumption is highly questionable from a theoretical perspective (Goldberg et al. 2005; Barber and Bellwood 2005) and so the robustness of this approach remains uncertain. In addition, distributional patterns of species within the Indo-Pacific remain poorly known, and even for better-sampled taxa such as reef fishes controversy exists regarding spatial patterns of endemism (Hughes, Bellwood, and Connolly 2002; Mora et al. 2003). While a consensus about the origin of the diversity gradient within the Indo-Pacific remains elusive, evolutionary relationships of Indo-Pacific taxa are increasingly being investigated using molecular markers (Barber and Bellwood 2005; McMillan and Palumbi 1995; Meyer 2003) and the resulting phylogenies in conjunction with better biogeographic and paleontological data should make it possible to evaluate the relative contributions of these hypotheses.

What Is Needed to Better Understand Spatial Patterns of Diversity in the Sea?

It is perhaps not surprising that we still do not have a full picture of how species richness of marine invertebrates varies along environmental gradients and along geography. After all, the nature of the habitat necessitates that most samples are collected using remote methods (e.g., dredge or grab samples from ships), many species are surprisingly rare (Bouchet et al. 2002), and systematics of most marine invertebrate groups remain poorly studied. What is surprising is how poorly sampled many parts of the world ocean still are, even for well-studied groups such as marine mollusks (Bouchet et al. 2002). Obviously, process-based hypotheses about marine biodiversity cannot be properly tested unless we have better estimates of local and regional diversities, especially in the tropics. Another limitation is that most descriptions of diversity gradients are based on narrow taxonomic groups, and the lack of diversity estimates of "whole" communities can prevent meaningful tests of ecological hypotheses based on biological interactions among species. Of particular importance here is the issue of spatial scale. Since diversity patterns are clearly scale dependent (Huston 1999; Willis and Whittaker 2002) macroecological hypotheses are not only best tested using data of comparable spatial grain, but comparison of richness patterns across scales can provide useful insights into the processes that are likely to generate them. For example, comparison of local and regional richness patterns suggest that for some marine invertebrate groups beta diversity also changes with latitude (Clarke and Lidgard 2000; Witman, Etter, and Smith 2004), a pattern that is most likely to result from historical processes rather than ecological mechanisms operating at local scales. Yet very few analyses of species richness in the ocean have been done at the scale of local communities and there is a clear need for data on how richness of local assemblages change along geographic and environmental gradients. Furthermore, the emerging recognition of the role of regional processes in shaping the diversity of local communities (Karlson, Cornell, and Hughes 2004; Witman, Etter, and Smith 2004) underscores the need for more research on how the regional pool effect is manifested at the scale of local sites (Russell, Wood, Allison, and Menge 2006).

Even though macroecology emphasizes comparative statistical analyses (Brown 1995), many macroecological hypotheses are amenable to experimental tests, at least on regional scales (Witman and Roy, this volume). At present such tests are largely lacking, but marine ecology has a long history of experimental studies, and there is a growing interest in conducting experiments across large spatial scales to test macroecological hypotheses (Sanford and Bertness; Connell and Irving; Witman and Roy; all this volume). In fact, as Brown (1995) pointed out, macroecology relies on statistical analyses of empirical observations simply for practical reasons—many of the macroecological questions involve spatial and temporal scales at which experimental manipulations would be difficult, impossible, or even immoral. Indeed, some macroecological hypotheses will always remain beyond the scope of experimental tests, but because of technological advances it is becoming increasingly feasible to deploy experiments over larger spatial scales. We view this as an exciting new development that would not only nicely complement the comparative statistical approach but also provide crucial experimental support for some of the key conceptual underpinnings of macroecological hypotheses.

While it is widely thought that historical processes are likely to have played an important role in shaping spatial patterns of species richness seen today (e.g., Fischer 1960; Ricklefs and Schluter 1993), estimating how rates of speciation and extinction vary along latitudinal and other environmental gradients have proven to be difficult. Many groups of marine invertebrates such as mollusks, foraminifera, and benthic ostracodes have an excellent fossil record that, in conjunction with increasing work on the molecular phylogenetics (Duda and Kohn 2005; Hellberg 1998; Latiolais et al. 2006; Meyer 2003), provide ideal systems for investigating the relative roles of ecological and evolutionary processes in generating spatial patterns of species richness. Paleontological data strongly suggests that net diversification rates (origination-extinction) of mollusks and foraminifera vary substantially along latitude (Buzas, Collins, and Culver 2002; Crame 2000; Flessa and Jablonski 1996; Stehli, Douglas, and Newell 1969) but relatively little work has been done so far to separate the effects of originations and extinctions as well as shifts in distributions of taxa over time. Teasing apart the contributions of these distinct processes would not only require the integration of paleontological and phylogenetic information, but also the development of spatially explicit models that take into account variations in all three of these parameters together (Goldberg et al. 2005; Jablonski, Roy, and Valentine 2006; Roy and Goldberg 2007).

REFERENCES

Allen, A. P., J. F. Gillooly, V. M. Savage, and J. H. Brown. 2006. Kinetic effects of temperature on rates of genetic divergence and speciation. *Proceedings of the National Academy of Sciences, USA* 103:9130–35.

Astorga, A., M. Fernandez, E. E. Boschi, and N. Lagos. 2003. Two oceans, two taxa and one mode of development: Latitudinal diversity patterns of South American crabs and test for possible causal processes. *Ecology Letters* 6:420–27.

Barber, P. H., and D. R. Bellwood. 2005. Biodiversity hotspots: Evolutionary origins of biodiversity in wrasses (*Halichoeres*: Labridae) in the Indo-Pacific and new world tropics. *Molecular Phylogenetics and Evolution* 35:235–53.

Bouchet, P., P. Lozouet, P. Maestrati, and V. Heros. 2002. Assessing the magnitude of species richness in tropical marine environments: Exceptionally high numbers of molluscs at a New Caledonia site. *Biol. J. Linn. Soc.* 75:421–36.

Briggs, J. C. 1999. Coincident biogeographic patterns: Indo-West Pacific ocean. *Evolution* 53:326–35.

———. 2004. A marine center of origin: reality and conservation, In *Frontiers of biogeography: New directions in the geography of nature,* ed. M. V. Lomolino and L. R. Heaney, 255–69. Sunderland, MA: Sinauer.

Brown, J. H. 1995. *Macroecology.* Chicago: University of Chicago Press.

Buzas, M. A., L. S. Collins, and S. J. Culver. 2002. Latitudinal difference in biodiversity caused by higher tropical rate of increase. *Proceedings of the National Academy of Sciences, USA* 99:7841–43.

Case, T. J., R. D. Holt, M. A. Mcpeek, and T. H. Keitt. 2005. The community context of species' borders: Ecological and evolutionary perspectives. *Oikos* 108:28–46.

Clarke, A. 1992. Is there a latitudinal diversity cline in the sea? *Trends in Ecology & Evolution* 7:286–87.

Clarke, A., and J. A. Crame. 1997. Diversity, latitude and time: Patterns in the shallow sea. In

Marine biodiversity: Patterns and processes, ed. R. F. G. Ormond, J. D. Gage, and M. V. Angel, 122–147. Cambridge: Cambridge University Press.

Clarke, A., and S. Lidgard. 2000. Spatial patterns of diversity in the sea: Bryozoan species richness in the North Atlantic. *Journal of Animal Ecology* 69:799–814.

Coates, M. 1998. A comparison of intertidal assemblages on exposed and sheltered tropical and temperate rocky shores. *Global Ecology and Biogeography Letters* 7:115–24.

Compton, T. J., M. J. A. Rijkenberg, J. Drent, and T. Piersma. 2007. Thermal tolerance ranges and climate variability: A comparison between bivalves from differing climates. *Journal of Experimental Marine Biology and Ecology* 352:200–211.

Connolly, S. R., D. R. Bellwood, and T. P. Hughes. 2003. Geographic ranges and species richness gradients: A re-evaluation of coral reef biogeography. *Ecology* 84:2178–90.

Crame, J. A. 2000. Evolution of taxonomic diversity gradients in the marine realm: Evidence from the composition of Recent bivalve faunas. *Paleobiology* 26:188–214.

———. 2002. Evolution of taxonomic diversity gradients in the marine realm: A comparison of Late Jurassic and Recent bivalve faunas. *Paleobiology* 28:184–207.

Currie, D. J. 1991. Energy and large scale patterns of animal and plant species richness. *American Naturalist* 137:27–49.

Currie, D. J., G. G. Mittelbach, H. V. Cornell, R. Field, J.-F. Guégan, B. A. Hawkins, D. M. Kaufman, et al. 2004. Predictions and tests of climate-based hypotheses of broad-scale variation in taxonomic richness. *Ecology Letters* 7:1121–34.

Dauvin, J.-C., and D. Bellan-Santini. 2004. Biodiversity and the biogeographic relationships of the Amphipoda: Gammaridea on the French coastline. *Journal of the Marine Biological Association U.K.* 84:621–28.

Duda, T. F. J., and A. J. Kohn. 2005. Species-level phylogeography and evolutionary history of the hyperdiverse marine gastropod genus *Conus. Molecular Phylogenetics and Evolution* 34:257–72.

Ekman, S. 1953. *Zoogeography of the sea.* London: Sidgwick and Jackson.

Fischer, A. G. 1960. Latitudinal variations in organic diversity. *Evolution* 14:64–81.

Flessa, K. W., and D. Jablonski. 1996. The geography of evolutionary turnover: A global analysis of extant bivalves. In *Evolutionary paleobiology,* ed. D. Jablonski, D. H. Erwin, and J. H. Lipps, 376–97. Chicago: University of Chicago Press.

Forbes, E. 1856. Map of the distribution of marine life. In *The physical atlas of natural phenomena,* ed. A. K. Johnston. Philadelphia: Lea and Blanchard.

Francis, A. P., and D. J. Currie. 1998. Global patterns of tree species richness in moist forests: Another look. *Oikos* 81:598–602.

———. 2003. A globally consistent richness-climate relationship for angiosperms. *American Naturalist* 161:523–36.

Fraser, R. H., and D. J. Currie. 1996. The species richness-energy hypothesis in a system where historical factors are thought to prevail: Coral reefs. *American Naturalist* 148:138–59.

Gaston, K. J. 2000. Global patterns in biodiversity. *Nature* 405:220–27.

Gaston, K. J., T. M. Blackburn, and J. I. Spicer. 1998. Rapoport's rule: Time for an epitaph? *Trends in Ecology & Evolution* 13:70–74.

Gaylord, B., and S. D. Gaines. 2000. Temperature or transport? Range limits in marine species mediated solely by flow. *American Naturalist* 155:769–89.

Giangrande, A., and M. Licciano. 2004. Factors influencing latitudinal pattern of biodiversity: An example using Sabellidae (Annelida, Polychaeta). *Biodiversity and Conservation* 13:1633–46.

Goldberg, E. E., K. Roy, R. Lande, and D. Jablonski. 2005. Diversity, endemism, and age distributions in macroevolutionary sources and sinks. *American Naturalist* 165:129–35.

Gray, J. S. 2002. Species richness of marine soft sediments. *Marine Ecology Progress Series* 244:285–97.

Hawkins, B. A., R. Field, H. V. Cornell, D. J. Currie, J.-F. Guégan, D. M. Kaufman, J. T. Kerr, et al. 2003. Energy, water and broad-scale geographic patterns of species richness. *Ecology* 84:3105–17.

Hayden, B. P., and R. Dolan. 1976. Coastal marine fauna and marine climates of the Americas. *Journal of Biogeography* 3:71–81.

Hellberg, M. E. 1998. Sympatric sea shells along the sea's shore: The geography of speciation in the marine gastropod *Tegula*. *Evolution* 52:1311–24.

Helmuth, B. C., D. G. Harley, P. M. Halpin, M. O'Donnell, G. E. Hofmann, and C. A. Blanchette. 2003. Climate change and latitudinal patterns of intertidal thermal stress. *Science* 298:1015–17.

Hillebrand, H. 2004a. On the generality of the latitudinal diversity gradient. *American Naturalist* 163:192–211.

———. 2004b. Strength, slope and variability of marine latitudinal gradients. *Marine Ecology Progress Series* 273:251–67.

Hughes, T. P., D. R. Bellwood, and S. R. Connolly. 2002. Biodiversity hotspots, centers of endemicity, and the conservation of coral reefs. *Ecology Letters* 5:775–84.

Huston, M. A. 1999. Local processes and regional patterns: Appropriate scales for understanding variation in the diversity of plants and animals. *Oikos* 86:393–401.

Jablonski, D. 1993. The tropics as a source of evolutionary novelty through geological time. *Nature* 364:142–44.

Jablonski, D., K. Roy, and J. W. Valentine. 2006. Out of the tropics: Evolutionary dynamics of the latitudinal diversity gradient. *Science* 314:102–6.

Karlson, R. H., H. V. Cornell, and T. P. Hughes. 2004. Coral communities are regionally enriched along an oceanic biodiversity gradient. *Nature* 429:867–870.

Kendall, M. A., and M. Aschan. 1993. Latitudinal gradients in the structure of macrobenthic communities: A comparison of Arctic, temperate and tropical sites. *Journal of Experimental Marine Biology and Ecology* 172:157–69.

Kerr, J. T. 1999. Weak links: "Rapoport's rule" and large-scale species richness patterns. *Global Ecology and Biogeography* 8:47–54.

Latiolais, J. M., M. S. Taylor, K. Roy, and M. E. Hellberg. 2006. A molecular phylogenetic analysis of strombid gastropod morphological diversity. *Molecular Phylogenetics and Evolution* 41:436–44.

Macpherson, E. 2002. Large-scale species richness gradients in the Atlantic Ocean. *Proceedings of the Royal Society of London Series B* 269:1715–20.

McKinney, M. L. 1997. Extinction vulnerability and selectivity: Combining ecological and paleontological views. *Annual Review of Ecology and Systematics* 28:495–516.

McMillan, W. O., and S. Palumbi. 1995. Concordant evolutionary patterns among Indo-West Pacific butterfly fishes. *Proceedings of the Royal Society of London Series B* 260:229–36.

Meyer, C. P. 2003. Molecular systematics of cowries (Gastropoda: Cypraeidae) and diversification patterns in the tropics. *Biological Journal of the Linnean Society* 79:401–59.

Mittelbach, G. G., D. W. Schemske, H. V. Cornell, A. P. Allen, J. M. Brown, M. Bush, S. P. Harrison, et al. 2007. Evolution and the Latitudinal Diversity Gradient: Speciation, extinction and biogeography. *Ecology Letters* 10:315–31.

Mora, C., P. M. Chittaro, P. F. Sale, J. P. Kritzer, and S. A. Ludsin. 2003. Patterns and processes in reef fish diversity. *Nature* 421:933–36.

Myers, A. A. 1996. Species and generic gamma-scale diversity in shallow-water marine amphipoda with particular reference to the Mediterranean. *Journal of the Marine Biological Association U.K.* 76:195–202.

Nee, S., E. C. Holmes, R. M. May, and P. H. Harvey. 1994. Extinction rates can be estimated from molecular phylogenies. *Philosophical Transactions of the Royal Society of London B* 344:77–82.

Palumbi, S. 1997. Molecular biogeography of the Pacific. *Coral Reefs* 16:S47–S52.

Pandolfi, J. M. 1992. Successive isolation rather than evolutionary centres for the origination of Indo-Pacific reef corals. *Journal of Biogeography* 19:593–609.

Paradis, E. 2004. Can extinction rates be estimated without fossils? *Journal of Theoretical Biology* 229:19–30.

Parmesan, C., S. D. Gaines, L. Gonzalez, D. M. Kaufman, J. G. Kingsolver, A. T. Peterson, and R. Sagarin. 2005. Empirical perspectives on species borders: From traditional biogeography to global change. *Oikos* 108:58–75.

Pianka, E. R. 1966. Latitudinal gradients in species diversity: A review of concepts. *American Naturalist* 100:33–46.

Quian, H., and R. E. Ricklefs. 2004. Taxon richness and climate in angiosperms: Is there a globally consistent relationship that precludes region effects? *American Naturalist* 163:773–79.

Ree, R. H., B. R. Moore, C. O. Webb, and M. J. Donoghue. 2005. A likelihood framework for inferring the evolution of geographic range on phylogenetic trees. *Evolution* 59:2299–2311.

Richardson, M. D., and J. W. Hedgpeth. 1977. Antarctic soft-bottom, macrobenthic community adaptations to a cold, stable, highly productive, glacially affected environment. In *Adaptations within Antarctic ecosystems,* ed. G. A. Llano, 181–95. Washington, DC: Smithsonian Institution.

Ricklefs, R. E. 1987. Community diversity: Relative roles of local and regional processes. *Science* 235:167–71.

———. 2004. A comprehensive framework for global patterns in biodiversity. *Ecology Letters* 7:1–15.

———. 2006. Evolutionary diversification and the origin of the diversity-environment relationship. *Ecology* 87:S3–S13.

Ricklefs, R. E., and D. Schluter. 1993. Species diversity: Regional and historical influences. In *Species diversity in ecological communities,* ed. R. E. Ricklefs and D. Schluter, 350–63. Chicago: University of Chicago Press.

Rivadeneira, M. M., M. Fernandez, and S. A. Navarrete. 2002. Latitudinal trends of species diversity in rocky intertidal herbivore assemblages: Spatial scale and the relationship between local and regional species richness. *Marine Ecology Progress Series* 245:123–31.

Rivadeneira, M. M. 2005. Macroecologia evolutiva de los bivalvos marinos de la costa Pacifica de sudamerica. Unpublished PhD thesis, Pontificia Universidad Catolica de Chile.

Rohde, K. 1999. Latitudinal gradients in species diversity and Rapoport's rule revisited: A review of recent work, and what can parasites teach us about the causes of the gradients? *Ecography* 22:593–613.

Rohde, K., and M. Heap. 1996. Latitudinal ranges of teleost fish in the Atlantic and Indo-Pacific oceans. *American Naturalist* 147:659–65.

Rohde, K., M. Heap, and D. Heap. 1993. Rapoport's Rule does not apply to marine teleosts and cannot explain latitudinal gradients in species richness. *American Naturalist* 142:1–16.

Rosen, B. R. 1988. Progress, problems and patterns in the biogeography of reef coral and other tropical marine organisms. *Helg. Meeres.* 42:269–301.

Roy, K., and E. E. Goldberg. 2007. Origination, extinction, and dispersal: Integrative models for understanding present-day diversity gradients. *American Naturalist* 170:S71–S85.

Roy, K., D. Jablonski, and J. W. Valentine. 1994. Eastern Pacific molluscan provinces and latitudinal diversity gradient: No evidence for "Rapoport's rule." *Proceedings of the National Academy of Sciences, USA* 91:8871–74.

———. 2000. Dissecting latitudinal diversity gradients: Functional groups and clades of marine bivalves. *Proceedings of the Royal Society Biological Sciences Series B* 267:293–99.

———. 2001. Climate change, species range limits and body size in marine bivalves. *Ecology Letters* 4:366–70.

Roy, K., D. Jablonski, J. W. Valentine, and G. Rosenberg. 1998. Marine latitudinal diversity gradients: Tests of causal hypotheses. *Proceedings of the National Academy of Science, USA* 95:3699–3702.

Roy, K., and J. M. Pandolfi. 2005. Responses of marine species and ecosystems to past climate change. In *Climate change and biodiversity,* ed. T. E. Lovejoy and L. Hannah, 160–75. New Haven, CT: Yale University Press.

Russell, R., S. A. Wood, G. Allison, and B. A. Menge. 2006. Scale, environment, and trophic status: The context dependency of community saturation in rocky intertidal communities. *American Naturalist* 167:E-158–E-170.

Sanders, H. L. 1968. Marine benthic diversity: A comparative study. *American Naturalist* 102:243–82.

Stanley, S. M., W. O. Addicott, and K. Chinzei. 1980. Lyellian curves in paleontology: Possibilities and limitations. *Geology* 8:422–26.

Stanley, S. M., K. L. Wetmore, and J. P. Kennett. 1988. Macroevolutionary differences between the two major clades of Neogene planktonic foraminifera. *Paleobiology* 14:235–49.

Stebbins, G. L. 1974. *Flowering plants: Evolution above the species level.* Cambridge, MA: Belknap Press.

Stehli, F. G., R. G. Douglas, and N. D. Newell. 1969. Generation and maintenance of gradients in taxonomic diversity. *Science* 164:947–49.

Stehli, F. G., and J. W. Wells. 1971. Diversity and age patterns in hermatypic corals. *Systematic Zoology* 20:115–26.

Stevens, G. C. 1989. The latitudinal gradient in geographical range: How so many species coexist in the tropics. *American Naturalist* 133:240–56.

Stillman, J. H., and G. N. Somero. 2000. A comparative analysis of the upper thermal tolerance limits of eastern Pacific porcelain crabs, genus Petrolisthes: Influences of latitude, vertical zonation, acclimation, and phylogeny. *Physiol. Biochem. Zool.* 73:200-208.

Thorson, G. 1957. Bottom communities (sublittoral or shallow shelf). *Geol. Soc. Am. Mem.* 67:461–534.

Todd, J. A., J. B. C. Jackson, K. G. Johnson, H. M. Fortunato, A. Heitz, M. Alvarez, and P. Jung. 2002. The ecology of extinction: Molluscan feeding and faunal turnover in the Caribbean Neogene. *Proceedings of the Royal Society of London B* 269:571–77.

Tomanek, L., and G. N. Somero. 1999. Evolutionary and acclimation-induced variation in the heat-shock responses of congeneric marine snails (genus Tegula) from different thermal habitats: Implications for limits of thermotolerance and biogeography. *Journal of Experimental Biology* 202:2925–36.

Valdovinos, C., S. A. Navarrete, and P. A. Marquet. 2003. Mollusk species diversity in the southeastern Pacific: Why are there more species towards the pole? *Ecography* 26:139–44.

Valentine, J. W. 1966. Numerical analysis of marine molluscan ranges on the extratropical north-eastern Pacific shelf. *Limnology and Oceanography* 11:198–211.

Vermeij, G. J. 1987. The dispersal barrier in the tropical Pacific: Implications for molluscan speciation and extinction. *Evolution* 41:1046–58.

Warwick, R. M., and Ruswahyuni. 1987. Comparative study of the structure of some tropical and temperate marine soft-bottom macrobenthic communities. *Marine Biology* 95:641–49.

Wei, K.-Y., and J. P. Kennett. 1983. Nonconstant extinction rates of Neogene planktonic foraminifera. *Nature* 305:218–20.

Willig, M. R., D. M. Kaufman, and R. D. Stevens. 2003. Latitudinal gradients of biodiversity: Pattern, process, scale, and synthesis. *Annual Review of Ecology and Systematics* 34:273–309.

Wiens, J. J., and M. J. Donoghue. 2004. Historical biogeography, ecology and species richness. *Trends in Ecology & Evolution* 19:639–44.

Willis, K. J., and R. J. Whittaker. 2002. Species diversity—scale matters. *Science* 295:1245–48.

Witman, J. D., R. J. Etter, and F. Smith. 2004. The relationship between regional and local species diversity in marine benthic communities: A global perspective. *Proceedings of the National Academy of Sciences, USA* 101:15664–69.

Witman, J. D., M. Cusson, P. Archambault, A. J. Pershing, and N. Mieszkowska. 2008. The relation between productivity and species diversity in temperate–arctic marine ecosystems. *Ecology* 89 (11) Supplement: S66–S80.

CHAPTER FIVE
MACROECOLOGICAL PATTERNS AMONG MARINE FISHES

ENRIQUE MACPHERSON, PHILIP A. HASTINGS, AND D. ROSS ROBERTSON

Introduction

Worldwide, there are over 28,000 species of fishes, which comprise more than 50 percent of all vertebrate species. Over 16,000 of these are marine for all or part of their lives, and fishes live in all parts of the world's oceans (Nelson 2006). "Fishes" represents a paraphyletic group that includes hagfishes (Myxini), lampreys (Petromyzontida), coelacanths, and lungfishes (Sarcopterygii), as well as the better known and much more diverse cartilaginous fishes (Chondrichthyes) and more recently evolved ray-finned fishes (Actinopterygii). They exhibit a considerable diversity of modes of reproduction, life cycles, and capacities for dispersal. The vast majority of marine ray-finned fishes have relatively small pelagic eggs and/or a pelagic larval phase, which, because it is spent in the water column away from adult habitat, enhances their capacity for dispersal and expansion of their geographic ranges (Leis and McCormick 2002). Cartilaginous fishes, in contrast, have large, benthic, or internally brooded eggs and lack a pelagic larval stage. However, their dispersal capabilities likely are enhanced by the large body size and mobility of adults, and these characteristics may have produced the large ranges many exhibit. The ancient origins of such species may also have given them more time than has been available to many recently derived teleost species to expand their ranges (Pyle 1999; Robertson, Grove, and

McCosker 2004). These fundamental differences in life histories of higher-level taxa can have important implications for distributional patterns and macroecology on both local and regional scales.

In marine environments, in contrast to terrestrial and freshwater environments, physical barriers to dispersal are thought to be weak and often absent (Briggs 1995). Coverage of most of the earth's surface by water, the continuity of such habitat over large distances, persistent transoceanic currents, intermittent large-scale oceanographic events (such as those due to the El Niño phenomenon) that produce extremes of variation in current flows, and the presence of pelagic larval stages should combine to increase dispersal potential, and produce high levels of gene flow and broad ranges for many species (Lessios and Robertson 2006). This combination of characteristics is likely to result in very different macroecological patterns among marine fishes compared to those observed in both terrestrial and freshwater species.

In this chapter, we briefly review current knowledge about patterns in the distribution and diversity of marine fish species in the context of the enormous spatial and temporal variation that results from oceanographic variability. We also discuss processes commonly implicated in controlling the distribution and regional diversity of marine fishes. This review necessarily depends on our current understanding of species-level diversity. Broader study of the genetics of putative species, especially widespread coral reef species (e.g., Muss et al. 2001) and open-ocean species (e.g., Miya and Nishida 1997) may reveal significantly greater cryptic species diversity. In addition, recent analysis indicates no decline in the rate of morphospecies descriptions of tropical shore-fishes in recent decades (Zapata and Robertson 2006). This review is also necessarily couched within the prevailing concept of species. Broader application of a phylogenetic species concept, recently advocated for coral-reef fishes (Gill 1999), would result in an increase in overall species diversity and a concomitant decrease in average species range (Agapow et al. 2004). Nonetheless, the predominant macroecological patterns for marine fishes discussed in this chapter are not expected to change significantly if that reorganization of the species concept were applied.

Global Patterns in Species Richness

Latitudinal Gradients

Latitudinal gradients in species richness represent the most widely argued about macroecological pattern relating to the large-scale spatial distributions of organisms (Willig, Kaufman, and Stevens 2003). In general, biological diversity increases from polar to equatorial latitudes, a trend documented in a large array of terrestrial and marine taxa (e.g., Stehli, McAlester, and Newell 1969; Roy et al. 1998; Rex et al. 1993, 2005; Gaston and Blackburn 2000; Macpherson 2002; Hillebrand 2004; see also Valentine; and McClain et al., Roy and Witman, this volume). Although some exceptions exist that are associated with depth or habitat characteristics (e.g., Clarke 1992; Rohde 1999; Gray 2001), this gradient is one of the most predominant macroecological patterns known.

This latitudinal pattern of a peak of richness at the equator also holds for many assemblages of marine fishes (figs. 5.1 and 5.2), including not only pelagic and benthic species on each side of the Atlantic (Macpherson 2002), but also tropical reef-fishes in the Indo-Pacific (Connolly, Bellwood, and Hughes 2003; Mora et al. 2003). In the Indo-Pacific, a convex diversity pattern with latitude holds for tropical reef fishes (and corals) (Connolly, Bellwood, and Hughes 2003; Mora et al. 2003; Bellwood et al. 2005; figure 5.1, panel B). However, the extent to which the central peak results from overlap of widely distributed species rather than a concentration of local endemics with narrow ranges remains unclear (see Hughes, Bellwood, and Connolly 2002a and Mora et al. 2003, for opposing views).

The tropical eastern Pacific is a biogeographically discrete region with a high level of endemism: about 79 percent of the resident coastal fishes are regional endemics (Robertson and Allen 2002; Zapata and Robertson 2006). Mora and Robertson (2005a) found that the richness of endemic coastal fishes living on the continental shore has a generally bell-shaped latitudinal distribution that peaks in the center of that region (fig. 5.2, panel B). Broadly

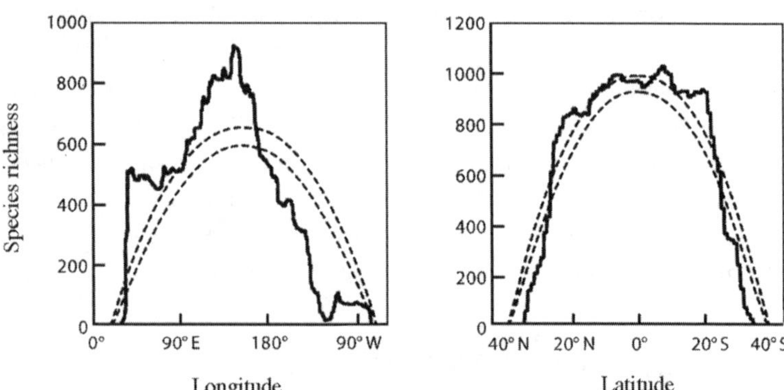

FIGURE 5.1 Longitudinal and latitudinal variation in the species-richness of reef fishes (n = 1,766) in the Indo-Pacific. Dotted lines represent the 99 percent of the values obtained from the randomizations (mid-domain model, reproduced with permission from Connolly, Bellwood, and Hughes 2003).

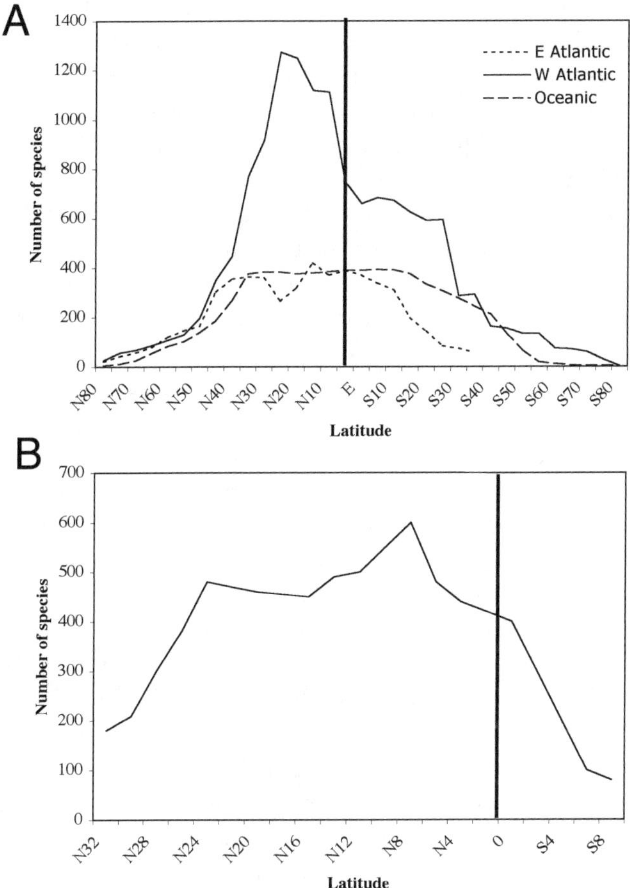

FIGURE 5.2 Latitudinal species richness gradients from (A) the Western and Eastern Atlantic fishes (solid line: coastal; dashed line: shelf-slope; dotted line: abyssal) and (B) Tropical Eastern Pacific shore-fishes (data from Macpherson, 2002, and Mora and Robertson, 2005a).

distributed species are largely responsible for that convex curve. Narrow-range species contribute the two small departures from that convexity, as they are concentrated in the two sections of the coast (Panama/Costa Rica and the Gulf of California) that have relatively large areas of habitat, an abundance of near-shore islands, and high habitat diversity.

In contrast to shore fishes, species assemblages in other habits, such as the continental slope and abyss, show a more homogeneous distribution of species richness with respect to latitude (Merrett and Haedrich 1997; Macpherson 2002; fig. 5.2, panel A). Departures from the general shallow-water pattern can also occur at finer scales, within biogeographically discrete regions; for example, in the tropical eastern Pacific, species richness of endemic shorefishes peaks at about 8° N and declines not only northward but also southward, toward the equator (Mora and Robertson 2005a; fig. 5.2, panel B). Further latitudinal patterns of species richness of coastal benthic species are strongly influenced by major local hydrographic features. These include large upwelling areas, such as those off Benguela (Mas et al. 1991), and the Sahara coast (Binet 1997), and the points at which equator-bound cold temperate currents turn westward away from the equator in the eastern Pacific (Mora and Robertson 2005a). Marked declines in richness of many groups are associated with such features. For example, the freshwater and sediment outflow from the Amazon and nearby Orinoco rivers largely eliminate reef development and shallow coastal habitat for reef fishes along ~2,300 km of the equatorial coast of south America (Collette and Rutzler 1977), which provides a significant barrier to dispersal of reef organisms and probably is responsible for most of the endemism found in southern Brazilian coastal habitats (Joyeux et al. 2001; Rocha 2003). Consequently, in this part of the Atlantic Ocean, richness of coastal fishes does not peak on the equator itself, but well to the north of it in the Caribbean Sea, between 9–20° N (Macpherson 2002; fig. 5.2, panel A). Pelagic fishes are usually widely distributed, and few large faunal regions based on the distributions of these fishes have been defined (Briggs 1995). In general, these regions follow large-scale circulation patterns and discontinuities in oceanographic characteristics and hence are not influenced by river outflows and coastal upwellings (Longhurst 1998; Briggs 2003). In the Atlantic Ocean, there are strong poleward decreases in species richness among pelagic species at around 50°N and 40°S (Angel 1997; Macpherson 2002). These faunal breakpoints coincide with well defined transition zones between biogeochemically distinct oceanic provinces (see Longhurst et al. 1995).

While the fish faunas of temperate and polar seas generally are comparatively low in total diversity, those faunas tend to be dominated by clades that have undergone significant radiations (Briggs 2003). Conspicuous examples among fishes in the North Pacific include the scorpaenoids (e.g., *Sebastes* rockfishes and related genera; Love, Yaklovich, and Thorsteinson 2002) and cottoids (Bolin 1944). Consequently, coastal fish communities in the NE temperate Pacific are dominated by acanthopterygian fishes from both temperately and tropically derived lineages (Hobson 1994). Similarly, the Southern Ocean fish fauna is characterized by an overall low diversity and a high percentage of endemics with two striking radiations, those of nototheniids on the continental shelf and of liparids on the continental slope (Andriashev 1965, 1986; Gon and Heemstra 1990; Miller 1993; Eastman and Clarke 1998; Clarke and Johnston 2003). Although the diversity of Antarctic fishes is relatively low, that is not the case for other benthic taxa in the same waters. For groups such as filter feeders (e.g., sponges, bryozoans), Antarctic diversity is similar to that described for tropical areas (Brey et al. 1994). Frequent fluctuations in habitat availability originating from fluctuations in the extent of the continental ice sheet, together with the effects of the long-term cooling of sea water (Clarke and Crame 1997) may have been the main causes of the absence of many teleost families (through extinction) and the differences in the latitudinal clines in species richness between groups (Clarke and Johnston 2003). Thatje, Hillenbrand, and Larter (2005) suggest that differences in species richness between taxa can also be related to the extreme conditions faced by pelagic early life-history stages, which limited the capacity for recolonization by many species (e.g., teleost fishes and decapods) after the isolation of Antarctica ca. 34 million years ago, resulting in the loss of major taxonomic groups.

Latitudinal gradients are observed in the trophic structure of fish assemblages as well as in species richness. There are sharp declines in the abundance of herbivorous fishes with increasing latitude (Harmelin-Vivien 2002; Floeter et al. 2005). In general, tropical fish assemblages show a larger trophic spectrum, characterized by increased use of low-quality food resources, that is, algae, sponges, cnidarians (Hobson 1994). This may be associated to higher water temperatures facilitating digestive processes of low-caloric diets (Ebeling and Hixon 1991). More recently, Frank, Petrie, and Schackell (2007), using data from exploited ecosystems, found pronounced geographical variation in the type of trophic forcing (top-down or bottom-up) that was related to species richness and temperature. Their results suggest that this relationship has a strong influence on resilience to fishing, with cold and species-poor ecosystems with top-down control succumbing more readily (and recovering more slowly) than species-rich ecosystems from warmer areas, which normally experience fluctuating levels of top-down and bottom-up control.

While the general latitudinal gradient in species richness is well established (Gaston and Blackburn 2000; Willig, Kaufman, and Stevens 2003; Hillebrand 2004), more than thirty hypotheses that attempt to explain the pattern have been proposed (Rohde 1992). No single causal mechanism has proven sufficient to explain the overall pattern across a range of terrestrial and marine faunas and floras, although geographic area, productivity, ambient energy supply, Rapoport-rescue, the rates of speciation and extinction, and geometric constraints have been those most widely considered (Rohde

1992; Willig, Kaufman, and Stevens 2003). Some studies have found that habitat area does not explain the latitudinal pattern in benthic fish species richness on a large spatial scale (Rohde 1992; Roy et al. 1998; Macpherson 2002), while others working with different systems have reported habitat-area effects (e.g., for Indo-Pacific reef-fishes see Hughes et al. 2002b; Bellwood et al. 2005, and for tropical eastern Pacific shore-fishes see Mora and Robertson 2005a).

In Atlantic fishes, significant correlations exist between some environmental factors that represent proxies of energy input (sea surface temperature, nitrates and chlorophyll) and species richness of fishes, not only of pelagic and shallow-water benthic species, but also of species living at greater depths (Macpherson 2002). Because these variables provide indirect measures of energy input (Rohde 1992; Fraser and Currie 1996), those results support the view that the level of energy entering ecosystems (from solar energy and/or oceanographic events), and the way that this energy is used, plays an important role in determining the latitudinal distribution of fish diversity. However, Mora and Robertson's (2005a) assessment of potential effects of various major potential determinants of the latitudinal diversity gradient of coastal fishes within a single tropical biogeographic region (the tropical eastern Pacific) produced different results. Their analysis identified the mid-domain effect (MDE), which predicts such a distribution when the species ranges are randomly placed within a bounded geographical domain (e.g., Colwell and Lees 2000) as the major determinant of the distribution of broad-range species and habitat abundance and diversity as the main determinant for narrow-range species. They found no effects of energy supply (as indicated by temperature and primary production) or environmental variability, on the fauna as a whole or any of its components that they considered. However, the use of mid-domain and other null-models in biogeographic analyses remains controversial and there is no consensus about what really constitutes an appropriate null model for such assessments (Zapata, Gaston, and Chown 2005). Recent analyses have included effects of environmental gradients as well as domain boundaries (Connolly 2005). For example, Bellwood et al. (2005) demonstrated that the MDE and habitat area are both predictor variables of reef-fish species richness in the Indo-Pacific. These studies highlight the need for the testing of various hypotheses together, in order to take into account effects of colinearity among predictive variables (e.g., temperature covaries with the mid-domain effect), and for the use of techniques that cope with statistical effects of spatial autocorrelation within variables (see Mora and Robertson 2005a).

Thus marine biogeography, and biogeography in general, still lacks an adequate general predictive theory of latitudinal gradients in species richness (Gaston and Blackburn 2000; Willig, Kaufman, and Stevens 2003, but see Allen, Brown and Gillooly 2002 and Gillooly et al. 2002, for a thermodynamic hypothesis for the species diversity gradient). Differences in the latitudinal patterns in species richness observed in Atlantic pelagic versus benthic fishes and in the relationship of each to different environmental factors indicate that a unique predictor of these gradients is unlikely (Macpherson 2002) and that gradients will be governed by different processes in different environments and parts of the world, as will gradients of different components of a single fauna. Indeed, Mora and Robertson (2005a) found quite different predictors of latitudinal patterns of species richness for tropical eastern Pacific shore-fishes with large and small geographic ranges. They pointed out that the so-called "Tropical Indo-Pacific" actually consists of a group of distinct subregions, with different processes likely influencing distributional patterns within each of those.

Longitudinal Patterns

Strong longitudinal variation in species richness is also evident both at the general level (Gaston and Blackburn 2000), and among tropical marine shore-fishes in particular (Briggs 1974; Bellwood and Hughes 2001; Connolly et al. 2003; Mora et al. 2003). The Indo-Australian Area (IAA) at the junction of the Pacific and Indian oceans has by far the greatest diversity of fishes of any part of the world (Briggs 1974; Randall 1998; Connolly, Bellwood, and Hughes 2003; Mora et al. 2003). In descending order of species richness, other global centers of diversity of tropical coastal fishes are the Greater Caribbean (within the tropical western Atlantic), the tropical eastern Pacific and the tropical eastern Atlantic (Briggs 1974).

Within the Indo-Pacific, species richness of reef-fish assemblages declines strongly with distance from the global hotspot of diversity in the IAA (Connolly, Bellwood, and Hughes 2003; Mora et al. 2003; fig. 5.1, panel A), which is centered on the Philippine Islands (Carpenter and Springer 2005). Two major processes contribute to this pattern. First, there are faunal losses due to reduction in habitat diversity that reflects the absence of continental habitats (e.g., estuaries, mangroves, large areas of shoreline influenced by river runoff), as one moves eastward onto the Pacific plate, where oceanic islands constitute the only habitat available (Randall 1998). Second, there is a decrease in species diversity within numerous lineages of coastal fishes (Springer 1982; Findley and Findley 2001; Allen 1975, 1979) that does not simply reflect declining habitat diversity. Much has been written concerning this latter pattern (e.g., Briggs 1974, 1995, 2003; Randall 1998; Bellwood and Hughes 2001; Mora et al. 2003; Connolly, Bellwood, and Hughes 2003). Prominent hypotheses that attempt to account for the longitudinal diversity gradients in coastal fishes of the Indo-Pacific include the following. (a) Effects of habitat area: The amount of shelf area available to support a diversity of species declines as one moves from the center of diversity (Bellwood and Hughes 2001; Bellwood et al. 2005). (b) Environmental stability: The relative environmental stability of the IAA, especially with regard to environmental changes associated with Pleistocene glaciation events, has facilitated species survival relative to other regions that were more adversely affected, such as the Greater Caribbean (Chenoweth et al. 1998). (c) Increased potential for allopatric speciation: The geographic complexity of the Indo-Australian area, coupled with repeated cycles of exposure and submergence of land barriers during eustatic sea-level fluctuations, has facilitated allopatric speciation most in the IAA (Springer 1982; Springer and Williams 1990; Carpenter and Springer 2005). (d) Dispersal from a center of origin: Centers of high diversity, such as the IAA, have traditionally been thought to be the centers of origin for most extant species for a particular region (Briggs 1974, 1995, 2003). Some phylogenetic and genetic evidence supports the notion of such a pattern (Briggs 1995) and the view that colonization of the Pacific occurs mostly in an easterly direction (Lavery et al. 2004). Mora et al. (2003) argue that the decline in species richness from the center of diversity can be accounted for by variation in dispersal ability (as mediated by pelagic larval duration) outwards from a center of origin. (e) Center of accumulation: Effects of westward geostrophic flow in major ocean currents on the largest-scale directionality of dispersal of pelagic propagules may produce accumulations of species at the western boundaries of both the Pacific and Indian oceans basins, the IAA and E Africa, respectively (Jokiel and Marinelli 1992; Connolly, Bellwood, and Hughes 2003).

It seems likely that more than one, and quite possibly all, of these factors have played a role in building and maintaining the extraordinary diversity of fishes (and other organisms) at the junction of the Indian and Pacific oceans in the center of the Indo-Pacific. Historical biogeographic analyses that employ genetic techniques to identify the sites of origin and patterns of subsequent spread of taxa throughout the tropics should help resolve issues that are raised by (but cannot be resolved by) description and correlation alone (e.g., Palumbi 1997; Bernardi et al. 2003; Robertson, Grove, and McCosker 2004).

The Indian and Pacific oceans both have the greatest concentrations of reef-fish diversity on their western boundaries (Connolly, Bellwood, and

Hughes 2003). A similar pattern exists among shorefishes in the Atlantic basin, where diversity is greater on the western than eastern side (Briggs 1974). This pattern is consistent with most of the hypotheses described previously. In contrast, relatively few analyses of the distribution of diversity within the center of diversity in the Atlantic—the Greater Caribbean—are available (e.g., Robins 1971). Recently Smith, Carpenter, and Waller (2002) found, for a subset of the ichthyofauna, the highest diversity in two areas: (a) the Florida peninsula, and northern Cuba, and (b) the northern coast of South America. Thus, in longitudinal terms, diversity within the Greater Caribbean shows no evidence of a western-boundary concentration that could reflect the action of an accumulation process, despite the existence of a large-scale circulation pattern that might be expected to produce such a boundary effect. The predominant large-scale circulation pattern in the Greater Caribbean is thought to facilitate dispersal and gene flow throughout that region (e.g., Shulman and Bermingham 1995). However, there are counter-examples to panmixia within the Greater Caribbean, in which deep genetic breaks both occur over very short distances and show little correspondence to circulation patterns (Carlin, Robertson, and Bowen 2003; Taylor and Hellberg 2003; Rocha et al. 2005a). Planes (2002) provides similar examples in the central Pacific.

The tropical eastern Pacific (TEP) exhibits a very different diversity pattern to that of the tropical Indo-Pacific and tropical Atlantic. The shore-fish fauna of that region, with ~1,285 species, has a very high level of endemism: about 79 percent of the resident species and 16 percent of the genera occur nowhere else. Rates of endemism in equivalent sized areas in the Indo-Pacific are much lower (Connolly, Bellwood, and Hughes 2003; Mora et al. 2003), and the level of endemism in the Greater Caribbean, the sister biogeographic region of the tropical eastern Pacific, is only about one third that in the TEP (see Smith, Carpenter, and Waller 2002 for Caribbean data). The TEP has a long history of both isolation from the central Pacific (Grigg and Hey 1992) and close association with the Greater Caribbean. The latter connection was broken only relatively recently (~3 mya—see Coates and Obando 1996) by the final closure of the Isthmus of Panama. As a result, the coastal fish faunas of those two regions have strong taxonomic affinities: about 35 percent of the genera in the tropical eastern Pacific are shared with, and only with, the tropical western Atlantic (Rosenblatt 1967). The neotropics have a remarkable abundance of blennioid fishes, several families of which make major contributions to both faunas and occur primarily in the new world (Rosenblatt 1967; Robertson 1998). The eastern Pacific barrier (EPB), the world's widest deep-water barrier (5,000–7,000 km), has isolated the TEP from the central Indo-Pacific for ~65 my (Grigg and Hey 1992). Normal currents across the barrier are sufficiently slow that transit times probably are beyond the larval durations of most species (Briggs 1961; Leis 1984). Consequently, demersal shore-fishes that have migrated eastward across the EPB constitute only 10 percent of the tropical eastern Pacific fauna. In contrast, almost all of the eastern Pacific epipelagic fishes and 25 percent of its inshore pelagic fishes have successfully crossed the EPB (Robertson, Grove, and McCosker 2004). Among the demersal fishes, the transpacific migrants tend to be concentrated on the oceanic islands in the eastern Pacific, which have habitats similar to the islands from which they originated in the central Pacific. More than any other tropical region, the TEP is affected by oceanographic effects of El Niño events. While it is thought that greatly increased eastward current flow across the EPB during such events enhances eastward migration (e.g., Richmond 1990), there is little direct evidence to support that view (Robertson, Grove, and McCosker 2004). During El Niño events, however, there are temporary range expansions by shore-fishes both within the TEP and beyond the normal latitudinal bounds of that region (Lea and Rosenblatt 2000; Victor et al. 2001; Robertson, Grove, and McCosker 2004; Mora and Robertson 2005b).

Patterns of Variation in Species Range Sizes

The geographic area occupied by a fish species depends not only on its biological characteristics, such as habitat preferences, dispersal capabilities of larvae and adults, and interactions with competitors and predators, but also on the processes of speciation, extinction, and range fragmentation as a result of vicariant events (Briggs 1974; Connolly, Bellwood, and Hughes 2003).

HABITATS Not surprisingly, range size tends generally to be larger in pelagic and deep-sea fishes than in benthic and shallow-water species (Merret and Haedrich 1997; Macpherson 2003; fig. 5.3). Mora and Robertson (2005b) examined the distributions of range sizes among four groups of shorefishes endemic to the tropical eastern Pacific (fig. 5.4). Insular species have the smallest ranges, and species with small ranges are predominantly island forms. The range-size frequency distribution of continental species is bimodal, with most species having moderate to large ranges. Those authors argue that this pattern reflects a combination of (a) the paucity of existing barriers to dispersal on the continental shore, (b) a lack of intermittent barriers that appear and disappear with changing sea levels (i.e., like those that have repeatedly fragmented ranges in the Indo-Australian Area), and (c) the fact that those continental barriers affect primarily reef-fishes, which represent only 38

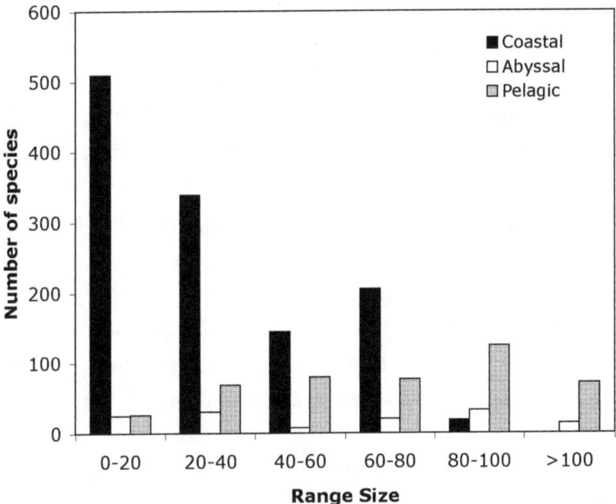

FIGURE 5.3 Frequency distributions of latitudinal range sizes for Western Atlantic coastal (0–100 m depth), abyssal (>1,000 m depth) and pelagic fishes (recorded in the water column). Range in degrees of latitude (see also Macpherson 2003).

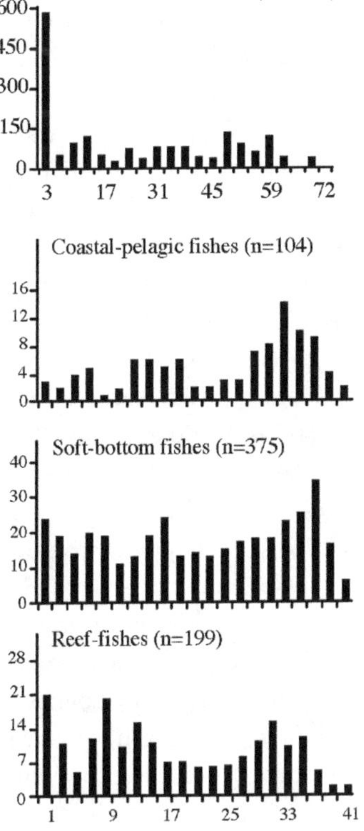

FIGURE 5.4 Frequency distributions (Y-axis, number of species) of latitudinal range sizes (X-axis, range size in degrees of latitude) three ecological groups of Tropical Eastern Pacific shore-fishes, plus Indo-Pacific reef fishes for comparison with the eastern Pacific reef fishes. (reproduced with permission from Mora and Robertson, 2005 B).

percent of the fauna. They found that, among continental species, range size decreases in the following hierarchy: pelagic species (continuous habitat) > benthic species living on soft bottoms (continuous habitat) > benthic species living on reefs (moderately discontinuous habitat) > species restricted to the oceanic islands (highly discontinuous habitat). They concluded that adult dispersal ability has the strongest effects on range size, and that isolation by open ocean barriers has stronger effects than isolation by habitat discontinuities on the continental shore. They also noted possible effects of variation in the geographic complexity of regions and the occurrence of intraregional barriers on the structure of range-size frequency distributions in shore-fishes in different parts of the Atlantic and Indo-Central Pacific.

BARRIERS Large-scale oceanographic discontinuities are generally responsible for producing different pelagic biogeographic provinces in the open oceans (Longhurst 1998). Such discontinuities evidently also represent a major factor limiting the distributions of coastal as well as pelagic fishes, because the range limits of both often coincide with such oceanographic breakpoints (Ekman 1953; Briggs 1974; Zezina 1997; Macpherson 2003; Mora and Robertson 2005a). However, different types of species do not necessarily exhibit the same patterns of geographic structure, probably as a result of different patterns of colonization and dispersal capabilities and different ecological requirements. The frequency of occurrence on both sides of the eastern Pacific Barrier is much higher for oceanic pelagic species than for shore-fishes, and transbarrier species constitute a much greater percentage of the TEP fauna of pelagics than is the case for shorefishes (Robertson, Grove, and McCosker 2004). Other barriers include recently developed permanent land barriers, such as the central American isthmus, which finally severed longstanding connections between the tropical biotas of the west Atlantic and eastern Pacific some 3 mya (e.g., Coates and Obando 1996). They also include eustatically variable land barriers such as those in the Indo-Australian Area (Springer 1982; Springer and Williams 1990; Carpenter and Springer 2005). Major oceanographic processes also act as barriers (e.g., upwelling areas, river discharges, principal currents, oceanographic fronts), and pronounced discontinuities in the temperature, salinity, and productivity characteristics of water masses. In the Atlantic Ocean, the boundaries of ranges of pelagic fishes tend to coincide with transition zones between oceanic domains that have distinct biogeochemical properties and plankton communities, domains that were described by Longhurst et al. (1995) and Longhurst (1998). The distributions of range end-points in benthic taxa are mainly influenced by the Sahara and Benguela upwellings in the eastern Atlantic, and by the boundaries of Labrador and Falkland currents and the Amazon/Orinoco river discharge in the western Atlantic. The zones on each side of these boundaries display marked differences in species richness. Furthermore, numerous species tend to become rarer toward their distributional limits, near these boundaries (Brown 1984; Macpherson 1989; Brown, Stevens, and Kaufman 1996). Rocha et al. (2002) showed how variation in the degree of genetic connectivity between the Caribbean and Brazilian populations of three members of a single genus is related to differences in habitat preferences that affect their ability to live within the 2,300 km wide Amazon barrier. The actions of such marine boundaries are also depth dependent because the effect of the oceanographic events marking biogeographic boundaries on the shelf and slope extends only as deep as the broad slope/rise. Consequently such boundaries are poorly defined on broad abyssal plain areas (Gordon and Duncan 1985; Haedrich and Merrett 1990; Macpherson 2003).

In the eastern Pacific, distributions of coastal species are strongly affected by pole-to-pole temperature gradients (Hubbs 1948), with notable breakpoints occurring where cold equator-bound currents turn westward and define the northern and southern limits of the centrally located tropical eastern Pacific (Briggs 1974; Hastings 2000; Mora and Robertson 2005b). Within the tropical part of the eastern Pacific the distribution of shallow coastal reef fishes is limited by effects of two large (300–750 km wide) "gaps" in the continental shore that lack reefs and consist entirely of sand and mud shorelines (Hastings 2000; Mora and Robertson 2005b).

ISLANDS AND SEAMOUNTS The few oceanic islands of the tropical Atlantic are widely scattered and support faunas that exhibit reduced species diversity, high levels of endemism, and other effects of isolation, including those due to greatly reduced habitat diversity (Briggs 1995; Robertson 2001). Considering that these islands vary in age and distance from adjacent coasts, comparative studies of their faunas can reveal considerable information about general patterns of evolution and distribution (Briggs 1995). Robertson (2001) analyzed the endemic shore-fish faunas of small, highly isolated tropical islands in the eastern Pacific and central Atlantic to assess whether they have unusual biological characteristics. He found that they have no particular characteristics in terms of body size, general dispersal capabilities, or taxonomic composition and concluded that shore-fishes in general are capable of maintaining persistent endemic populations on such islands if they can disperse to them.

The recent and continuing accumulation of genetic data on the shore-fish faunas of those islands is providing insights into the origins and ages of island species and the extent of ongoing connections between island and mainland faunas (see Bernardi et al. 1999; Muss et al. 2001; Bowen et al. 2001; Rocha et al. 2002; Carlin, Robertson, and Bowen 2003; Rocha et al. 2005a, b; Robertson et al. 2006). Such work should eventually lead to a greatly enhanced appreciation of general patterns and processes governing the structure of island faunas and the extent to which they interact with mainland faunas, as well as the action of islands as stepping stones for transoceanic migration.

Seamounts tend to be dominated by species inhabiting neighboring areas (Rogers 1994). However, seamount faunas also show high levels of endemism (Wilson and Kaufmann 1987; Parin, Mironov, and Nesis 1997), and exhibit previously unsuspected high diversity (Richer de Forges, Koslow, and Poore 2000). Although there are few studies of seamount fish communities, results from various invertebrate groups indicate that seamounts likely are sites with high rates of speciation, as a result of reproductive and genetic isolation resulting from their geographic isolation both from other seamounts and continental shelf areas, and from hydrographic conditions that trap larvae that originated on a seamount and promote self-recruitment and sustaining local populations (Parker and Tunnicliffe 1994).

DEEP SEA There has been considerable improvement in our knowledge of the deep fish fauna in the last few decades. While those findings have provided some useful insights, logistical difficulties are such that the biota of the deep seas (>1,000 m) remains much more poorly known than that of shallower marine habitats (Haedrich 1997; Merret and Haedrich 1997). Differences in deep-sea community structure are thought to be associated with variation in productivity and levels of seasonal organic enrichment from sinking phytodetritus (Merret and Haedrich 1997; Rex, Stuart, and Coyne 2000). Despite their apparent isolation from immediate surface events, environmental changes associated with climatic fluctuations can also have significant affects on community structure in deep-sea habitats (Ruhl and Smith 2004). Deep pelagic species often associated with particular water masses, and, as a consequence, species composition and/or abundances often change rapidly along fronts between water masses (e.g., Backus, Craddock, and Shores 1969; Figueroa, Díaz de Astarloa, and Martos 1998). Among such fishes species diversity tends to be highest in mixing areas where species from neighboring water masses co-occur (e.g., Beamish et al. 1999).

The deep sea has unusual physico-chemical activity not found in shallow environments that has unique effects on associated biological communities.

Both hot and cold deep-sea hydrothermal vents, which are small, relatively short-lived and scattered, support unusual fish communities that are based on chemoautotrophic primary production (Van Dover 2000). There appears to be some interchange among these fish communities separated by long distances (ca. 1000 km), associated with high dispersal capabilities (Hashimoto et al., 1995), although the degree of local endemism is also high (Tunnicliffe 1991; Tunnicliffe and Fowler 1996). Mechanisms of colonization by vent organisms remain largely unknown for most of the mid-ocean ridge systems, although recent studies indicate that dispersal of invertebrate larvae occurs via deep-ocean currents (Van Dover et al. 2002).

The greatest diversity of such fishes occurs at mid-depths, near 1,500 m for demersal species (Rex et al. 1993; Haedrich 1997). However, depth trends in diversity vary geographically and among fish groups. At Porcupine Seabight (N Atlantic), for example, species richness peaks around 1,000 m and falls away steadily to the 2,400 m level, where it increases again (Haedrich and Merret 1997). Diversity in the NE Pacific is also bimodal, with peaks at 600–700 m and 2,000–2,100 m (Pearcy, Stein, and Carney 1982). Like the deep-sea benthic fishes, the greatest diversity of deep-sea pelagic fishes occurs at mid depths (Ebeling 1967; Haedrich 1997).

Although some demersal deep-sea fishes are widely distributed, many species appear to have small ranges (Haedrich and Merret 1988, 1990). As a consequence, similarities in demersal fish faunas between different areas within the deep sea usually are low (Haedrich 1997; Merret and Haedrich 1997). For example, the composition of deep-sea fish faunas in the N Atlantic and N Pacific are quite different, with <10 percent of species being shared (Haedrich and Merret 1988; Pearcy, Stein, and Carney 1982). The deep-sea fish fauna of SE Australia is more similar to that of the N Atlantic than that of the N Pacific (Koslow, Bulman, and Lyle 1994). While many pelagic deep-sea fishes are thought to have broad distributions encompassing one or more ocean basins (Haedrich 1997), detailed morphological analyses (e.g., Gibbs 1986) and recent genetic studies have revealed previously unrecognized cryptic diversity (e.g., Miya and Nishida 1997) calling this assumption into question.

RAPOPORT'S RULE The effect of latitude on species' range-size, known as Rapoport's rule, has been examined for many different groups since Stevens (1989) first discerned the tendency for range size to increase with latitude (Gaston, Blackburn, and Spicer 1998; Rohde 1999; Gaston and Blackburn 2000). Rapoport's rule is based on the rationale that greater environmental variation at higher latitudes selects for broad tolerance (and hence large ranges) species, which can also live at less variable lower latitudes, where reduced variability also promotes the evolution of narrow tolerance (small range) species (Stevens 1989). Distributions of Atlantic fishes suggest that Rapoport's rule does hold, as species with distributions that reach latitudes nearest the poles do have the broadest ranges; however, this trend is relatively weak, suggesting that this rule is not the primary factor responsible for latitudinal patterns in range size (Macpherson 2003; see the following). Other studies (Rohde, Heap, and Heap 1993; Rohde and Heap 1996) indicate that Pacific fishes do not follow Rapoport's rule on a broad scale, though such a relationship may exist on a smaller scale. Mora and Robertson (2005a) found that latitudinal patterns of variation in range-size among regionally endemic tropical eastern Pacific shorefishes are consistent with Rapoport's Rule (fig. 5.5). However, they concluded that those patterns arise simply as corollaries of the mid-domain effect, which is the major determinant of the latitudinal distribution of species richness within that region. They also found opposite latitudinal patterns of variation in range size depending on whether average range-size in a latitudinal band was measured using either (a) the mean size of the ranges of all species present in that band (the original method of measurement—see Stevens 1989), or (b) the mean size of the ranges whose midpoints occur in that band (an alternative method developed by Rhode et al. 1993 to cope with a lack of statistical independence among measurements produced by the former method).

Many authors (e.g., Roy, Jablonski, and Valentine 1994; Gaston and Blackburn 2000) have considered various mechanisms that might explain latitudinal gradients in range size: climatic variability, area size, extinction rate, competition, and biogeographic boundaries. Pole-to-pole studies of latitudinal patterns in range size examine patterns that span multiple biogeographic provinces. Hence these largest-possible-scale latitudinal gradients may reflect the distinct environmental differences that separate adjacent provinces (e.g., temperature, salinity, and productivity), while there is greater constancy of environmental conditions within a given biogeographic province (Rutherford, D'Hondt, and Prell 1999). Consequently, the location of these oceanographic boundaries, the environmental conditions at the boundaries, and the ability of species to cross them may be the main factors that account for such largest-scale latitudinal gradients in range size (Roy, Jablonski, and Valentine 1994; Macpherson 2003; but see Mora and Robertson 2005a).

LARVAL DISPERSAL There is no clear relationship between range-size and larval dispersal capabilities among either tropical reef or temperate shore-

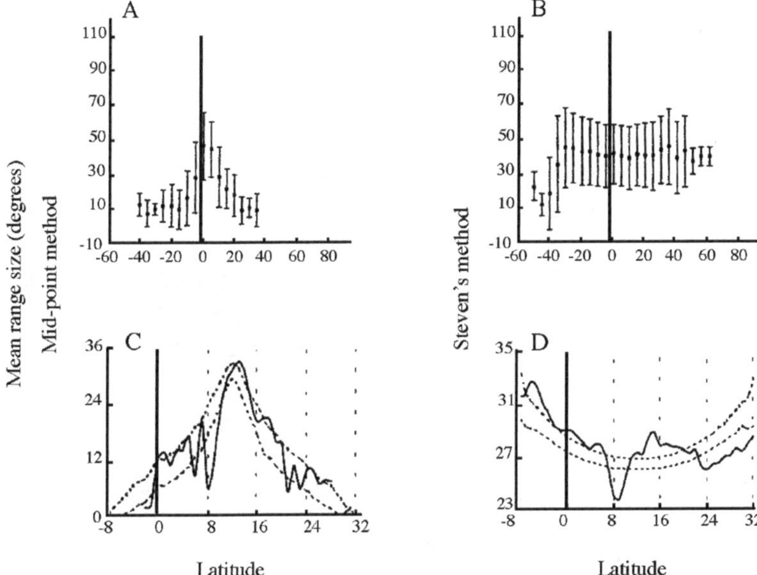

FIGURE 5.5 Rapoport's rule. A–B. Trends in mean range-size of teleost fishes from the Indo-Pacific. C–D. The same for the shorefishes of the Tropical Eastern Pacific. A and C: Mid-point method, using the midpoint of each species' latitudinal range as a single value, thus yielding a set of independent data points. B and D: Steven's method, where the mean distribution ranges of the species present in each 5° bin are calculated, and latitude is regressed on the mean range in each bin. Bars in figs. A–B are standard deviations; dotted lines in figs. C–D are the 95 percent confidence limits of the mean range size distributions generated by the mid-domain model. Equator marked by a vertical line. (A–B reproduced with permission from Rohde and Heap 1996, and C–D from Mora and Robertson 2005 A).

fishes, at least as measured by variation in the length of the pelagic larval life (e.g., Victor and Wellington 2000; Lester and Ruttenberg 2005; but see Mora et al. 2003). This suggests that range size is not set by a single mechanism, such as the length of the larval life. However, dispersal potential of such fishes may be influenced not only by larval durations, but also by whether larvae are restricted to nearshore habitats or range further offshore, and by their spawning characteristics (e.g., benthic or pelagic eggs, season of planktonic life; Shanks and Eckert 2005; Macpherson and Raventos 2006). Recent climate changes have affected the distributional pattern of numerous tropical and temperate marine fishes (e.g., Rocha et al. 2005b), to the extent that, in some cases, the position of a species' center of distribution has been significantly modified (Stenseth et al. 2002; Walther et al. 2002; Genner et al. 2004). It would be useful to analyze not only larval dispersal capabilities but various other biological characteristics of such species, in order to better understand processes that facilitate such patterns of colonization and range modification.

DEPTH Geographic range size-frequency distributions vary strongly in relation to depth. In general, coastal species have smaller geographic ranges than species inhabiting the continental slope or abyssal plains (Briggs 1974; Stevens 1996; Macpherson 2003). Furthermore, there is a tendency for depth range to increase with depth of occurrence (e.g., Ekman 1953; Pineda 1993; Stevens 1996; Haedrich 1997; Merrett and Haedrich 1997; Smith and Brown 2002). These depth-range distributions of species are the bases for the recognition of depth provinces, the boundaries of which tend to lie around the edge of the continental shelf (300 m depth), the upper continental slope (300–1,000 m) and the abyssal domain (>1,000 m; Haedrich 1997; Merrett and Haedrich 1997; Macpherson 2003). However, separations between these depth provinces are less evident than those between latitudinal provinces. This is not surprising given that processes regulating the latitudinal distributions of species are unlikely to be the same as those regulating depth distributions (Macpherson 2003). The depth-domain boundaries tend to be related to a suite of oceanographic parameters, including discontinuities in temperature, discontinuities in productivity, sedimentary features, and hydrodynamics (Gordon and Duncan 1985; Haedrich and Merrett 1990). The extent of species depth ranges changes with latitude, because environmental conditions tend to be more uniform over depth at higher latitudes (Stevens 1996; Zezina 1997; Longhurst 1998). As a consequence, the depth-range frequency distributions of species are right-skewed near the equator, where most species have small depth ranges, and left-skewed near the poles, where most species have large depth ranges.

CLIMATE AND CHANGES IN DISTRIBUTION Sporadic shifts in the distributions of fish species have frequently been documented. In the tropical eastern Pacific two types of range-changes occur during El Niño events, when a surge of heated water moves eastward across the Eastern Pacific Barrier. The effects of that surge extend beyond the usual northern and southern limits of the region, which are defined by the westward turning points of cold equator-bound currents. First, many tropical species extend their ranges temporarily into adjacent temperate areas (northward extensions: Hubbs 1948; Lea and Rosenblatt 2000; southward extensions: Chirichigno and Velez 1998). Second, among species restricted entirely to the tropical eastern Pacific there are temporary range expansions within that area (Victor et al. 2001). A number of recent shifts in species ranges and changes in community structure in temperate areas have been attributed to climatic changes in both the north Pacific (e.g., Fields et al. 1993; McFarlane, King, and Beamish 2000; Zhang et al. 2000; Brooks, Schmitt, and Holbrook 2002; Beamish et al. 2004) and the north Atlantic and its sub-basins (e.g., Francour et al. 1994; Genner et al. 2004). Invasion of the tropical Atlantic by reef fishes has been linked to such environmental variation, with such changes providing insight into the effects of ongoing global climate change (Rocha et al. 2005b). Implications of global warming for tropical reef fishes have only begun to be explored (see Mora and Ospina 2001; Rocha et al. 2005a). In addition, the affects of climate change on even the remotest abyssal communities may be significant but remain poorly understood (Ruhl and Smith 2004).

Body Size Distributions

Many biological characteristics of species are related to, and often dependent on, body size (Peters 1983). Consequently, the distribution of body sizes of the species inhabiting an area has the potential to provide insights into mechanisms that determine the species composition of fish assemblages. Species' body-size distributions are mostly right-skewed (most species are small) in terrestrial animal communities, although skewness can change with geographical scale: from right-skewed at large scales to a variety of different shapes at local or regional scales (see Gaston and Blackburn 2000; Roy et al. 2000).

Body size is also related to metabolic rate, and large species consume more energy than small species, although they require less energy per gram of body weight (Peters 1983). The relationship between body size and abundance is one aspect of macroecology that has been assessed in numerous terrestrial organisms (e.g., Gaston and Blackburn 2000), but poorly studied in marine fishes. However, body size is a poor predictor of species abundance for SW Atlantic fishes (Macpherson 1989), as well as for Indo-Pacific reef fishes (Munday and Jones 1998; fig. 5.6). Furthermore, body-size and density relationships are different for smallest and largest size classes, probably associated with different patterns of resource acquisition (Ackerman, Bellwood, and Brown 2004). Biomass/size spectra (i.e., the biomass density of organisms belonging to different size classes—see Cyr, Peters, and Downing 1997) have been commonly employed in aquatic studies, to compare the structure of different nonfish communities (e.g., phyto- and zooplankton: Sheldon, Prakash, and Sutcliffe 1972; Rodriguez et al. 2001) and assess possible effects of system productivity on that size-spectral structure (Sprules and Munawar

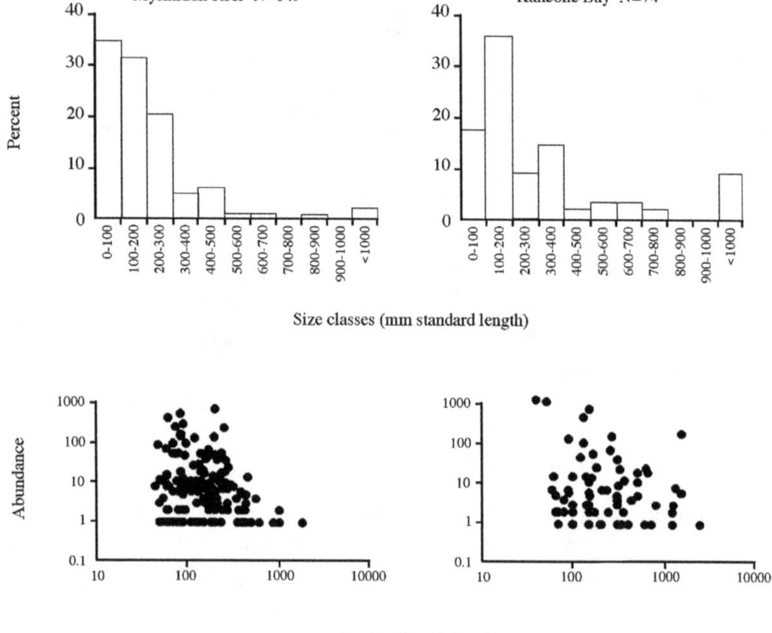

FIGURE 5.6 Body size distributions (upper panel) and body size versus abundance of each species (lower panel) of reef fishes from Hawaii and the Australian Great Barrier Reef (reproduced with permission from Munday and Jones 1996).

1986). Differences in the slopes of biomass/size spectra have been found to be associated with productivity: higher slope values tend to occur in areas of high productivity. That is, whereas high productivity areas have greater concentrations of biomass in smaller species (which have high turnover rates), in oligotrophic areas biomass is distributed more evenly among the different size classes, or may even be skewed toward the larger size classes (Sprules and Munawar 1986). There are few similar studies of body size distributions among fish communities (e.g., Merrett and Haedrich 1997; Munday and Jones 1998; Ackerman, Bellwood, and Brown 2004). As with the plankton, indications are that, among those fishes, there is greater biomass of large organisms in the more oligotrophic abyssal zone compared to the upwelling areas on the continental shelf, where the opposite relationship prevails (Macpherson and Gordoa 1996). Furthermore, interesting patterns of variation in the diversity of large oceanic predators have recently been identified, with diversity being highest at intermediate latitudes (20–30°N and S), as well as near reefs, shelf breaks, and seamounts that can enhance local productivity and food supply (Worm, Lotze, and Myers 2003).

Unfortunately, industrial fisheries have produced significant disturbances to the relationship between size and abundance in many marine fish communities (Jennings and Kaiser 1998), as well as in life-history parameters and population characteristics such as average body size (Tittensor et al., this volume). These relationships may be more strongly altered in the deep sea (>1,000 m), an area virtually unknown, where new fishing technologies can have devastating consequences (Roberts 2002). Hence, future studies of natural size-distribution patterns among marine fishes would only be possible for unexploited communities in extremely large protected areas that have been established for a sufficiently long time that the ecosystem has recovered to an approximation of its original state (Marquet, Navarrete, and Castilla 1990; Jackson 2001), However, in some cases, highly disturbed marine communities may become irretrievably altered by settling into new stable states that are very different from the predisturbance condition (Knowlton 2004). Recently, fish body-size spectra and predator/prey body-mass ratios have been used to predict the original (pre-fishing) fish abundance and size-structure in the intensively fished North Sea. The results of that work suggest that the long-term depletion of large fishes through exploitation exceeds the level of depletion indicated by many short-term studies (Jennings and Blanchard 2004).

Conclusions

The availability of data on macroecological trends and processes in marine fishes is much as it is in other groups of marine organisms, with many large gaps (McClain et al., this volume; Santelices et al., this volume). Although the latitudinal and longitudinal distributions of species richness of fishes and invertebrates are largely concordant (e.g., Smith, Carpenter, and Waller 2002), species richness of some important nonfish groups, such as marine macroalgae, shows quite different latitudinal trends to that among marine fishes (Kerswell 2006; Santelices et al., this volume). Even so, it is likely that some mechanisms have similar influences on richness and range-size patterns among macroalgae, corals, and reef fishes. For example, distributions of diversity peaks in all three groups are consistent with predictions of accumulation patterns brought about by dispersal on large-scale ocean-current systems (Connolly, Bellwood, and Hughes 2003; Kerswell 2006). These studies provide new perspectives of drivers of observed diversity patterns. Further, rather than simply explaining declines in richness from the center by invoking processes that enhance richness in the center, Connolly, Bellwood, and Hughes (2003) examined those declines from the perspective of how environmental factors produce disproportionate changes in the centers and endpoints of species distributions. They concluded that the distributions (directions and positions) of major ocean currents within both the Indian and Pacific basins explained much of both longitudinal and latitudinal patterns in corals and reef fishes. However, while mechanisms such as this are relevant to explaining patterns within ocean basins, for between-basin differences we must examine quite different processes. At that scale, major differences in historical processes become of primary importance (Briggs 1995; Rex et al. 2005). The global center of tropical marine diversity in the IAA at the western edge of the Pacific has been much more stable, and thus amenable to tropical organisms through evolutionary time than has the distinctly less diverse tropical west Atlantic, which was strongly influenced by repeated ice ages and is known to have suffered major extinction events of benthic taxa during the Pleistocene (Budd, Johnson, and Semann 1994).

The existence of similar spatial trends in diversity across a range of taxonomically distant groups of organisms that have very different larval and adult dispersal capabilities, that inhabit depth zones from the coastal zone down to the abyssal zone, and that include both pelagic and benthic forms, indicates that there are some general causal mechanisms that contribute to global macroecological gradients and patterns. That said, unfortunately, sufficient complexity has emerged from studies of macroecological patterns among marine fishes, other marine taxa, and terrestrial organisms over the last decade that biogeographers recognize that they are a long way from being able to claim we actually understand those causal mechanisms (Gaston and Blackburn 2000).

REFERENCES

Ackerman, J. L., D. R. Bellwood, and J. H. Brown. 2004. The contribution of small individuals to density-body size relationships: Examination of energetic equivalence in reef fishes. *Oecologia* 139:568–71.

Agapow, P. M., O. R. P. Bininda-Edmonds, K. A. Crandall, J. L. Gittleman, G. M. Mace, J. C. Marshall, and A. Purvis. 2004. The impact of species concept on biodiversity studies. *Quarterly Review of Biology* 79:161–79.

Allen, A. P., J. H. Brown, and J. F. Gillooly. 2002. Global biodiversity, biochemical kinetics, and the energetic-equivalence rule. *Science* 297:1545–48.

Allen, G. R. 1975. *Damselfishes of the South Seas.* Neptune City, NJ: TFH Publications.

———. 1979. *Butterfly and angelfishes of the World 2.* MERGUS, Hans A. Baensch, Melle, Germany.

Andriashev, A. P. 1965. A general review of the Antarctic fish fauna. In *Biogeogaphy and ecology in Antarctica, Mongraphy Biology, XV,* ed. P. van Oye and J. van Mieghem, 491–550. The Hague: Junk.

———. 1986. *Review of the snailfish genus* Paraliparis *(Scorpaeniformes: Liparididae) of the Southern Ocean.* Theses Zool. 7. Koenigstein: Koeltz Scientific Books.

Angel, M. V. 1997. Pelagic biodiversity. In *Marine biodiversity: Patterns and processes,* ed. R. F. G. Ormond, J. D. Gage, and M. V. Angel, 35–68. Cambridge: Cambridge University Press.

Backus, R. H., J. E. Craddock, and D. L. Shores. 1969. Mesopelagic fishes and thermal fronts in the western Sargasso Sea. *Marine Biology* 3:87–106.

Beamish, R. J., A. J. Benson, R. M. Sweeting, and C. M. Neville. 2004. Regimes and the history of the major fisheries off Canada's west coast. *Progress in Oceanography* 60:355–85.

Beamish, R. J., K. D. Leask, O. A. Ivanov, A. A. Balanov, A. M. Orlov, and B. Sinclair. 1999. The ecology, distribution, and abundances of midwater fishes of the subarctic Pacific gyres. *Progress in Oceanography* 43:399–442.

Bellwood, D. R., and T. P. Hughes. 2001. Regional-scale assembly rules and biodiversity of coral reefs. *Science* 292:1532–34.

Bellwood, D. R., T. P. Hughes, S. R. Connolly, and J. Tanner. 2005. Environmental and geometric constraints on Indo-Pacific coral reef biodiversity. *Ecology Letters* 8:643–51.

Bernardi, G., D. R. Robertson, K. E. Clifton, and E. Azzuro. 1999. Ecology, biogeography and evolution of the Atlantic parrotfish genus *Sparisoma* inferred from mtDNA sequences. *Molecular Phylogenetics and Evolution* 15:292–300.

Bernardi, G., G. Bucciarelli, D. Costagliola, D. R. Robertson, and J. B. Heiser. 2003. Ecology and evolution of the coral reef fish genus *Thalassoma* (Labridae): 1. Molecular phylogeny and biogeography. *Marine Biology* 144:369–75.

Binet, D. 1997. Climate and pelagic fisheries in the Canary and Guinea currents 1964-1993: The role of trade winds and the southern oscillation. *Oceanologica Acta* 20:177–90.

Bolin, R. L. 1944. A review of the marine cottid fishes of California. *Stanford Ichthyological Bulletin* 3:1–135.

Bowen, B. W., A. L. Bass, A. I. Garcia-Rodriguez, L. A. Rocha, and D. R. Robertson. 2001. Phylogeography of the trumpetfish (*Aulostomus* spp.): A ring species complex on a global scale. *Evolution* 55:1029–39.

Brey, T., M. Klages, C. Dahm, M. Gomy, J. Gutt, S. Hain, M. Stiller, W. E. Arntz, J. W. Wägele, and A. Zimmermann. 1994. Antarctic benthic diversity. *Nature* 368:297.

Briggs, J. C. 1961. The East Pacific barrier and the distribution of marine shore fishes. *Evolution* 15:545–54.

———. 1974. *Marine zoography.* New York: McGraw-Hill.

———. 1995. *Global biogeography.* Amsterdam: Elsevier.

———. 2003. Marine centres of origin as evolutionary engines. *Journal of Biogeography* 30:1–18.

Brooks, A. J., R. J. Schmitt, and S. J. Holbrook. 2002. Declines in regional fish populations: Have species responded similarly to environmental change? *Marine and Freshwater Research* 53:189–98.

Brown, J. H. 1984. On the relationship between abundance and distributions of species. *The American Naturalist* 124:255–79.

Brown, J. H., G. C. Stevens, and D. M. Kaufman. 1996. The geographic range: Size, shape, boundaries and internal structure. *Annual Review of Ecology and Systematics* 27:597–623.

Budd, A. F., K. G. Johnson, and T. A. Stemann. 1994. Plio-Pleistocene extinctions and the origin of the modern Caribbean coral reef fauna. In *Global aspects of coral reefs: Health, hazards and history,* ed. R. N. Gingsburg, 7–13. Miami: University of Miami Publications, Miami.

Carlin, J. L., D. R. Robertson, and B. W. Bowen. 2003 Ancient vicariance and recent dispersal in the tropical Atlantic reef fishes *Epinephelus adscensionis* and *Rypticus saponaceus* (Percoidei: Serranidae). *Marine Biology* 143:1057–69.

Carpenter, K. E., and V. G. Springer. 2005. The center of the center of marine shore fish biodiversity: The Philippine Islands. *Environmental Biology of Fishes* 72:467–480.

Chenoweth, S. F., J. M. Hughes, C. P. Keenan, and S. Lavery. 1998. When oceans meet: A teleost shows secondary intergradation at an Indian-Pacific interface. *Proceedings of the Royal Society of London, B* 265 (1394): 415–20.

Chirichigno, N. F., and J. D. Velez. 1998. Clave para indentificar los peces marinos del Perú. Perú: Instituto del Mar del Perú.

Clarke, A. 1992. Is there a latitudinal diversity cline in the sea? *Trends in Ecology and Evolution* 7:286–87.

Clarke, A., and J. A. Crame. 1997. Diversity, latitude and time: patterns in shallow sea. In *Marine biodiversity: Patterns and processes,* ed. R. F. G. Ormond, J. G. Gage, and M. V. Angel, 122–47. Cambridge: Cambridge University Press.

Clarke, A., and N. M. Johnston. 2003. Antarctic marine benthic diversity. *Oceanography and Marine Biology. Annual Review* 41:47–114.

Coates, A. G., and J. A. Obando. 1996. The geologic evolution of the Central American Isthmus. In *Evolution and environments in tropical America,* ed. J. B. C. Jackson, A. F. Budd, and A. G. Coastes, 21–56. Chicago: University of Chicago Press, Chicago.

Collette, B. B. and K. Rutzler. 1977. Reef fishes over sponge bottoms off the mouth of the Amazon River. Miami: Third International Coral Reef Symposium.

Colwell, R. K., and D. C. Lees. 2000. The mid-domain effect: geometric constraints on the geography of species richness. *Trends in Ecology and Evolution* 15:70–76.

Connolly, S. R. 2005. Process-based models of species distributions and the mid-domain effect. *The American Naturalist* 166:1–11.

Connolly, S. R., D. R. Bellwood, and T. Hughes. 2003. Indo-Pacific biodiversity of coral reefs: Deviations from a mid-domain model. *Ecology* 84:2178–90.

Cyr, H., R. H. Peters, and J. A. Downing. 1997. Population density and community size structure: Comparison of aquatic and terrestrial systems. *Oikos* 80:139–49.

Darwin, C. 1872. *The origin of species by means of natural selection, 6th ed.* New York: Random House.

Eastman, J. T., and A. Clarke. 1998. A comparison of adaptative radiations of Antarctic fish with those of non-Antarctic fish. In *Fishes of Antarctica: a biological overview,* ed. G. di Prisco et al., 3–26. Berlin: Springer-Verlag.

Ebeling, A. W. 1967. Zoogeography of tropical deep-sea animals. *Studies in Tropical Oceanography* 5:593–613.

Ebeling, A. W., and M. A. Hixon. 1991. Tropical and temperate reef fishes: comparisons of community structures. In *The ecology of fishes on coral reefs,* ed. P. F. Sale, 509–63. London: Academic Press.

Ekman, S. 1953. *Zoogeography of the sea.* London: Sidgwick and Jackson.

Fields, P. A., J. B. Graham, R. H. Rosenblatt, and G. N. Somero. 1993. Effects of expected global climate change on marine faunas. *Trends in Ecology and Evolution* 8:361–67.

Figueroa, D. E., J. M. Díaz de Astarloa, and P. Martos. 1998. Mesopelagic fish distribution in the southwest Atlantic in relation to water masses. *Deep-Sea Research* I 45:317–32.

Findley, J. S., and M. T. Findley. 2001. Global, regional, and local pattern in species richness and abundance of butterflyfishes. *Ecological Monographs* 71:69–91.

Floeter, S. R., M. D. Behrens, C. E. L. Ferreira, M. J. Paddack, and M. H. Horn. 2005. Geographical gradients of marine herbivorous fishes: patterns and processes. *Marine Biology* 147:1435–47.

Francour, P., F. C. Boudouresque, J. G. Harmelin, M. Harmelin-Vivien, and J. P. Quingnard. 1994. Are the Mediterranean waters becoming warmer? Information from biological indicators. *Marine Pollution Bulletin* 28:523–26.

Frank, K. T., B. Petrie, and N. L. Schackell. 2007. The ups and downs of trophic control in continental shelf ecosystems. *Trends in Ecology and Evolution* 22:236–42.

Fraser, R. H., and D. J. Currie. 1996. The species richness-energy hypothesis in a system where historical factors are thought to prevail: Coral reefs. *The American Naturalist* 148:138–59.

Gaston, K. J., and T. M. Blackburn. 2000. *Pattern and process in macroecology.* London: Blackwell Science.

Gaston, K. J., T. M. Blackburn, and J. I. Spicer. 1998. Rapoport's rule: Time for an epitaph? *Trends in Ecology and Evolution* 13:70–74.

Genner, M. J., D. W. Sims, V. J. Wearmouth, E. J. Southall, A. J. Southward, P. A. Henderson, and S. J. Hawkins. 2004. Regional climatic warming drives long-term community changes of British marine fish. *Proceedings of the Royal Society of London, B* 271:655–61.

Gibbs, R. J., Jr. 1986. The stomiid fish genus *Eustomias* and the oceanic species concept. Pelagic Biogeography. *UNESCO Technical Papers in Marine Science* 49:98–103.

Gill, A. C. 1999. Subspecies, geographic forms and widespread Indo-Pacific coral-reef fishes species: A call for change in taxonomic practice. In *Proceedings of the 5th Indo-Pacific Fish Conference, Nouméa* ed. B. Séret and J.-Y. Sire, 79–87. Paris: Societé Française d' Ichyologie, 1999.

Gillooly, J. F., E. L. Charnov, G. B. West, V. M. Savage, and J. H. Brown. 2002. Effects of size and temperature on developmental time. *Nature* 417:70–73.

Gon, O., and P. C. Heemstra, eds. 1990 *Fishes of the Southern Ocean.* Grahamstown, South Africa: CTP Book Printers.

Gordon, J. D. M., and J. A. R. Duncan. 1985. The ecology of the deep-sea benthic and benthopelagic fish on the slopes of the Rockall Trough, Northeastern Atlantic. *Progress in Oceanography* 15:37–69.

Gray, J. S. 2001. Marine diversity: The paradigms in patterns of species richness examined. *Scientia Marina* 65, supplement 2:41–56.

Grigg, R. W., and R. Hey. 1992. Paleoceanography of the tropical eastern Pacific Ocean. *Science* 255:172–78.

Haedrich, R. L. 1997. Distribution and population ecology. In *Deep-sea fishes,* ed. D. J. Randall and A. P. Farrell, 79–114. Vol. 16. Fish Physiology series. London: Academic Press.

Haedrich, R. L., and N. R. Merrett. 1988. Summary atlas of deep-living demersal fishes in the North Atlantic. *Journal of Natural History* 22:1325–62.

———. 1990. Little evidence for faunal zonation or communities in deep demersal fish fauna. *Progress in Oceanography* 24:239–50.

Harmelin-Vivien, M. 2002. Energetics and fish diversity on coral reefs. In *Coral reef fishes. Dynamics and diversity in a complex ecosystem,* ed. P. F. Sale, 265–74. Amsterdam: Academic Press.

Hashimoto, J., S. Ohta, K. Fujikura, and T. Miura. 1995. Microdistribution pattern and biogeography of the hydrothermal vent communities of the Minami-Ensei Knoll in the Mid-Okinawa trough, western Pacific. *Deep-sea Research I* 42:577–98.

Hastings, P. A. 2000. Biogeography of the tropical eastern Pacific: Distribution and phylogeny of chaenopsid fishes. *Zoological Journal of the Linnean Society* 128:319–35.

Hillebrand, H. 2004. Strength, slope and variability of marine latitudinal gradients. *Marine Ecology Progress Series* 273:251–67.

Hobson, E. S. 1994. Ecological relations in the evolution of acanthopterygian fishes in warm-temperate communities of the northeastern Pacific. *Environmental Biology of Fishes* 40:49–90.

Hubbs, C. L. 1948. Changes in the fish fauna of western North America correlated with changes in ocean temperature. *Journal of Marine Research* 7:459–82.

Hughes, T. P., D. R. Bellwood, and S. R. Connolly. 2002a. Biodiversity hotspots, centres of endemicity, and the conservation of coral reefs. *Ecology Letters* 5:775–84.

Hughes, T. P., A. H. Baird, E. A. Dinsdale, V. J. Harriott, N. A. Moltschaniwskyj, M. S. Pratchett, J. E. Tanner, and B. L. Willis. 2002b. Detecting regional variation using meta-analysis and large-scale sampling: Latitudinal patterns in recruitment. *Ecology* 83:436–51.

Jackson, J. B. C. 2001. What was natural in the coastal oceans? *Proceedings of the National Academy of Sciences* 98:5411–18.

Jennings, S., and J. L. Blanchard. 2004. Fish abundance with no fishing: Predictions based on macroecological theory. *Journal of Animal Ecology* 73:632–42.

Jennings, S., and M. J. Kaiser. 1998. The effects of fishing on marine ecosystems. *Advances in Marine Biology* 34:203–352.

Jokiel, P., and F. J. Marinelli. 1992. The vortex model of coral reef biogeography. *Journal of Biogeography* 19:449–58.

Joyeux, J. C., S. R. Floeter, C. E. L. Ferreira, and J. L. Gasparini. 2001. Biogeography of tropical reef fishes: The South Atlantic puzzle. *Journal of Biogeography* 28:831–41.

Kerswell, A. P. 2006. Global biodiversity patterns of benthic marine algae. *Ecology* 87:2479–88.

Knowlton, N. 2004. Multiple "stable" states and the conservation of marine ecosystems. *Progress in Oceanography* 60:387–96.

Koslow, J. A., C. M. Bulman, and J. M. Lyle. 1994. The mid-slope demersal fish community off Southeastern Australia. *Deep-Sea Research I* 41:113–41.

Lavery, S., T. Y. Chan, Y. K. Tam, and K. H. Chu. 2004. Phylogenetic relationships and evolutionary history of the shrimp genus *Penaeus* s.l. derived from mitochondrial DNA. *Molecular Phylogenetics and Evolution* 31:39–49.

Lea, R. N., and R. H. Rosenblatt. 2000. Observations on fishes associated with the 1997–98 El Niño off California. *CalCOFI Report* 41:117–29.

Leis, J. M. 1984. Larval fish dispersal and the East Pacific barrier. *Oceanography Tropical* 19:181–92.

Leis, J. M., and M. I. McCormick. 2002. The biology, behavior and ecology of the pelagic, larval stage of coral reef fishes. In *Coral reef fishes. Dynamics and diversity in a complex ecosystem,* ed. P. F. Sale, 171–99. London: Academic Press.

Lessios, H. A., and D. R. Robertson. 2006 Crossing the impassable: Genetic connections in 20 reef fishes across the Eastern Pacific Barrier. *Proceedings of the Royal Society of London, B* 273:2201–8.

Lester, S. E., and B. I. Ruttenberg. 2005. The relationship between pelagic larval duration and range size in tropical reef fishes: A synthetic analysis. *Proceedings of the Royal Society of London B* 272:585–91.

Longhurst, A. 1998. *Ecological geography of the sea.* San Diego: Academic Press.

Longhurst, A., S. Sathyendranath, T. Platt, and C. Caverhill. 1995 An estimate of global primary production in the ocean from satellite radiometer data. *Journal of Plankton Research* 17:1245–71.

Love, M. S., M. Yaklovich, and L. Thorsteinson. 2002. The rockfishes of the Northeast Pacific. Berkeley, CA: University of California Press.

Macpherson, E. 1989. Influence of geographical distribution, body size and diet on population density of benthic fishes off Namibia (South West Africa). *Marine Ecology Progress Series* 50:295–99.

———. 2002. Large-scale species-richness gradients in the Atlantic Ocean. *Proceedings of the Royal Society of London B* 269:1715–20.

———. 2003. Species range size distributions for some marine taxa in the Atlantic Ocean. Effect of latitude and depth. *Biological Journal of the Linnean Society* 80:437–55.

Macpherson, E., and A. Gordoa. 1996. Biomass spectra in benthic fish assemblages in the Benguela system. *Marine Ecology Progress Series* 138:27–32.

Macpherson, E., and N. Raventos. 2006. Relationship between pelagic larval duration and geographic distribution in Mediterranean littoral fishes. *Marine Ecology Progress Series* 327:257–65.

McFarlane, G. A., J. R. King, and R. J. Beamish. 2000. Have there been recent changes in climate? Ask the fish. *Progress in Oceanography* 47:147–69.

Marquet, P. A., S. A. Navarrete, and J. C. Castilla. 1990. Scaling population density to body size in rocky intertidal communities. *Science* 250:1125–27.

Mas-Riera, J., A. Lombarte, A. Gordoa, and E. Macpherson. 1990. Influence of Benguela upwelling on the structure of demersal fish populations off Namibia. *Marine Biology* 104:175–82.

Merret, N. R., and R. L. Haedrich. 1997. *Deep-sea demersal fish and fisheries.* London: Chapman and Hall.

Miller, R. G. 1993. *History and atlas of the fishes of the Antarctic Ocean.* Carson City, Nevada: Foresta Institute.

Miya, M., and M. Nishida. 1997. Speciation in the open ocean. *Nature* 389:803–4.

Mora, C., P. M. Chittaro, P. F. Sale, J. P. Kritzer, and S. A. Ludsin. 2003. Patterns and processes in reef fish diversity. *Nature* 421:933–36.

Mora, C., and A. F. Ospina. 2001. Tolerance to high temperatures and potential impact of sea warming on reef fishes of Gorgona Island (tropical eastern Pacific). *Marine Biology* 139:765–69.

Mora, C., and D. R. Robertson. 2005a. Causes of latitudinal gradients in species richness: A test with the endemic shorefishes of the Tropical Eastern Pacific. *Ecology* 86:1771–82.

———. 2005b. Factors shaping the range-size frequency distribution of the endemic fish fauna of the Tropical Eastern Pacific. *Journal of Biogeography* 32:277–86.

Munday, P. L., and G. P. Jones. 1998. The ecological implications of small body size among coral-reef fishes. *Oceanography and Marine Biology. Annual Review* 36:373–411.

Muss, A., D. R. Robertson, C. A. Stepien, P. Wirtz, and B. W. Bowen. 2001. Phylogeography of *Ophioblennius*: The role of ocean currents and geography in reef fish evolution. *Evolution* 55:561–72.

Nelson, J. S. 2006. *Fishes of the world, 4th ed.* New York: Wiley.

Palumbi, S. R. 1997. Molecular biogeography of the Pacific. *Coral Reefs* 16 (Suppl.): 47–52.

Parin, N. V., A. N. Mironov, and K. N. Nesis. 1997. Biology of the Nazca and Sala y Gomez Submarine Ridges, an outpost of the Indo-West Pacific fauna in the eastern Pacific Ocean: Composition and distribution of the fauna, its communities and history. *Advances in Marine Biology* 32:145–242.

Parker, T., and V. Tunnicliffe. 1994. Dispersal strategies of the biota on an oceanic seamount: Implications for ecology and biogeography. *Biological Bulletin* 187:336–45.

Pearcy, W. G., D. L. Stein, and R. S. Carney. 1982. The deep-sea benthic fish fauna of the northeastern Pacific Ocean on Cascadia and Tufts abyssal plains and adjoining continental slopes. *Biological Oceanography* 1:375–428.

Peters, R. H. 1983. *The ecological implications of body size.* Cambridge: Cambridge University Press.

Pineda, J. 1993. Boundary effects on the vertical ranges of deep-sea benthic species. *Deep-Sea Research I* 40:2179–92.

Planes, S. 2002. Biogeography and larval dispersal inferred from population genetic analysis. In *Coral reef fishes. Dynamics and diversity in a complex ecosystem*, ed. P. F. Sale, 201–20. Amsterdam: Academic Press.

Pyle, R. L. 1999. Patterns of Pacific reef and shore fish biodiversity. In *Marine and coastal biodiversity in the tropical island Pacific region*, Volume 2: Population, development, and conservation priorities, ed. L. G. Eldredge, J. E. Maragos, P. F. Holthus, and H. F. Takeuchi, 157–75. Honolulu: Pacific Science Association.

Randall, J. E. 1998. Zoogeography of shore fishes of the Indo-Pacific region. *Zoological Studies* 37:227–68.

Rex, M. A., J. A. Crame, C. T. Stuart, and A. Clarke. 2005. Large-scale biogeographic patterns in marine molluscs: A confluence of history and productivity? *Ecology* 86:2288–97.

Rex, M. A., C. T. Stuart, and G. Coyne. 2000. Latitudinal gradients of species richness in the deep-sea benthos of the North-Atlantic. *Proceedings of the National Academy of Sciences* 97:4082–85.

Rex, M. A., C. T. Stuart, R. R. Hessler, J. A. Allen, H. L. Sanders, and G. D. F. Wilson. 1993. Global-scale latitudinal patterns of species diversity in the deep-sea benthos. *Nature* 365:636–39.

Richer de Forges, B., J. A. Koslow, and G. C. B. Poore. 2000. Diversity and endemism of the benthic seamount fauna in the southwest Pacific. *Nature* 405:944–47.

Richmond, R. H. 1990. The effects of El Niño/Southern Oscillation on the dispersal of corals and other marine organisms. In *Global ecological consequences of the 1982-83 El Niño Southern Oscillation*, 153–90. Amsterdam: Elsevier.

Roberts, C. M. 2002. Deep impact: The rising toll of fishing in the deep sea. *Trends in Ecology and Evolution* 17:242–45.

Robertson, D. R. 1998. Do coral-reef fish faunas have a distinctive taxonomic structure? *Coral Reefs* 17:179–86.

———. 2001. Population maintenance among tropical reef-fishes: Inferences from the biology of small-island endemics. *Proceedings of the National Academy of Sciences* 98:5668–70.

Robertson, D. R., and G. R. Allen. 2002. Shorefishes of the tropical eastern Pacific: An information system. CD-ROM. Balboa, Panama: Smithsonian Tropical Research Institute.

Robertson, D. R., J. S. Grove, and J. E. McCosker. 2004. Tropical transpacific shorefishes. *Pacific Science* 4:507–65.

Robertson, D. R., F. Karg, R. Moura, B. C. Victor, and G. Bernardi. 2006. Mechanisms of speciation and faunal enrichment in Atlantic parrotfishes. *Molecular Phylogenetics and Evolution* 40:795–807.

Robins, C. R. 1971. Distributional patterns of fishes from coastal and shelf waters of the tropical western Atlantic. *FAO Fisheries Report* 72-2:249–255.

Rocha, L. A. 2003. Patterns of distribution and processes of speciation in Brazilian reef fishes. *Journal of Biogeography* 30:1161–71.

Rocha, L. A., A. Bass, D. R. Robertson, and B. W. Bowen. 2002. Adult habitat preferences, larval dispersal and the comparative phylogeography of three Atlantic surgeonfishes (Teleostei: Acanthuridae). *Molecular Evolution* 11:243–52.

Rocha, L. A., D. R. Robertson, J. Roman, and B. W. Bowen. 2005a. Ecological speciation in tropical reef fishes. *Proceedings of the Royal Society of London B* 272:573–79.

Rocha, L. A., D. R. Robertson, C. R. Rocha, J. L. Van Tassell, M. Craig, and B. W. Bowen. 2005b. Recent colonization of the Atlantic by an Indo-Pacific reef fish. *Molecular Ecology* 14:3921–28.

Rodriguez, J., J. Tintore, J. T. Allen, J. M. Blanco, D. Gomis, A. Reul, J. Ruiz, V. Rodriguez,

F. Echevarria, and F. Jimenez-Gomez. 2001. Mesoscale vertical motion and the size structure of phytoplankton in the ocean. *Nature* 410:360–63.

Rogers, A. D. 1994. The biology of seamounts. *Advances in Marine Biology* 30:305–50.

Rohde, K. 1992. Latitudinal gradients in species diversity: The search for the primary cause. *Oikos* 65:514–27.

———. 1999. Latitudinal gradients in species diversity and Rapoport's rule revisited: A review of recent work and what can parasites teach us about the causes of the gradients? *Ecography* 22:593–613.

Rohde, K., and M. Heap. 1996. Latitudinal ranges of teleost fish in the Atlantic and Indo-Pacific oceans. *The American Naturalist* 147:659–65.

Rohde, K., M. Heap, and D. Heap. 1993. Rapoport's rule does not apply to marine teleosts and cannot explain latitudinal gradients in species richness. *The American Naturalist* 142:1–16.

Rosenblatt, R. H. 1967. The zoogeographic relationships of the marine shore fishes of tropical America. *Studies of Tropical Oceanography* 5:579–92.

Roy, K., D. Jablonski, and J. W. Valentine. 1994. Eastern Pacific molluscan provinces and latitudinal diversity gradient: No evidence for "Rapoport's rule." *Proceedings of the National Academy of Sciences* 91:8871–74.

———. 2000 Dissecting latitudinal diversity gradients: Functional groups and clades of marine bivalves. *Proceedings of the Royal Society of London B* 267:293–99.

Roy, K., D. Jablonski, J. W. Valentine, and G. Rosenberg. 1998 Marine latitudinal diversity gradients: Tests of causal hypotheses. *Proceedings of the National Academy of Sciences* 95:3699–3702.

Ruhl, H. A., and K. L. Smith, Jr. 2004. Shifts in deep-sea community structure linked to climate and food supply. *Science* 305:513–15.

Rutherford, S., S. D'Hondt, and W. Prell. 1999. Environmental controls on the geographic distribution of zooplankton diversity. *Nature* 400:749–53.

Shanks, A. L., and G. L. Eckert. 2005. Population persistence of California current fishes and benthic crustaceans: A marine drift paradox. *Ecological Monographs* 75:505–24.

Sheldon, R. W., A. Prakash, and W. H. Sutcliffe. 1972. The size distribution of particles in the ocean. *Limnology and Oceanography* 17:327–40.

Shulman M. J., and E. Bermingham. 1995. Early-life histories, ocean currents, and the population-genetics of Caribbean reef fishes. *Evolution* 49:897–910.

Smith, K. F., and J. H. Brown. 2002. Patterns of diversity, depth range and body size among pelagic fishes along a gradient of depth. *Global Ecology and Biogeography* 11:313–22.

Smith, M. L., K. E. Carpenter, and R. K. Waller. 2002. An introduction to the oceanography, geology, biogeography, and fisheries of the tropical and subtropical western central Atlantic. In *The living marine resources of the western central Atlantic. Vol 1: Introduction, molluscs, crustaceans, hagfishes, sharks, batoid fishes, and chimaeras,* ed. K. E. Carpenter, 1–23. Rome: FAO.

Springer, V. G. 1982. Pacific plate biogeography, with special reference to shorefishes. *Smithsonian Contribution of Zoology* 465:1–182.

Springer, V. G., and J. T. Williams. 1990. Widely distributed Pacific plate endemics and lowered sea-level. *Bulletin of Marine Science* 47:631–40.

Sprules, W. G., and M. Munawar. 1986. Plankton size spectra in relation to ecosystem productivity, size and perturbation. *Canadian Journal of Fishery and Aquatic Sciences* 43:1789–94.

Stehli, F. G., A. L. McAlester, and C. E. Newell. 1969. Generation and maintenance of gradients of taxonomic diversity. *Science* 164:947–49.

Stenseth, N. C., A. Mysterud, G. Ottersen, J. W. Hurrell, K. S. Chan, and M. Lima. 2002. Ecological effects of climate fluctuations. *Science* 297:1292–96.

Stevens, G. C. 1989. The latitudinal gradient in geographical range: How so many species coexist in the tropics. *The American Naturalist* 133:240–56.

———. 1996. Extending Rapoport's rule to Pacific marine fishes. *Journal of Biogeography* 23:149–54.

Taylor, M. S., and M. E. Hellberg. 2003. Genetic evidence for local retention of pelagic larvae in a Caribbean reef fish. *Science* 299:107–9.

Thatje, S., C. D. Hillenbrand, and R. Larter. 2005. On the origin of Antarctic marine benthic community structure. *Trends in Ecology and Evolution* 20:534–40.

Tunnicliffe, V. 1991. The biology of hydrothermal vents: Ecology and evolution. *Oceanography and Marine Biology Annual Reviews* 29:319–407.

Tunnicliffe, V., and C. M. R. Fowler. 1996. Influence of sea-floor spreading on the global hydrothermal vent fauna. *Nature* 379:531–33.

Van Dover, C. L. 2000. *The ecology of deep-sea hyrothermal vents.* Princeton, NJ: Princeton University Press.

Van Dover, C. L., C. R. German, K. G. Speer, L. M. Parson, and R. C. Vrijenhoek. 2002. Evolution and biogeography of deep-sea vent and seep invertebrates. *Science* 295:1253–57.

Victor, B. C., and G. M. Wellington. 2000. Endemism and the pelagic larval duration of reef fishes in the eastern Pacific Ocean. *Marine Ecology Progress Series* 205:241–48.

Victor, B. C., G. M. Wellington, D. R. Robertson, and B. I. Ruttenberg. 2001. The effect of El Niño-Southern Oscillation event on the distribution of reef-associated labrid fishes in the eastern Pacific Ocean. *Bulletin of Marine Science* 69:279–88.

Walther, G. R., E. Post, P. Convey, A. Menzel, C. Parmesan, T. J. C. Beebee, J. M. Fromentin, O. Hoegh-Guldberg, and F. Bairlein. 2002. Ecological responses to recent climate change. *Nature* 416:389–395.

Willig, M. R., D. M. Kaufman, and R. D. Stevens. 2003. Latitudinal gradients of biodiversity: Pattern, process, scale and synthesis. *Annual Review of Ecology and Systematics* 34:273–309.

Wilson, R. R., and R. S. Kaufmann. 1987. Seamount biota and biogeography. In *Seamounts, islands and atolls,* ed. B. H. Keating, P. Fryer, R. Batiza, and G. W. Boehlert, 319–34. Geophysical Monograph 43. Washington, DC: American Geophysical Union.

Worm, B., H. K. Lotze, and R. A. Myers. 2003. Predator diversity hotspots in the blue ocean. *Proceedings of the National Academy of Sciences* 100:9884–88.

Zapata, F. A., K. J. Gaston, and S. L. Chown. 2005. The mid-domain effect revisited. *The American Naturalist* 166:144–48.

Zapata, F. A., and D. R. Robertson. 2006. How many species of shore fishes are there in the Topical Eastern Pacific? *Journal of Biogeography* 34:38–51.

Zezina, O. N. 1997. Biogeography of the bathyal zone. *Advances in Marine Biology* 32:389–426.

Zhang, C. I., J. B. Lee, S. Kim, and J. Oh. 2000. Climatic regime shifts and their impacts on marine ecosystem and fisheries resources in Korean waters. *Progress in Oceanography* 47:171–90.

CHAPTER SIX
MARINE ALGAL COMMUNITIES

BERNABÉ SANTELICES, JOHN J. BOLTON, AND
ISABEL MENESES

Introduction

Macroecology is understood as the search for major statistical patterns in the types, distributions, abundances, and richness of species, from local to global scales, and the development and testing of underlying theoretical explanations of these patterns (Brown and Maurer 1989; Brown 1995; Lawton 1999). As defined by Lawton (1999), it is a blend of ecology, biogeography, and evolution, leading to interdisciplinary explanations of large-scale patterns.

Although it is now recognized that scientists have been doing macroecology for decades (e.g., Preston 1962; Pianka 1966; McArthur and Wilson 1967), the global macroalgal diversity patterns have not been well studied, with most diversity comparisons being regional and restricted to the Northern Hemisphere. It is true that phycologists have accumulated taxonomic and geographic information on algal species for some time. However, most of these studies have taken the form of descriptions of regional flora, analysis of floras in terms of their biogeographic components, comparisons of floras from different areas, delineation of floristic regions and provinces, and relationships between floras and provinces with the environment (e.g., currents, temperature, or salinity patterns and others). Less often have authors addressed the problem of global diversity (e.g., Pielou 1977, 1978; van den Hoek 1984; Lüning 1990; Silva 1992; Bolton 1994) or of large-scale distribution patterns of some algal attributes (e.g., Gaines and Lubchenco 1982) such as size, morphology, growth forms, or life history strategies.

Macroalgal assemblages are major components of intertidal and subtidal communities, contributing significantly to marine littoral primary production and structuring important and often exclusive habitats for a diverse benthic fauna and nursery grounds for invertebrates and fish. Traditionally the emphasis in intertidal and subtidal communities has focused on processes operating at small scales (e.g., the abiotic environment, Doty 1946; the biotic interactions, Connell 1961) and many of such findings can not be extrapolated freely into macroecological theory. Only recently (e.g., Menge et al. 1997) has attention been placed on larger-scale phenomena, such as the oceanographic processes that drive the delivery of nutrients and propagules into these communities. Following this new approach, ecological analysis is being conducted at a variety of spatial scales, the largest of which approaches those of macroecology.

Macroalgal assemblages have been one of the major test grounds of community ecology, with many ecologically important concepts being developed or tested by experimentalists in these assemblages. It is now clear that numerous biotic and abiotic factors may limit the small-scale macroalgal distribution patterns. Those include, among others, the abiotic extremes (e.g., Lobban and Harrison 1994), grazing (Duffy and Hay 2001), competition (Witman and Dayton 2001), facilitation (Bruno and Bertness 2001) and the relative importance of nutrient enhancement and herbivory (Menge and Branch 2001). It is suspected that these same factors and their interplay may also be having a role at larger scales, determining macroecological patterns, but the experimental evidence is missing. Therefore, in the seaweed, many macroecological patterns remain unexplored and the few already known remain unexplained. In this chapter we attempt to produce a synthesis of macroscale patterns in algal diversity and distribution as a first step to the study of this subject with seaweeds. For this we reviewed eight such patterns, including latitudinal gradients of total species richness and of the three major phylogenetic components (Rhodophyta, Chlorophyta, Phaeophyta, as the (R+C)/P ratio), latitudinal patterns of geographic range (Rapoport's rule) and species richness as function of coastline length. In addition, we examine longitudinal patterns of species richness, the effects of upwelling on diversity and morphology, and latitudinal patterns of morphologies and growth styles. The scarce data available in the literature for each of these patterns was complemented by new data gathered by the authors during the preparation of this chapter.

In order to complement previous work we used a basic data set, which consisted of published numbers of macroalgal species for eighty-five particular sections of coastline (either islands, parts of countries, countries, or larger regions). These were correlated against coastline length and latitude to investigate global patterns of both species diversity and the proportions of red, green, and brown algae in any particular coastline (references indicated in table 6.1). Coastlines that were considered poorly collected were omitted. Some of the available species numbers for regions or islands were not separated into red (R), green (C), or brown (P) in six original references. Therefore the analysis of ratios of species numbers in those groups were done with a slightly smaller (seventy-nine) number of floras. For large-scale patterns of major regions, the species lists from smaller islands were not considered. Thus, we used a smaller data set, with sixty-three localities for those comparisons. For comparative studies of latitudinal species richness, patterns of morphologies and growth forms in these coastlines, a more detailed data set with the marine flora of the temperate western coast of South America and the Western and Eastern coasts of South Africa was used. Additional data for specific comparisons were gathered from selected references, as indicated in each following section.

Latitudinal Patterns of Total Species Richness

Perhaps the most widely cited global-scale pattern in species diversity is that species numbers decrease moving away from the equator in both hemispheres (Pianka 1966). In their respective reviews, Rosenzweig (1995) and Willig, Kaufman, and Stevens (2003) mention that the pattern is found in plants and animals; in vertebrates, aerial and quadrupedal; warm-blooded and cold; in invertebrates (see also Roy and Witman, this volume); in both aquatic and terrestrial environments; and even among fossil Foraminifera dating back to the Cenozoic. However, there are a few examples of major groups of organisms that do not follow this pattern. These exceptions are very well represented by the macroalgae and other aquatic plants (fig. 6.1). As early as 1980, Santelices found that species richness of the seaweed flora of temperate Pacific South America increased between 10°S and 55°S, a pattern later supported by additional floristic studies (fig. 6.1, panel A; see review in Santelices and Meneses 2000). Two years later, Gaines and Lubchenco (1982) found that total algal richness in North and Central America was greatest at midlatitudes on the west coast (fig. 6.1, panel B) and mid- to low-latitudes on the east coast (fig. 6.1, panel C). Later research (Santelices and Marquet 1998, this study) indicated that macroalgae may exhibit a variable pattern of species richness in different areas. The seaweed flora on the

TABLE 6.1 Local and regional floras used in this study

Locality	References
Adriatic Sea	Furnari, Cormaci, and Serio 1999
Aeolian Islands	Cormaci et al.1990
Alaska	Stekoll 1998
Antarctic	Wienke and Clayton 2002
Aruba and Bonaire (Caribbean)	Vroman and Stegenga 1988
Azores	Tittley and Neto 1995
Brazil–Bahia	http://www.ib.usp.br/algamare-br/
Brazil–Ceara	http://www.ib.usp.br/algamare-br/
Brazil–Espirito Santo	http://www.ib.usp.br/algamare-br/
Brazil–Sao Paulo	http://www.ib.usp.br/algamare-br/
British Isles	Parke and Dixon 1976
British Columbia, Washington, Oregon	Scagel et al. 1989
Cape Hatteras to Cape Canaveral, USA	Schneider and Searles 1991
California	Abbott and Hollenberg 1976
Canadian arctic	Lee 1990
Cape Verde Is.	John and Lawson 1997
Cayman Is.	Tittley 1994
Chile 13-33°S	I. Meneses (data set)
Chile 34-42°S	I. Meneses (data set)
Chile 4-12°S	I. Meneses (data set)
Chile 43-56°S	I. Meneses (data set)
Congo-Angola	John and Lawson 1997
Danish West Indies	Vroman and Stegenga 1988
Easter Is.	Santelices and Abbott 1987
England and Wales	Guiry and Hessian 1998
Fiji	South and Skelton 2003
Filicudi Is. (Isole Eolie)	Giaccone et al. 2000
France	Guiry and Hessian 1998
French Polynesia	South and Skelton 2003
Ghana	John and Lawson 1997
India	Mairh et al. 1998
Inhaca Island	Critchley et al. 1997
Ireland	Guiry and Hessian 1998
Isolie dei Cyclopi	Giaccone and Pizzuto 2001
Jamaica	Tittley 1994
Japan	Womersley 1981
Low countries	Tittley and Neto 1995
Macquarie	Ricker 1987
Maine	Tittley and Neto 1995
Malaysia/Indonesia	Womersley 1981
Maritime Provinces	Tittley and Neto 1995
Mauretania-Senegal (subtrop)	John and Lawson 1997
Micronesia	South and Skelton 2003
North Atlantic Warm temperate (Canaries)	John and Lawson 1997
N France/Belgium	Coppejans 1995
Namibia	Lluch 2002
New Caledonia	South and Skelton 2003
New Zealand	Knox 1963
Newfoundland	Tittley and Neto 1995
Norfolk Island	Millar 1999

TABLE 6.1 Continued

Locality	References
Northern Norway	Tittley and Neto 1995
Northern Spain	Tittley and Neto 1995
Norway	Jensen 1998
New South Wales and Lord Howe Is.	South and Skelton 2003
NZ Auckland Is.	Brown 1998
NZ Campbell Is.	Brown 1998
NZ Chatham Is.	Brown 1998
NZ Kermadec	Brown 1998
NZ North Is.	Brown 1998
NZ South Is.	Brown 1998
NZ Stewart Is.	Brown 1998
NZ Three Kings	Brown 1998
Oregon	Hansen 1997
Philippines	Silva et al. 1987
Portugal	Sousa-Pinto 1998
Puglia (Italy)	Cormaci et al. 2001
Salina Is. (Isole Eolie)	Cormaci et al. 1990
Samoan Archipelago	South et al. 2001
Sao Tome	John and Lawson 1997
Scotland	Guiry and Hessian 1998
Senegal	John and Lawson 1997
Shetland	Tittley and Neto 1995
Solomon Is.	Womersley and Bailey 1970
South Africa: Cape Agulhas to Kei Mouth	J. J. Bolton (data set)
South Africa: Kwazulu-Natal	J. J. Bolton (data set)
South Africa: Orange R. to Cape Peninsula	J. J. Bolton (data set)
South Africa (total)	J. J. Bolton (data set)
South Georgia	John et al. 1994
Southern Australia	Womersley 1981
Southern Norway	Tittley and Neto 1995
Southern Spain/Portugal	Tittley and Neto 1995
Tremiti Islands (Adriatic)	Cormaci et al. 2000
Trinidad	Richardson 1975
Gambia-Gabon	John and Lawson 1997
Vietnam	Nang and Dinh 1998
West Baltic	Tittley and Neto 1995

Atlantic coast of Europe shows the classical pattern of species richness increasing to the Equator (fig. 6.1, panel D). Although with a smaller latitudinal range, the floras of the west and east coast of South Africa (fig. 6.1, panels E and F) also show opposite trends, with species number increasing with latitude on the west coast and decreasing with latitude on the east coast. Thus, seaweed floras in different coastlines may increase in species richness to the equator, decrease or peak at mid latitude.

A different approach to this problem was invoked by Bolton (1994). Rather than plotting species numbers along a latitudinal gradient, Bol-

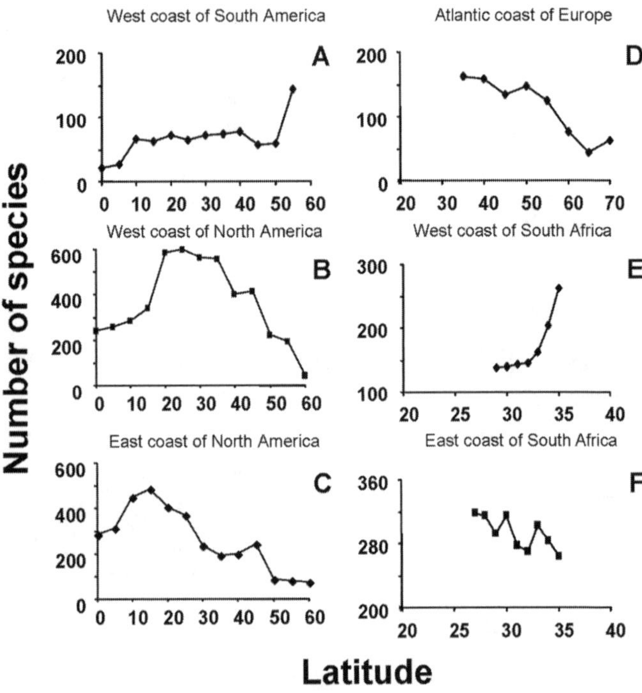

FIGURE 6.1 Latitudinal patterns of macroalgal species richness in six regional floras. Data for the Atlantic coast of Europe are taken from van den Hoeck (1975), those from the West coast of South America from Santelices (1980), and Santelices and Meneses (2000). Data for the east and west coast of South Africa were compiled by J. J. Bolton (*in litteris*), while the values for east and west North America are taken from Gaines and Lubchenco (1982).

ton (1994) studied the total species richness from twenty-nine distinct regions representing diverse marine climates, in order to evaluate if maximal richness occurred at low latitudes. There was no evidence of a peak in species numbers in tropical latitudes, as poor and rich seaweed flora occurred throughout temperate and tropical regions. For this chapter we have added the respective values of thirty-four floras from other regions resulting in a data set involving sixty-three floras (table 6.1). In order to compare the diversity of coastlines of different lengths, the unit of comparisons in this study was the log of the number of species per kilometer. Results indicate that the Antarctic and Arctic floras are considerably species poor (fig. 6.2), as has been noted by previous authors (van den Hoek 1984; Lüning 1990; Bolton 1994). There is, however, no other pattern in the data, with relatively rich and poor floras occurring throughout the globe from 60° N to 60° S. In fact, the statistical value of the regression equation on fig. 6.2 becomes non-

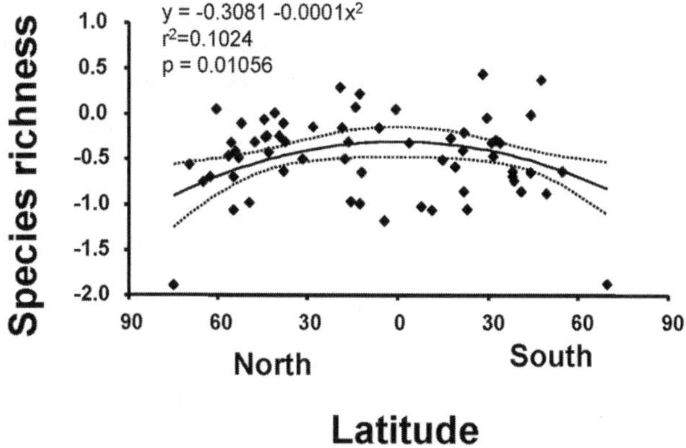

FIGURE 6.2 Macroalgal species richness as a function of latitude for sixty-three published seaweed floras. Species richness is expressed as the log of the number of species per km of coastline. Latitude corresponds to average latitude and the lines in the graph are 95 percent confidence intervals.

significant ($p = 0.6438$; $r^2 = 0.0036$; $\hat{y} = -0.4002 - 0.0002x^2$) if the two points with lowest species richness (Arctic and Antarctic) are not considered. Thus, this evidence additionally supports that, apart from lower species richness at the Poles, seaweed diversity is not related to latitude. Furthermore, several temperate regions have the potential to achieve algal species numbers at least as high as those in the tropics.

There is no clear explanation for the lack of a consistent latitudinal pattern of macroalgal species richness at a global scale. Given the relative abundance of taxonomic studies in temperate and tropical latitudes, it is unlikely that this pattern results from a lack of taxonomic studies in the tropics. In fact, Bolton (1994) estimated that in order to produce a significant peak in species numbers in the tropics, the Philippine flora would have to double and the Caribbean flora triple.

Several other factors have been suggested to explain the atypical patterns of species richness, either at a global or more local scales. Massive increases in herbivory by fish and invertebrates has been invoked (Gaines and Lubchenco 1982) as a general explanation for the lack of significant increases in species richness with decreasing latitudes, and experimental studies tend to support this hypothesis (reviewed in Hay 1997). Herbivores commonly remove almost all seaweed biomass on shallow fore reefs, leaving primarily encrusting corallines, which are resistant to herbivore removal, and small, rapidly growing filamentous form, that tolerate herbivory by rapidly replacing lost tissues. Rates of herbivory on coral reefs is faster than in any other habitat measured, either terrestrial or marine, generated by a high density and diversity or reef herbivores (fish and invertebrates) that may exhibit high metabolic rates, high densities, or both. It is not surprising, then, that the majority of seaweed secondary metabolites appear to be produced by genera that are predominantly tropical or subtropical in distribution, and in the few cases in which species from these genera also occur in temperate seas, the temperate representative appear to contain less defensive compounds than do their tropical counter parts (Hay 1988). However, not all species are able to protect themselves from herbivory, and experimental or natural herbivore removal result in significant increments of algal biomass and diversity, sometimes involving the replacement of corals by benthic algae (see McCook 1999; Díaz-Pulido and McCook 2003, for reviews). Floristic studies also have shown than the "turf" algal communities occurring on tropical shores have an extremely high α diversity (Stuercke and McDermid 2004; Anderson et al. 2005).

At regional levels, scientists have identified specific factors limiting diversity in medium and low latitude areas. Thus, the lack of hard substratum along the expansive sedimentary coasts, poor underwater light due to sediment suspension, and the discharges of large rivers are the accepted explanations for the reduced macroalgal richness along tropical West Africa (Lawson 1978; Lawson and John 1982; John 1986). The tropical western Atlantic, on the other hand, is a relatively narrow tropical region, which may have lost many warm-water species when temperatures declined at the end of the Tertiary and Pleistocene (Lüning 1990). The relative low species diversity on the Chile/Peru coastline has been explained by the presence of the Humboldt Current, which isolates this land mass from migrations from the Central Pacific and the Tropics (Santelices 1980). In addition, Peru and the northern parts of Chile are periodically affected by El Niño, such that the warm water is probably restricting the northward extension of species coming from the south (van den Hoeck 1984). The coastline of Namibia shares a seaweed flora with the west coast of South Africa (Stegenga, Bolton, and Anderson 1997; Lluch 2002). It has been previously documented that any 50 km section of the west coast of South Africa has around half the seaweed species of any 50 km section of the south and east coasts of South Africa (Bolton and Stegenga 2002). This west coast flora is related to a major upwelling region (the Benguela), and it has been hypothesized that major upwelling may reduce species diversity (Bolton 1994). This is not the case with minor coastal upwelling, where short stretches of coastline with upwelled conditions may add extra species to the flora of a general region (see section six, following).

Since the latitudinal patterns of species richness exhibited by the macroalgae in many areas are one of the few exceptions to the broadly accepted latitudinal increase of species to the tropics, this is a most promising area of future research in macroecology of macroalgae.

Latitudinal Patterns of Species Richness in the Three Major Macroalgal Divisions and the (R+C)/P Ratio

Macroalgae form a rather heterogeneous assemblage of primary producers, including at least three algal divisions: the Rhodophyta (red algae), the Chlorophyta (green algae), and the Phaeophyta (the brown algae), each with distinct patterns of distribution around the globe, and evolution in different climates and times. Given these differences, it is important to look at their respective patterns of distribution and to the relative importance of each of these components in the flora from different regions.

As early as 1937, Feldmann demonstrated that the North Atlantic Ocean habitats characterized by warmer water had fewer brown algal species and more red algae than cooler water habitats. Thus, he introduced the ratio of red-to-brown seaweed species in a flora (the R/P ratio) as an indicator of the temperature affinities of that flora. Values of 1.0–2.0 are common in cold temperate regions, rising to values of 3.5–4.3 in tropical regions (Feldmann 1937; Lüning 1990). Forty years later, Cheney (1977) modified Feldmann's R/P ratio, relating the sum of red and green seaweed species numbers to the number of browns ("(R+C)/P"). In the North Atlantic, the latter ratio ranges from figures of <3 in cold water floras to >6 in tropical floras. Bolton (1986) pointed out that this scale does not work in the same manner in southern Africa, where figures in floras ranging from cool temperate to subtropical waters are around 5–7 (Farrell et al. 1993). This discrepancy has also been demonstrated in warm temperate regions of the Arabian Sea (Schils and Coppejans 2003) with upwelling communities in cooler waters having higher ratios. To date, however, there have been no global studies of these ratios, nor of other ratios, such as the relative proportions of green seaweed species compared to browns and reds. In this section we use the floristic composition of seventy-nine regions (those flora in table 6.1 that have readily available figures for species numbers in the three groups) to compare Feldmann's R/P ratio with Cheney's (R+C)/P ratio and with the proportion between green and brown species (C/P ratio). As coastlines of the same length may span very different ranges of latitude, depending on orientation, in these data we decided to plot species diversity against the average latitude of a section of coastline, measured as the midpoint of latitude between the southernmost and northernmost points.

Results indicate a clear similarity between the R/P and the R+C/P ratios (fig. 6.3, panels A and B). Both exhibit low index values at higher latitudes in both hemispheres. The pattern is most defined in the Northern Hemisphere,

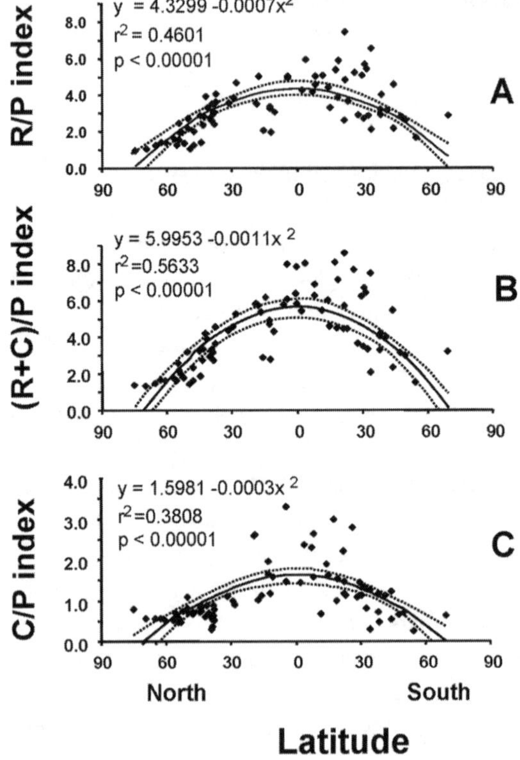

FIGURE 6.3 Latitudinal patterns of species proportion indices of Rhodophyta (R), Phaeophyta (P) and Chlorophyta (C) in seventy-nine local floras from different regions. The respective index has been plotted as a function of mid-point latitude for each area. Lines in the graph are 95 percent confidence intervals.

particularly from latitudes above 30 degrees, with figures of 4–5 around 30 degrees of latitude, dropping to around 2 at 70 degrees of latitude. A few tropical floras have anomalously low figures, but most tropical floras have an R/P of 5 to 6. Data in the Southern Hemisphere is more dispersed, with both indices reaching their highest values (6–8) in latitudes between 10 and 25°S. Thus, both indices provide an indication of tropicality in the respective flora. In fact, the regressions on the R/P and (R+C)/P indices are statistically significant. Researchers, however, should be aware of the data dispersal, and of the possibilities of finding high values in warm-temperate and subtropical floras, particularly in the Southern Hemisphere. Due to its larger r^2 value, the (R+C)/P index should be preferred over the R/P index.

Application of the ratio of green-to-brown seaweed species of the same flora (fig. 6.3, panel C) results in most floras showing slightly higher figures in lower latitudes in both hemispheres, ranging from 1–1.5 at around 30 degrees to around 0.5 above 50 degrees. In tropical floras, however, there is much more variation, with many tropical floras having very high C/P ratios ranging from 1.5–3. Thus, this ratio doesn't seem appropriate to characterize flora.

The global patterns shown by these various ratios can be explained by an analysis of the distribution of the percentages of red, green, and brown seaweed species in the various floras at different latitudes. There are relatively more brown seaweeds species in floras from higher latitudes in both hemispheres (fig. 6.4, panel A). This is the clearest of all the patterns, and it appears to work very similarly in both hemispheres. This is predictive and, apart from very few exceptions, it appears possible to rather accurately estimate the proportion of brown algal species in a seaweed flora from the midpoint of its latitudinal range. The exceptions are the recorded flora of the Antarctic, which has a lower proportion of browns (23 percent) than would be predicted (over 35 percent), and two tropical floras, which have many more browns (33 to37 percent) than expected (10 to 20 percent). The reduced species diversity of brown algae in warmer waters does not apply to all groups of brown algae. The Dictyotales and some members of the Fucales

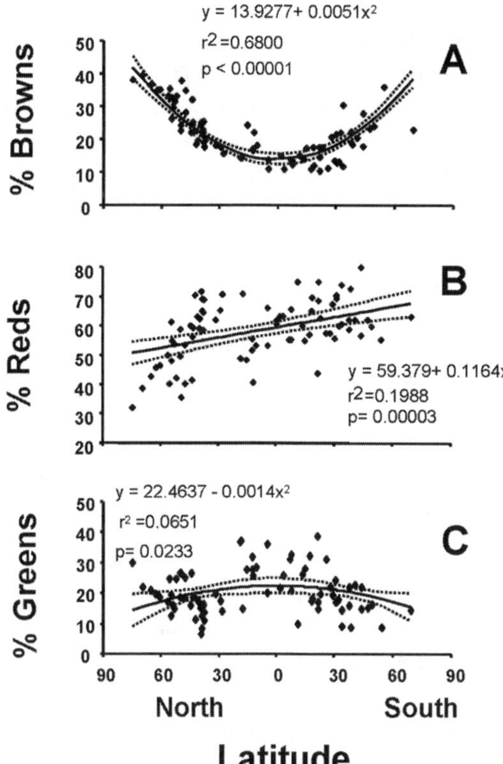

FIGURE 6.4 Relative abundance of brown, red, and green macroalgal species expressed as the percentage of total seaweeds in seventy-nine local floras as a function of latitude. Dotted lines are 95 percent confidence limits.

(e.g., *Sargassum*) tend to be more species rich in the tropics, while Laminariales, other Fucales, and other brown algal groups are more diverse in temperate or cold waters.

The percentages of red algae (fig. 6.4, panel B) reveal an interesting difference between hemispheres. In the Northern Hemisphere the relative abundance of red algae increases from the poles to the tropics, while in the Southern Hemisphere their relative abundance increases from the tropics to the poles. Thus, in temperate regions of the Northern Hemisphere there is a tendency for fewer red algae at higher latitudes, with 60 to 70 percent reds around 40 degrees of latitude, dropping to 30 to 40 percent reds in the Arctic. In the Southern Hemisphere, by contrast, not a single flora has less than 50 percent red algal species, reaching 52 to 60 percent closer to the Antarctic.

The global pattern of the percentage of green algal species (fig. 6.4, panel C) in a seaweed flora is much more variable than that for the other two phyla, although there is an obvious tendency for a pattern opposite to that of the brown algae. Tropical floras tend to have a larger proportion of green species than temperate flora in both hemispheres.

To summarize this section on the relative abundance of species of red, brown, and green algae in the various seaweed flora, it is clear that there are distinct global patterns. Brown seaweeds are predominantly temperate and polar, as shown by the proportion of the seaweed species present. Red seaweeds increase from the Arctic to the tropics to the Subantarctic. The green algae show a great deal of variation, but with a tendency to form a greater proportion of the species in floras of warmer waters at lower latitudes. These data suggest that the evolution and diversification of the brown algae as a whole was in cooler seas. Judging by the greater predominance of green algae in current tropical floras, it is likely that at least the major groups of green algae arose in tropical waters. Red algae are distributed more evenly, although there is a suggestion of a greater diversification of cooler-water red algal species in the Southern rather than the Northern Hemisphere.

As explained previously, the differences in proportions of green, brown, and reds in regional floras originally was associated to temperature differences (Feldman 1937; Cheney 1977). However, the data in figures 6.3 and 6.4 suggest a much more complex situation. Sites of origin and areas of diversification appear to be the overall controlling factors. The opposing distributional patterns of green and brown algae might suggest competitive interactions. Kelps and fucoids well represented in cold and temperate waters are among the taxa most frequently described as successful competitors in intertidal and subtidal habitats (see Witman and Dayton [2001] and Santelices [2004], for reviews) often excluding a variety of smaller green and red algal species. Although algal competitive interactions are recognized to have the potential to locally exclude other species, seldom have they been expanded to explain large-scale patterns of distribution.

Latitudinal Pattern of Geographic Ranges (Rapoport's Rule)

Rapoport's rule (Rapoport 1975, 1982) describes a pattern in which the sizes of species' distributional ranges become smaller toward the tropics. The rule supposes that high-latitude environments have a greater annual range of climatic conditions, thereby favoring the evolution of eurytolerant species with larger geographic ranges than those that evolved in less variable, tropical environments. If low-latitude species exhibit small geographic ranges, then most low-latitude localities will have relatively more species near the edges of their geographic ranges than high-latitude sites. Those edge populations, assumed to be poorly adapted to local conditions, persist thanks to the continual supply of individuals from nearby favorable areas (the "rescue" process), thereby maintaining greater species richness. Since many taxa exhibit both—the latitudinal pattern of species richness and Rapoport's rule—it was suggested that both patterns were linked and probably resulted from the same phenomenon (Rapoport 1982; Stevens 1989, 1992).

A great diversity of taxa from aquatic and terrestrial environments exhibit the pattern described previously (see reviews in Willig, Kaufman, and Stevens [2003] and Roy and Witman, this volume). However, numerous groups of species have failed to show the proposed pattern, and there are suggestions indicating the rule might be an artifact, derived from nonindependent sampling (Rhode, Heap, and Heap 1993) or inadequate consideration of province-scale biogeography (Roy, Jablonski, and Valentine 1994), or that the rule is only applicable under a very restricted set of circumstances.

Since low and high latitudes can be areas of high diversity for macroalgae from different coastlines (see section beginning on p. 155), macroalgae can be used to evaluate if species in high richness areas exhibit smaller latitudinal ranges than species in low richness areas, independent of climate. For this, the latitudinal distribution patterns of four regional floras (Santelices and Marquet 1998 and this study) have been studied. They include the marine flora from Chile, the Atlantic coast of Europe (van den Hoek 1975), and the east and the west coasts of South Africa (Bolton, this study). A direct testing of the relationship between number of species and their mean range at the various latitudes cannot be done, because the two data sets would lack independence. Therefore, we have independently tested (fig. 6.5) number of species and mean range against latitude.

As described earlier, the floras from the east coast of South Africa (fig.

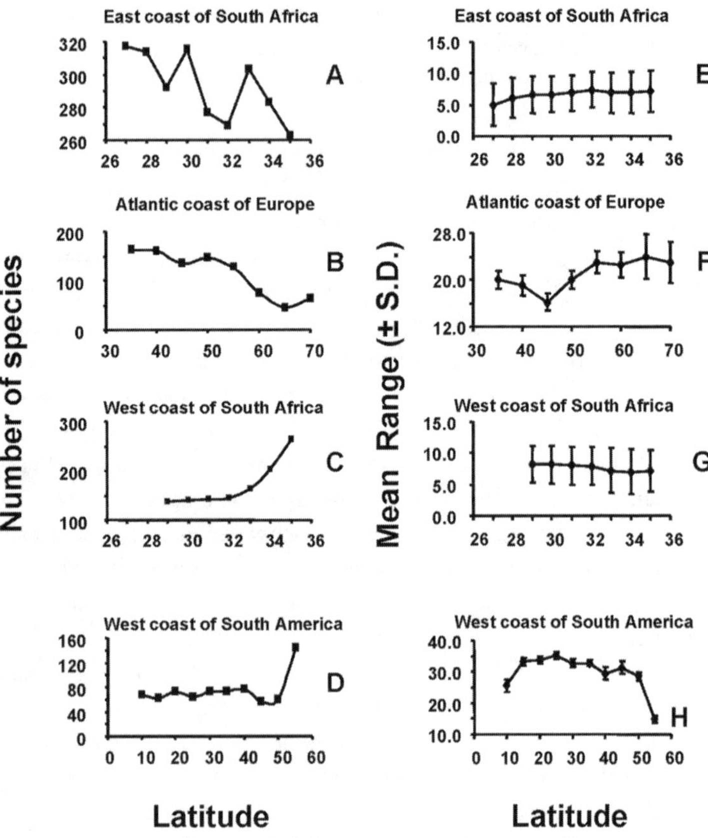

FIGURE 6.5 Species richness and mean geographic range size as a function of latitude in four regional floras. The regression equations are $y = 464.61 - 5.55x$, $r^2 = 0.5385$, $p = 0.024$ for the East coast of South Africa (panel A) and $y = 298.44 - 3.5193x$, $r^2 = 0.8592$, $p = 0.0003$ for the Atlantic coast of Europe (panel B); $y = 426.14 + 18.643x$, $r^2 = 0.7400$, $p = 0.013$ for the West coast of South Africa (panel C) and $y = 66.64 + 1.0 \times 10^{-22} e^x$, $r = 0.93$, $p = 0.00001$ for the West coast of South America (panel D). The corresponding values for mean distributional ranges and latitudes are $r = 0.8071$, $p = 0.009$, for the East coast of South Africa (panel E), $r = 0.7523$, $p = 0.031$, for the Atlantic coast of Europe (panel F), $r = -0.9218$, $p = 0.003$, for the West coast of South Africa (panel G) and $r = -0.4915$, $p = 0.149$ for the West coast of South Africa (panel H).

6.5, panel A) and the Atlantic coast of Europe (fig. 6.5, panel B) exhibit the classical pattern of species richness increasing to the equator. The floras of the west coast of South Africa (fig. 6.5, panel C) and the west coast of South America (fig. 6.5, panel D) exhibit the reverse pattern. In all these cases the correlations, positive or negative, between number of species and latitudes are significant (figs. 6.5, panel A, to 6.5, panel D). On the other hand, the correlations between mean distributional range and latitudes are positive and significant for the East coast of South Africa and the Atlantic coast of Europe, while they are negative for the west coast of South Africa and the west coast of South America. This last correlation, however, is statistically nonsignificant due to the reduced mean range of the flora at 10°S. If that data point is not considered, the respective correlation value becomes significant ($r = 0.7454, p = 0.021$). Since the marine flora of Ecuador was not included in this study, the northernmost distributional range of the marine flora considered at this latitude probably is artificially limited to the geopolitical limits of Peru, yielding a shorter mean range than real for the flora at 10°S.

Thus, the data on the four floras studied suggest opposite tendencies for species richness and mean distributional ranges with latitude, independent of the latitudinal pattern of species richness. These results support the prediction of an inverse relationship between latitudinal species richness and mean latitudinal range in general, and not only in relation to the pole-to-tropic species richness gradient. Furthermore, studies with the macroalgae from temperate Pacific South America (Santelices and Marquet 1998) as well as cursory evaluation of those in fig. 6.5 indicate that the higher diversity areas are not necessarily related to environments with smaller annual range of climatic condition. Therefore, the Rapoport-rescue hypothesis, as applied to macroalgae, may have a predictive value on the relationship between species richness and mean range of distribution, but it is not able to explain the mechanisms producing latitudinal gradients of species richness or the mechanisms determining the inverse relationships between mean distributional range and species richness.

Species Richness and Coastline Length

Seaweeds grow in a narrow strip around islands and continents, limited by substratum availability and light penetration. Relatively few species grow below 50 m depth on most coasts. Thus, it is reasonable to use coastline length as a surrogate for area in the species/area comparison. However, obtaining a measure of coastline length is problematic in that coastlines are fractal in nature and the smaller the scale used to measure, the longer the coastline. If too small a scale is used, then very indented coastlines, particularly those with fjords such as Norway or southern Chile, have enormous coastline lengths compared to countries with generally straight coastlines (e.g., the coastlines of African countries). These coastline indentations appear to add little to coastal diversity: for example, Jensen (1998) commented that despite the fact that Norway spans 13 degrees of latitude and has 10,000 km^2 of substrate suitable for the growth of macroalgae, the main environmental factors and the algal flora are rather homogeneous along the outer coast. Many marine coastal biogeographers use sections of coastline of 50 or 100 km to compare species distributions, as this is considered to include a range of microhabitats. For the purpose of this study, coastline lengths are calculated in 100 km segments along a given coast, measured with dividers using maps from the Times Concise Atlas of the World (1979). The coastline length is expressed in km throughout, although estimated in 100 km segments, as explained previously. Since the relationship between diversity and area is nonlinear (Rosenzweig 1995), comparisons of diversity of coastlines with different lengths are done relating log of species number to log of coastline length. Plots of coastline length against log species number were shown to be more strongly correlated than plots of linear values in a large-scale study of Indian Ocean seaweed floras (Price et al. 2006).

Results suggest (fig. 6.6, panel A) a significant correlation between the log of coastline length and the log of species richness. However, variability is great and there is a rather large number of coastlines with species richness values below the confidence limits of the equation. Explanations for the low species richness in relation to coastline length of thirteen of these floras (within the circle in fig. 6.6, panel A) have been put forward in the literature. These include the flora of the Antarctic and the Arctic, three sub-polar floras (Newfoundland, Alaska, and South Georgia), three sections of tropical West Africa, four points along the Chile-Peru coastline, and the flora of Namibia. An energy argument could be made for the low diversity of polar and sub-polar floras, as has been done by previous authors for various groups of organisms. The low species richness of tropical West African seaweeds, those on the Namibia coast, and those along Chile-Peru have been discussed previously (see section 1). The low numbers can be explained by individual or combinations of abiotic factors acting on the large-scale pattern and limiting macroalgal richness in each region.

If the data from the above thirteen low-richness floras are removed (fig. 6.6, panel B), the correlation between species number and coastline length increases in significance as does the explicative ability of the correlation (r^2). Therefore, the general pattern of greater species richness with increasingly greater areas (Rosenzweig 1995) also applies to macroalgae.

It is interesting to note that the slope of the curve in fig. 6.6, panel B is less than one, suggesting that area increments are not followed in equal proportions by increments in species richness. Bolton (*in litteris*) has found an exponentially decreasing function in the number of species per km of coastline calculated over coastlines of different lengths (fig. 6.6, panel C) Similar nonlinear relationships have also been obtained with terrestrial taxa (Rosenzweig 1995) and with several species-area relationships.

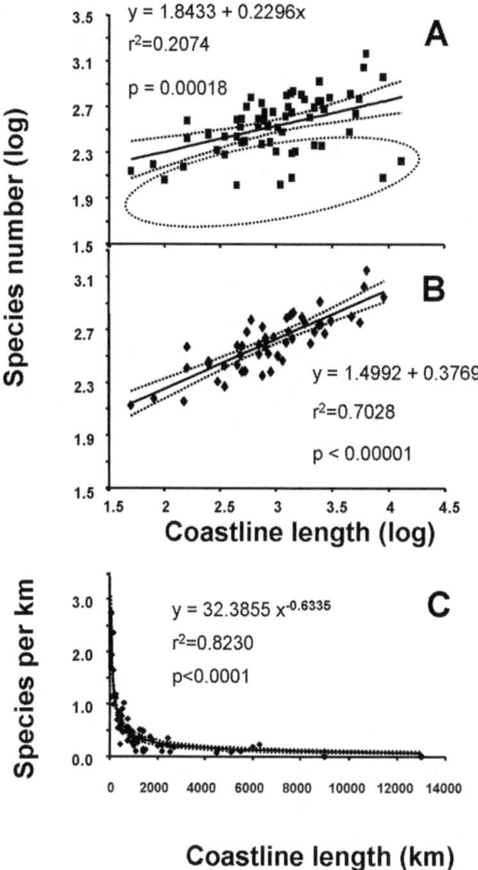

FIGURE 6.6 Relationships between coastline length and species number. Panels A and B indicate log of species number as a function of log of coastline length. Panel A includes (in the dotted circle) and panel B excludes thirteen local floras with low number of species. The total number of regional floras considered in panel A is sixty-three. Panel C shows the number of species per km of coastline as a function of coastline length ($n = 63$).

Longitudinal Patterns of Species Richness

There are a few long coastlines around the world running east-west and with a wide range of longitudes over a small range of latitudes. The marine flora of some of these coastlines have been studied in enough detail to search for macro-patterns in them.

Some of the coastlines, such as the Mediterranean (Lüning 1990) or, on a smaller scale, the South African south coast (Bolton and Stegenga 2002) experience such gradient of water temperature regimes and species turnover along them that comparisons are difficult and results may be misleading. For that reason those two coasts were not considered.

One coastline with a minimum of change in seawater temperature regime over a very long distance is the coast of southern Australia, which allows for investigations of the effect of coastal distance on macroalgal patterns, with minimal overriding temperature effects. Bolton (1996) analyzed Womersley's (1987) species distribution data of brown algae along 1,000 km of the coastline of Victoria, southern Australia. This showed that there was a considerable reduction in brown algal species richness in the easternmost 400 km (around half the species as in the westernmost sections, with an even greater proportional reduction in species of the numerically dominant Fucales). It is likely that this may be caused partly by a lack of available rocky coast in this region, possibly exacerbated by reduced seaweed studies (Bolton 1996).

Experimental studies along the southern Australian coastline are showing that changes in the major canopy-forming species has effects both on species richness patterns and in ecological interactions. The subtidal communities of the western coasts are dominated by the kelp *Ecklonia radiata* (C. Ag.) J. Ag., whereas on the southern Australian coast the communities are dominated by a mixed fucalean community of species of *Sargassum, Cystophora,* and *Platythalia* (Kendrick, Lavery, and Phillips 1999; Kendrick et al. 2004; Goldberg and Kendrick 2004; Collings and Cheshire 1998). The reefs dominated by *Ecklonia* have a considerably lower species richness (<27 spp. m^{-2}) than those with fucalean overstory (43 spp. m^{-2}).

Not only the structure of the canopy-forming dominant may affect the species-richness patterns of the understory. Irving, Connell, and Gillanders (2004) have compared the canopy-benthos associations of *Ecklonia*-dominated kelps in western, southern, and eastern Australia and northern New Zealand. They found significant differences in benthic assemblages among monospecific, mixed, and open stands, indicating that failure to distinguish between superficially similar habitats can lead to over-generalizations about the ecology of kelp forests. Similar findings had been reported earlier by Foster and Vanblaricom (2001) for *Macrocystis* beds from Central California. Comparing the kelp forest structure from different depths, they concluded on the occurrence of two general "types" of kelp forest in the region: one with abundant understory kelps and coralline algae and the other with an understory dominated by sessile invertebrates. The studies in Australia and New Zealand also found (Irving, Connell, and Gillanders 2004) that the benthic patterns in western Australia were more similar to each other but were distinct from eastern Australia and New Zealand, which also differed among them. Thus, interregional differences in species make-up along a longitudinal gradient and the type and morphology of the canopy dominants and their density may produce significant differences in the relative abundance, species composition, and species richness of the understory assemblages.

Diversity and Morphological Effects of Upwelling

The effects of upwelling on macroalgal diversity and distribution have been repeatedly discussed in the phycological literature over the last sixty years. It is now known that upwelling may affect macroalgal species richness, biomass, and productivity, and the relative representation of different algal morphologies. These three effects are discussed in the following.

PATTERNS OF DIVERSITY AND DISTRIBUTION The first emphasis of the studies on upwelling effects on macroalgae was on geographic distribution patterns (Dawson 1945, 1950, 1951), especially of some cold-temperate species extending their presence into warmer-water areas and sometimes exhibiting otherwise unexplained disjunct patterns of distribution. The realization of the occurrence of a high nutrient load in upwelled waters later led to the expectation of high macroalgal stocks and productivity in upwelling areas and to the presence of species with large-sized individuals (e.g., kelps; Dawson, Neushul and Wildman 1961). The habitat modifications induced by the upwelling process and the presence of these large-sized plants in turn led to the expectation of a taxonomically richer flora.

Comparative observations in different upwelling areas around the world now indicate that the net effect of upwelling on macroalgal diversity and species richness depends on the strength and permanence of the upwelling process and on the marine climate of the area affected by upwelling. Localized or diffuse upwelling zones in cold-temperate areas (e.g., the coast of Chile) do not necessarily increase macroalgal diversity because there are no major temperature modifications or habitat diversifications, and the greater nutrient loads of the upwelling areas may be used in growth and production by the species already occurring in the area (Santelices 1991).

Strength and permanence of the upwelling zones are important in determining macroalgal diversity when the upwelling-waters occur in warm-water areas. Strong and permanent upwelling in tropical and subtropical coastlines decreases rather than increases species richness. The best-known example of this is the west coast of South Africa. Another example is the tropical coast of Africa, where the species-poor flora is explained by the cooling effects of strong upwelling combined with a lack of firm substratum, low light penetration, and historical reasons (John and Lawson 1991).

Weak and seasonal upwelling in warmer-water regions, on the other hand, means exposure to a broader range of oceanographic conditions and the possibility to increase the diversity of thermal regimes and macroalgal diversity. Perhaps the best-known example is the flora of the California Channel Islands. There, the coastlines of some islands can be under the influence of the warm waters of the southern California Counter Current, while others are under the influence of the cold California Current, and many islands receive the influence of seasonal coastal upwelling (Neushul, Calrke, and Brown 1967, Murray and Littler 1981). The flora of this area is remarkably diverse and when ordinated according to biogeographic units, it consists in five temperature-determined groups, ranging from warm-water to cold-water species (Murray and Littler 1981). Similar studies on the seaweeds of the Arabian Sea (Schils and Coppejans 2003) found the highest species richness in upwelling-affected communities, with this latter biotope also including species-rich overlap communities.

PRODUCTIVITY AND BIOMASS For many years, phycologists have assumed that the upwelling of deep nutrient-laden waters would account for high algal productivity and biotic richness in the coastal areas washed by upwelled waters (e.g., Dawson 1966). Experimental testing of this idea has been rather recent, infrequent, and the few results gathered so far have not always agreed with the expectations. In a series of studies with intertidal communities, Bosman and Hockey (1986) and Bosman et al. (1986) found that nutrient enrichment of intertidal and nearshore waters by dissolved seabird guano could modify the intertidal community structure on the shores of seabird breeding islands. Those islands exhibited greater algal production and greater biomass of limpets. In order to determine whether the process of nutrient enrichment through upwelling could lead to similar modifications, Bosman, Hockey, and Siegfried (1987) compared midshore community structure in areas with and without upwelling in Chile, southern Africa, and the Canary Islands. As expected, they found that algal cover and the biomass of herbivorous limpets supported per-unit area on rocky shores were significantly greater in regions of coastal upwelling than in regions where upwelling did not occur, suggesting that enhanced algal production was a factor distinguishing the two types of region. Almost ten years later Bustamante et al. (1995) compared productivity and average standing stock of macroalgae in the west and the east coasts of South Africa, expecting to find differences due to the existence of strong upwelling on the west coast and its virtual absence on the east coast. Bustamante et al. (1995) found clear overall differences in productivity between both coasts. However, average macroalgal standing stocks on the west and the east coasts were not significantly different, suggesting that standing stock was a poor indicator of productivity and that growth and dominance of intertidal primary producers at the community level was not only determined by the concentrations of nutrients in the surrounding waters but also by the rates of herbivory. They concluded that at the local scale, biotic interactions may determine the algal landscape and abundance, while at larger spatial scales, over hundreds and thousands of kilometers, nutrient concentration may control primary production.

MORPHOLOGICAL PATTERNS The idea of a preferential representation of some algal morphologies associated to upwelling areas has been in the literature since Dawson's suggestions (1950, 1951) of kelp distribution associated to upwelling areas along the coast of Baja California, USA. Expansion of these ideas to other algal groups is recent and is restricted to a handful of studies done in South Africa and Chile. In their comparison of the west and east coasts of South Africa, Bustamante et al. (1995) found that the stocks of macroalgae found on the west coast localities (with upwelling) was dominated by filamentous, foliose, and corticated forms, while articulated and crustose corallines and noncoralline turfs were much less abundant. By contrast, on the east coast the dominant forms were crustose coralline forms and noncoralline turfs, while filamentous, foliose, and corticated algal forms were much less abundant. The high representation of these latter algal morphologies on the east coast relative to the low importance of foliose and corticated forms is explained (Bustamante et al. 1995) by the lower nutrient concentrations and high fish grazing pressure on the intertidal and shallow subtidal systems of that area. However, the high representation of the corticated and foliose morphologies on the west side could not be explained because, although the nutrient loads on that coastline are high, grazing pressure was also extremely high.

More recently, Broitman et al. (2001) compared the mid-intertidal macroalgal assemblages in zones with and without upwelling along the Chilean coastline, between 20°S and 36°S. In general, the low intertidal of these wave-exposed habitats are dominated by crustose coralline algae and kelps, some cover of corticated forms, and very few ephemerals. In the mid-intertidal, mussels and corticated seaweeds dominate the primary space, followed in abundance by ephemeral algae and crustose forms. Broitman et al. (2001) found that low intertidal areas directly influenced by upwelling exhibited a primary cover of kelps significantly higher than in nonupwelling areas, while in the mid-intertidal the sites with direct influence of upwelling had higher cover of corticated algae, and ephemeral forms showed a significantly lower cover than in sites without upwelling. The abundance of other functional groups was not significantly different between sites with different influences of upwelling. The explanation for these differences seems not only to involve between-site differences in nutrients, but also grazing and competitive interactions. Through experimental manipulation, Nielsen and Navarrete (2004) concluded that at high-upwelling sites, corticated algae grew faster, attained higher biomass, apparently suppressed ephemerals and, at least in their study sites, herbivores played a minor role. At low-upwelling sites, by contrast, the growth of corticated algae was reduced, ephemeral algae attained higher biomass, and herbivores could to some extent control the abundance of ephemerals. The selective removal of key-hole limpets by fishermen along these coasts was suggested (Nielsen and Navarrete 2004) as an explanation for the reduced grazing pressure on corticated algae.

It is as yet unknown if similar kinds of interactions determine the dominant algal morphology in other upwelling regions. If so, the general pattern would be similar to those suggested by Bustamante et al. (1995) in relation to productivity: within the diversity of forms allowed by the high nutrient availability, the dominant algal morphologies would be determined by biotic and abiotic interactions at the local scale.

Latitudinal Patterns of Algal Morphologies

Historically, external morphology has been one of the most important characters allowing the categorization of seaweeds. This character has been used in taxonomic and systematic ordination, grouping of life history cycles (isomorphic versus heteromorphic cycles) and more recently to predict functional responses of a given morphology in a given environment (the "form-function" hypothesis; Littler and Littler 1980).

The "form-function" hypothesis originated from the assumption that the external morphology of macroalgae integrates several algal functions (Neushul 1972) and is, therefore, simultaneously related to several environmental factors. Seaweed-dominated inshore communities are composed of a mixture of species from the three main phyla (red, green, and brown), many of which exhibit strong convergence of form types. Therefore, external morphologies common to phylogenetically different algal groups are understood as convergent adaptations to a given environment, while differences between morphologies would represent divergent responses to a given selection factor. Thus, environment and habitat requirements of species with convergent morphologies would be expected to be more similar among themselves than the requirements of species with different morphologies. Within this framework, the "form-function" hypothesis distinguished seven groups of algal morphologies (Littler and Littler 1980) that would respond differently to photosynthesis and productivity, herbivory, successional stages of the community, and desiccation stress.

Experimental work developed after the original formulation of the "form-function" hypothesis has found consistent results in relation to photosynthesis and productivity. Attempts to relate the model to other environmental parameters, such as herbivory, successional stages of the community, and desiccation stress (Littler and Littler 1980; Littler and Arnold 1982; Steneck and Watling 1982) were either unsuccessful, attained solely for species-specific interactions, or were explained by factors other than seaweed morphology (Padilla 1985; Paul and Hay 1986; Padilla and Allen 2000).

In spite of these results, the "form-function" model proposed by Littler and Littler (1980), and modified by Steneck and Watling (1982) is the most widely used and forms the basis for the macroalgal grouping in this section. The only comparable study known to us is the study of phytogeographic morphological patterns done (Gaines and Lubchenco 1982) with the floras of the east and west coasts of the United States.

In the present study we distinguished the following ten morphological categories (fig. 6.7):

1. Thin sheetlike forms (fig. 6.7, panel A): these include species with laminar fronds that do not have a medulla nor a cortex, but just a few layers of cells or are thicker in a costa or central zone of the frond only. They have a high area/volume ratio. Species included here are *Porphyra* spp., *Ulva* spp. and members of the family Delesseriaceae.
2. Thick sheetlike forms or corticated sheetlike forms (fig. 6.7, panel B): These are all species with ample lamina-type fronds, with a differentiated medulla and cortex. They include a large number of species such as Gigartinales and several Cryptonemiales among the Rhodophyta, and are represented by genera such as *Desmarestia, Dictyota, Padina,* and kelps.
3. Thin filaments (fig. 6.7, panel C): These are either uniseriate filamentous green (*Enteromorpha, Cladophora, Rhizoclonium, Chaetomorpha*, etc.) or red algae (*Acrochaetium, Griffithsia, Ballia, Antithamnionella*, etc.). All Ectocarpoids are included in this category.
4. Corticated filaments (fig. 6.7, panel D): These forms also are filamentous but have differentiation at the cortical level. This category includes species of *Sphacelaria, Halopteris, Bostrychia, Ceramium, Centroceras,* and all corticated Ceramiales.
5. Cylinder-like forms (fig. 6.7, panel E): Includes those seaweeds with a clearly differentiated medulla and cortex and a more or less cylindrical shape. Most *Gracilaria, Gymnogongrus, Gelidium, Ahnfeltia* species are Rodophytes that belong to this category, *Scytosiphon, Myrogloia, Chordaria* are some of the brown algae included in this category. Chlorophytes

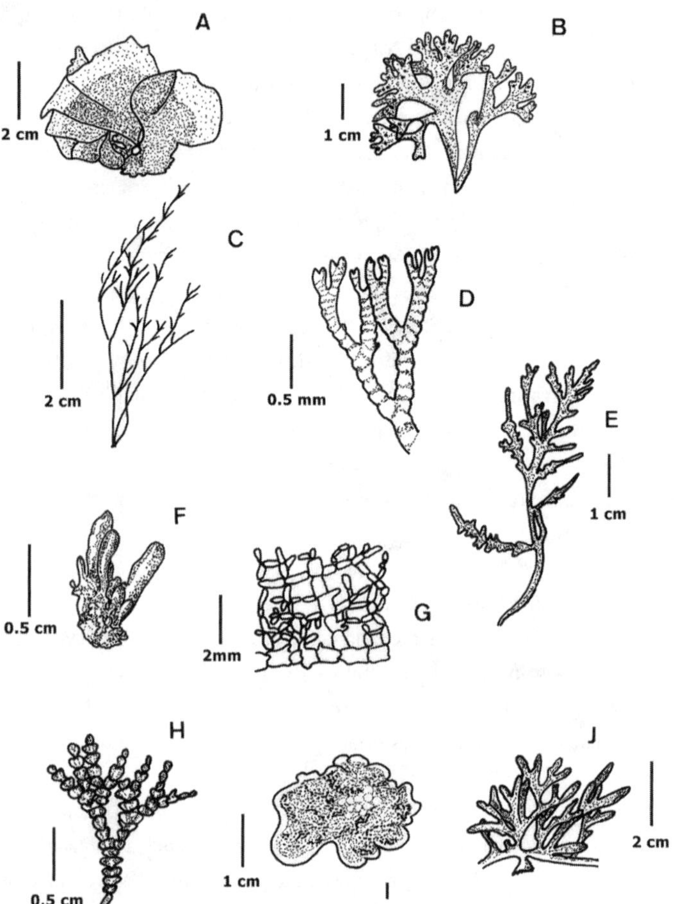

FIGURE 6.7 Morphological types used in this study to explore distributional patterns. (A) Thin sheetlike forms, (B) thick sheetlike forms or corticated sheetlike forms, (C) thin filaments, (D) corticated filaments, (E) cylinder-like forms, (F) saccate or cushionlike forms, (G) netlike forms, (H) jointed calcareous forms, (I) crustose forms, (J) coenocytic upright forms.

included in this category are represented by some species of *Caulerpa* and *Codium*.

6. Saccate or cushionlike forms (fig. 6.7, panel F): Prostrate thick forms either flat or globose (e.g., *Codium, Colpomenia, Dyctiosphaeria*). In figures 6.8 and 6.9 these are indicated as "mats."
7. Netlike forms (fig. 6.7, panel G): structurally thin filaments, that rather than growing freely, they grow forming a net-looking form. Green algae such as *Boodlea* or *Microdyction* are included here.

8. Jointed calcareous forms (fig. 6.7, panel H): Mostly geniculate Corallinaceae and calcified greens (*Halimeda* spp.) are included here.
9. Crustose (fig. 6.7. panel I): Calcified and noncalcified species of crustose shape belong to this category such as: crustose Corallinaceae, *Petrocelis, Hildenbrandia, Ralfsia.*
10. Coenocytic upright forms (fig. 6.7, panel J): Those forms have internal coenocytic structure, like most species of *Codium* and *Caulerpa,* and grow as an upright thallus with one or several main axes and branches. These are somehow related to some of the species belonging to the saccate, cushion-like morphologies but differ externally in not being prostrate.

Once the species were grouped into these categories, the latitudinal distribution of each group was plotted by adding all species of each category for each degree of latitude. Best-fit curve adjustments were performed using SSPS Regression Models of SSPS Base 10.0 (SSPS Applications).

In the case of the flora from temperate Pacific South America we deleted those species that have been recorded in a single locality (the one-degree forms; Santelices and Meneses 2000; Meneses and Santelices 2000) to gain major taxonomic certainty working with repeatedly collected taxa.

In this flora, only five of the morphological groups distinguished had enough representation in the areas (fig. 6.8, panel A). These included filaments (thin and thick), sheetlike forms (thin and thick), crustose forms, cylinders, and mats (saccate and cushionlike forms). Filamentous and sheetlike forms are the most abundant in number of species, while mats and calcified forms are the less abundant and exhibit scarce latitudinal variation in richness. Crustose species, on the other hand, do increase moving southward. However, many of the species recorded for the southern tip of South America are calcareous crusts and were described during the nineteenth century. Given the morphological variability shown within species in this group and since the species have not been revised recently, the number of crustose species in fig. 6.8, panel A is probably overestimated.

The remaining groups exhibit rather well-defined latitudinal patterns of distribution. Thin and thick filaments have contrasting patterns of distribution (fig. 6.8, panel B), a situation that also occurs with thin and thick sheetlike forms (fig. 6.8, panel C). Thin filamentous forms are poorly represented in the north and reach their maximum specific richness in the extreme south (forty-five species). On the other hand, thick filaments are present at the northernmost latitudes, have their maximum abundance between 20 and 35°S and then decrease drastically southward (no species recorded at 56°S).

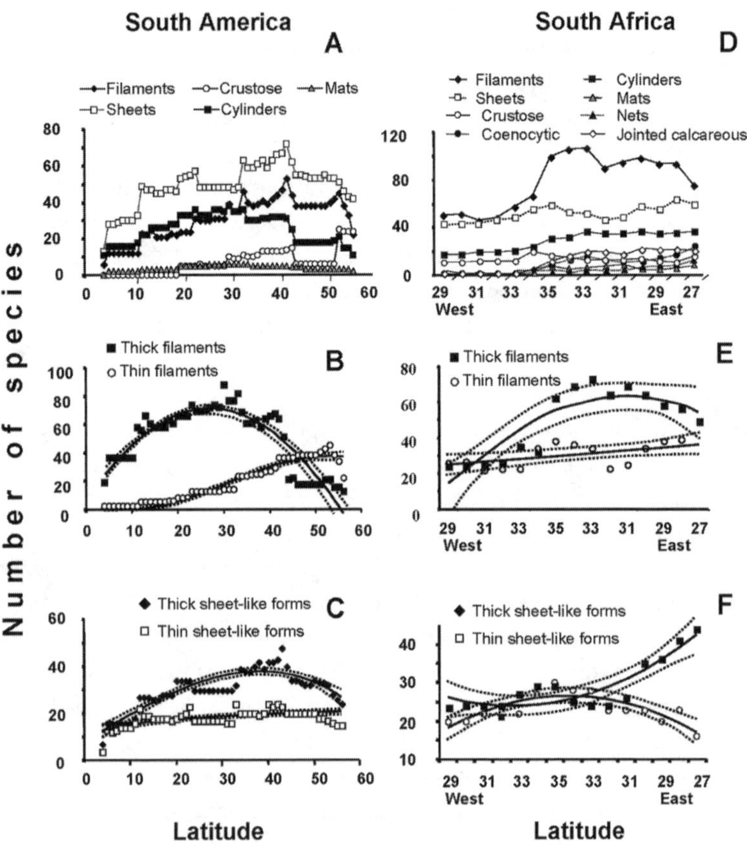

FIGURE 6.8 Number of species of different thallus morphologies in temperate Pacific South America and on the West and East coasts of South Africa. Panels A and D show the general abundance of different algal types in both places. Panels B and E indicate the species richness of thin and thick filamentous forms, and panels C and F show the species richness of thin and thick sheetlike forms. The correlation coefficients of scatter plot data are $r^2 = 0.94$; $p = 0.0001$ for thin and $r^2 = 0.82$; $p = 0.001$ for thick filaments in panel B; $r^2 = 0.38$ and $p = 0.001$ for thin and $r^2 = 0.80$, $p = 9.0001$ for thick sheetlike forms in panel C; $r^2 = 0.34$, $p = 0.022$ for thin and $r^2 = 0.38$ and $p = 0.00015$ for thick filaments in panel E and $r^2 = 0.81$, $p = 0.000005$ for thin and $r^2 = 0.68$ and $p = 0.001$ for thick sheet like forms in panel F.

Similar to the thin filamentous forms, thin sheetlike forms are also poorly represented in northern South America, reaching a maximum of approximately twenty species between 20 and 30°S. These thallus forms include species of *Ulva* and *Porphyra,* with Delesseriaceae being added to the south of Chile. Thick corticated sheetlike forms do show a different distribution: they are always more abundant than thin sheetlike forms, and they increase in number of species to 40°S and then decrease slightly further south, but not as drastically as thick filaments do.

In the case of the South African flora (fig. 6.8, panel D), eight of the ten categories of seaweed morphologies distinguished previously occurs. Allowing for minor differences in the curves describing the respective specific richness, seven of them share a general pattern of distribution. All these groups, except net-forming species, increase in species richness in warm waters; the increment is most evident in coenocytic and cylinder-like forms. The presence of net-forming algae, saccate forms (mats), and the abundance of coenocytic and crustose forms constitute significant differences between this flora and that of South America. However, both have in common high abundances of filaments and sheetlike morphologies.

When the filamentous species are separated into thick and thin filaments, two different situations emerge (fig. 6.8, panel E). Thin filamentous forms exhibit an overall pattern of increase in species richness from the northern west coast and then northward along the east coast. Thick and corticated filaments increase in abundance toward the southern tip of South Africa, then slightly decreasing along the warm waters of the east coast.

The respective situation of thin and thick sheetlike forms is quite different (fig. 6.8, panel F). As with the thick filaments, thin sheetlike forms are most abundant in the south of South Africa, decreasing in abundance along both coastlines, while the number of thick, corticated sheetlike forms is very low in the west and south coasts of South Africa, increasing exponentially along the warmer eastern coast.

It is difficult to explain these latitudinal patterns in morphology. Looking at the patterns in South America, it could be argued that thin thalli (thin filaments and thin sheetlike forms) have a higher surface-to-volume ratio than thick-corticated thalli. Therefore, thin filaments and thick sheetlike forms are more sensitive to desiccation than their thick counterparts. Along South America, the increasingly higher levels of desiccation toward the north would comparatively affect thin morphologies more than thick morphologies, a pattern coherent with their reduced representations in lower latitudes. Along these lines, there would be a reduction in desiccation toward the south, coupled with the higher productivity of finely dissected forms in those low temperature waters, and enough sheltered habitats could explain the higher number of thin filaments and sheetlike forms there. However, when applied to the flora of South Africa, the explanation would hold for thin sheetlike forms only, which increase in the cold waters of the south west and decreases along the warm east coast. But in South Africa, thick filaments (and not the thin ones) are the kind increasing southward along the west coast and slightly decreasing along the east coast. The patterns exhibited by the thin filamentous forms and the thick sheetlike forms along the west coast of South Africa are completely different to those exhibited by species with similar morphologies along the west coast of South America.

As commented previously, it seems that only Gaines and Lubchenco (1982) have previously attempted to trace biogeographic patterns of algal morphology. Although they found some consistent latitudinal patterns with the macroalgae from the east and west coasts of the United States, the explanations for the patterns were much more elusive. They recognized that the patterns may result from interactions among factors, which, also, may exhibit latitudinal changes. It seems that much more work is needed in this area. As discussed earlier, the morphological groups distinguished so far (Litter and Littler 1980; Steneck and Watling 1982; Padilla and Allen 2000) have the capacity to predict intermorphological differences in productivity, but not with other selective factors. Until we understand the selective advantages of different morphologies with respect to major interactions between factors and factor compensations, it will be most difficult to explain the biogeographic patterns of morphological distribution.

Latitudinal Patterns of Growth Forms

A recent review (Santelices 2004) distinguished three categories of seaweeds according to their growth types, regeneration capabilities, and genetic make-up. The three groups were clonal, aclonal, and coalescing seaweeds. The three groups exhibit different ecological responses and alternative strategies for colonizing a given environment. Since variations in selective regimes are expected along latitudinal gradients, it is attractive to look at the latitudinal patterns of distribution of these different growth types.

There seem to be some typical morphologies and sizes related to each growth type, although they are not exclusive. Aclonal species tend to have a main axis (of variable shape: frondose, cylindrical, plumose, etc.) with primary branches attached to it. The main axis is usually attached to the substrate at a single point. It should be noted that larger sizes are more common among aclonal species. Clonal species typically have a prostrate, creeping axis attached to the substrate by several points and a short, branched, upright mass of axes. Usually, erect and prostrate portions of the plant are able to be cut off and to continue growing as an entirely different plant. The coalescing species are attached to the substrate by a discoid holdfast from which one or several branched or unbranched axes arise. The latter one may be flattened or cylindrical.

In the following analysis we have used the floras of the west coast of

South America and the east and west coasts of South Africa. Clonal species are considered as those able to propagate genetically identical individuals from the parental plant by thallus fragmentation due to mechanical forces that could be abiotic (i.e., wave action) or biotic (i.e., herbivory). Nevertheless, if we consider this as a form of vegetative propagation, further examples should be considered in which special devices of the plant are able to act as propagules. Table 6.2 indicates some of the species that have been reported as having asexual propagation of some kind and includes the means by which propagation occurs. Few of these examples have actually been tested in the field, and more examples could probably be found by digging further

TABLE 6.2 Macroalgal species in which clonal growth has been observed, means of aclonal propagation, and morphology type of the species.

Species Name	Propagation	Morphology type	Reference
Acanthophora	Thallus fragments	Cylinder	Collado-Vides 2002
Antithamnionella sarniensis	Thallus fragments	Thin filament	L'Hardy-Halos 1986
Ascophyllum nodosum	Piece of holdfast with entire branches	Corticated sheet	Lazo and Chapman 1998
Bostrychia	Thallus fragments	Thin filament	Collado-Vides 1997
Caulerpa prolifera	Basal system (stolon)	Coenocytic upright	Meinesz 1979
Chondracanthus chamissoi	Thallus fragments	Corticated sheet	Bulboa and Machiavello 2001
Cladophoropsis membranacea	Thallus fragments	Thin filament	Van der Strate et al. 2002
Corallina officinalis	Thallus fragments	Jointed calcareous	Littler and Kauker 1984
Deucalion levringii	Three-celled multinucleate ovoid propagules	Thin sheet	Huisman and Kraft 1982
Ectocarpoids	Thallus fragments	Thin filament	Santelices 1992
Gelidium sesquipedale	Thallus fragments	Corticated sheet	Seoane-Camba 1969; Santos 1993
Glossophora kunthii	"Ligules"—leaflets upon the blade surface	Corticated sheet	García 1996
Gracilaria chilensis	Thallus fragments	Cylinder	Santelices 2004
Halimeda discoidea	Thallus fragments	Jointed calcareous	Walters and Smith 1994
Monosporus spp	Possibly two-celled monosporangia	Thin filament	Baldock 1976
Phyllophora californica	Deciduous "leaflets" along the edge and surface of upper segments (may act as vegetative gemmules)	Thin sheet	Dawson 1958
Sargassum natans	Thallus fragments	Corticated sheet	Børgesen 1924
Sphacelaria californica	Modified short branches	Corticated filament	Setchell and Gardner 1925

into the literature. We did not consider in the list species such as *Enteromorpha*, *Delesseria*, and *Ulva*, which have been shown to be able to regrow from fragments under laboratory conditions, but actually have low probabilities of becoming attached in the field.

Coalescing species may be clonal or aclonal; however, evidence for the number of coalescing species is lacking because the ecological importance of coalescence has only recently been realized (Santelices et al. 1999; Santelices 2004). Therefore, species of each flora were separated into clonal and aclonal growth forms only, and their relative importance calculated as the percentage of the total numbers of species.

The relative abundance of clonal and aclonal forms along temperate Pacific South America changes between 25–30°S (fig. 6.9, panel A). To the north of this region, there is a difference in the relative abundance of both algal types, with aclonal species being more abundant than clonal forms (60% vs. 40%). This numerical difference gradually decreases to the south. Beyond 30°S, the relative representation of both growth forms is relatively similar.

The dominance of aclonal over clonal forms is more marked along the west coast of South Africa (80 percent versus 20 percent; fig. 6.9, panel B), with just a slight decrease in the southernmost tip of South Africa. The increase of clonal species occurs mainly along the eastern coast of South Africa, reaching a value close to 40 percent; the latitudinal change, however, is not statistically significant for the clonal or the aclonal species (fig. 6.9, panel B).

Among the clonal forms, red algae form the dominant group both in South America and South Africa (figs. 6.9, panel C and 6.9, panel D). In South America and along the west coast of South Africa the relative representation of red clones decreases to the south. However, while the decrease in relative abundance along South America is steep (95 percent to 40 percent), the reduction in South Africa is very gradual (80 percent to 70 percent at 34°S). In both places, the decreasing abundance of red clonal forms is replaced by increments of the abundance of brown algae, which are more evident at the southern tip of South America. Along the east coast of South Africa, the relative abundance of clonal red algae continues to decrease, being replaced by both brown and green clonal species.

The relative abundance of aclonal forms differs markedly between South Africa and South America (figs. 6.9, panel E and 6.9, panel F). Most aclonal forms along the west coast of temperate South America are green algae, whose relative abundance decreases steadily toward the south. By contrast, most aclonal species in South Africa are red algae, for which the relative

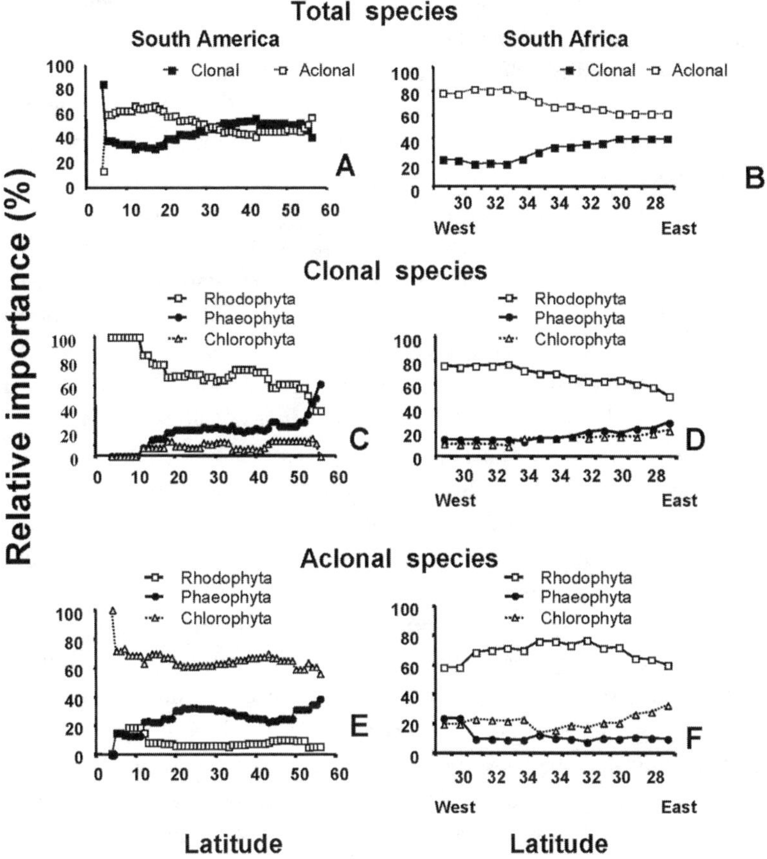

FIGURE 6.9 Relative importance of clonal and aclonal growth forms in the floras from temperate Pacific South America and the East and West coasts of South Africa (panels A and B). Data in lower figures illustrate the relative importance of green, brown, and red algal species among the clonal (panels C and D) and aclonal (panels E and F) species. The correlation coefficient of scatter plot data are $r^2 = 0.52$, $p < 0.001$ for clonal and $r^2 = 0.51$, $p < 0.001$ for aclonal species in South America (panel A); $r^2 = 0.39$, $p = 0.14$ for clonal and $r^2 = 0.39$, $p = 0.14$ for aclonal species in South Africa (panel B); $r^2 = 0.85$, $p < 0.001$ for clonal Rhodophyta, $r^2 = 0.83$, $p < 0.001$ for clonal brown algae and $r^2 = 0.60$, $p < 0.001$ for clonal green algae in South America (panel C); $r^2 = 0.79$, $p < 0.001$ for clonal red algae, $r^2 = 0.28$, $p = 0.31$ for clonal brown and $r^2 = 0.76$, $p = 0.001$ for clonal green species in South Africa (panel D). The respective values for aclonal species are $r^2 = 0.38$, $p = 0.005$ for red, $r^2 = 0.62$, $p < 0.0001$ for brown, and $r^2 = 0.51$, $p < 0.001$ for green seaweeds in South America (panel E) and $r^2 = 0.52$, $p = 0.045$ for reds, $r^2 = 0.65$, $p = 0.008$ for browns, and $r^2 = 0.31$, $p = 0.26$ for green in South Africa.

abundance remains more or less constant along the west coast and slightly declines northwardly along the east coast. The replacement of the declining representation of aclonal green species in South America is by brown algae, while in South Africa it is the green aclonal forms the ones becoming comparatively more abundant whenever the abundance of the red algae decreases.

The preceding results suggest a few generalizations that could be advanced as working hypotheses for future work. The most striking pattern found with growth forms is the gradual increase in clonal representation north along the warmer eastern coast of South Africa. This is consistent with the subtropical type of habitat that occurs there and the network-type of interactions that allow the coexistence and growth of a large number of clonal species in that kind of habitat (Santelices 2004). Dominance of aclonal forms in cold-temperate waters was also expected, since in those environments kelps, saccates, sheetlike forms, and other types of aclonal seaweeds are abundant.

Besides climate, phylogenetic components also seem important in the relationships illustrated in figure 6.9. On the east and west coasts of South Africa, red algae are numerically dominant among both clonal and aclonal species. The increase in clonal forms along the east coast is due primarily to increases in the representation of green and brown clones. By contrast, green aclonal species replace the red aclonal species along the east coast. Along temperate Pacific South America, both red and green algae have radiated, but the red as clonal forms and the green as aclonal forms. The declining southward representation of these two forms is replaced by brown algae that, from the cool south, are radiating both as clonal and aclonal forms. Additional data is needed to evaluate the influences of climate and phylognetic constraints on the distribution and abundance of clonal versus aclonal forms on other coasts.

Conclusions

The evidence here discussed allows for a first characterization of macroecological patterns in macroalgae. There is now little doubt that macroalgal species richness does not always increase toward the tropics, as is the case for most kinds of macroscopic organisms in terrestrial and marine habitats. Brown algal species richness increases and green algal species richness decreases toward higher latitudes while red algal species richness increases from the Arctic to the Tropics to the Subantarctic. In all the floras examined, there is a tendency for species to have smaller mean latitudinal ranges in the areas of greater species richness, but this is independent of the latitudinal patterns of species richness. For the macroalgae, using coastline length as a surrogate for area, there is a significant correlation between coastline length and species richness. Both local species richness and the productivity and morphological effects of upwelled waters seem determined by regional factors as well as local interactions. Although latitudinal patterns of macroalgal morphologies can be traced in many areas, the explanations of such patterns remain elusive, as we have little understanding of the selective advantages of the different types of macroalgal morphologies. Latitudinal patterns of growth types, on the other hand, suggest increases in clonal species richness in warmer waters.

These patterns should be taken as working hypotheses rather than well-established generalizations. Some of them (e.g., upwelling effects on productivity and morphologies) are based on just a few studies. Others, using larger data sets with significant dispersal of the data, may suggest a number of exceptions to each of these generalizations. In any case, additional work, some in experimental ecology, some in biogeography and evolution, is needed to more firmly establish the validity of the above general patterns and to search for explanations to these patterns.

Future work should not only be restricted to these ideas. There are many other algal responses and relations that constitute important avenues for future analysis. Global patterns of sexual and asexual reproduction, latitudinal patterns of productivity, abundance and herbivory, macroecology of body size, patterns of speciation and endemisms, dominant habitat-forming species and habitat associations, interactions between algae and invertebrates, large-scale trends in chemical defences are, among others, macroecological patterns awaiting future studies.

ACKNOWLEDGMENTS

We appreciate the important help of Marcelo Bobadilla with data analysis and illustrations, and the comments by Randy Finke and two anonymous reviewers to earlier drafts of this manuscript. This study was supported by FONDAP 1501001-1 Program 7 and grants FONDECYT 1060474 (to Bernabe Santelices), 1030524 (to Isabel Meneses), and by funding from the National Research Foundation and Department of Environment and Tourism (South Africa: to John Bolton).

REFERENCES

Abbott, I. A., and G. J. Hollenberg. 1976. *Marine algae of California*. Stanford, CA: Stanford University Press,

Anderson, R. J., C. McKune, J. J. Bolton, O. De Clerck, and E. Tronchin. 2005. Patterns in subtidal seaweed communities on coral-dominated reefs on the KwaZulu-Natal coast, South Africa. *African Journal of Marine Science* 27:529–37.

Baldock, R. N. 1976. The Griffithsieae group of the Ceramiaceae (Rhodophyta) and its Southern representatives. *Australian Journal of Botany* 24:509–93.

Bolton, J. J. 1986. Marine phytogeography of the Benguela upwelling system on the west coast of southern Africa: A temperature dependent approach. *Botanica Marina* 29:251–56.

———. 1994. Global seaweed diversity: Patterns and anomalies. *Botanica Marina* 36:241–46.

———. 1996. Patterns of species diversity and endemism in comparable temperate brown algal floras. *Hydrobiologia* 326/327:173–78.

Bolton, J. J., and H. Stegenga. 2002. Seaweed biodiversity in South Africa. *South African Journal of Marine Science* 24:9–18.

Bosman, A. L., and P. A. R. Hockey. 1986. Seabird guano as a determinant of rocky intertidal community structure. *Marine Ecology Progress Series* 32:247–57.

Bosman, A. L., J. T. du Toit, P. A. R. Hockey, and G. M. Branch. 1986. A field experiment demonstrating the influence of seabird guano on intertidal primary production. *Estuarine, Coastal and Shelf Science* 23:283–94.

Bosman, A. L., P. A. R. Hockey, and W. R. Siegfried. 1987. The influence of coastal upwelling on the functional structure of rocky intertidal communities. *Oecologia* 72:226–32.

Broitman, B. R., S. A. Navarrete, F. Smich, and S. D. Gaines. 2001. Geographic variation of southeastern Pacific intertidal communities. *Marine Ecology Progress Series* 224:21–34.

Brown, J. H. 1995. *Macroecology.* Chicago: University of Chicago Press.

Brown, J. H., and B. Maurer. 1989. Macroecology: The division of food and space among species and continents. *Science* 243:1145–50.

Brown, M. T. 1998. The seaweed resources of New Zealand. In *Seaweed resources of the world,* ed. A. T. Critchley and M. Ohno, 127–37. Yokosuka, Japan: Japan International Cooperation Agency.

Bruno, J. F., and M. D. Bertness. 2001. Habitat modification and facilitation in benthic marine communities. In *Marine community ecology,* ed. M. D. Bertness, S. D. Gaines, and M. E. Hay, 201–20. Sunderland, MA : Sinauer.

Bulboa, C. R., and J. E. Macchiavello. 2001. The effects of light and temperature on different phases of the life cycle in the carrageenan producing alga *Chodracanthus chamissoi* (Rhodophyta, Gigartinales). *Botanica Marina* 44:371–74.

Bustamante, R. H., G. M. Branch, S. Eekshout, B. Robertson, P. Zoutendyk, and M. Schleyer. 1995. Grandients of intertidal primary productivity around the coasts of South Africa and their relationship with consumer biomass. *Oecologia* 102:189–201.

Cheney, D. F. 1977. R & C/P, a new and improved ratio for comparing seaweed floras. *Journal of Phycology* (Suppl.):12.

Collado-Vides, L. 2002. Clonal architecture in marine macroalgae: Ecological and evolutionary perspectives. *Evolutionary Ecology* 15:531–45.

Collado-Vides, L., G. Gómez, V. Gómez, and G. Lechuga. 1997. Simulation of the clonal growth of *Bostrichya radicans* (Ceramiales, Rhodophyta) using L-Systems. *ByoSystems* 42:19–27.

Collings, G. J., and A. C. Cheshire. 1998. Composition of subtidal macroalgal communities of the lower gulf waters of South Australia, with reference to water movement and geographical separation. *Australian Journal of Botany* 46:657–59.

Connell, J. 1961. The influence of interspecific competition and other factors on the distribution of the barnacle *Chthamallus stellatus. Ecology* 42:710–23.

Coppejans, E. 1995. Flora algologique des côtes du Nord de la France et de la Belgique. *Scripta Botanica Belgica* 9:1–454.

Cormaci, M., G. Furnari, D. Serio, and F. Pizzuto. 1990. Osservazioni sulle fitocenosi bentoniche dell'isola di Salina (Isole Eolie). In *2° Workshop Progellot Strategico. Clima Ambiente e Territorio nel Mezzogiorno,* ed. S. M. Salina (Isole Eolie), 28–30 Maggio 1990. Parchi Naturali e Aree Protette. 339-365.

Cormaci, M., G. Furnari, G. Alongi, M. Catra, and D. Serio. 2000. The benthic algal flora on rocky substrata of the Tremiti Islands (Adriatic Sea). *Plant Biosystems* 134:133–52.

Cormaci, M., G. Furnari, G. Aloni, D. Serio, A. Petrocelli, and E. Cecere. 2001. Censimento delle macroalghe marine bentoniche delle coste pugliesi. *Thallassia Salentina* 25:1–158.

Critchley, A. T., M. E. Aken, S. O. Bandeira, and M. Kalk. 1997. A revised checklist for the seaweeds of Inhaca Island, Mozambique. *South African Journal of Botany* 63:426–36.

Dawson, E. Y. 1945. Marine algae associated with upwelling along the Northwestern coast of Baja California, Mexico. *Bulletin of the Southern California Academy of Sciences* 44:57–71.

———. 1950. A note on the vegetation of a new coastal upwelling area of Baja California. *Sears Foundation for Marine Research, Journal of Marine Research* 9:65–68.

———. 1951. A further study of upwelling and associated vegetation along Pacific Baja California, Mexico. *Sears Foundation for Marine Research, Journal of Marine Research* 10:39–58.

———. 1958. Notes on Pacific Coast marine algae VII. Bulletin of the Southern California Academy of Sciences 57 (2): 65–80.

———. 1966. *Marine botany: An introduction.* New York: Holt, Rinehart, and Winston.

Dawson, E. Y., M. Neushul, and R. D. Wildman. 1961. Seaweeds associated with kelp beds along Southern California and Northwestern Mexico. *Pacific Naturalist* 1 (14): 1–81.

Diaz-Pulido, G., and L. J. McCook. 2003. Relative roles of herbivory and nutrients in the recruitment of coral-reef seaweeds. *Ecology* 84:2026–33.

Doty, M. S. 1946. Critical tide factors that are correlated with the vertical distribution of marine algae and other organisms along the Pacific coast. *Ecology* 27:315–28.

Duffy, J. E., and M. Hay. 2001. The ecology and evolution of marine consumers-prey interactions. In *Marine community ecology,* ed. M. D. Bertness, S. D. Gaines, and M. E. Hay, 131–59. Sunderland, MA: Sinauer.

Farrell, E. G., A. T. Critchley, M. E. Aken, and R. N. Pienaar. 1993. The intertidal algal flora of Isipingo Beach, Natal, South Africa and its phytogeographical affinities. *Helgol. Meeresunters* 47:145–60.

Feldmann, J. 1937. Recherches sur la végétation marine de la Méditerranée. La côte des Albères. *Revue Algologique* 10:1–339.

Foster, M. S., and G. R. Vanblaricom. 2001. Spatial variation in kelp forest communities along the Big Sur coast of central California, U.S.A. *Cryptogamie Algologie* 22:173–86.

Furnari, G., M. Cormaci, and D. Serio. 1999. Catalogue of the benthic marine macroalgae of the Italian coast of the Adriatic Sea. *Bocconea* 12:1–214.

Gaines, S. A., and J. Lubchenco. 1982. A unified approach to marine plant-herbivore interactions. 2. Biogeography. *Annual Review of Ecology and Systematics* 13:111–38.

García, J. C. 1996. Fenología y ciclo de vida de *Glosophora kunthii* (C.A.) J. Agardh de poblaciones intermareales y submareales en Puerto Aldea IV Región. Tesis para optar al título de Biólogo Marino, Universidad Católica del Norte.

Giaccone, G., M. Cormaci, G. Furnari, B. Scammacca, G. Alongi, M. Catra, V. Di Martino, G. Marino, and D. Serio. 2000. Biodiversità vegetale marina dell'arcipelago "Isole Eolie." *Boll. Accademia Gioenia di Scienze Naturali* 32:191–242.

Giaccone, G., and F. Pizzuto. 2001. Stato delle conoscenze sulla biodiversità algale marina delle Isole dei Ciclopi (Catania, Sicilia orientale). *Boll. Accademia Gioenia di Scienze Naturali* 34:5–22.

Goldberg, N. A., and G. A. Kendrick. 2004. Effects of island groups, depth, and exposure to ocean waves on subtidal macroalgal assemblages in the Recherche Arcipelago, Western Australia. *Journal of Phycology* 40:631–41.

Guiry, M. D., and C. C. Hessian. 1998. The seaweed resources of Ireland. In *Seaweed resources of the world,* ed. A. T. Critchley and M. Ohno, 210–16. Yokosuka, Japan: Japan International Cooperation Agency.

Hansen, G. I. 1997. A revised checklist and preliminary assessment of the macrobenthic marine algae and seagrasses of Oregon. In *Conservation and management of native flora and fungi,* ed. T. N. Kaye, A. Liston, R. M. Love, D. L. Luoma, R. J. Meinke, and M. V. Wilson, 175–200. Corvallis, OR: Native Plant Society of Oregon.

Hay, M. E. 1988. Marine plant-herbivore interaction: The ecology of chemical defences. *Annual Review of Ecology and Systematics* 19:111–45.

———. 1997. The ecology and evolution of seaweed-herbivore interaction on coral reefs. *Coral Reefs* 16, Supplement:S67–S76.

Huisman, J. M., and G. T. Kraft. 1982. *Deucalion* gen. nov. and *Anisoschizus* gen. nov. (Ceramiaceae, Ceramiales), two new propagule-forming red algae from Southern Australia. *Journal of Phycology* 18:177–92.

Irving, A. D., S. D. Connell, and B. M. Gillanders. 2004. Local complexity in patterns of canopy-benthos associations produces regional patterns across temperate Australasia. *Marine Biology* 144:361–68.

Jensen, A. 1998. The seaweed resources of Norway. In *Seaweed resources of the world,* ed. A. T. Critchley and M. Ohno, 200–209. Yokosuka, Japan: Japan International Cooperation Agency.

John, D. M. 1986. Coastal vegetation. Littoral and sub-littoral marine vegetation. In *Plant ecology in West Africa,* ed. G. W. Lawson, 195–246. Chichester: Wiley.

John, D. M., and G. W. Lawson. 1991. Littoral ecosystems of tropical Western Africa. In *Intertidal and littoral ecosystems,* ed. A. Mathieson, C. Mathieson, and P. H. Nienhuis, 297–320. Amsterdam: Elsevier.

———. 1997. Seaweed biodiversity in West Africa: A criterion for designating Marine Protected Areas. In *The coastal zone of West Africa: Problems and management,* ed. S. M. Evans, C. J. Vanderpuye, and A. K. Armah, 111–13. Sunderland, UK: Penshaw Press.

John, D. M., I. Tittley, G. W. Lawson, and P. J. A. Pugh. 1994. Distribution of seaweed floras in the Southern Ocean. *Botanica Marina* 36:235–40.

Kendrick, G. A., E. S. Harvey, T. Wernberg, N. Harman, and N. Goldberg. 2004. The role of disturbance in maintaining diversity of benthic macroalgal assemblages in southwestern Australia. *Japanese Journal of Phycology* 52 (suppl.): 5–9.

Kendrick, G. A., P. S. Lavery, and J. A. Phillips. 1999. Influence of *Ecklonia radiata* kelp canopy on structure of macroalgal assemblages in Marmion lagoon, Western Australia. *Hydrobiologia* 398/399:275–83.

Knox, G. A. 1963. The biogeography and ecology of the Australasian coasts. *Marine Biology Annual Review* 1:341–404.

Lawson, G. W. 1978. The distribution of seaweed floras in the tropical and subtropical Atlantic Ocean: A quantitative approach. *Botanical Journal of the Linnean Society, London* 76:177–93.

Lawson, G. W., and D. M. John. 1982. The marine algae and coastal environment of tropical West Africa. *Nova Hedwigia (Beih.)* 70:1–455.

Lawton, J. H. 1996. Patterns in ecology. *Oikos* 75:145–47.

———. 1999. Are there general laws in ecology? *Oikos* 84:177–92.

Lazo, M. L., and A. R. O. Chapman. 1998. Components of crowding in a modular seaweed: Sorting through the contradictions. *Marine Ecology Progress Series* 174:257–67.

Lee, R. K. S. 1990. A catalogue of the marine algae of the Canadian Arctic. *National Museum of Canada Publications in Botany* 9:1–83.

L'Hardy-Halos, M. T. 1986. Observations on two species of *Antithamnionella* from the coast of Brittany. *Botanica Marina* 29:37–42.

Littler, M. M., and K. E. Arnold. 1982. Primary productivity of marine macroalgal functional-form groups from southwestern North America. *Journal of Phycology* 18:307–11.

Littler, M. M., and B. J. Kauker. 1984. Heterotrichy and survival strategies in the red algae *Corallina officinalis. L. Botanica Marina* 27:37–44.

Littler, M. M., and D. S. Littler. 1980. The evolution of thallus form and survival strategies in benthic marine macroalgae: Field and laboratory tests of a functional-form model. *American Naturalist* 116:25–44.

Lluch, J. 2002. Marine benthic algae of Namibia. *Scientia Marina* 66 (Supplement 3): 5–256.

Lobban, C. S., and P. J. Harrison. 1994. *Seaweed ecology and physiology.* New York: Cambridge University Press.

Lüning, K. 1990. Seaweeds: Their environment, biogeography, and ecophysiology. New York: Wiley.

Mairh, O. P., C. R. K. Reddy, and G. Raja Krishna Kumar. 1998. The seaweed resources of India. In *Seaweed resources of the world,* ed. A. T. Critchley and M. Ohno, 110–26. Yokosuka, Japan: Japan International Cooperation Agency.

McArthur, R. H., and E. O. Wilson. 1967. *The theory of island biogeography.* Princeton, NJ: Princeton University Press.

McCook, L. J. 1999. Macroalgae, nutrients, and phase shifts on coral reefs: Scientific issues and management consequences for the Great Barrier Reef. *Coral Reefs* 18:357–67.

Meinesz, A. 1979. Contribution al'étude de *Caulerpa prolifera* (Førsskäl) Lamouroux (Chlorophycée, Caulerpale). I. Morphogénèseet croissance dens une station des côtes continentales Françaises de la Méditerranée. *Botanica Marina* 22:27–39.

Meneses, I., and B. Santelices. 2000. Patterns and breaking points in the distribution of benthic algae along the temperate Pacific coast of South America. *Revista Chilena de Historia Natural* 73:615–23.

Menge, B. A., and G. M. Branch. 2001. Rocky intertidal communities. In *Marine community ecology,* ed. M. D. Bertness, S. D. Gaines, and M. E. Hay, 221–51. Sunderland, MA: Sinauer.

Menge, B. A., B. A. Daley, P. A. Wheeler, E. Dahlhoff, E. Sanford, and P. T. Strub. 1997. Benthic-pelagic links and rocky intertidal communities: Bottom-up effects on top-down control? *Proceedings of the National Academy of Sciences, USA* 94:14530–35.

Millar, A. J. K. 1999. Marine benthic algae of Norfolk Island, South Pacific. *Australian Systematic Botany* 12:479–547.

Murray, S. N., and M. M. Littler. 1981. Biogeographical analysis of intertidal macrophyte floras of southern California. *Journal of Biogeography* 8:339–51.

Nang, H. Q., and N. H. Dinh. 1998. The seaweed resources of Vietnam. In *Seaweed resources of the world,* ed. A. T. Critchley and M. Ohno, 62–69. Yokosuka, Japan: Japan International Cooperation Agency.

Neushul, M. 1972. Functional interpretation of benthic marine algal morphology. In *Contributions to the systematics of benthic marine algae of the Northern Pacific,* ed. I. A. Abbott and M. Kurogi, 47–74. Kobe, Japan: Japanese Society of Phycologists.

Neushul, M., W. D. Clarke, and D. W. Brown. 1967. Subtidal plant and animal communities of the Southern California Islands. In *Proceedings of the symposium on the biology of the*

California Islands, ed. R. N. Philbrick, 37–55. Santa Barbara, CA: Santa Barbara Botanical Garden.

Nielsen, K. J., and S. A. Navarrete. 2004. Mesoscale regulation comes from bottom-up: Intertidal interactions between consumers and upwelling. *Ecology Letters* 7:31–41.

Padilla, D. K. 1985. Structural resistance of algae to herbivores. A biomechanical approach. *Marine Biology* 90:103–9.

Padilla, D. K., and B. J. Allen. 2000. Paradigm lost: Reconsidering functional form and group hypotheses in marine ecology. *Journal of Experimental Marine Biology and Ecology* 250:207–21.

Parke, M., and P. S. Dixon. 1976. Check-list of British marine algae—third revision. *Journal of the Marine Biological Association of the UK* 56:527–94.

Paul, V. J., and M. E. Hay. 1986. Seaweed susceptibility to herbivory: Chemical and morphological correlates. *Marine Ecology Progress Series* 33:255–64.

Pianka, E. R. 1966. Latitudinal gradients in species diversity: A review of concepts. *American Naturalist* 100:33–46.

Pielou, E. C. 1977. The latitudinal spans of seaweed species and their pattern of overlap. *Journal of Biogeography* 4:299–311.

———. 1978. Latitudinal overlap of seaweed species: Evidence for quasi-sympatric speciation. *Journal of Biogeography* 5:227–38.

Preston, F. W. 1962. The canonical distribution of commonness and rarity. *Ecology* 43:185–215.

Price, A. R. G., L. P. A. Vincent, A. J. Venkatachalam, J. J. Bolton, and P. W. Bason. 2006. Concordance between different measures of biodiversity in Indian Ocean macroalgae. *Marine Ecology Progress Series* 319:85–91.

Rapoport, E. H. 1975. *Aerografía: Estrategias geográfica de especies.* México City, DF: Fundo Cult. Econ.

———. 1982. *Aerography: Geographic strategies of species.* New York: Pergamon.

Richardson, W. D. 1975. The marine algae of Trinidad, West Indies. *Bull. Br. Mus. (Nat. Hist.) Botany* 5 (3): 1–143, + pls. 16–27.

Ricker, R. W. 1987. *Taxonomy and biogeography of Macquarie Island seaweeds.* London: British Museum (Natural History).

Rohde, K., M. Heap, and D. Heap. 1993. Rapoport's rule does not apply to marine teleosts and cannot explain latitudinal gradients in species richness. American Naturalist 142:1–16.

Rosenzweig, M. L. 1995. Species diversity in space and time. Cambridge: Cambridge University Press.

Roy, K., D. Jablonsky, and J. W. Valentine. 1994. Eastern Pacific molluscan provinces and latitudinal diversity gradients: No evidence for "Rapoport's rule." *Proceedings of the National Academy of Science, USA* 91:8871–74.

Roy, K., and J. D. Witman. 2008. Spatial patterns of species diversity in the shallow marine invertebrates: Patterns, processes and prospects (this volume).

Santelices, B. 1980. Phytogeographic characterization of the temperate coast of Pacific South America. *Phycologia* 19:1–12.

———. 1990. Patterns of reproduction, dispersal and recruitment in seaweeds. *Oceanography and Marine Biology Annual Review* 28:177–276.

———. 1991. Littoral and sublittoral communities of continental Chile. Chapter 14, pp. 347-369. In *Intertidal and littoral ecosystem of the world,* vol. 24 in the series "Ecosystems of the World," ed. A. C. Mathieson and P. H. Nienhuis, 347–69. New York: Elsevier Scientific Publishing.

———. 1992. Digestion survival in seaweeds: An overview. In *Plant-animal interactions in the marine benthos,* Systematics Association, vol. 46, ed. D. M. John, S. J. Hawkins, and J. H. Price, 363–84. Oxford: Clarendon Press.

———. 2004. A comparison of ecological response among aclonal (unitary), clonal and coalescing macroalgae. *Journal of Experimental Marine Biology and Ecology* 300:31–64.

Santelices, B., and I. A. Abbott. 1987. Geographic and marine isolation: An assessment of the marine algae of Easter Island. *Pacific Science* 41:1–20.

Santelices, B., and P. Marquet. 1998. Seaweeds, latitudinal diversity patterns and the Rapoport's rule. *Diversity and Distributions* 4:71–75.

Santelices, B., and I. Meneses. 2000. A reassesment of the phytogeographic characterization of temperate Pacific South America. *Revista Chilena de Historia Natural* 73:605–14.

Santelices, B., J. A. Correa, D. Aedo, V. Flores, M. Hormazábal, and P. Sánchez. 1999. Convergent biological processes in coalescing Rhodophyta. *Journal of Phycology* 35:1127–49.

Santelices, B., P. Skelton, and G. South. 2004. Observations on *Gelidium samoense* from the Fiji Islands. In *Taxonomy of economic seaweeds with reference to some Pacific species, vol. 9,* ed. I. A. Abbott and K. J. McDermid, 119–29. San Diego: California Sea Grant College Program.

Santos, R. 1993. Plucking or cutting *Gelidium sesquipedale*? A demographic simulation of harvest impact using a population projection matrix model. *Hydrobiologia* 260/261:269–76.

Scagel, R. F., P. W. Gabrielson, D. J. Garbary, L. Golden, M. W. Hawkes, S. C. Lindstrom, J. C. Oliveira, and T. B.Widdowson. 1989. A synopsis of the benthic marine algae of British Columbia, Southeast Alaska, Washington and Oregon. Vancouver, Canada: Phycological Contribution No. 3, University of British Columbia.

Schils, T., and E. Coppejans. 2003. Phytogeography in upwelling areas of the Arabian Sea. *Journal of Biogeography* 30:1339–56.

Schneider, C. W., and R. B. Searles. 1991. Seaweeds of the Southeastern United States: Cape Hatteras to Cape Canaveral. Durham, NC: Duke University Press.

Seoane-Camba, J. 1969. Crecimiento, reproducción y desprendimiento de biomasa en *Gelidium sesquipedale* (Chem.) *Thuret. Proceedings of International Seaweed Symposium* 6:365–74.

Setchell, W. A., and N. L. Gardner. 1925. The marine algae of the Pacific coast of North America. Part III. Melanophyceae. *University of California Publications in Botany* 8:383–898.

Silva, P. C. 1992. Geographic patterns of diversity in benthic marine algae. *Pacific Science* 46:429–37.

Silva, P. C., E. H. Menez, and R. L. Moe. 1987. Catalogue of the benthic algae of the Philippines. *Smithsonian Contribution to Marine Science* 27:1–179.

Sousa-Pinto, I. 1998. The seaweed resources of Portugal. In *Seaweed resources of the world,* ed. A. T. Critchley and M. Ohno, 176–84. Yokosuka, Japan: Japan International Cooperation Agency.

South, G. R., and P. A. Skelton. 2003. Catalogue of marine benthic macroalgae of the Fiji Islands, South Pacific. *Australian Systematic Botany* 16:647–98.

South, G. R., P. A. Skelton, and A. Yoshinaga. 2001. Subtidal benthic marine algae of the Phoenix Islands, Republico of Kiribati, Central Pacific. *Botanica Marina* 44:559–70.

Stegenga, H., J. J. Bolton, and R. J. Anderson. 1997. Seaweeds of the South African west coast. *Contributions from the Bolus Herbarium.* 18:1–655.

Stekoll, M. S. 1998. The seaweed resources of Alaska. In *Seaweed resources of the world,* ed. A. T. Critchley and M. Ohno, 258–65. Yokosuka, Japan: Japan International Cooperation Agency.

Steneck, R. S., and L. Watling. 1982. Feeding capabilities and limitations of herbivorous molluscs: A functional-form approach. *Marine Biology* 68:299–319.

Stevens, G. C. 1989. The latitudinal gradient in geographical ranges: How so many species coexist in the tropics. *American Naturalist* 133:240–56.

———. 1992. The elevational gradient in altitudinal range: An extension of Rapoport's latitudinal rule to altitude. *American Naturalist* 140:893–911.

Stuercke, B., and K. J. McDermid. 2004. Variation in algal turf species composition and abundance on two Hawaiian shallow subtidal reefs. *Cryptogamie Algolgie* 25:353–65.

Tittley, I. 1994. Marine algae of the Cayman Islands: A preliminary account. In *The Cayman Islands: Natural history and biogeography,* ed. M. A. Brunt and J. E. Davies, 125–30. Amsterdam: Kluwer Academic.

Tittley, I., and A. I. Neto. 1995. The marine algal flora of the Azores and its biogeographical affinities. *Bolletim do Museu Municipal do Funchal,* Supplement no. 4:747–66.

UK HarperCollins. 1979. *The Times concise atlas of the world, rev. ed.* Edinburgh: John Bartholomew and Son Ltd.

van den Hoek, C. 1975. Phytogeographic provinces along the coast of the northern Atlantic Ocean. *Phycologia* 14:317–30.

———. 1984. Worldwide longitudinal seaweed distribution patterns and their possible causes, as illustrated by the distribution of rhodophytan genera. *Helgoländer Meeresunters* 41:261–72.

van der Strate, H. J., L. Van de Zande, W. T. Stam, and J. L. Olsen. 2002. The contribution of haploids, diploids and clones to fine-scale population structure in the seaweed *Cladophoropsis membranacea* (Chlorophyta). *Molecular Ecology* 11:329–45.

Vroman, M., and H. Stegenga. 1988. An annotated checklist of the marine algae of the Caribbean islands of Aruba and Bonaire. *Nova Hedwigia* 46:433–80.

Walters, L. J., and C. M. Smith. 1994. Rapid rhizoid production in *Halimeda discoidea* Decaisne (Chlorophyta, Caulerpales) fragments: A mechanism for survival after separation from adult thalli. *Journal of Experimental Marine Biology and Ecology* 175:105–20.

Wienke, C., and M. N. Clayton. 2002. *Biology of Antarctic seaweeds.* Liechtenstein: ARG Gantner Verlag KG.

Willig, M. R., D. M. Kaufman, and R. D. Stevens. 2003. Latitudinal gradients of biodiversity: Pattern, process, scale and synthesis. *Annual Review of Ecology and Systematics* 34:273–309.

Witman, J. D., and P. K. Dayton. 2001. Rocky subtidal communities. In *Marine community ecology,* ed. M. D. Bertness, S. D. Gaines, and M. E. Hay, 339–93. Sunderland, MA: Sinauer.

Womersley, H. B. S. 1981. Biogeography of Australasian marine macroalgae. In *Marine botany: An Australasian perspective,* ed. M. N. Clayton and R. J. King, 292–307. Melbourne: Longman Cheshire.

———. 1987. *The marine benthic flora of southern Australia, II. Phaeophyta.* Adelaide, South Australia: Government Printer.

Womersley, H. B. S., and A. Bailey. 1970. Marine algae of the Solomon Islands. *Philosophical Transactions of the Royal Society of London* 259:257–352.

PART TWO

PROCESSES UNDERLYING MACROECOLOGICAL PATTERNS

CHAPTER SEVEN

PLANKTON—NOT SO PARADOXICAL AFTER ALL

SEAN NEE AND GRAHAM STONE

Introduction

The paradox of the plankton is, essentially, an apparently massive violation of the principle of competitive exclusion (Hardin 1960), or the "one species, one niche" requirement for coexistence. In "The paradox of the plankton" Hutchinson (1961) writes:

> The problem that is presented by the phytoplankton is essentially how it is possible for a number of species to coexist in a relatively isotropic or unstructured environment all competing for the same sorts of materials.

Hutchinson entertained the idea that the possibility did exist for coexistence by conventional niche differentiation exploiting, for example, the light gradient in the water column. He discounted this, however, being of the view that water is too well mixed. "It is hard to believe that in turbulent open water many physical opportunities for niche diversification exist." He and MacArthur were also aware that specialist predators or parasites could facilitate coexistence, but appears not to have placed much store in this mechanism.

It is quite extraordinary, the extent to which this world view of homogenous water with little opportunity for niche diversification has been completely overturned by the discoveries of recent years, so it is appropriate to return for another look at the paradox of the plankton. In fact, our understanding of marine life itself has been transformed beyond recognition. What would Hutchinson think of the discovery in 1977—that most of the biomass in the ocean consists of bacteria at concentrations of about 10^6 ml^{-1}; that is, that most plankton are actually bacteria and are responsible for most of the ocean's metabolism (Hobbie, Daley, and Jasper 1977; Azam 1998)? What would he think of the fact that the numerically most abundant life form in the oceans is *viral* (Azam and Worden 2004) with viruses occurring at densities of 10^7 ml^{-1}: these would not even be discovered until thirty years after his seminal paper (Bergh et al. 1989; Proctor and Fuhrman 1990)? And finally, what would Hutchinson make of the fact that, from the point of view of these life-forms, water is not a well-stirred fluid but "is structured with cross-linked polymers, colloids and nano- and microgels, creating an organic matter continuum and a wealth of surfaces displaying activity and biodiversity hotspots" (Azam and Worden 2004, 1623).

In the first part of this chapter we will see if the paradox of the plankton can be resolved by invoking the neutral theory of biodiversity. Having concluded that it cannot, we will then revisit the paradox in the light of recent discoveries and suggest that the paradox is readily resolved by a combination of niche differentiation and specialist enemies. In particular, specialist parasites: the marine environment is remarkably species poor in primary consumers compared to the land.

The Neutral Theory of Biodiversity

In addition to the existence of many new discoveries, another reason to revisit the paradox is the current interest in the neutral theory of biodiversity, since such a theory was presented as a resolution of the paradox many years ago. The neutral theory of biodiversity (Bell 2001; Hubbell 2001) was originally proposed as a candidate explanation of tropical tree diversity (Hubbell 1979). The enormous tree diversity in the tropical forests seems to present a similar puzzle to the ecologist, as do the plankton. A tree just needs a patch of ground, some light, air, and water, so how can such a huge number of species coexist? On the neutral view, species *do* all have the same niche, and competitive exclusion just takes a very long time. This appears to have been proposed in the context of the paradox of the plankton by Riley (1963), cited in Colinvaux (1993). On this view, the population sizes of the species simply fluctuate at random. Over time, species follow a random walk to extinction while other species, appearing by the random process of speciation, increase in abundance. This is the exact equivalent of the neutral theory of molecular evolution, originally proposed as the solution to what was felt to be the puzzle of enormous amounts of molecular population genetic variability (Kimura 1983). Nee (2005) presents a brief collection of basic quantitative results of neutral theory derived by Hubbell and others.

A critical role is played in this theory by the "fundamental biodiversity number" usually denoted by θ:

$$\theta = 2Nu$$

where N is the sum total number of all individuals of all species and u is the per-individual/per generation speciation rate. It is natural that θ should play a central role in the neutral theory as it measures the flux of new species into the "metacommunity," which is simply the entire tropical forest. There is a slight complication, however, which we need to set aside. The factor of 2 in the definition of θ only occurs if we derive our results using a model of discrete, nonoverlapping generations. If, as might be thought more appropriate for trees and plankton, we derive our results assuming overlapping generations (more specifically, the verbal model described by Hubbell put into mathematical form, as distinct from his mathematical models), then θ is simply Nu. A remarkably concise set of derivations of results from an overlapping generations model can be found in Appendix 8.2 of Leigh (1999). But, having noted this, we will continue to use Hubbell's definition to avoid confusion.

With a random sample of individuals from a community ("metacommunity" in the language of neutral theory), it is straightforward to come up with a point estimate of θ in the community. Let p_i be the relative abundance of the ith species in the sample. Then,

$$\frac{1}{\sum_1^s p_i^2} = \hat{\theta} + 1,$$

with the "hat" on θ indicating that it is an estimate. Using data tabulated in Patrick (1968), we can use this formula to infer that, in the diatom community she was sampling, $\theta = 9$, that is, that about nine new species appear per generation. We can also put confidence intervals on this estimate using the sampling variance of the Simpson diversity index (e.g., Lande, Engen, and Saethler 2003).

If this number seems rather large, recall that the fate of most new species is swift extinction: the probability that a species will survive for n generations is, for large n, close to $2/n$ (Fisher 1958, 84). So this large estimate of θ is not a quantitative problem for neutral theory.

However, in the next few pages of his book, Fisher developed the mathematics of neutral theory further and derived a quantitative result that creates a difficulty for neutral theory as an explanation of the abundance of tree (or plankton species), as was pointed out by Leigh (1999: chapter 8). Consider a species that arose *n* generations ago. What is the probability, under neutral drift, that its abundance exceeds *k* individuals today? The probability is:

$$\Pr(\text{abundance} > \mathrm{k}) = \frac{2}{n} e^{-2k/n}.$$

Fisher (1958, 87) described the significance of this result in words: "An inference of some interest is that in the absence of favorable selection, the number of individuals [of a species] . . . cannot greatly exceed the number of generations since its occurrence. Actually, the chance is less than 1 in 1,000 that [the number of individuals] should exceed $3n/2$." (Fisher appears to have omitted the phrase "Assuming the [species] arose at least 100 generations ago.")

The significance of this is that it would take an impossibly long time for a new species to get to the colossal population sizes achieved by plankton, even with their short generation times of, very roughly, 100 generations per year.

It is tempting to invoke the following "fix," which we believe is not a straw man but a real possibility that must be examined. Many groups of organism do not achieve high abundances until some considerable time after their first appearance. For example, diatoms appeared in the Jurassic but did not begin their ascent to numerical predominance until the Cretaceous (Damste et al. 2004). During their period of relatively low abundance they perhaps could achieve the diversity we observe after their release.

Unfortunately, this fix does not work, because the timescale problem arises whichever direction you look at time. The time taken for a new species to become common is the same as the time taken for a common species to become extinct. Lande, Engen, and Saether (2003, 166) describe this problem as follows:

> For extremely abundant communities, such as oceanic plankton, tropical insects and even tropical trees, this predicts that extinction of common species is not expected to occur within the age of the earth, whereas species observed in the fossil record become extinct within a few to several million years.

Other possible "fixes" get "broken" in Nee (2005).

However, it is important to note that the apparent fact that neutral theory cannot account for plankton diversity does not in any way compromise its utility as a null model for analyzing ecological data. An excellent example of this use is provided by Leigh et al. (1993), which is also discussed in Nee (2005).

Hutchinson (1961) himself used neutral theory to make a timescale calculation concerning the importance of random extinction. Taking the basic mathematical results from the great mathematical ecologist Skellam, he calculated that lakes would have to be impossibly old for the probability of the random-walk extinction of a plankton species to become nonnegligible. This is the same insight as that of Lande, Engen, and Saether (2003) at a smaller scale.

Niche Diversification

We are in the middle of an extraordinary period of discovery of ecological diversification in plankton. We will discuss this mainly using results for *Prochlorococcus* and *Synechococcus,* two closely related genera of plankton. This is partly because these photosynthesising cyanobacteria are the numerically predominant marine plankton and produce a substantial fraction of the Earth's oxygen. But also, they dramatically illustrate the huge potential for niche diversification in plankton.

Prochlorococcus and *Synechococcus* are tiny—a few percent of the size of *E. coli*. They have so few genes—around 2,000—they are considered a "minimal life unit" (Fuhrman 2003; Rocap et al. 2003). It is thought that they economize on DNA as an adaptation to life in an environment typically poor in nutrients like nitrogen and phosphorous (Fuhrman 2003). They actually turn the paradox of the plankton on its head. *Prochlorococcus* strains have such similar sequences at their small subunit RNA genes that they would be considered the same species by the normal criteria used by microbiologists. However, the strains exhibit such ecological diversity that they would be considered different species by normal ecological criteria. Hence, the paradox is, in a way, created by its own resolution! First we will discuss the partitioning of the light column, analagous to the story of MacArthur's warblers partitioning trees by height (MacArthur 1958). Then we will discuss the enormous potential for niche diversification in nutrient exploitation.

Different strains of *Prochlorococcus* are adapted to exploit different levels of light. This is true to such an extent that one strain may not grow at all in the optimal level for a different strain (Moore, Rocap, and Chisholm 1998). Lower light-adapted *Prochlorococcus* have a larger number, as well as more specialised, *pcb* genes, which encode the major light harvesting proteins (Bibby et al. 2003).

But not only do plankton partition light intensity, they partition the electromagnetic spectrum! The following account is based on Ting at al. (2002). *Prochlorococcus* uses a light-harvesting system that is better at absorbing blue wavelengths, whereas *Synechococcus* uses systems with absorption maxima in the blue-green region. Different wavelengths of light penetrate to different depths, with blue enriched at greater depths. Also, the photosynthetic system of *Synechococcus* is considered to be more expensive than that of *Prochlorococcus* in terms of nutrients. Together, this explains why *Prochlorococcus* is dominant in deeper ocean waters, which are nutrient poor, whereas *Synechococcus* is dominant in coastal, nutrient rich, surface waters. Coastal waters are also enriched with green light by phytoplankton.

Different strains of *Synechococcus* have been found to be partitioning the light spectrum in the Baltic Sea (Stomp et al. 2004). A striking visual illustration of this fact is seen in the color of glass flasks of the two strains—red or green—reflecting which wavelengths they are not using. A cyanobacterium of a different genus, *Tolypothrix tenuis*, was discovered coexisting with both *Synechococcus* strains: *Tolypothrix* is able to adapt its photosynthetic system to exploit the wavelengths not being used by its neighbors, presumably incurring some cost for this flexibility.

Marine environments are typically poor in one or more essential nutrients except in regions of coastal upwelling. Such nutrients that do exist are available in a variety of forms. Consider iron, 99% of which is bound to organic molecules in the sea (Gelder 1999). These molecules may be iron binding compounds such as porphyrins, mechanically released from cells that have, for example, been lysed by viruses. They may also be iron-binding compounds actively secreted by cells to facilitate iron uptake, like siderophores. *Synechococcus* and other cyanobacteria produce siderophores, whereas eukaryotic phytoplankton do not. Since it goes to the expense of manufacturing syderophores, it is not surprising that *Synechococcus* is much more effective at the uptake of iron bound to them rather than porphyrins, whereas the opposite is the case for diatoms (Hutchins et al. 1999). Hence, we have a niche partition.

Other nutrients are similarly partitioned. For example, strains of *Prochlorococcus* differ in their abilities to use nitrate, nitrite and cyanate as nitrogen sources (Rocap et al. 2003). They also differ in their abilities to exploit organic sources of phosphorous, and they differ from *Synechococcus* in their abilities to exploit iron sources (Rocap et al. 2003).

Trade-offs allowing niche partitioning may include combinations of light and nutrient axes, dramatically increasing the scope of possible niche partitioning. For example, marine diatoms (eukaryotes) use a photosynthetic system that requires much less iron than that used by coastal diatoms, where iron limitation is less of a problem. The marine diatoms can photosynthesise at the same rates as coastal diatoms, but may be less able to adapt rapidly to changing light intensities (Strzepek and Harrison 2004).

Niche diversification in cyanobacteria like *Prochlorococcus* and *Synechococcus* may be facilitated by the fact that they are under such strong pressure to economise on DNA. To exploit a resource requires the genetic information specific to that particular exploitation system. Perhaps each ecologically distinct strain—ecotype—is a particular "minimal life unit" out of the many possible, with the tradeoffs imposed by restrictions on genome size.

The Janzen/Connell Hypothesis

Under this hypothesis, diversity is maintained by specialist parasites and predators, giving rare species an advantage. It has been proposed in a variety of contexts. Perhaps the earliest was in the context of polymorphism at genetic loci associated with infectious disease resistance by J. B. S. Haldane. Leigh (1999), for example, discusses the evidence for it as an explanation of tropical forest diversity: the hypthesis is named after those who proposed it in this context.

As we saw, Hutchinson did not think much of this as an explanation of plankton diversity. But what we have learned about marine viruses, with the first discoveries of the existence of marine viruses coming thirty years *after* Hutchinson's seminal paper (Bergh et al. 1989; Proctor and Fuhrman 1990; Suttle, Chan, and Cottrell 1990), necessitates a rethink.

Viruses are the most abundant life forms in the sea (Wilhelm and Suttle 1999) with new families continuing to be discovered (Culley, Lang, and Suttle 2003). Planktonic mortality from viral lysis is comparable in importance to mortality from predation (Fuhrman 1999), contributing significantly to turnover of populations and nutrients (Wilhelm and Suttle 1999). Although there are generalist viruses infecting many *Prochlorococcus* strains as well as *Synechococcus*, there are also known to be extremely strain-specific viruses (Sullivan, Waterbury, and Chisholm 2003): the story is similar with algal viruses (Suttle pers. com.). Because of their ubiquity and high levels of activity, it is hardly surprising that viruses have been proposed as a solution to the paradox of the plankton (Fuhrman and Suttle 1993; Weinbauer and Rassoulzadegan 2004).

A striking and important contrast between marine and terrestrial foodwebs and diversity patterns may result entirely from a lack of specialist predation in the former. In marine systems, the primary producers are photosynthetic bacteria and unicellular plants that live fast and die young. They invest low resources in somatic maintainance, and high resources in reproduction—hence the phenomenon of ephemeral algal blooms. Algae are fed on by a relatively low diversity of primary consumers—predominantly crustaceans. These include familiar krill (Euphausiaceans, such as *Euphausia superba*) and calanoid copepods. These groups have low species richness (for example, there are only around ninety different types of krill in the world, (Mauchline 1980). These primary consumers feed by filtering unicellular plants out of the water, and select food plants by size rather than by species. There are thus fewer species of primary consumer than of primary producer. As a result of this low specificity, there is little correlation between plant species and herbivore species richness in pelagic marine ecosystems (Irigoien, Huisman, and Harris 2004).

In contrast, plant biomass in terrestrial ecosystems is dominated by contributions from large, long-lived multicellular plants. These invest considerable resources in somatic maintainance and defense against herbivores, resulting in an evolutionary arms race between metabolic plant defenses (including toxins such as strychnine, caffeine, and nicotine) and insect detoxification systems. Many insect countermeasures are highly specific to the defenses of particular plant taxa, resulting in high specificity of primary consumers to specific sets of plants, and well-documented patterns of insect-plant coevolution and co-cladogenesis (Thomson 1994). The inversion of the relative sizes of plants and arthropod herbivores on land (large plants, small insects, rather than small plants, large crustaceans) means that, in contrast to marine systems, each terrestrial plant commonly supports multiple species of primary consumer. Though relatively undiverse morphologically in comparison with marine crustaceans, terrestrial insects are extremely species rich, for example, the weevil family Curculionidae, with 48,000 species in 4,300 genera presently described (Anderson 1998), is the most species-rich family of organisms known. As a result of this herbivore-host plant specificity, there is a strong positive correlation between plant species and herbivore species richness on land (Irigoien, Huisman, and Harris 2004).

Discussion

The cyanobacteria *Prochlorococcus* had not even been discovered at the time Hutchinson wrote his seminal paper, "The paradox of the plankton" (Hutchinson 1961) in spite of the fact that, as we now know, they are major contributors to global oxygen. At the time, Hutchinson felt that the species diversity of the plankton was problematic from the point of view of classical ecological ideas concerning the coexistence of species. It is, perhaps, ironic that the many recent discoveries and investigations of the marine planktonic world, made primarily by modern molecular biology, should seem to so readily reintegrate plankton into the classical world view of subdivided niche space.

Another recent discovery—the superabundance of viruses in the sea—also integrates the plankton into an increasingly "classical" view that diversity can be maintained by specialist enemies. It might be thought odd that the natural history comes *after* the ideas required to explain it. But the ideas could be developed in the context of other biological systems: the natural history of most of life on Earth—invisible life— has had to wait for the advance of modern molecular biology.

ACKNOWLEDGMENTS

We are grateful to Bert Leigh for explaining the origin of the factor of two in the common definition of the fundamental biodiversity number. Curt Suttle kindly brought us up to speed on viruses in the sea.

REFERENCES

Anderson, R. S. 1998. Weevils (curculionoidea). In *Assessment of species diversity in the montane cordillera ecozone,* ed. I. M. Smith and G. G. E. Scudder. Burlington: Ecological Monitoring and Assessment Network.

Azam, F. 1998. Microbial control of oceanic carbon flux: The plot thickens. *Science* 280:694–96.

Azam, F., and A. Z. Worden. 2004. Microbes, molecules and marine ecosystems. *Science* 303:1622–24.

Bell, G. 2001. Neutral macroecology. *Science* 293:2413–17.

Bergh, O., K. Y. Borsheim, G. Bratbak, and M. Heldal. 1989. High abundance of viruses found in aquatic environments. *Nature* 340 (6233):467–68.

Bibby, T. S., I. Mary, J. Nield, F. Partensky, and J. Barber. 2003. Low-light-adapted prochlorococcus species possess specific antennae for each photosystem. *Nature* 424 (6952): 1051–54.

Colinvaux, P. 1993. *Ecology 2.* New York: Wiley.

Culley, A. I., A. S. Lang, and C. A. Suttle. 2003. High diversity of unknown picorna-like viruses in the sea. *Nature* 424:1054–57.

Damste, J. S. S., G. Muyzer, B. Abbas, S. W. Rampen, G. Masse, W. G. Allard, S. T. Belt, et al. 2004. The rise of the rhizosolenid diatoms. *Science* 304 (5670): 584–87.

Fisher, R. A. 1958. The genetical theory of natural selection. New York: Dover.

Fuhrman, J. A. 1999. Marine viruses and their biogeochemical and ecological effects. *Nature* 399:541–48.

———. 2003. Genome sequences from the sea. *Nature* 424:1001–1002.

Fuhrman, J. A., and C. A. Suttle. 1993. Viruses in marine planktonic systems. *Oceanography* 6:51–63.

Gelder, R. J. 1999. Complex lessons of iron uptake. *Nature* 400:815–16.

Hardin, G. 1960. The competitive exclusion principle. *Science* 131:1292–97.

Hobbie, J. E., R. J. Daley, and S. Jasper. 1977. Use of nucleopore filters for counting bacteria by flourescence microscopy. *Applied and Environmental Microbiology* 33:1225–28.

Hubbell, S. P. 1979. Tree dispersion, abundance and diversity in a tropical dry forest. *Science* 203:1299–309.

———. 2001. *The unified neutral theory of biodiversity and biogeography.* Princeton, NJ: Princeton University Press.

Hutchins, D. A., A. E. Witter, A. Butler, and G. W. Luther. 1999. Competition among marine phytoplankton for different chelated iron species. *Nature*: 858–61.

Hutchinson, G. E. 1961. The paradox of the plankton. *American Naturalist* 95:137–145.

Irigoien, X., J. Huisman, and R. P. Harris. 2004. Global biodiversity patterns of marine phytoplankton and zooplankton. *Nature* 429:863–67.

Kimura, M., 1983. *The neutral allele theory of molecular evolution.* Cambridge: Cambridge University Press.

Lande, R., S. Engen, and B. E. Saether. 2003. *Stochastic populations dynamics in ecology and conservation.* Oxford: Oxford University Press.

Leigh, E. G. 1999. *Tropical forest ecology.* Oxford: Oxford University Press.

Leigh, E. G., S. J. Wright, E. A. Herre, and F. E. Putz. 1993. The decline of tree diversity on newly isolated tropical islands—A test of a null hypothesis and some implications. *Evolutionary Ecology* 7 (1): 76–102.

MacArthur, R. H. 1958. Population ecology of some warblers of northeastern coniferous forests. *Ecology* 39:599–619.

Mauchline, J. 1980. The biology of mysids and euphausiids. *Advances in Marine Biology* 18:1–61.

Moore, L. R., G. Rocap, and S. W. Chisholm. 1998. Physiology and molecular phylogeny of coexisting prochlorococcus ecotypes. *Nature* 393:464–67.

Nee, S. 2005. The neutral theory of biodiversity: Do the numbers add up? *Functional Ecology* 19:173–176.

Patrick, R. 1968. The structure of diatom communities in similar ecological conditions. *American Naturalist* 102:173–83.

Proctor, L. M., and J. A. Fuhrman. 1990. Viral mortality of marine-bacteria and cyanobacteria. *Nature* 343 (6253): 60–62.

Rocap, G., F. W. Larimer, J. Lamerdin, S. Malfatti, P. Chain, N. A. Ahlgren, A. Arellano, et al. 2003. Genome divergence in two prochlorococcus ecotypes reflects oceanic niche differentiation. *Nature* 424 (6952): 1042–47.

Stomp, M., J. Huisman, F. de Jongh, A. J. Veraart, D. Gerla, M. Rijkeboer, B. W. Ibelings, U. I. A. Wollenzien, and L. J. Stal. 2004. Adaptive divergence in pigment composition promotes phytoplankton biodiversity. *Nature* 432:104–6.

Strzepek, R. F., and P. J. Harrison. 2004. Photosynthetic architecture differs in coastal and oceanic diatoms. *Nature* 431:689–92.

Sullivan, M. B., J. B. Waterbury, and S. W. Chisholm. 2003. Cyanophages infecting the oceanic cyanobacterium prochlorococcus. *Nature* 424 (6952): 1047–51.

Suttle, C. A., A. M. Chan, and M. T. Cottrell. 1990. Infection of phytoplankton by viruses and reduction of primary productivity. *Nature* 347:467–69.

Ting, C. S., G. Rocap, J. King, and S. W. Chisholm. 2002. Cyanobacterial photosynthesis in the oceans: The origins and significance of divergent light-harvesting strategies. *Trends in Microbiology* 10 (3): 134–42.

Weinbauer, M. G., and F. Rassoulzadegan. 2004. Are viruses driving microbial diversification and diversity? *Environmental Microbiology* 6 (1): 1–11.

Wilhelm, S. W., and C. A. Suttle. 1999. Viruses and nutrient cycles in the sea. *Bioscience* 49:781–88.

CHAPTER EIGHT

BASIN-SCALE OCEANOGRAPHIC INFLUENCES ON MARINE MACROECOLOGICAL PATTERNS

JAMES J. LEICHTER AND JON D. WITMAN

Macroecology can be broadly described as the study of patterns of biological processes such as species distribution, diversity, and abundance that become emergent when viewed at regional and larger spatial scales (Brown 1995; Denny et al. 2004). As such, the scales of macroecological patterns in marine environments necessarily overlap those of regional physical variability structured by a range of oceanographic climate and forcing processes. Macroecological patterns in individual biological parameters such as body size or in relationships among parameters such as abundance and diversity arise from processes acting at the scales of individual organisms and communities, but become evident only when examined at the much larger spatial scales over which structuring environmental processes vary. The goal of this chapter is to provide an overview of known and potential links between oceanographic forcing mechanisms, biological processes acting at scales of individual organisms to communities, and the resulting macroecological patterns. We provide an overview of a range of physical processes that may influence and modify macroecological patterns, and hope to suggest promising avenues for future research. We pay particular attention to phenomena acting on large spatial scales of ocean basins (>1,000 km), as it is at these scales that macroecological patterns are most likely to occur and be recognized.

Interest in the mechanisms producing macroecological pattern in marine systems leads to recognition that these patterns are likely to be strongly influenced by oceanographic dynamics. Conceptually, oceanographic mechanisms acting across a range of spatial and temporal scales can be viewed as a base level of environmental structure that influences a suite of biological processes that in turn influence the formation of macroecological patterns, both directly and indirectly (fig. 8.1). For purposes of this illustration and the following discussion, we subdivide oceanographic processes that in reality span a continuous range of space and time into discrete scales encompassing climate processes acting at ocean basin scales, advective processes associated with open ocean and coastal currents, and diffusive processes associated with mixing. Environmental forcing associated with these underlying oceanographic scales influences key biological processes including productivity, dispersal, and retention that in turn contribute to macroecological patterns such as variation in body size, species ranges and abundance, and diversity at regional scales. Feedbacks among processes and patterns acting at a variety of scales are likely to increase overall complexity.

Discovering emergent relationships between body size, abundance, diversity, and range size are a central theme of macroecology (Gaston and Blackburn 2000). While macroecology represents a relatively new concep-

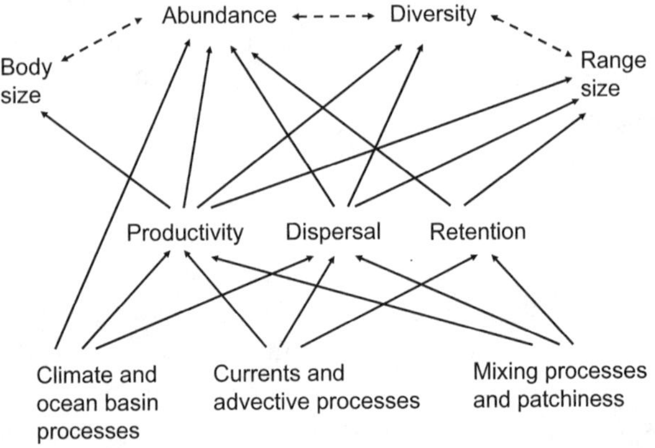

FIGURE 8.1 Diagram of pathways of oceanographic influence on macroecological patterns. Oceanographic processes on the (bottom level) are linked to biological processes (intermediate level) and ultimately to macroecological patterns and relationships (top level). The oceanographic processes can be viewed as a base level or environmental structure that influences a suite of biological processes, which in turn influence the formation of macroecological patterns both directly and indirectly (dotted arrows). Climatic and ocean basin processes are ENSO, PDO, NAO, Rossby and Kelvin waves, and storminess. Currents and advective processes are all transport mechanisms. Mixing processes are derived from tidal energy changes and upwelling and downwelling. Elucidating emergent relationships between body size, abundance, diversity, and range size is a central theme of macroecology.

tual mode for the interpretation of broad-scale ecological patterns, recognition that ocean dynamics can strongly influence biological processes at large scales has a long history in biological oceanography. For example, at least one set of marine macroecological patterns—spatial and temporal variation in fish recruitment at broad regional scales—has been interpreted at least partially in terms of oceanographic variability since the early twentieth century. Hjort (1914, 1926) proposed a set of general hypotheses to explain ways in which the timing of zooplankton prey availability might affect the survival and recruitment of larval fishes. The subsequent development and tests of the so-called match/mismatch hypothesis (Cushing 1975, 1990; Jackson and Johnson 2000; Reid et al. 2003) explicitly emphasized the roles of physical dynamics in the temporal and spatial overlap or nonoverlap of larval fishes and their zooplankton prey.

More recently, the last twenty to thirty years has seen an increasing number of marine ecological studies that directly incorporate measurements of physical dynamics at scales of tens to hundreds of km across environmental gradients (see Sanford and Bertness, this volume). Much of the resulting interaction between oceanographers and ecologists has focused on population-level processes, such as recruitment. For instance, a series of invertebrate recruitment studies in coastal systems in the 1980s pushed a growing trend of directly measuring coastal currents and associated variation in temperature, salinity, and wind forcing, along with more traditional ecological metrics such as settlement and recruitment. In central California, work by Roughgarden and associates (Gaines, Brown, and Roughgarden 1985; Gaines and Roughgarden 1985; Roughgarden, Gaines, and Possingham 1988, Farrell, Bracher, and Roughgarden 1991; Connolly, Menge, and Roughgarden 2001) led to an intentionally simplified conceptual picture of how variation in coastal upwelling may contribute to pulsed settlement of intertidal barnacles and spatial patterns in recruitment. Studies by Wing et al. (1995a, 1995b, 1998) produced a more complex view of the roles of coastal buoyancy currents associated with upwelling relaxation in the settlement of subtidal crabs and sea urchins. Due in large part to technological advances associated with the reduced cost of sophisticated electronic instruments and inexpensive digital memory, it has become increasingly common since about 1990 for marine ecologists to make relatively detailed physical measurements and to collaborate with coastal physical oceanographers on primarily ecological studies. A consistent finding in these types of efforts is the extent and potential importance of physical variability at high frequencies and small spatial scales (Denny 1995; Denny et al. 2004; Leichter, Deane, and Stokes 2005).

With this background and the increasing focus on physical dynamics as an important component of marine ecological studies, it comes as no surprise that the processes producing macroecological patterns can be strongly influenced by oceanographic dynamics. In fact, given the range of patterns and spatial scales over which they occur, it would be much more surprising to find that these patterns were *not*, in general, influenced by oceanographic dynamics than to find that they are. However, detailed knowledge of the links between oceanographic dynamics and macroecological patterns have received only limited direct investigation to date, and most of those studies have focused on a small subset of population-level processes associated with dispersal, coastal boundary currents, and settlement. Given that benthic communities offshore and along the shoreline depend on water transport to supply food and larvae and to remove metabolic wastes, it would be impossible to understand large-scale patterns in the ocean, their causes, and how they are changing in a human-dominated world without considering the links between oceanographic processes and ecological pattern. There is an enormous amount of important macroecological research yet to be done, at all levels of ecological organization, by integrating oceanographic and ecological perspectives. One of our goals is to highlight potentially interesting research directions.

We organize the following discussion around decreasing spatial scales, and emphasize the importance of episodic events and the interactions of processes across temporal scales—specifically the potential for broad-scale, low-frequency and episodic processes to modulate mechanisms acting at much higher frequencies. For example, we discuss recent evidence for the Galápagos Islands that suggests the passage of Kelvin waves associated with basin-scale forcing has a strong effect on the higher-frequency dynamics of internal wave motions, which in turn affect invertebrate recruitment rates. It is useful to take a broad view of macroecological patterns to include ecological patterns resulting from or occurring at regional and larger scales. For example, upwelling is a major oceanographic process operating on basin spatial scales with a demonstrated influence on recruitment and food webs in the rocky intertidal zone. Since these large-scale effects of upwelling have been reviewed elsewhere (Roughgarden, Gaines, and Possingham 1988; Menge et al. 1997; Connolly, Menge, and Roughgarden 2001; Menge et al. 2004; Navarrete et al. 2005; Witman 2007; Sanford and Bertness [this volume]) they are not covered here. While any list of the macroecological patterns that are potentially influenced by oceanographic processes is likely to be incomplete, it is worthwhile to consider some of the patterns likely to be strongly influenced. Following table 1.1 in Gaston and Blackburn (2000), we expect the following macroecological patterns are very likely to be modulated by oceanographic dynamics: species-area relationships; species-richness-isolation relationships; species richness-energy relationships; longitudinal gradient in species richness; vertical (depth) gradients in species richness; species-range size distributions; geographical range structure; latitudinal gradients in geographical range size (Rapoport's rule); latitudinal gradients in abundance and body size (Bergmann's rule); species-body size distributions, and population density-body size relationships.

Mechanisms Acting at Basin Scales

Significant advances have been made in the past two decades in understanding the dynamics of broad-scale forcing at ocean basin scales. Elucidating the connections between these large-scale processes and local marine populations promises to be one of the exciting outcomes of the intersection of climate and ecological research in coming years. Studies of the links between climate, large-scale ocean dynamics, and ecological patterns are directly relevant to macroecological pattern. The major basin-wide ocean circulations are driven by global winds ultimately arising from latitudinal variation in solar heating, and strongly influenced by seasonal variability and Coriolis effects. As such, the major features, such as basin-scale circulations, formation of the central gyres, and intensification of western boundary currents, are reasonably well understood by physical oceanographers. While these features clearly respond to large-scale climate forcing, they also appear to have existed over long enough periods of earth history to allow the development of complex biological responses at scales from the life history characteristics of individual species to broad biogeographic patterns such as the basin-scale distribution of stony corals. The migrations of many large pelagic species appear to be in some sense "tuned" to ocean basin circulation patterns (Bakun 1996). For example, in the North Atlantic, migrations associated with completion of life cycles of large vertebrates such as turtles, tuna, and eels, as well as complex invertebrates such as squid, map well onto the patterns of the North Atlantic gyre and the western boundary current, the Gulf Stream. Poleward of the subtropical gyres in the North Atlantic and North Pacific, the subpolar gyres represent regions of intense, seasonal primary productivity. Large-scale surveys of water column productivity reveal striking macroecological patterns of chlorophyll concentration in relation to temperature, density, and other hydrographic parameters. For example, the abundance of shallow North Atlantic picoplankton is positively related to temperature, while the cytometric diversity of phtytoplankton decreases with temperature in this region (Li, this volume).

Major components of oceanographic variability at large spatial and temporal scales are the basin-wide oscillations such as the El Niño Southern Oscillation (ENSO), the North Atlantic Oscillation (NAO), and the Pacific Decadal Oscillation (PDO). Through the dramatic influence on climate at the scales of entire ocean basins, these coupled ocean-atmosphere processes can have large effects on both terrestrial and marine ecosystems (Glynn 1988; Beaugrand et al. 2002; Stenseth et al. 2002; Stenseth et al. 2003). A variety of indices of these large-scale processes have been developed and ecologists have explored a range of correlations with species-specific, community, and ecosystem-wide patterns. In addition to variation in areas directly affected by the ocean dynamics of these processes, so-called teleconnections can also exist, for example, among tropical and temperature environments. Fluctuations in the both the NAO and ENSO are associated with large-scale shifts in surface and higher atmospheric winds and are associated with measurable changes in surface currents and temperature (Greatbach 2000; Stenseth et al. 2002). The expression of ENSO in the eastern tropical Pacific has long been recognized as having a profound effect on local biological and ecological patterns such as along the Peruvian coast (Barber and Chavez 1983; Glynn 1985; Glynn 1988). An important aspect of the coupled ocean and ecosystem dynamics associated with basin-scale oscillations is a temporal dependence on the degree of correlation among events, as well as lags between forcing and biological responses. In some cases, strong links among forcing processes can exist at one time period, but the relationships can break down in others (Ottersen et al. 2001). Clearly, such variation through time in the relationships among processes make the results of broad-scale investigations sensitive to the time and timing of particular studies. These basin-scale processes can influence the scaling of population density and body mass, M, figure 8.2, population density via competitive relationships among algae in kelp forests (Dayton et al. 1999) and as well as link ecological processes at large spatial scales causing, for example, synchronous population fluctuations among independent populations (Moran 1953; Ranta, Kaitala, and Lindstrom 1999).

One of the general effects of basin-scale processes such as ENSO on marine organisms is severe food limitation during one phase of the oscillation (Glynn 1988; Dayton et al. 1999). Consequently, one of the major implications of ENSOs for macroecological patterns of body size and abundance are manifested through starvation-induced changes in body size, mortality, and failures of reproduction and recruitment. For example, a well-known macroecological relationship indicates that population density and body size (mass) are inversely related with an exponent of –3/4 (Damuth 1981). The

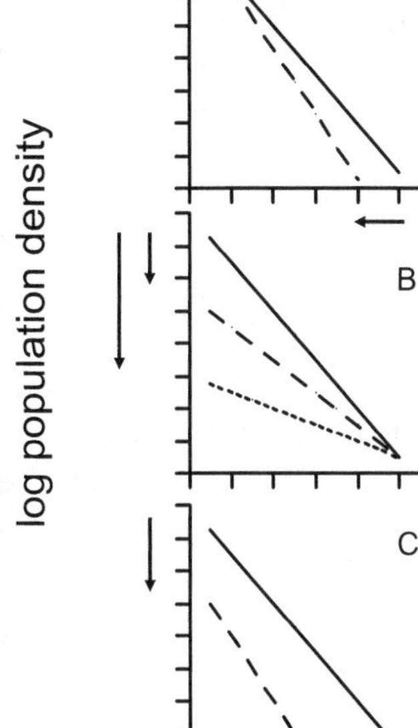

FIGURE 8.2 Conceptual model predicting how basin-scale oceanographic events such as ENSO would modify the scaling of body mass to population density in marine populations. In panel A, the size range of organisms is truncated by ENSO induced starvation and shrinkage of larger individuals, or by the mortality of large individuals (panel A). Consequently, the slope of the regression line would be increased, resulting in a larger negative exponent than the typical –3/4 value reported for a wide variety of organisms. Reproductive or recruitment failures associated with ENSO would impact the density-body mass scaling (panel B) by lowering the elevation of the intercepts of the regression lines, making the slope less steep and shifting the exponent toward smaller negative numbers. Both types of El Nino impacts may simultaneously occur (panel C); one effect would be to decrease the elevation of the regression line and increase the slope of the scaling relationship. One possible result would yield an exponent of –1.0 (panel C). The overall result of the model is that by limiting food resources, basin-wide climatic events may change the scaling of population density and body mass so that energetic equivalence is less likely to be an emergent macroecological property of post-ENSO marine populations.

–3/4 scaling between density and body mass holds for vertebrates (Damuth 1981; Blackburn et al. 1993) invertebrates (Marquet, Navarrete, and Castilla 1990; Marquet et al. 2005), phytoplankton (Ottersen et al. 2001; Li 2002) and terrestrial plants (Ottersen et al. 2001). The "energetic equivalence rule" was developed from the cross product of the scaling of whole-organism metabolic rate to body mass ($M^{3/4}$) and the ($M^{-3/4}$) scaling between density and body mass, yielding the exponent of M^0 (Damuth 1981). The rule indicates that the energy use of a population is independent of body size, and is similarly constrained for terrestrial and marine organisms (Damuth 1981; Ottersen et al. 2001). Figure 8.2 predicts how the scaling of population density and body mass would be altered by ENSO's and other basin-wide climatic events. Truncating the size range of organisms in the population by ENSO-induced starvation and shrinkage of larger individuals to smaller size classes or by the mortality of large individuals (fig. 8.2, panel A) would increase the slope of the regression line, resulting in a larger negative exponent than –3/4. The effect of reduced recruitment (fewer small individuals into the population) caused by reproductive or recruitment failures associated with El Niño—on the density mass scaling (fig 8.2, panel B) would be to lower the elevation of the intercepts of the regression lines, making the slope less steep and shifting the exponent toward smaller negative numbers. Both types of El Niño impacts (reduction in recruitment and reduction in food supply, and thus body mass) may occur together; one effect would be to decrease the elevation of the regression line and increase the slope of the scaling relationship. One possible result would yield an exponent of –1.0 (fig 8.2, panel C). The upshot of the model is that by causing food limitation, basin-wide climatic events may change the scaling of population density and body mass (either increasing or decreasing the exponent) so that the energetic equivalence is less likely to be an emergent macroecological property marine communities following El Niño events.

Another macroecological effect of ENSO, the PDO, and NAO is that they change the spatial distributions of marine organisms on basin-wide scales by affecting dispersal. This was evident by the dramatic range extension of a species of wrasse, apparently transported eastward from the Indo West Pacific by currents associated with the 1997–98 ENSO (Victor et al. 2001). Indeed, Glynn (1988) suggested that increased west-to-east current flows during ENSO years may generally serve as a conveyor of fish larvae from the tropical western Pacific to the eastern Pacific. In the North Atlantic Basin, major changes in copepod biodiversity and biogeography, such as a 10-degree northern latitudinal extension of warm-water species, have occurred in response to NAO events and increasing ocean temperatures (Beaugrand et al. 2002). Thus, a general effect of basin-wide climatic events is that they modify species ranges (Gaines et al., this volume), and consequently range-abundance relationships (Blackburn et al. 1993). By affecting long-distance dispersal and causing migration from one region to another, the ENSO, NAO, and PDO may change macroecological relationships between regional and local diversity (Ricklefs 1987) and increase the diversity of local communities embedded in the recipient region.

Major Currents Meanders, Frontal Eddies, and Spin-Off Rings

The major western boundary currents, such as the Gulf Stream or Kuroshio Current, tend to meander as they progress poleward. These meanders can remain part of the main flow, and are often associated with frontal eddies, or the meanders can pinch off of the main flow, forming isolated, rotating gyres or rings. In the case of the Gulf Stream, marked thermal fronts tend to occur between the relatively warm, poleward-flowing current and the cooler inshore waters. Considerable attention has been given to the effects of Gulf Stream/Florida Current meanders and frontal eddies along the south

Florida and the U.S. Atlantic Bight (Lee and Mayer 1977; Lee et al. 1992; Lee et al. 1994). Lee et al (1992, 1994) suggested that upwelling of isotherms associated with the passage of Florida Current (FC) frontal eddies could provide a major source of subsurface nutrients to the shallow water shelf of south east Florida. Sponagule et al. (2005) have recently shown periods of sharply elevated fish recruitment co-occurring with changes in the alongshore currents, indicative of the passage of FC frontal eddies. The passage of frontal eddies can also modulate the high-frequency variability associated with the thermocline in this and other regions. Along the Florida Keys reef tract, temperature time series show strong modulation of high-frequency variability over periods of five to seven days that appear to be at least partially associated with the passage of Florida Current eddies (Leichter, Stewart, and Miller 2003; Leichter, Deane, and Stokes 2005).

Cowen and colleagues have recently explored the effects of basin-scale circulation patterns in the Caribbean for the dispersal of fish larvae produced in close proximity to islands. Both the large-scale trajectory of currents and smaller-scale dynamics associated with island-specific flows can have an influence on predicted dispersal trajectories (Cowen et al. 2006; Purcell et al. 2006). However, even more important appears to be the specific swimming behaviors of these fish larvae even at very early stages in their development (e.g., within five to ten days). Migrations to deeper depths out of the major surface flows can dramatically increase the potential for retention near natal sites (Cowen, Paris, and Srinivasin 2006). The influence of such variation on macroecological patterns of abundance, body size and diversity (fig. 8.1) is likely to be exerted through the effects on cohort strength of at settlement and synchrony of recruitment.

In deeper waters farther from the shelf, oscillations of western boundary currents lead to large instabilities in current direction and velocity. These instabilities can generate spin-off rings representing semiclosed parcels of water, with the thermal and chemical signatures of the water found on the opposite sides of the currents (reviewed in Mann and Lazier 1996). In the North Atlantic, so-called cold core rings with scales of 100 to 300 km continue to rotate cyclonically and can maintain their integrity for up to a year. The cyclonic circulation of these rings produces upwelling of isotherms and nutrients toward the interior and associated primary production. Warm-core rings, generally smaller on scales of tens to 100 km, by contrast rotate anticyclonically on the western side of the Gulf Stream and often propagate inshore on the shelf, producing periods of anomalous warming and associated transport of subtropical organisms, for example, to the New England coastline in summer. On the order of ten cold-core rings typically form in any given year (Joyce et al. 1984), and as the waters within the rings age, distinct biological communities develop within the rings. From an ecological perspective, cold-core rings are of sufficient scale, with their own internal dynamics, that they represent self-contained, "replicate" ecosystems, or metacommunities (Holyoak, Leibold, and Holt 2005) perhaps an ideal setting in which to observe patterns giving rise to pelagic macroecological patterns. Gulf Stream rings can collide with shore, dissipate, or, commonly, make contact with the Gulf Stream and be reabsorbed into it.

Cold-core rings in the Gulf Stream system effectively transport large volumes of cool, relatively nutrient-rich waters into the nutrient-deplete surrounding Sargasso Sea. Rapid phytoplankton blooms can support locally elevated populations of planktivorous fishes, as well as large fish and seabirds that migrate into the food-rich patches. Over the course of months, as the surface waters within cold-core rings warm and nutrients are depleted, the primary production lessens and tends to move to deeper water. Thus, entire pelagic communities distinct from those in the surrounding Sargasso Sea form, with internal dynamics in cold-core rings. For warm-core rings, a somewhat more complicated pattern develops (Mann and Lazier 1996). Warm-core rings can have biological impacts in both the pelagic realm and, in cases where they move onto the shelf, on nearshore and benthic habitats. While the water initially contained in these rings tends to be nutrient deplete, the circulation at the periphery of the rings is favorable to upwelling of subsurface waters and associated nutrients. Stratification at the interior of these rings tends to be favorable for the development and maintenance of plankton blooms. Thus, warm-core rings, like their cold-core counterparts can, in fact, be areas of enhanced production relative to surrounding shelf waters, even though the water originally contained in the warm-core rings originates on the more oligotrophic side of the Gulf Stream. On the continental shelf, warm-core rings have the potential to move inshore, and at times they collide with the coastline. During such events, the onshore movement of warm water can be accompanied by transport of exotic warm-water species occasionally found, for example, in New England and as far north as the Scotian Shelf. The movement of warm water onto the shelf is accompanied by offshore advection of ambient-shelf waters, and this mechanism may be associated with offshore losses of locally produced shelf larvae. Thus, the collision of warm-core rings with shelf waters can lead to subsequent low recruitment. These processes are likely to influence both the timing and year-class strength of recruitment in fish and invertebrates with pelagic larvae and thus could influence macroecological patterns through cascading ecosystem effects on scales of tens to 100 km and time scales of weeks to months. Clearly, spin-off rings have a large effect on the size of species ranges, especially if the exotic migrants survive in the new environment long enough to reproduce.

Rossby and Kelvin Waves

Another large-scale feature of basin-wide circulation is the generation of Rossby and Kelvin waves. Rossby waves occur as large (hundreds of km) horizontal deflections of open ocean currents, slowly propagating across ocean basins. The occurrence of Rossby waves is due to equatorial-to-polar gradients in the Coriolis effect, associated with variation in the earth's rotation (Mann and Lazier 1996). Either depressions (local low pressure) or elevations (local high pressure) in a major open-ocean current, such as the equatorial current in the north Pacific, will slowly propagate in a westward direction. If embedded in an eastward current of faster mean velocity than the phase velocity of the Rossby wave, the wave features can have a net movement to the east, or if the phase velocity of the Rossby wave is exactly offset by an eastward current the wave feature can appear stationary, as so-called "arrested" Rossby waves (Bakun 1996). Estimated phase speeds for Rossby waves in the Pacific and the dependence of this speed on latitude suggest a time to cross the Pacific basin of nearly a decade at 30°, two years at 15°, and only seven months near the equator (Bakun 1996).

The dynamics of traveling Rossby waves change when they reach either eastern or western ocean boundaries. At these boundaries, the waves become "trapped" along the coastline in the form of faster-propagating Kelvin waves. At western boundaries, Kelvin waves of both depression or elevation of sea level are constrained to travel toward the equator in both the northern and southern hemispheres, whereas at eastern boundaries these features travel toward the poles in both hemispheres. For the western boundary cases, when coastally trapped Kelvin waves encounter the equator they propagate eastward along the equator as equatorial Kelvin waves along what has been termed the *equatorial wave guide* (see Bakun 1996). Kelvin waves can also become trapped around large oceanic islands where they tend to continuously circle the coastline.

The macroecological consequences of both Rossby and Kelvin are unknown but are likely to include influences on dispersal and recruitment. Indeed, Kelvin waves traveling eastward along the equator often signal the onset of an El Niño. The small changes in sea level (centimeters to tens of centimeters) caused by the waves are probably of little ecological significance, but there may be important effects associated with significant vertical displacements of the subsurface pycnocline/thermocline generated by these horizontally large features. For example, an unusual warming event occurred in the central Galápagos Islands, where temperatures rose abruptly to over 28°C in March 2005 and were sustained for one to two weeks, then decreased (fig 8.2, panel B). At one site (Rocas Gordon), the maximum temperature at 6 m depth was 29.4°C, which was above the 1982–1983 ENSO temperature maxima reported for the Galápagos (Glynn 1985; Glynn 1988, fig 8.2, panel B). At this time, there was a failure of barnacle recruitment, with *Megabalanus peninsularis* recruit densities lower than in any other of six other measurement periods in 2002–2005 (J. Witman, M. Brandt, unpublished data, Witman et al., in press). Witman and colleagues have hypothesized that the abrupt warming event was caused by an offshore warm-water mass traveling through the central Galápagos region as a Kelvin wave. Large-scale measurements of sea surface height taken around the temperature sampling (fig. 8.2, panel A) are consistent with this interpretation, as they show significantly elevated sea surface in the Galápagos region in March and early April of 2005, as would be expected for Kelvin waves (http://topex-www.jpl.nasa.gov/science/jason1-quick-look/). Local barnacle recruitment may have been depressed because the offshore, warm water associated with Kelvin waves was likely devoid of barnacle larvae.

Changes in mean thermocline depth associated with Kelvin waves can significantly affect higher-frequency processes, including internal waves associated with vertical water column stratification. When the episodic Kelvin wavelike phenomena are absent in the Galápagos, flows associated with internal waves (Witman and Smith 2003) appear to have a major influence on the recruitment of the barnacle *Megabalanus peninsularis,* a common prey of predatory fish and invertebrates. For example, barnacle recruitment to the subtidal zone was consistently higher at the top five-ranked upwelling sites than at sites with weak or no upwelling in 2002, 2003, and 2004 (Witman et al., unpublished ms). Barnacle recruitment is predicted by the vertical and horizontal flows associated with internal waves, which spread barnacle larvae up and down vertical rock walls, increasing the depth range and abundance of adult barnacles at strong upwelling sites. (Witman et al. in press). Pineda and Lopez (2002) examined patterns of water column stratification, temperature variability, and barnacle settlement at two sites separated by 100 km along the California coastline in Mexico and the United States. In addition to showing higher overall settlement at the northern site, where internal tidal activity was greatest, they suggested that local stratification is influenced by a combination of local wind forced upwelling and south-to-north propagation of thermocline shoaling events in coastally trapped waves. A

A.

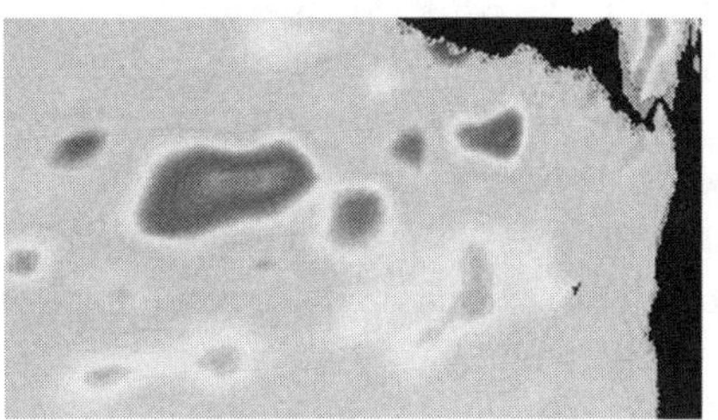

B.

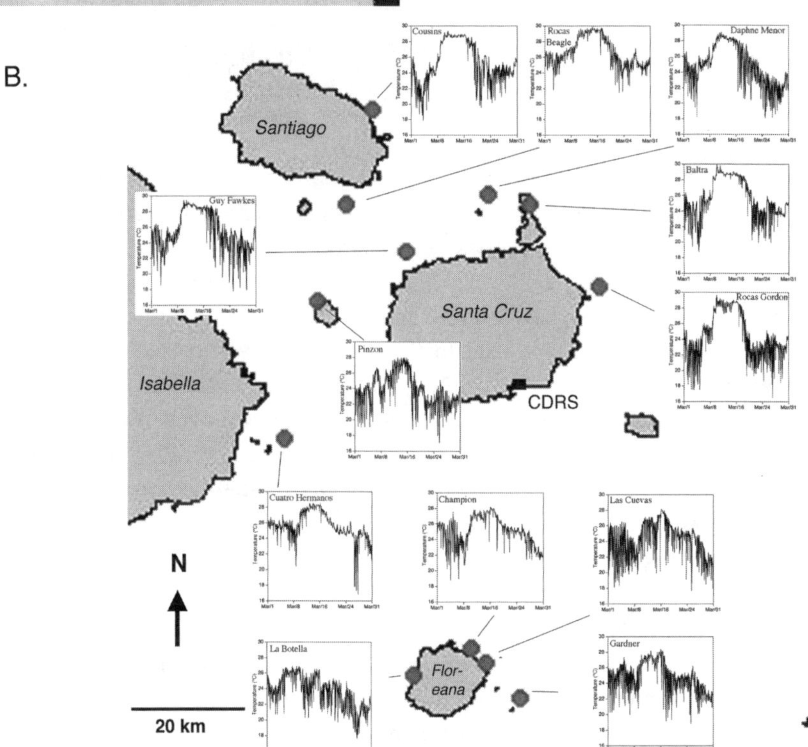

FIGURE 8.3 Panel A: Anomalies in the elevation of the sea surface in the eastern Pacific measured by the TOPEX- Poseidon and Jason satellites averaged over a ten-day period from March 31–April 9, 2005. Note the light-colored area extending eastward to the Galápagos (black dot off coast of South America at right.) Sea surface is approximately 20 mm higher than normal, which is consistent with the presence of Kelvin waves. Panel B: Graph of temperature records from rock walls at twelve sites (6 m depth) in the central Galápagos. Note that there was a rapid increase in temperature across the region for 1–2 weeks in March 2005. We hypothesize that they represent the temperature signal of a Kelvin wave passing through the archipelago. The macrecological effects of this phenomenon were apparently to reduce the subtidal recruitment of the barnacle, *Megabalanus peninsularis*.

number of other reports have pointed to the potential importance of coastally trapped waves in this region (Lentz and Winant 1986; Hickey, Dobbins, and Allen 2003). Pringel and Riser (2003) showed lagged correlation between alongshore winds in Baja California and nearshore temperature variability hundreds of km to the north in kelp beds near San Diego, with the lag consistent with predicted phase speed of first-mode coastally trapped waves. Coastally trapped waves are also thought to carry the signal of ENSO from north along the west coast of North America (Bakun 1996).

Dynamics Associated with Density Stratification—Internal Waves and Subsurface Chlorophyll Maximum (SCM) Layers

In areas of the ocean and at times of vertical water column density stratification (usually resulting from vertical temperature stratification with a smaller component of salinity stratification), internal waves can propagate either horizontally along density interfaces (interfacial waves) or at a range of angles in continuously stratified water columns. In coastal systems, the vertical density gradients tend to be concentrated within distinct pycnoclines, with a mean depth typically from 10 to 50 m depth, and one result is that most of the internal wave activity is likely to consist of interfacial waves. The dynamics and generation of internal waves have been well treated (Pond and Pickard 1983; Mann and Lazier 1996; Garrett 2001). Macroecological consequences of internal waves can be found within and near the pycnocline/thermocline, at locations where these density interfaces intersect the bottom (Witman et al. 2004), and also well away from the interfaces at the ocean surface (Shanks 1995). The surface slicks associated with subsurface internal waves can often be detected visually from ships and from shore, as well as from aircraft and satellites. A recent atlas of internal waves photographed from space shows that internal waves are prevalent in nearly all oceans and coastal areas of the world (http://www.internalwaveatlas.com/Atlas_index.html). Synthetic aperture radar is also a powerful tool for detecting internal wave slicks in coastal settings. Surface slicks have been implicated in the aggregation and transport of invertebrate larvae (Kingsford and Choat 1986; Shanks 1995). In order for significant transport to occur, however, it appears likely the slicks must be formed over nonlinear internal waves. Nonlinear internal waves are likely in many coastal settings because the vertical thickness (height) of the density layers above and below the pycnocline are not large relative to the amplitude and wave length of the incident internal waves, as required for interfacial waves to behave according to linear wave theory. Instead, the waves can become asymmetrical, can steepen, and may break. Often the leading front of the internal tide contains a packet of much higher-frequency internal waves (Leichter et al. 1996; Leichter et al. 1998). Upon running into shallow water these waves can break, forming internal bores with strong onshore transport. There can also subsequently be strong offshore transport. Along the coast of southern California, Pineda proposed a two-phase process, whereby the initial onshore flow of a subsurface cool layer is followed by offshore flow near the bottom and strong onshore flow of a surface, a warm bore that may transport larvae (Pineda 1991; Pineda 1994).

Internal waves in coastal settings are likely to cause vertical and horizontal transport of dissolved nutrients and suspended particles, including phytoplankton and zooplankton (Pingree and Mardell 1981; Shea and Broenkow 1982; Sandstrom and Elliott 1984; Leichter et al. 1998; Leichter, Stewart, and Miller 2003). Internal waves can also have dramatic influences on near-bottom turbulence and sediment transport (Cacchione and Drake 1986; Cacchione, Pratson, and Ogston 2002). The hypothesized biological significance of internal waves includes enhanced water column productivity in the region of the shelf break. The impact of internal waves has been suggested as an explanation for enhanced productivity at shelf-break zones. Often at the shelf break a zone of significantly enhanced chlorophyll concentration can be delineated along with elevated concentrations of dissolved macronutrients. Recently, Leichter et al. (1996) described the impact of internal waves for the coral reef tract of the Florida Keys and showed (Leichter, Stewart, and Miller 2003) significantly elevated nutrient fluxes associated with this mechanism.

Another consequence of vertical temperature stratification is that it influences the location and development of a subsurface layer of concentrated phytoplankton, known as the Subsurface Chlorophyll Maximum (SCM) layer (Cullen 1982). This is because the thermocline represents a transition between the warm, sunlit, shallow-surface layer, which is often low in nutrients and the deeper cold, nutrient-rich waters (Mann and Lazier 1996). The vertical position of phytoplankton represents, in some sense, a compromise between the requirement for high irradiance near the surface to maximize photosynthesis and the requirement for nutrients concentrated in deeper waters. Consequently, the greatest concentration of phytoplankton typically occurs at the interface between the two conditions at the base of the thermocline. It is not unusual for chlorophyll *a* concentrations to be ten-fold greater in the SCM layer than above or below it (fig. 8.4; Witman et al. 1993). SCM layers can be spatially continuous for hundreds of kilometers (Townsend,

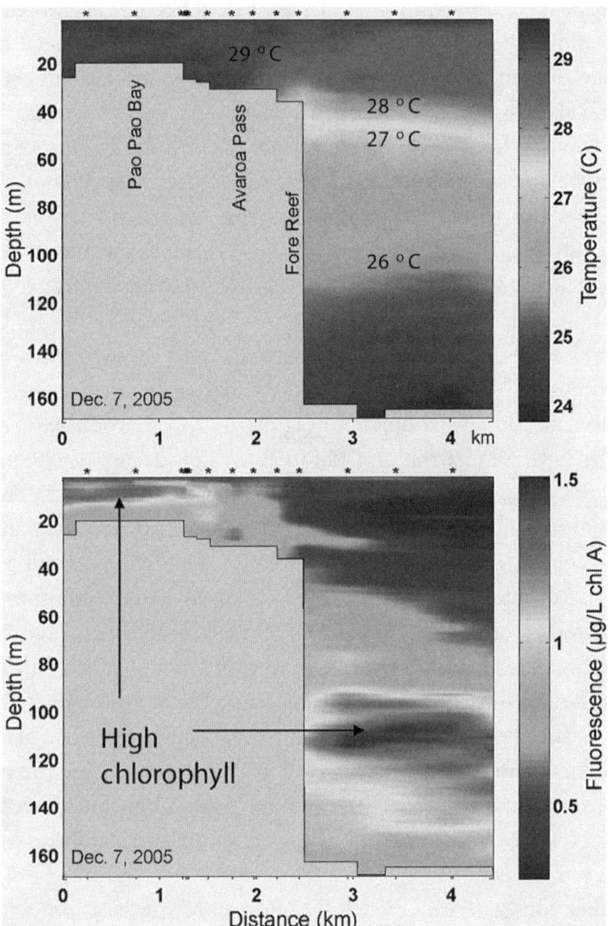

FIGURE 8.4 Section plots of temperature (top) and chlorophyll-*a* concentration (bottom) on the north shore of Moorea, French Polynesia. Sampling is from the head of PaoPao Bay out to a distance of 5 km offshore. Points along top axes indicate CTD station locations. Thermal variation between the bay and ocean and vertical stratification offshore are evident. The top of the offshore subsurface chlorophyll maximum (SCM) layer is associated with the 26° C isotherm. An inshore region of high chlorophyll concentration is associated with elevated nutrient levels and high productivity in the inner bay.

Cucci, and Berman 1984), and they are common in many areas of the world's oceans when the water column is vertically stratified (fig. 8.4). Productivity at the subsurface chlorophyll-maximum layer can be enhanced by internal waves, both by the transport of nutrients above the pycnocline (if there is turbulence generated at the interface) and also by transiently elevating the entire layer into shallower depths, where light availability is greater. On rocky pinnacles in the central Gulf of Maine, Witman et al. (1993, 2004b) described predictable vertical displacements of a marked SCM layer by internal waves creating sharp increases in the fluxes of phytoplankton (measured as chlorophyll *a*) to benthic suspension feeders at 30 m depth

Since SCM layers are regions of concentrated productivity that are widely distributed, they likely have many unexplored effects on macroecological patterns and processes on the bottom and in the water column. The supply of phytoplankton food and other nutrient resources should be higher where SCM layers impinge on the bottom, either in the highly dynamic environment of internal waves (Zimmerman and Kremer 1984; Leichter et al. 1996; Witman and Smith 2003) or where they are horizontally advected from the water column to the bottom habitats at the same depth as the SCM. It is logical that large areas of the bottom that routinely receive up to ten times more food due to the influence of SCM layers support communities with different productivity-driven macroecological relationships than other areas. From published locations of SCM layers, areas that are likely influenced by SCM food supply include: intermediate depths (10 to 30 m) in the rocky subtidal zone, the shelf-break zone, depths of perhaps 35 to 80 m on coral reefs, and other topographically high areas of the bottom that generate and receive internal waves. If the greater food supply in SCM-influenced areas resulted in greater body size than non-SCM areas, then any of the macroecological patterns related to body size, such as range size-body size, population density-body size, and species richness-body-size relationships would be expected to differ between SCM-influenced and noninfluenced areas. Testing the hypothesis that the body sizes of species that consume food resources of SCM layers are larger in benthic habitats influenced by the SCM than in those where the SCM is not present would be a logical first step toward testing the associated macroecological relationships. There are few tests of species-energy theory (Mittlebach et al. 2001) in marine communities (Shinn et al. 1989; Chown and Gaston 1999; Roy, Jablonski, and Valentine 2000; Witman et al. 2008). As productivity is often used as an energy variable to predict diversity, a testable hypothesis would be that productivity controls on diversity vary between areas with and without the influence of the SCM layer.

Summary and Future Directions

As vectors of food and larvae, oceanographic processes are fundamental to macroecological patterns and relationships in the ocean. Enough is known about physical phenomena at basin-wide scales to pose testable hypotheses about how they drive macroecological relationships and how these relationships may change with changing environmental conditions including global climate changes. Our detailed and conceptual knowledge in this area is limited however, to the effects of current mediated dispersal on the size of populations in the rocky intertidal zone, the effects of upwelling on some components of intertidal food webs, and to ENSO-driven mortalities of marine organisms. We suggest that it is time to broaden the perspective on oceanographic forcing of marine populations, communities and ecosystems. Exciting progress could be made by investigating basin-scale oceanographic processes and their effect on macroecological patterns in the following areas:

- Studying food and recruitment limitation effects of ENSO, as surrogates for climate change, on macroecological patterns involving body size, distribution, and abundance and diversity.
- Oceanographic influences on species diversity, in particular, are grossly understudied, yet they have great potential to enhance local diversity by increasing the supply of larvae and propagules, to reduce diversity as agents of mortality, and to affect the magnitude of diversity as linear or nonlinear functions of productivity.
- Exploring specific effects of Rossby and Kelvin waves on dispersal and recruitment.
- Studying regional effects of cold- and warm-core rings on species ranges and range-abundance relationships.
- Investigating the macroecological effects of productivity that is vertically concentrated in the water column on body size, species diversity, and abundance

Since many of these oceanographic phenomena are episodic, their imprint on macroecological patterns will be difficult to resolve unless time series data are available from fixed locations or from remote sensing across spatial scales from one to thousands of kilometers.

ACKNOWLEDGMENTS

Our work on macroecological effects of oceanographic processes was supported by NOAA's National Undersea Research Program and by the Biological Oceanography Program of the U.S. National Science Foundation.

REFERENCES

Bakun, A. 1996. *Patterns in the ocean.* Monterey, CA: California Sea Grant.
Barber, R. T., and F. P. Chavez. 1983. Biological consequences of El Nino. *Science* 222:1203–10.

Beaugrand, G., P. C. Reid, F. Ibanez, J. A. Lindley, and M. Edwards. 2002. Reorganization of North Atlantic marine copepod biodiversity and climate. *Science* 296:1692–94.

Blackburn, T. M., V. K. Brown, B. M. Doube, J. J. D. Greenwood, J. H. Lawton, and N. E. Stork. 1993. The relationship between abundance and body-size in natural animal assemblages. *Journal of Animal Ecology* 62:519–28.

Brown, J. H. 1995. *Macroecology.* Chicago: University of Chicago Press.

Cacchione, D. A., and D. E. Drake. 1986. Nepheloid layers and internal waves over continental shelves and slopes. *Geo-Marine Letters* 6:147–52.

Cacchione, D. A., L. F. Pratson, and A. S. Ogston. 2002. The shaping of continental slopes by internal tides. *Science* 296:724–27.

Chown, S. L., and K. J. Gaston. 1999. Patterns in procellariiform diversity as a test of species energy theory. *Evolutionary Ecology Research* 1:365–73.

Connolly, S. R., B. A. Menge, and J. Roughgarden. 2001. A latitudinal gradient in recruitment of intertidal invertebrates in the northeast Pacific Ocean. *Ecology* 82:1799–1813.

Cowen, R. K., C. B. Paris, and A. Srinivasan. 2006. Scaling of connectivity in marine populations. *Science* 311:522–27.

Cullen, J. J. 1982. The deep chlorophyll maximum: Comparing vertical profiles of chlorophyll a. *Canadian Journal of Fisheries and Aquatic Science* 39:791–803.

Cushing, D. H. 1975. *Marine ecology and fisheries.* Cambridge: Cambridge University Press.

———. 1990. Plankton production and year-class strength in fish populations: An update of the match/mismatch hypothesis. *Adv. Mar. Biol.* 26:249–93.

Damuth, J. 1981. Population-density and body size in mammals. *Nature* 290:699–700.

Dayton, P. K., M. J. Tegner, P. B. Edwards, and K. L. Riser. 1999. Temporal and spatial scales of kelp demography: The role of oceanographic climate. *Ecological Monographs* 69:219–50.

Denny, M. 1995. Predicting physical disturbance: Mechanistic approaches to the study of survivorship on wave-swept shores. *Ecological Monographs* 65:371–418.

Denny, M., V. Brown, E. Carrington, G. Kraemer, and A. Miller. 1989. Fracture mechanics and the survival of wave-swept macroalgae. *Journal of Experimental Marine Biology and Ecology* 127:211–28.

Denny, M. W., B. Helmuth, G. H. Leonard, C. D. G. Harley, L. J. H. Hunt, and E. K. Nelson. 2004. Quantifying scale in ecology: Lessons from a wave-swept shore. *Ecological Monographs* 74:513–32.

Farrell, T. M., D. Bracher, and J. Roughgarden. 1991. Cross-shelf transport causes recruitment to intertidal populations in central California [USA]. *Limnology and Oceanography* 36:279–88.

Gaines, S., S. Brown, and J. Roughgarden. 1985. Spatial variation in larval concentrations as a cause of spatial variation in settlement for the barnacle, *Balanus glandula. Oecologia* (Berlin) 67:267–72.

Gaines, S., and J. Roughgarden. 1985. Larval settlement rate: A leading determinant of structure in an ecological community of the marine intertidal zone. *Proceedings of the National Academy of Sciences, USA* 82:3707–11.

Garrett, C. 2001. Internal waves. In *Encyclopedia of ocean sciences* ed. J. H. Steel, S. A. Thorpe, and K. K. Turekian. Academic Press, San Diego.

Gaston, K. J., and T. M. Blackburn. 2000. *Pattern and process in macroecology.* Malden, MA: Blackwell Science.

Glynn, P. W. 1985. El-Nino-associated disturbance to coral reefs and post disturbance mortality by Acanthaster-Planci. *Marine Ecology-Progress Series* 26:295–300.

———. 1988. El Nino–Southern Oscillation 1982–1983: Nearshore population, community, and ecosystem responses. *Ann. Rev. Ecol. Syst.* 19:309–45.

Greatbach, R. J. 2000. The North Atlantic Oscilllation. *Stochastic Environmental Research and Risk Assessment* 14:213–42.

Hickey, B. M., E. L. Dobbins, and S. E. Allen. 2003. Local and remote forcing of currents and temperature in the central Southern California Bight. *Journal of Geophysical Research-Oceans* 108.

Hjort, J. 1914. Fluctuations in the great fisheries of northern Europe. *Rapports, Conceil Permanent International pour l'Exploration de la Mer* 20:1–227.

———. 1926. Fluctuations in the year classes of important food fishes. *J. Cons. Int. Explor. Mer* 1:5–18.

Holyoak, M., M. A. Leibold, and R. D. Holt. 2005. *Metacommunities: Spatial dynamics and ecological communities.* Chicago: University of Chicago Press.

Jackson, J. B. C., and K. G. Johnson. 2000. Life in the last few million years. *Paleobiology* 26:221–35.

Joyce, T., R. Backus, K. Baker, P. Blackwelder, O. Brown, T. Cowles, R. Evans, et al. 1984. Rapid evolution of a Gulf-Stream warm-core ring. *Nature* 308:837–40.

Kingsford, M. J., and J. H. Choat. 1986. Influence of surface slicks on the distribution and onshore movements of small fish. *Marine Biology* (Berlin) 91:161–72.

Lee, T. N., M. E. Clarke, E. Williams, A. F. Szmant, and T. Berger. 1994. Evolution of the Tortugas gyre and its influence on recruitment in the Florida Keys. *Bulletin of Marine Science* 54:621–46.

Lee, T. N., and D. A. Mayer. 1977. Low-frequency current variability and spin-off eddies along shelf off southeast Florida. *Journal of Marine Research* 35:193–220.

Lee, T. N., C. Rooth, E. Williams, M. McGowan, A. M. Szmant, and M. E. Clarke. 1992. Influence of Florida Current, gyres and wind-driven circulation on transport of larvae and recruitment in the Florida Keys coral reefs. *Cont. Shelf Res.* 12:971–1002.

Leichter, J. J., G. B. Deane, and M. D. Stokes. 2005. Spatial and temporal variability of internal wave forcing on a coral reef. *Journal of Physical Oceanography* 35:1945–62.

Leichter, J. J., G. Shellenbarger, S. J. Genovese, and S. R. Wing. 1998. Breaking internal waves on a Florida (USA) coral reef: A plankton pump at work? *Marine Ecology Progress Series* 166:83–97.

Leichter, J. J., H. L. Stewart, and S. L. Miller. 2003. Episodic nutrient transport to Florida coral reefs. *Limnology and Oceanography* 48:1394–1407.

Leichter, J. J., S. R. Wing, S. L. Miller, and M. W. Denny. 1996. Pulsed delivery of subthermocline water to Conch Reef (Florida Keys) by internal tidal bores. *Limnology and Oceanography* 41:1490–1501.

Lentz, S. J., and C. D. Winant. 1986. Subinertial currents on the Southern-California shelf. *Journal of Physical Oceanography* 16:1737–50.

Li, W. K. 2002. Macroecological patterns of phytoplankton in the northwestern Atlantic Ocean. *Nature* 419:154–57.

Mann, K. H., and J. R. N. Lazier. 1996. *Dynamics of marine ecosystems: Biological-physical interactions in the oceans.* Cambridge, MA: Blackwell Science.

Marquet, P. A., S. A. Navarrete, and J. C. Castilla. 1990. Scaling population-density to body size in rocky intertidal communities. *Science* 250:1125–27.

Marquet, P. A., R. A. Quinones, S. Abades, F. Labra, M. Tognelli, M. Arim, and M. Rivadeneira. 2005. Scaling and power-laws in ecological systems. *Journal of Experimental Biology* 208:1749–69.

Menge, B. A., C. Blanchette, P. Raimondi, T. Freidenburg, S. Gaines, J. Lubchenco, D. Lohse, G. Hudson, M. Foley, and J. Pamplin. 2004. Species interaction strength: Testing model predictions along an upwelling gradient. *Ecological Monographs* 74:663–84.

Menge, B. A., B. A. Daley, P. A. Wheeler, E. Dahlhoff, E. Sanford, and P. T. Strub. 1997. Benthic-pelagic links and rocky intertidal communities: Bottom-up effects on top-down control? *Proceedings of the National Academy of Sciences, USA* 94:14530–35.

Mittlebach, G. G., C. F. Steiner, S. M. Scheiner, K. L. Gross, H. L. Reynolds, R. B. Waide, M. R. Willig, S. I. Dodson, and L. Gough. 2001. What is the observed relationship between species richness and productivity? *Ecology* 82:2381–96.

Moran, P. A. P. 1953. The statistical analysis of the Canadian lynx cycle II. Synchronization and meteorology. *Australian Journal of Zoology* 1:291–98.

Navarrete, S. A., E. A. Wieters, B. R. Broitman, and J. C. Castilla. 2005. Scales of benthic-pelagic and the intensity of species interactions: From recruitment limitation to top-down control. *Proceedings of the National Academy of Sciences, USA* 102:18046–51.

Ottersen, G., B. Planque, A. Belgrano, E. Post, P. C. Reid, and N. C. Stenseth. 2001. Ecological effects of the North Atlantic Oscillation. *Oecologia* 128:1–14.

Pineda, J. 1991. Predictable upwelling and the shoreward transport of planktonic larvae by internal tidal bores. *Science* 253:548–51.

———. 1994. Internal tidal bores in the nearshore: Warm-water fronts, seaward gravity currents and the onshore transport of neustonic larvae. *Journal of Marine Research* 52:427–58.

Pineda, J., and M. Lopez. 2002. Temperature, stratification and barnacle larval settlement in two Californian sites. *Continental Shelf Research* 22:1183–98.

Pingree, R. D., and G. T. Mardell. 1981. Slope turbulence, internal waves and phytoplankton growth at the Celtic Sea Shelf-Break. *Philosophical Transactions of the Royal Society of London Series A—Mathematical Physical and Engineering Sciences* 302:663–78.

Pond, S., and G. L. Pickard. 1983. *Introductory dynamical oceanography,* 2nd ed. Oxford: Pergamon.

Pringle, J. M., and K. Riser. 2003. Remotely forced nearshore upwelling in Southern California. *Journal of Geophysical Research-Oceans* 108 (C4): 3131.

Purcell, J. F. H., R. K. Cowen, C. R. Hughes, and D. A. Williams. 2006. Weak genetic structure indicates strong dispersal limits: A tale of two coral reef fish. *Proceedings of the Royal Society of London B* 273:1483–90.

Ranta, E., V. Kaitala, and J. Lindstrom. 1999. Spatially autocorrelated disturbances and patterns in population synchrony. *Proceedings of the Royal Society of London B* 266:1851–56.

Reid, P. C., M. Edwards, G. Beaugrand, M. Skogen, and D. Stevens. 2003. Periodic changes in the zooplankton of the North Sea during the twentieth century linked to oceanic inflow. *Fisheries Oceanography* 12:260–69.

Ricklefs, R. E. 1987. Community diversity—Relative roles of local and regional processes. *Science* 235:167–71.

Roughgarden, J., S. Gaines, and H. Possingham. 1988. Recruitment dynamics in complex life cycles. *Science* 241:1460–66.

Roy, K., D. Jablonski, and J. W. Valentine. 2000. Dissecting latitudinal diversity gradients: Functional groups and clades of marine bivalves. *Proceedings of the Royal Society of London B* 267:293–99.

Sandstrom, H., and J. A. Elliott. 1984. Internal tide and solitons on the Scotian shelf: A nutrient pump at work. *Journal of Geophysical Research* 89:6415–26.

Shanks, A. L. 1995. Mechanisms of cross-shelf dispersal of larvale invertebrates and fish. In *Ecology of marine invertebrate larvae,* ed. L. McEdward, 323–67. Boca Raton, FL: CRC.

Shea, R. E., and W. W. Broenkow. 1982. The Role of internal tides in the nutrient enrichment of Monterey Bay, California. *Estuarine Coastal and Shelf Science* 15:57–66.

Shinn, E. A., B. H. Lidz, J. L. Kindinger, J. H. Hudson, and R. B. Halley. 1989. *Reefs of Florida and the Dry Tortugas: A guide to the modern carbonate environments of the Florida Keys and the Dry Tortugas.* St. Petersburg, FL: U.S. Geological Survey.

Sponaugle, S., T. Lee, V. Kourafalou, and D. Pinkard. 2005. Florida current frontal eddies and the settlement of coral reef fishes. *Limnology and Oceanography* 50:1033–48.

Stenseth, N. C., A. Mysterud, G. Ottersen, J. W. Hurrell, K. S. Chan, and M. Lima. 2002. Ecological effects of climate fluctuations. *Science* 297:1292–96.

Stenseth, N. C., G. Ottersen, J. W. Hurrell, A. Mysterud, M. Lima, K. S. Chan, N. G. Yoccoz, and B. Adlandsvik. 2003. Studying climate effects on ecology through the use of climate indices: the North Atlantic Oscillation, El Nino Southern Oscillation and beyond. *Proceedings of the Royal Society of London B* 270:2087–96.

Townsend, D. W., T. L. Cucci, and T. Berman. 1984. Subsurface chlorophyll maxima and vertical-distribution of zooplankton in the Gulf of Maine. *Journal of Plankton Research* 6:793–802.

Victor, B. C., G. M. Wellington, D. R. Robertson, and B. I. Ruttenberg. 2001. The effect of the El Nino-Southern Oscillation event on the distribution of reef-associated labrid fishes in the eastern Pacific Ocean. *Bulletin of Marine Science* 69:279–88.

Wing, S. R., L. W. Botsford, J. L. Largier, and L. E. Morgan. 1995a. Spatial structure of relaxation events and crab settlement in the northern California upwelling system. *Marine Ecology Progress Series* 128:199–211.

Wing, S. R., L. W. Botsford, S. V. Ralston, and J. L. Largier. 1998. Meroplanktonic distribution and circulation in a coastal retention zone of the northern California upwelling system. *Limnology and Oceanography* 43:1710–21.

Wing, S. R., J. L. Largier, L. W. Botsford, and J. F. Quinn. 1995b. Settlement and transport of benthic invertebrates in an intermittent upwelling region. *Limnology and Oceanography* 40:316–29.

Witman, J. D. 2007. Benthic pelagic coupling. In *Encyclopedia of tidepools and rocky shores,* ed. M. W. Denny and S. D. Gaines, 68–71. Berkeley, CA: University of California Press.

Witman, J. D., M. Brandt, and F. Smith. In press. Coupling between subtidal prey and consumers along a mesoscale upswelling gradient in the Galápagos Islands. *Ecological Monographs.*

Witman, J. D., M. Cusson, P. Archambault, A. J. Pershing, and N. Mieszkowska. 2008. The relation between productivity and species diversity in temperate–arctic marine ecosystems. *Ecology* 89 (11) Supplement: S66–S80.

Witman, J. D., J. J. Leichter, S. J. Genovese, and D. A. Brooks. 1993. Pulsed phytoplankton supply to the rocky subtidal zone: Influence of internal waves. *Proceedings of the National Academy of Sciences, USA* 90:1686–90.

Witman, J. D., M. R. Patterson, and S. J. Genovese. 2004. Benthic pelagic linkages in subtidal communities: Influence of food subsidy by internal waves. In *Food webs at the landscape level,* ed. G. A. Polis, M. E. Power, and G. R. Huxel, 133–53. Chicago: University of Chicago Press.

Witman, J. D., and F. Smith. 2003. Rapid community change at a tropical upwelling site in the Galapagos Marine Reserve. *Biodiversity and Conservation* 12:25–45.

Zimmerman, R. C., and J. N. Kremer. 1984. Episodic nutrient supply to a kelp forest ecosystem in Southern California. *Journal of Marine Research* 42:591–604.

CHAPTER NINE

DISPERSAL AND GEOGRAPHIC RANGES IN THE SEA

STEVEN D. GAINES, SARAH E. LESTER, GINNY ECKERT, BRIAN P. KINLAN, RAFE SAGARIN, AND BRIAN GAYLORD

One of the most fundamental characteristics of any species is its geographic range—the collection of all locations where the species occurs. Although the study of geographic ranges has been a historically important component of the field of biogeography, it has emerged in recent decades as it own discipline—areography (Rapoport 1982). Geographic ranges vary dramatically in size, shape, and location. This variation has been obvious for centuries and has yielded important insight into a wide range of fundamental issues in ecology and evolution (Gaston 2003). There is no shortage of hypothesized explanations for areographic variation.

One factor that has garnered considerable attention as a potential driver of species' ranges is dispersal. Darwin (1859) wrote extensively on the potential influences of dispersal and barriers to dispersal on the extent of species' distributions. He even performed experiments on the potential for seeds and dried materials from scores of species to disperse by floating in the sea. Moreover, his observations that the fauna of a given location cannot be entirely explained by environmental conditions and that striking faunistic differences exist between the New and Old World suggest an important role for dispersal shaping species' distributions.

> In considering the distribution of organic beings over the face of the globe, the first great fact which strikes us is, that neither the similarity nor the dissimilar-

> ity of the inhabitants of various regions can be wholly accounted for by climatal and other physical conditions. Of late, almost every author who has studied the subject has come to this conclusion. The case of America alone would almost suffice to prove its truth; for if we exclude the arctic and northern temperate parts, all authors agree that one of the most fundamental divisions in geographical distribution is that between the New and Old Worlds; yet if we travel over the vast American continent, from the central parts of the United States to its extreme southern point, we meet with the most diversified conditions; humid districts, arid deserts, lofty mountains, grassy plains, forests, marshes, lakes, and great rivers, under almost every temperature. There is hardly a climate or condition in the Old World which cannot be paralleled in the New—at least as closely as the same species generally require. No doubt small areas can be pointed out in the Old World hotter than any in the New World; but these are not inhabited by a fauna different from that of the surrounding districts; for it is rare to find a group of organisms confined to a small area, of which the conditions are peculiar in only a slight degree. Notwithstanding this general parallelism in the conditions of the Old and New Worlds, how widely different are their living productions!

Similarly, by the early twentieth century, Grinnell (1922) clearly recognized the role of rare dispersal events on species' distributions.

> These *pioneers* are of exceeding importance to the species in that they are continually being centrifuged off on scouting expeditions (to mix the metaphor), to seek new country which may prove fit for occupancy. The vast majority of such individuals, 99 out of every hundred perhaps, are foredoomed to early destruction without any opportunity of breeding. Some few individuals may get back to the metropolis of the species. In the relatively rare case two birds comprising a pair, of greater hardihood, possibly, than the average, will find themselves a little beyond the confines of the metropolis of the species, where they will rear a brood successfully and thus establish a new outpost. Or, having gone farther yet, such a pair may even stumble upon a combination of conditions in a new locality the same as in its parent metropolis, and there start a new detached colony of the species.

Despite centuries of interest, definitive connections between dispersal scale and many biogeographical patterns have remained speculative. One of the key problems has been the difficulty of studying dispersal. Unbiased quantitative estimates of dispersal distances are rare across the range of variation within and among species in most natural ecosystems. The ultimate goal is to characterize the probability that individuals will disperse any given distance and see how this dispersal kernel varies among species, locations, and times. In practice, techniques for measuring dispersal distances generally offer a very narrow window on the full kernel. For example, mark recapture studies can definitively show that an individual moved between two locations, but they frequently underestimate the long-distance tails of dispersal distributions, since the likelihood of recapture declines with distance. Moreover, many species have dispersing life stages that are difficult to tag at all. This is especially true in the sea, where most species have a microscopic planktonic life stage that can disperse on ocean currents for days to months, depending on the species.

However, while measuring dispersal in the ocean may be difficult, dispersal data are particularly valuable for marine systems, where species show remarkable variation in both dispersal capability (Kinlan and Gaines 2003; Shanks, Grantham, and Carr 2003; Siegel et al. 2003; Kinlan, Gaines, and Lester 2005) and biogeographic patterns (Lester and Ruttenberg 2005; Lester et al. 2007).

Fortunately, in the last few years, the window on dispersal in the sea has begun to open with advances in archival and broadcasting tags (Block et al. 1998; Block et al. 2001; Boustany et al. 2002; Block et al. 2005), syntheses of rates of spread of exotic species (Kinlan and Gaines 2003; Shanks, Grantham, and Carr 2003; Kinlan and Hastings 2005), syntheses of genetic estimates of average dispersal distance (Wares, Gaines, and Cunningham 2001; Kinlan and Gaines 2003; Palumbi 2003; Kinlan, Gaines, and Lester 2005), new tagging techniques to identify sites of natal origin (Levin et al. 1993; Swearer et al. 1999; Thorrold et al. 2001; Thorrold et al. 2002; Palumbi et al. 2003; Zacherl et al. 2003; Jones, Planes, and Thorrold 2005), and new models of particle transport (Siegel et al. 2003; Cowen, Paris, and Srinivasan 2006; Gaylord et al. 2006).

As a result of progress in quantifying dispersal in the sea, there has been a resurgence of interest in exploring the consequences of variation in dispersal distance on characteristics of marine species' ranges. Here we examine recent advances in these connections for three characteristics of species ranges in the sea: (a) the size of geographic ranges, (b) the location of species borders, and (c) the distribution of individuals within a species' range. We chose these three characteristics because they each highlight notable findings. For the first two, the emerging results differ greatly from prior expectations. For the third, we are at the incipient stages of developing a strong mechanistic link between the marine biogeographic patterns and dispersal.

The issues we explore are not restricted in any way to marine systems.

Indeed, the connections between dispersal and these three macroecological characteristics of species' distributions are of broad ecological importance in all habitats (see general discussions in Gaston 2003). Marine species, however, offer several advantages for characterizing both the patterns and the potential underlying mechanisms. First, a broad diversity of marine animals and plants have relatively sedentary adults and disperse only as larval propagules. The isolation of dispersal within this early life stage helps separate the roles of dispersal from migration and other more directed forms of adult movement. Second, the range of variation in propagule dispersal is enormous. Average dispersal distance varies by more than seven orders of magnitude among species (Kinlan and Gaines 2003; Shanks, Grantham, and Carr 2003; Kinlan, Gaines, and Lester 2005). Third, this broad range of dispersal distances occurs within many distantly related taxonomic groups, which affords better opportunities to separate the effects of dispersal from other phylogenetically confounded factors (Kinlan and Gaines 2003; Kinlan, Gaines, and Lester 2005). Finally, for shallow-water coastal species, ranges have a simplified geometry. Since the depth component of the range is typically miniscule for such species relative to their latitudinal or longitudinal extent, coastal species essentially have a one-dimensional range with only two boundaries. Compared to the two-dimensional boundary that circumscribes most terrestrial species' ranges, this geometrical simplification greatly facilitates the exploration of a wide range of macroecological issues (Sagarin, Gaines, and Gaylord 2006).

The Influence of Dispersal on Range Size

Geographic ranges vary dramatically in size. They can be as small as a single reef or bay and as large as all of the world's oceans. The underlying causes of this variation are surely myriad and include both ecological and evolutionary factors. An organism's ability to disperse is one of the more commonly cited causes of variation in range size (Hanski et al. 1993; Kunin and Gaston 1993; Brown, Stevens, and Kaufman 1996; Gaston 1996). Examples of this claim abound for both marine (Shuto 1974; Hansen 1978; Hansen 1980; Jablonski 1982; Perron and Kohn 1985; Jablonski 1986; Scheltema 1989; Emlet 1995; Bonhomme and Planes 2000; Victor and Wellington 2000; Bradbury and Snelgrove 2001) and terrestrial species (insects—Juliano 1983; Gutierrez and Menendez 1997; Dennis et al. 2000; birds—Duncan, Blackburn, and Veltman 1999; and plants—Edwards and Westoby 1996; Thompson, Gaston, and Band 1999; Clarke, Kerrigan, and Westphal 2001; Kessler 2002).

Although the specific rationale for a connection between dispersal scale and range size is rarely stated explicitly, there are three broad classes of mechanistic hypotheses proposed to account for such a relationship (Lester et al. 2007):

- Site colonization hypotheses
- Speciation-rate hypotheses
- Selection hypotheses

Site Colonization: If you cannot get there, it will not be part of your range. This is the simple logic behind a set of hypotheses coupling range size to dispersal scale. For one, species with limited dispersal ability may have more geographically restricted ranges, simply because they fail to reach as many sites (Juliano 1983; Wellington and Victor 1989; Gutierrez and Menendez 1997; Thompson, Hodgson, and Gaston 1998). This logic is at the heart of the earlier comments by Darwin and Grinnell. A second formulation of the site colonization hypothesis originates from the theory of metapopulation dynamics (Levins 1969; Hanski et al. 1993). If local populations at the periphery of the range occasionally go extinct, the species' range diminishes in size until sites are recolonized. In such a dynamic setting, time to recolonization plays a critical role, and species with limited dispersal may therefore occupy smaller geographic ranges, since sites will remain unoccupied for longer periods of time. A special case of the site colonization hypothesis follows from the concept of the "rescue effect" (Edwards and Westoby 1996; Duncan, Blackburn, and Veltman 1999), where fringe populations at the edge of the range are demographic sinks that would otherwise go extinct without regular immigration from populations elsewhere in the range (Brown and Kodrick-Brown 1977; Gotelli 1991). If such a rescue effect is operating, the degree of range expansion should scale with dispersal distance; short distance dispersers can only "rescue" nearby sink populations.

Speciation Rate: Species with limited dispersal may experience greater isolation and lower gene flow, and thus a greater potential for local adaptation. Hence, restricted dispersal may enhance rates of speciation (Jackson 1974; Shuto 1974; Scheltema 1977; Hansen 1980; Hansen 1982; Jablonski 1986; Palumbi 1992). A higher rate of speciation at the margin of a species' range can decrease average range size by two mechanisms: (a) speciation cleaves off a piece of the historical range, and the new species may restrict expansion of the parental species, and (b) new species may have had insufficient time to expand their ranges (Hansen 1980; Oakwood et al. 1993). Thus, higher speciation rates could yield a distribution of range sizes that is skewed to smaller sizes.

Selection: In addition to the potential for dispersal influencing range size, range size could determine dispersal distance. If there is a cost, or at least no benefit, to long-distance dispersal when range size is small, species with small geographic ranges might experience selection for restricted dispersal (Gutierrez and Menendez 1997; Thompson, Gaston, and Band 1999). The hypothesis relies on an assumption that more geographically restricted species have a narrower range of tolerances, are more ecologically specialized, or occupy a restricted, isolated or infrequently disturbed habitat type, so that the costs of broad dispersal exceed any potential benefits. This may be true when the range size of a species is close to its average dispersal distance, as for small-island endemics; in this case, the costs of broad dispersal are extreme (Baskett, Weitz, and Levin 2007).

Many studies claim or assume a correlation between dispersal distance and range size based on these types of arguments. One would think that the presumed association between dispersal and range size would have been well tested and that these hypotheses would have been critically evaluated. However, this has not been the case, in part because the relationship has been difficult to test in any quantitative way. Since estimates of average dispersal distances are rarely available, dispersal ability has generally been classified by a proxy: for example, reproductive strategy (Kessler 2002) or seed size (Aizen and Patterson 1990; Oakwood et al. 1993; Edwards and Westoby 1996) in plants, developmental mode (planktonic versus nonplanktonic larvae) in marine gastropods (Hansen 1980; Perron and Kohn 1985; Scheltema 1989), and flight ability (flightless versus flight-capable) in insects (Juliano 1983; Gutierrez and Menendez 1997). Proxies, however, have inherent problems if they mistakenly characterize dispersal potential (e.g., because other independent traits also affect dispersal distances—Kinlan and Gaines 2003; Kinlan, Gaines, and Lester 2005). In addition, since they typically include a very small number of classes of species, it is impossible to characterize the functional relationship between dispersal and range size, even when the qualitative groups differ significantly (Lester and Ruttenberg 2005).

Recent advances in techniques for estimating dispersal in the sea have provided rapidly expanding quantitative data sets of the distribution of dispersal distances of marine species, which can be used to explore how range size scales with dispersal distance. To illustrate the new insight garnered by having more direct estimates of dispersal distances, consider a comparison among a taxonomically and geographically diverse group of more than thirty species of marine invertebrates, using a common proxy for dispersal distance—mode of larval development (Hansen 1980; Perron and Kohn 1985; Scheltema 1989). As has been found in other studies examining various marine invertebrate taxonomic groups, range size (here defined as the maximum linear distance within the range, in km; see Lester et al. 2007 for a description of data set and more detailed methods) is larger for species with planktotrophic (feeding planktonic larvae) larval development (fig. 9.1). Given that species with this mode of development spend longer periods, on average, drifting in the plankton compared to nonfeeding larvae or to species with direct development (no planktonic phase), they presumably also have larger average dispersal distances (but see Shanks and Eckert 2005 for a range of ways that larval behavior might diminish these differences). As a result, dispersal is commonly cited as an important component of the variation in range size found across taxa with different modes of development.

These thirty-five species, however, provide an opportunity to probe this issue further, since we have estimates of their average dispersal distance from genetic isolation by distance slopes (fig. 9.2, Kinlan and Gaines 2003; Palumbi 2003). When the same range size data are plotted against quantitative rather than categorical estimates of dispersal scale, we see a very different pattern (fig. 9.3). Although the mean range size for species with feeding planktonic larvae is larger than for species with the other two development modes, dispersal distance seems to play little role in generating this pattern. All three groups have substantial variation in dispersal distance, but

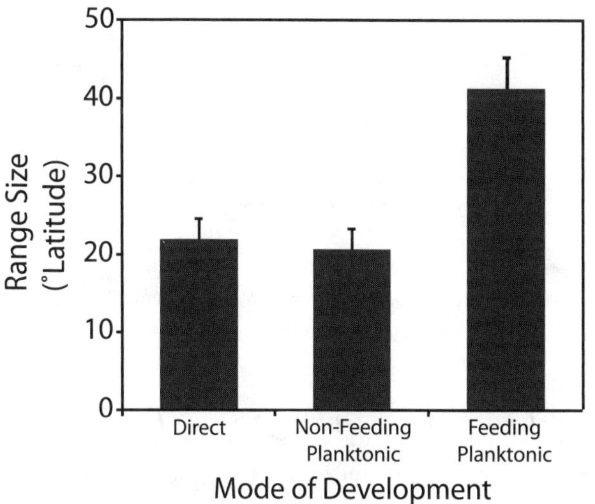

FIGURE 9.1 Average range size (measured here as maximum linear distance within the range, in km) for marine invertebrate taxa classified by a commonly used proxy for disperal—mode of larval development. Direct developers have no planktonic dispersal, since young develop at their natal site. Time in the plankton is on average much larger for species with feeding larvae than for species with nonfeeding larvae. This data set includes a diverse set of invertebrate species, representing five phyla from around the world. See Lester et al. (2007) for more details on this data set and a complete description of how range size was calculated.

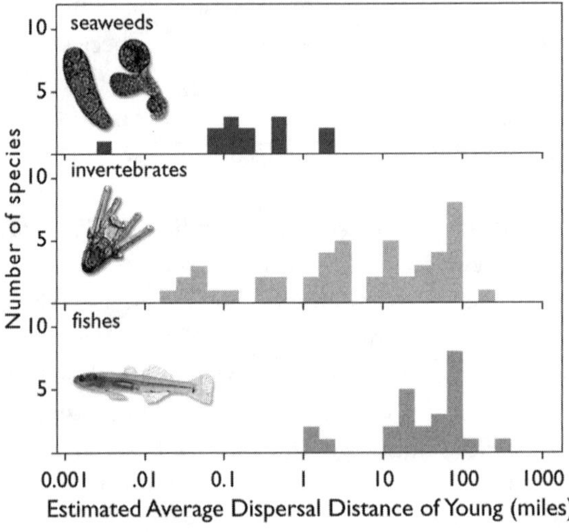

FIGURE 9.2 Frequency distribution of dispersal distances estimated from genetic measures of isolation by distance. (redrawn from Kinlan and Gaines 2003)

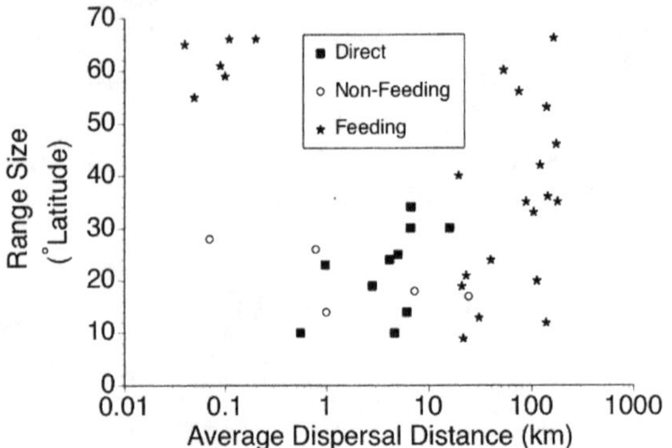

FIGURE 9.3 Average range size (measured here as maximum linear distance within the range, in km) for the same marine invertebrate taxa in fig. 9.1, as a function of average dispersal distance estimated by slopes of genetic isolation by distance.

there is no correlation for any group between this variation and range size. Range size is entirely uncorrelated with four orders of magnitude variation in dispersal.

Much broader evaluations of this connection have reached similar conclusions. There is strikingly little correspondence between range size and dispersal distance. Lester and Ruttenberg (2005) examined tropical reef fish from a wide diversity of families and geographical settings and found that dispersal scale (inferred from pelagic larval duration) only appears to influence range size in settings where there are enormous gaps between suitable habitat (e.g., in the tropical Pacific). When reef habitats are arrayed in more closely spaced stepping stones, range size is independent of dispersal scale for all fish families examined. Similarly, Lester et al. (2007) expand this comparison to include seaweeds, invertebrates, and marine fish from higher latitudes. They similarly find that dispersal scale correlates with range size only in selective situations. Given that there are several intuitively appealing hypotheses drawing a connection between dispersal and geographic range size, these findings suggest we need to reconsider the logic behind these hypotheses (Lester et al. 2007) and refocus on identifying the actual mechanisms underlying the enormous variation in species' range size.

The Influence of Dispersal on the Location of Range Boundaries

A second critical feature of a species' range is where it ends. If the causes of range boundaries were idiosyncratic to the unique tolerances, traits, and interactions of different species, we might expect that species' range boundaries would be distributed somewhat randomly along coastlines. However, known marine species distributions strongly suggest this is not the case in the sea. Striking clusters of species boundaries occur within some relatively short stretches of coastlines on most continental margins (Briggs 1974; Lüning 1990). These relatively abrupt latitudinal shifts in species composition make it possible to define some boundaries of marine biogeographic provinces with general consensus (Dana 1853; Ekman 1953; Valentine 1966; Briggs 1974; Pielou 1979).

Does the common location of range boundaries imply common causality? It has long been noted that clusters of range boundaries of marine species are typically associated with major coastal headlands or points that are characterized by distinctive oceanographic features (e.g., the convergence of two current systems, mesoscale eddies, or gyres; Dana 1853). Two classes of hypotheses have been proposed to account for the clustering of range boundaries at these prominent points:

> The evidence seems overwhelming that the boundaries of [marine] biotic provinces are determined by modern abiotic factors. . . . One of two possible [explanations] is that each offers unique environmental conditions, to which species from other provinces are unadapted; intruders therefore cannot establish themselves in a "wrong" province although nothing prevents their entering it. The other possibility is that actual barriers to dispersal exist that are difficult to cross. . . . Where such barriers to dispersal coincide with boundaries between differing environments, it is difficult to judge the relative importance of the two factors in maintaining the distinctness of biotic provinces (Pielou 1979).

Pielou's last sentence poses the fundamental problem. There are two classes of causes—one based on mortality outside the species' range, either due to physical or biological causes (hereafter the *mortality hypothesis*), the other based on barriers to larval dispersal (hereafter the *dispersal barrier hypothesis*). Unfortunately, the underlying climatic mechanisms potentially responsible for these two causes of range limits—steep physical gradients versus hydrographic barriers to dispersal—are typically confounded in space. Steep gradients in ocean temperature or other physical parameters cannot be generated without anomalous circulation patterns (e.g., convergent currents pushing water offshore), which tend to restrict along-coast larval dispersal.

Although the possible roles of both physical gradients and circulation have been noted in most marine biogeographic studies of the past century, the emphasis has been placed disproportionately on the physical gradients per se (generally temperature) as the ultimate cause (Clarke, chapter 10, this volume). This stems in part from the influence of the Hutchinsonian niche concept (Hutchinson 1957) on thinking in biogeography, whereby the geographic range is viewed as a "spatial reflection" of a species' niche (Brown and Lomolino 1998), emphasizing the role of environmental conditions in setting species' distributions. However, the evidence to support such a bias is not particularly compelling. As noted previously, correlations between the position of species boundaries and thermal parameters (e.g., maximum temperature, minimum temperature, temperature range) are necessarily confounded, because the isotherms themselves are correlated with changes in the pattern of circulation. In addition, correlations of the number of species range limits with thermal parameters rarely show strong statistical relationships (Valentine 1966; Doyle 1985; Lüning and Freshwater 1988). Furthermore, experiments that transplant marine species beyond their normal range, although extremely rare, do not commonly support the mortality hypothesis (Crisp 1950; Yamada 1977; Doyle 1985; Gilman 2006). Despite the lack of a clear causal connection between coastal marine provincial boundaries and corresponding gradients in physical conditions, only a few field studies of single species have advocated the dispersal hypothesis as the primary determinant of range limits (Crisp 1950; Yamada 1977; Cowen 1985; Doyle 1985).

One way to gain insight into the relative roles of mortality and dispersal barrier hypotheses is to use the life history variation within marine species as an exploratory tool. As noted earlier, marine invertebrates are a diverse group with considerable variation in their mode of reproduction and scales of dispersal. With respect to the clustering of range boundaries at particular locations, we can gain insight into the roles of mortality- versus dispersal-based hypotheses by considering two dispersal classes. The majority of marine invertebrate larvae develop for weeks or months in the plankton and currents may transport them far from their parents (hereafter *broad dispersers*). The direction and distance they disperse depend on patterns of circulation, potentially modified by the swimming behavior of the larvae (e.g., Botsford et al. 1994; Pineda 1991; Shanks and McCulloch 2003; Cowen, Paris, and Srinivasan 2006). The remaining species, however, spend little (minutes to hours) or no time developing in the plankton and do not disperse far from their parents (hereafter *limited dispersers*). Thus, they are affected less directly by patterns of coastal circulation.

Unlike the previous section, where we explored range size relative to quantitative estimates of dispersal distance, this simple dispersal dichotomy may provide considerable insight for this particular macroecological pattern. The reason is that only one of these two groups (broad dispersers) will be *directly* influenced by the pattern of circulation at biogeographic breaks, while both groups will be *indirectly* affected by the physical and biological gradients the circulation patterns create. Therefore, if we focus on invertebrates with relatively sessile adults, one test of the role of dispersal barriers is to ask if species that reproduce via broadly dispersing larvae are more likely to have a range boundary at a biogeographic boundary than species with limited-dispersing larvae. If hydrographic barriers to dispersal play an important role in clustering species range limits at major points and headlands, we would predict that species with broadly dispersing larvae should be more likely to have a range limit at these headlands than species with nondispersing larvae. By contrast, if steep physical gradients (e.g., in temperature) are the primary cause of the clustering of range limits, we would predict that both groups of invertebrates are equally likely to have range limits at these points.

To test such contrasting predictions, we assembled data on the range limits of intertidal invertebrates from the Pacific coast of North America (Morris, Abbott, and Haderlie 1980; Eckert 1999). Figure 9.4 plots the percentage of range limits (northern and southern limits plotted separately) for species with broadly dispersing versus limited-dispersing larvae in 0.5° increments of latitude. Sharp peaks indicate that the ranges of a large percentage of species end within a short stretch of coastline. Note that there is little correspondence between the latitudinal distribution of range limits for the two groups ($r^2 < 0.05$ between species with dispersing versus limited dispersing larvae for both northern and southern range limits). The differences are especially noteworthy for some prominent biogeographic breaks. For example, Point Conception, California (34.5°) is a clear northern boundary for species with broadly dispersing larvae but not for species with limited dispersing larvae (fig. 9.4). Similarly, two prominent headlands in Baja California

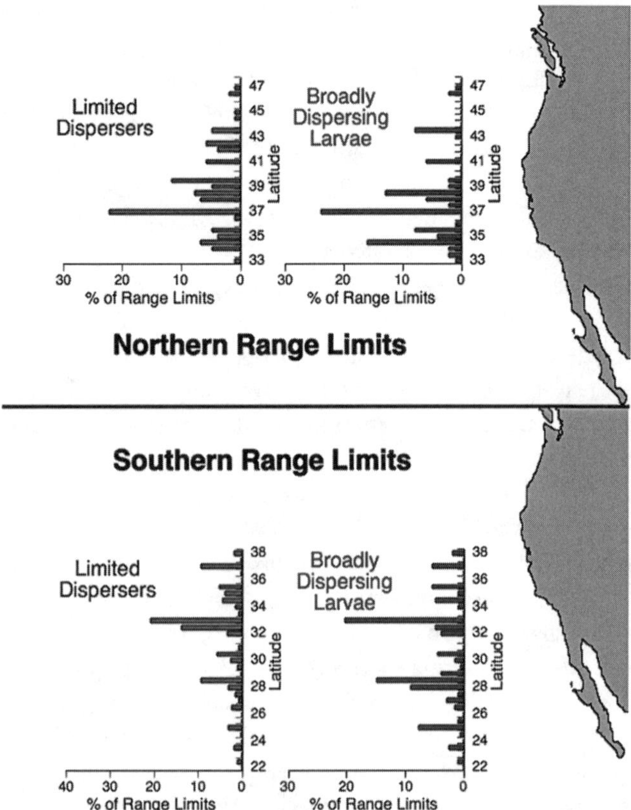

FIGURE 9.4 Distribution of species' borders for marine intertidal invertebrates with relatively sessile adults. Species are placed in two groups, based on their larval development. Those species whose larvae spend either no time in the plankton (direct developers) or only minutes to hours are classed as limited dispersals. Species whose larvae develop for many days to months in plankton are classed as broad dispersers. Range boundaries are from Morris, Abbott, and Haderlie (1980) with extensive updates from the literature (see Eckert 1999 for details).

are clear southern boundaries for species with dispersing larvae, but not for species with nondispersing larvae (fig. 9.4). Along this stretch of coastline, only the Monterey Bay region (36.5°) seems to be a common boundary for species with both larval types (fig. 9.4).

Further evidence for a connection between dispersal and the location of range boundaries comes from comparing northern versus southern range limits rather than comparing across species with different life histories. Figure 9.4 shows that some headlands represent "one-way" boundaries. Most noteworthy, Point Conception is a prominent northern boundary for species with broadly dispersing larvae, but it is not a southern boundary. This pattern is difficult to reconcile with hypotheses based upon physical or biological gradients that affect mortality, which would predict clustering of both northern and southern range limits. By contrast, the unidirectional nature of this boundary for broad dispersers follows directly from the general patterns of circulation and their likely effects on dispersal (Gaylord and Gaines 2000). Point Conception represents a point of convergence between a large coastal current (the California Current) and a large, seasonal mesoscale eddy (Lasker, Pelaez, and Laurs 1981; Husby and Nelson 1982; Doyle 1985). The pattern of flow is such that larval dispersal may be much more difficult from south to north around the point than the reverse. In support of this hypothesis, genetic analyses within populations of intertidal barnacles that span Point Conception show differential rates of gene flow (north to south > south to north; Wares, Gaines, and Cunningham 2001). In addition, recent range expansions of species that previously had northern range limits at Point Conception are consistent with a breakdown in dispersal barriers under altered flow conditions (e.g., during El Niño—Zacherl et al. 2003).

Simple models of larval dispersal under commonly observed flow conditions can create range boundaries surprisingly easily with convergent flows (Gaylord and Gaines 2000; Byers and Pringle 2006) even when there is no spatial variation in mortality in the adult habitat. The irony is that species with longer potential for dispersal are more susceptible to such flow-induced barriers than species with restricted propagule movement. Historically, dispersal barriers have been viewed more as a problem for species with quite limited dispersal who were unable to cross the barrier (see summary in Gaston 2003). For such hypothetical flow-induced boundaries, however, the absence of sufficient larvae that do not disperse is fundamental to the establishment of the species border (Gaylord and Gaines 2000). Even when the barrier to dispersal is quite leaky, a range boundary can be maintained, because larvae that settle and survive to adulthood beyond the boundary produce larvae that disperse disproportionately back toward the region of convergence (Gaylord and Gaines 2000). The observation that species of invertebrates with longer-lived planktonic larvae are the primary species with range boundaries at two sites along the west coast of North America with convergent flows (Point Conception and Punta Eugenia) suggests that the role of dispersal barriers in setting range boundaries in the sea warrants considerably more attention. Although alternative explanations to the dispersal barrier hypothesis (e.g., enhanced gene flow in species with long-distance dispersal may swamp local adaptation to changes in physical or biological conditions or differential larval mortality during development on opposite sides of the boundary) can account for some of the pattern data observed along the west coast of North America, they all include a fundamentally important role for larval dispersal. These issues are ripe for exploration of both the generality of the macroecological patterns at other biogeographic boundaries and for detailed experimental studies at the edges of species' ranges (e.g., see Gilman 2006 for an excellent example).

The Influence of Dispersal on Abundance across a Species' Range

So far, we have focused on the impact of dispersal on issues related to range edges. Such boundary definitions simplify the evaluation of population size to a problem of presence/absence. As a result, we have been able to ignore population size during all of this discussion. However, patterns of abundance across a species' range can have important consequences for a wide range of issues such as gene flow, species interactions, and responses to harvesting. Thus, it is critical to document geographic patterns of population abundance and understand the factors determining these patterns. Although a wide range of ecological theory and empirical studies have examined connections between dispersal and population size (e.g., Pulliam 1988; Boyce 1992; Lande, Engen, and Saether 1999), two emerging bodies of research suggest this connection may be strong across broad geographical scales in marine populations.

First, the modeling studies of the link between dispersal and range boundaries discussed briefly previously also consider the consequences of dispersal under different flow conditions to patterns of population size across species' ranges (Gaylord and Gaines 2000; Byers and Pringle 2006; see also Siegel et al. 2003 for new modeling approaches to dispersal in turbulent flows). Such models generate a diverse set of abundance patterns across geographic scales in the presence of different oceanographic flow fields (e.g., see fig. 9.5 and multiple figures in Gaylord and Gaines 2000 for simple examples).

In parallel with these modeling efforts, the last decade has seen a great expansion of studies of actual population sizes across entire species ranges (see Sagarin and Gaines 2002a; Sagarin and Gaines 2002b; Gaston 2003 for recent reviews). Contrary to the simple and ubiquitous biogeographic presumption that species are typically most abundant at sites near the center of their geographic range, these large-scale ecological studies have found a rich diversity of distributions of abundance with a surprisingly large number of cases where peaks in species' abundance occur relatively close to range boundaries rather than near range centers (see fig. 9.6 for a few examples

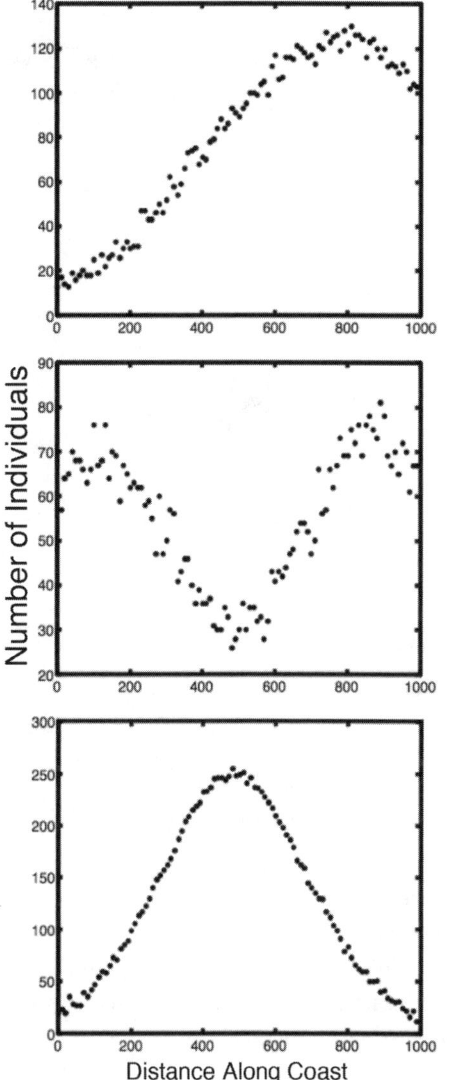

FIGURE 9.5 Population abundance across a species' geographic range with different patterns of coastal circulation and larval dispersal (sensu Siegel et al. 2003). Top panel: Unidirectional flow to right, mean flow = 5 cm/s, std. dev. = 15 cm/s, PLD = fourteen days, gaussian dispersal kernel. Middle panel: Diverging flow at midpoint, mean flow = +/–5 cm/s, std. dev. = 15 cm/s, PLD = fourteen days, gaussian dispersal kernel. Bottom panel: Converging flow toward midpoint, mean flow = –/+5 cm/s, std. dev. = 15 cm/s, PLD = fourteen days, gaussian dispersal kernel.

from the larger range of patterns summarized in Sagarin and Gaines 2002a; Sagarin and Gaines 2002b; Defeo and Cardoso 2004; Sagarin, Gaines, and Gaylord 2006).

Since the presumption of an abundant center is at the core of a number of ecological, evolutionary, biogeographic, and conservation theories and frameworks (see review in Sagarin, Gaines, and Gaylord 2006), these empirical findings and syntheses call into question a number of results and

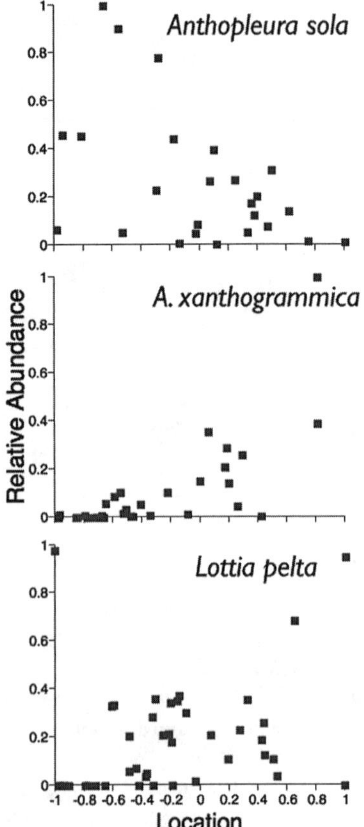

FIGURE 9.6 Selected examples of abundance distributions for marine species that show peaks of abundance near the range boundary (redrafted from Sagarin and Gaines 2002).

approaches. As a consequence, recent studies have begun to examine how such issues as genetic population structure (Vucetich and Waite 2003), habitat conservation (Hampe and Petit 2005), and species responses to climate change (Helmuth, Kingsolver, and Carrington 2005) might be altered by different patterns of abundance across species' ranges.

Further modeling work connects these empirical patterns of abundance to the theoretical findings demonstrating the geographic influence of dispersal and oceanography on population sizes. This work shows that a number of circulation scenarios lead to theoretical predictions of abundance patterns with peak abundances near the edge of species' ranges (e.g., see the middle panel of fig. 9.5 and numerous examples in Gaylord and Gaines 2000). Although other hypotheses could undoubtedly account for skewed abundance distributions (e.g., nonlinear physical gradients—Helmuth and Hofmann 2001; Helmuth et al. 2002), the connections with patterns of dispersal warrant more focused attention. This conclusion is supported by the observation that many of the species observed to have peaks of abundance at the edge of the range have these abundant edges at prominent biogeographic boundaries with convergent flows (e.g., Punta Eugenia in Baja California, Mexico—see Sagarin and Gaines 2002b, Sagarin, Gaines, and Gaylord 2006).

Although the number of species with detailed data on abundance across their geographical range does not currently permit the kinds of larger macroecological comparisons across species with different life-history traits or patterns of dispersal that were possible as discussed earlier, the fact that abundant edge distributions for some species coincide with locations that have disproportionate numbers of range boundaries for species with broadly dispersing planktonic development (fig. 9.4) suggests a significant role of dispersal. Whether this is just coincidence or an important new finding awaits more detailed studies at other locations and field experiments that test among competing hypotheses more rigorously. Furthermore, the question of whether abundance patterns tell us anything about the mechanisms setting range boundaries (Caughley et al. 1988; Sagarin and Gaines 2002a; Sagarin, Gaines, and Gaylord 2006) remains currently unresolved, but there are hints that they may be an important source of insight for some larger-scale dynamics in marine populations.

Conclusions

Dispersal redistributes individuals in space. Thus, it is not surprising that it is functionally linked to a wide range of large-scale issues in biogeography and macroecology. Here we have explored recent advances in our understanding of how dispersal might structure marine populations at large biogeographic scales. We considered three issues (size of ranges, location of range boundaries, and distribution of individuals across entire ranges) that illustrate both differences in approach and differences in conclusion.

In the first case (range size), dispersal was long suspected to play a disproportionately large role in determining the size of species' geographic distributions. New syntheses, however, that include more detailed and/or quantitative estimates of dispersal distances are tempering that conclusion. Dispersal may play a smaller role in setting the range size of marine species than previously suspected. By contrast, the role of dispersal barriers in setting the location of range boundaries in the sea has received comparatively little attention. Although dispersal barriers are usually listed as one of several hypothetical causes for the location of range boundaries, it is commonly assumed that gradients in environmental conditions are the primary determinant of species' range boundaries, particularly when species' boundaries cluster at a given location. Comparisons across taxa with different life histories, and thus different dispersal characteristics, suggest that the role of dispersal barriers in establishing species' range limits may be far more important than previously suspected, particularly in certain oceanographic settings (e.g., convergent flows).

Finally, explorations of range edges have been historically somewhat divorced from studies of geographical variation in population size. The emerging data on patterns of abundance across entire species' ranges and modeling work predicting patterns of abundance under different dispersal and oceanographic scenarios both suggest that it could be fruitful to further study the potential for dispersal to influence geographic patterns of population size. In examining these three issues, we stress the value of a multifaceted approach to macroecological studies. These insights were gained by a combination of empirical documentation of large-scale patterns for different life history groups, modeling efforts, and critical examinations of mechanistic hypotheses.

REFERENCES

Aizen, M. A., and W. A. Patterson, III. 1990. Acorn size and geographical range in the North American oaks (*Quercus L.*). *Journal of Biogeography* 17:327–32.

Baskett, M. L., J. S. Weitz, and S. A. Levin. 2007. The evolution of dispersal in reserve networks. *American Naturalist* 170:59–78.

Block, B. A., S. L. H. Teo, A. Walli, A. Boustany, M. J. W. Stokesbury, C. J. Farwell, K. C. Weng, H. Dewar, and T. D. Williams. 2005. Electronic tagging and population structure of Atlantic bluefin tuna. *Nature* 434:1121–27.

Block, B. A., H. Dewar, S. B. Blackwell, T. D. Williams, E. D. Prince, C. J. Farwell, A. Boustany, et al. 2001. Migratory movements, depth preferences, and thermal biology of Atlantic bluefin tuna. *Science* 293:1310–14.

Block, B. A., H. Dewar, C. Farwell, and E. D. Prince. 1998. A new satellite technology for tracking the movements of Atlantic bluefin tuna. *PNAS* 95:9384–89.

Bonhomme, F., and S. Planes. 2000. Some evolutionary arguments about what maintains the pelagic interval in reef fishes. *Environmental Biology of Fishes* 59:365–83.

Botsford, L. W., C. L. Moloney, A. Hastings, J. L. Largier, T. M. Powell, K. Higgins, and J. F. Quinn. 1994. The influence of spatially and temporally varying oceanographic conditions on meroplanktonic metapopulations. *Deep Sea Research* 41:107-145.

Boustany, A. M., S. F. Davis, P. Pyle, S. D. Anderson, B. J. Le Boeuf, and B. A. Block. 2002. Satellite tagging: Expanded niche for white sharks. *Nature* 415:35–36.

Boyce, M. S. 1992. Population viability analysis. *Annual Review of Ecology and Systematics* 23:481–97.

Bradbury, I. R., and P. V. R. Snelgrove. 2001. Contrasting larval transport in demersal fish and benthic invertebrates: The roles of behaviour and advective processes in determining spatial pattern. *Canadian Journal of Fisheries and Aquatic Sciences* 58:811–23.

Briggs, J. C. 1974. Marine Zoogeography. New York: McGraw-Hill.

Brown, J. H., and A. Kodrick-Brown. 1977. Turnover rates in insular biogeography: Effect of immigration on extinction. *Ecology* 58:445–49.

Brown, J. H., and M. V. Lomolino. 1998. *Biogeography*, 2nd ed. Sunderland, MA: Sinauer Associates.

Brown, J. H., G. C. Stevens, and D. M. Kaufman. 1996. The geographic range: Size, shape, boundaries and internal structure. *Annual Review of Ecology and Systematics* 27:597–623.

Byers, J. E., and J. M. Pringle. 2006. Going against the flow: Retention, range limits and invasions in advective environments. *Marine Ecology Progress Series* 313:27–41.

Caughley, G., D. Grice, R. Barker, and B. Brown. 1988. The edge of the range. *Journal of Animal Ecology* 57:771–85.

Clarke, P., R. A. Kerrigan, and C. J. Westphal. 2001. Dispersal potential and early growth in 14 tropical mangroves: Do early life history traits correlate with patterns of adult distribution? *Journal of Ecology* 89:648–59.

Cowen, R. K. 1985. Large scale pattern of recruitment by the labrid, *Semicossyphus pulcher*: Causes and implications. *Journal of Marine Research* 43:719–42.

Cowen, R. K., C. B. Paris, and A. Srinivasan. 2006. Scaling of connectivity in marine populations. *Science* 311:522–27.

Crisp, D. J. 1950. Breeding and distribution of *Chthamalus stellatus*. *Nature* 166:311–12.

Dana, J. D. 1853. On an isothermal oceanic chart, illustrating the geographical distribution of marine animals. *American Journal of Science and Art, Second Series* 16:153–67.

Darwin, C. 1859. *On the origin of species by means of natural selection, or the preservation of favoured races in the struggle for life*. London: John Murray.

Defeo, O., and R. S. Cardoso. 2004. Latitudinal patterns in abundance and life-history traits in the mole crab *Emerita brasiliensis* on South American sandy beaches. *Diversity and Distributions* 10:89–98.

Dennis, R. L., H. B. Donato, T. H. Sparks, and E. Pollard. 2000. Ecological correlates of island incidence and geographic range among British butterflies. *Biodiversity and Conservation* 9:343–59.

Doyle, R. F. 1985. Biogeographical studies of rocky shores near Point Conception, California. PhD diss., University of California, Santa Barbara.

Duncan, R. P., T. M. Blackburn, and C. J. Veltman. 1999. Determinants of geographical range sizes: A test using introduced New Zealand birds. *Journal of Animal Ecology* 68:963–75.

Eckert, G. L. 1999. Consequences of diverse reproductive strategies in marine invertebrates. PhD diss., Department of Ecology, Evolution, and Marine Biology, University of California, Santa Barbara.

Ekman, S. 1953. *Zoogeography of the Sea*. London: Sidgwick and Jackson.

Edwards, W., and M. Westoby. 1996. Reserve mass and dispersal investment in relation to geographic range of plant species: Phylogenetically independent contrasts. *Journal of Biogeography* 23:329–38.

Emlet, R. B. 1995. Developmental mode and species geographic range in regular sea urchins (Echinodermata: Echinoidea). *Evolution* 49:476–89.

Gaston, K. J. 1996. Species-range-size distributions: Patterns, mechanisms and implications. *Trends in Ecology & Evolution* 11:197–201.

———. 2003. *The structure and dynamics of geographic ranges*. Oxford: Oxford University Press.

Gaylord, B., and S. D. Gaines. 2000. Temperature or transport? Range limits in marine species mediated solely by flow. *American Naturalist* 155:769–89.

Gaylord, B., D. C. Reed, P. T. Raimondi, and L. Washburn. 2006. Macroalgal spore dispersal in coastal environments: Mechanistic insights revealed by theory and experiment. *Ecological Monographs* 76:481–502.

Gilman, S. E. 2006. Life at the edge: An experimental study of a poleward range boundary. *Oecologia* 148:270–79.

Gotelli, N. J. 1991. Metapopulation models: The rescue effect, the propagule rain, and the core-satellite hypothesis. *American Naturalist* 138:768–76.

Grinnell, J. 1922. On the role of the accidental. *Auk* 39:373–80.

Gutierrez, D., and R. Menendez. 1997. Patterns in the distribution, abundance and body size of carabid beetles (Coleoptera: Caraboidea) in relation to dispersal ability. *Journal of Biogeography* 24:903–14.

Hampe, A., and R. J. Petit. 2005. Conserving biodiversity under climate change: The rear edge matters. *Ecology Letters* 8:461–67.

Hansen, T. A. 1978. Larval dispersal and species longevity in lower tertiary gastropods. *Science* 199:885–87.

———. 1980. Influence of larval dispersal and geographic distribution on species longevity in neogastropods. *Paleobiology* 6:193–207.

———. 1982. Modes of larval development in early Tertiary neogastropods. *Paleobiology* 8:367–77.

Hanski, I., J. Kouki, and A. Halkka. 1993. Three explanations for the positive relationship between distribution and abundance of species. In *Species diversity in ecological communities: historical and geographical perspectives,* ed. R. E. Ricklefs and D. Schluter, 108–16. Chicago: University of Chicago Press.

Hanski, I., P. Turchin, E. Korpimaki, and H. Hentonnen. 1993. Population oscillations of boreal rodents: Regulation of mustelid predators leads to chaos. *Nature* 364:232–35.

Helmuth, B., C. D. G. Harley, P. M. Halpin, M. O'Donnell, G. E. Hofmann, and C. A. Blanchette. 2002. Climate change and latitudinal patterns of intertidal thermal stress. *Science* 298:1015–17.

Helmuth, B., J. G. Kingsolver, and E. Carrington. 2005. Biophysics, physiological ecology and climate change: Does mechanism matter? *Annual Review of Physiology* 67:177–201.

Helmuth, B. S. T., and G. E. Hofmann. 2001. Microhabitats, thermal heterogeneity, and patterns of physiological stress in the Rocky Intertidal Zone. *Biological Bulletin* 201:374–84.

Husby, D. M., and C. S. Nelson. 1982. Turbulence and vertical stability in the California Current. *CalCOFI Reports* 23:113–29.

Hutchinson, G. E. 1957. Concluding remarks. *Cold Spring Symposia on Quantitative Biology* 22:415–27.

Jablonski, D. 1982. Evolutionary rates and models in late Cretaceous gastropods: Role of larval ecology. *Proceedings of the Third North American Paleontological Convention* 1:257–62.

———. 1986. Larval ecology and macroevolution in marine invertebrates. *Bulletin of Marine Science* 39:565–87.

Jackson, J. B. C. 1974. Biogeographic consequences of eurytopy and stenotopy among marine bivalves and their evolution significance. *American Naturalist* 108:541–60.

Jones, G. P., S. Planes, and S. R. Thorrold. 2005. Coral reef fish larvae settle close to home. *Curr. Biol.* 15:1314–18.

Juliano, S. A. 1983. Body size, dispersal ability, and range size in North American species of *Brachinus* (Coleoptera: Carabidae). *Coleopterists Bulletin* 37:232–38.

Kessler, M. 2002. Range size and its ecological correlates among the pteridophytes of Carrasco National Park, Bolivia. *Global Ecology and Biogeography* 11:89–102.

Kinlan, B., and S. D. Gaines. 2003. Propagule dispersal in marine and terrestrial environments: A community perspective. *Ecology* 84:2007–20.

Kinlan, B., and A. Hastings. 2005. Rates of population spread and geographic range expansion: what exotic species tell us. In *Rates of population spread and geographic range expansion: What exotic species tell us,* ed. D. F. Sax, J. J. Stachowicz, and S. D. Gaines, 381–19. Sunderland, MA: Sinauer.

Kinlan, B. P., S. D. Gaines, and S. E. Lester. 2005. Propagule dispersal and the scales of marine community process. *Diversity and Distributions* 11:139–48.

Kunin, W. E., and K. J. Gaston. 1993. The biology of rarity: Patterns, causes and consequences. *Trends in Ecology & Evolution* 8:298–301.

Lande, R., S. Engen, and B.-E. Sæther. 1999. Spatial scale of population synchrony: Environmental correlation versus dispersal and density regulation. *American Naturalist* 154:271–81.

Lasker, R., J. Pelaez, and R. M. Laurs. 1981. The use of satellite infrared imagery for describing ocean processes in relation to spawning of the northern anchovy (*Engraulis mordax*). *Remote Sensing of the Environment* 11:439–53.

Lester, S., and B. I. Ruttenberg. 2005. The relationship between pelagic larval duration and range size in tropical reef fishes: A synthetic analysis. *Proceedings of the Royal Society of London B* 272:585–91.

Lester, S., B. I. Ruttenberg, S. D. Gaines, and B. Kinlan. 2007. The relationship between dispersal ability and geographic range size. *Ecology Letters* 10:745–58.

Levin, L., D. Huggett, P. Myers, T. Bridges, and J. Weaver. 1993. Rare-earth tagging methods for the study of larval dispersal by marine invertebrates. *Limnology and Oceanography* 38:346–60.

Levins, R. 1969. Some demographic and genetic consequences of environmental heterogeneity for biological control. *Bulletin of the Entomological Society of America* 15:237–40.

Lüning, K. 1990. *Seaweeds: Their environment, biogeography and ecophysiology.* New York: Wiley.

Lüning, K., and W. Freshwater. 1988. Temperature tolerance of Northeast Pacific marine algae. *Journal of Phycology* 24:310–15.

Morris, R. H., D. P. Abbott, and E. C. Haderlie. 1980. *Intertidal invertebrates of California.* Stanford, CA: Stanford University Press.

Oakwood, M., E. Jurado, M. Leishman, and M. Westoby. 1993. Geographic ranges of plant species in relation to dispersal morphology, growth form and diaspore weight. *Journal of Biogeography* 20:563–72.

Palumbi, S. R. 2003. Population genetics, demographic connectivity, and the design of marine reserves. *Ecological Applications* 13:S146–S158.

———. 1992. Marine speciation on a small planet. *Trends in Ecology and Evolution* 7:114–17.

Palumbi, S. R., S. D. Gaines, H. Leslie, and R. R. Warner. 2003. New wave: High-tech tools to help marine reserve research. *Frontiers in Ecology and the Environment* 1:73–79.

Perron, F. E., and A. J. Kohn. 1985. Larval dispersal and geographic distribution in coral reef gastropods of the genus *Conus. Proceedings of the 5th International Coral Reef Congress, Tahiti* 4:95–100.

Pielou, E. C. 1979. *Biogeography.* New York: Wiley.

Pineda, J. 1991. Predictable upwelling and the shoreward transport of planktonic larvae by internal tidal bores. *Science* 253:548–51.

Pulliam, H. R. 1988. Sources, sinks, and population regulation. *American Naturalist* 132:652–61.

Rapoport, E. H. 1982. *Areography: geographical strategies of species.* Oxford: Pergamon.

Sagarin, R. D., and S. D. Gaines. 2002a. The "abundant centre" distribution: To what extent is it a biogeographic rule? *Ecology Letters* 5:137–47.

———. 2002b. Geographical abundance distributions of coastal invertebrates using 1-dimensional ranges to test biogeographic hypotheses. *Journal of Biogeography* 29:985–98.

Sagarin, R. D., S. D. Gaines, and B. Gaylord. 2006. Moving beyond assumptions to understand abundance distributions across the ranges of species. *Trends in Ecology & Evolution* 21:524–29.

Scheltema, R. S. 1989. Planktonic and non-planktonic development among prosobranch gastropods and its relationship to the geographic range of species. Reproduction, genetics and distributions of marine organisms. *23rd European Marine Biology Symposium* 183–88.

———. 1977. Dispersal of marine invertebrate organisms: Paleobiogeographic and biostratigraphic implications. In *Dispersal of marine invertebrate organisms: Paleobiogeographic and biostratigraphic implications,* ed. E. G. Kauffman and J. E. Hazel, 73–108. Stroudsburg, PA: Dowden, Hutchinson, and Ross.

Shanks, A. L., and G. L. Eckert. 2005. Population persistence of California current fishes and benthic crustaceans: a marine drift paradox. *Ecological Monographs* 75:505–24.

Shanks, A. L., B. A. Grantham, and M. H. Carr. 2003. Propagule dispersal distance and the size and spacing of marine reserves. *Ecological Applications* 13:S159–S169.

Shanks, A. L., and A. McCulloch. 2003. Topographically generated fronts, very nearshore oceanography, and the distribution of chlorophyll, detritus, and selected diatom and dinoflagellate taxa. *Marine Biology* 143:969–80.

Shuto, T. 1974. Larval ecology of prosobranch gastropods and its bearing on biogeography and paleontology. *Lethat* 7:239–57.

Siegel, D., B. P. Kinlan, B. Gaylord, and S. D. Gaines. 2003. Lagrangian descriptions of marine larval dispersion. *Marine Ecology Progress Series* 260:83–96.

Swearer, S. E., J. E. Caselle, D. W. Lea, and R. R. Warner. 1999. Larval retention and recruitment in an island population of a coral-reef fish. *Nature* 402:799.

Thompson, K., K. J. Gaston, and S. R. Band. 1999. Range size, dispersal and niche breadth in the herbaceous flora of central England. *Journal of Ecology* 87:150–55.

Thompson, K., J. G. Hodgson, and K. J. Gaston. 1998. Abundance-range size relationships in the herbaceous flora of central England. *Journal of Ecology* 86:439–48.

Thorrold, S. R., G. P. Jones, M. E. Hellberg, R. S. Burton, S. E. Swearer, J. E. Neigel, S. G. Morgan, and R. R. Warner. 2002. Quantifying larval retention and connectivity in marine populations with artificial and natural markers. *Bull. Mar. Sci.* 70:291–308.

Thorrold, S. R., C. Latkoczy, P. K. Swart, and C. M. Jones. 2001. Natal homing in a marine fish metapopulation. *Science* 291:297–99.

Valentine, J. W. 1966. Numerical analysis of marine molluscan ranges on the extratropical northeastern Pacific shelf. *Limnology and Oceanography* 11:198–211.

Yamada, S. B. 1977. Geographic range limitation of the intertidal gastropods *Littorina sitkana* and *L. planaxis*. *Marine Biology* 39:61–65.

Victor, B. C., and G. M. Wellington. 2000. Endemism and the pelagic larval duration of reef fishes in the eastern Pacific Ocean. *Marine Ecology Progress Series* 205:241–48.

Vucetich, J. A., and T. A. Waite. 2003. Spatial patterns of demography and genetic processes across the species range: Null hypotheses for landscape conservation genetics. *Cons. For. Genet.* 4:639–45.

Wares, J. P., S. D. Gaines, and C. W. Cunningham. 2001. A comparative study of asymmetric migration events across a marine biogeographic boundary. *Evolution* 55:295–306.

Wellington, G. M., and B. C. Victor. 1989. Planktonic larval duration of one hundred species of Pacific and Atlantic damselfishes (Pomacentridae). *Marine Biology* 101:557–68.

Zacherl, D., S. D. Gaines, and S. I. Lonhart. 2003. The limits to biogeographical distributions: Insights from the northward range extension of the marine snail, *Kelletia kelletii* (Forbes, 1852). *Journal of Biogeography* 30:913–24.

Zacherl, D. C., P. H. Manríquez, G. Paradis, R. W. Day, J. C. Castilla, R. R. Warner, D. W. Lea, and S. D. Gaines. 2003. Trace elemental fingerprinting of gastropod statoliths to study larval dispersal strategies. *Marine Ecology Progress Series* 248:297–303.

CHAPTER TEN
TEMPERATURE AND MARINE MACROECOLOGY

ANDREW CLARKE

Introduction

> With the ratification of long tradition, the biologist goes forth, thermometer in hand, and measures the effect of temperature on every parameter of life. Lack of sophistication poses no barrier; heat storage and exchange may be ignored or Arrhenius abused; but temperature is, after time, our favourite abscissa. One doesn't have to be a card-carrying thermodynamicist to wield a thermometer. (Vogel 1981, 1)

Temperature is one of the most important physical factors affecting organisms. It is also relatively simple to measure accurately and precisely although, as Vogel (1981) has pointed out in a few memorable sentences, its relationship to ecology and physiology is often complex. Temperature, typically combined with other environmental factors as climate, has long been regarded as a key factor regulating the diversity of organisms, and from the time of the earliest naturalists the perceived favorableness of climate has been regarded as a key factor in determining how many and what kinds of organisms could live in a given place. Early discussions were concerned predominantly with the terrestrial environment, and it was not really until Thorson (1957) first described latitudinal clines in the diversity of North Atlantic marine organisms, and Hutchinson's classic studies of diversity (Hutchinson 1959, 1961) that the discussion was extended to the sea. Connell and Orias (1964) reviewed previous ideas of the role of environmental rigorousness in controlling diversity, and Stevens (1989) reinvigorated the modern debate over the influence of climate (and its variability) on diversity.

Although many of these ideas were developed for the terrestrial realm, they have frequently been extended to the sea. Marine environments have many environmental features in common with the terrestrial realm, including movement, light, photoperiod, and temperature, but there are also some more specifically aquatic factors such as nutrient concentration and salinity. Most attention, however, has been directed at temperature.

Temperature in the Sea

Surface temperatures show a clear latitudinal cline from polar waters, with a mean annual temperature around zero to a broad zone of tropical waters in the range of 25 to 30°C (fig. 10.1). Although the pattern is similar in the two major ocean basins, there are important differences in detail. In the western Atlantic and the western Pacific, warm water extends farther north under the influence of western boundary currents. In several places, coastal upwellings lead to colder waters in lower latitudes, and the terrestrial environment immediately inland is frequently highly arid; important examples here are the Benguela off Namibia, the Peruvian upwelling, and California.

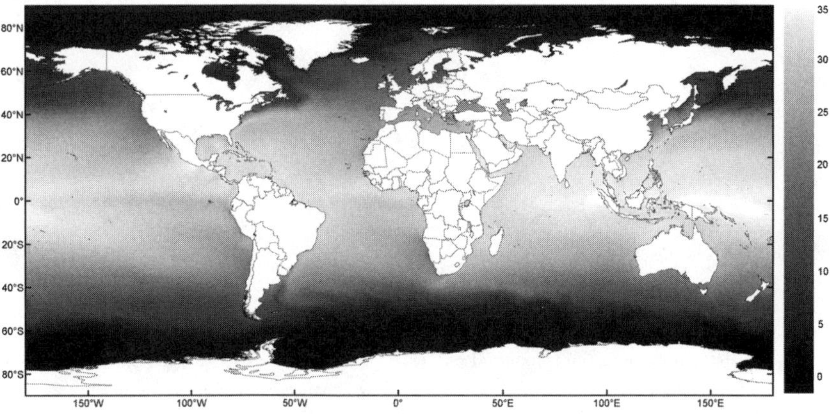

FIGURE 10.1 The global distribution of sea surface temperature. The map shows the mean surface temperature for the period 2001 to 2007, as determined by the NASA Aqua MODIS satellite instrument. The original resolution is 4 km, and units are degrees centigrade. Data was provided by Ocean Color Web, NASA Goddard Space Flight Center (http://oceancolor.gsfc.nasa.gov).

Tropical zones extend to higher latitudes in the Pacific compared with the Atlantic, particularly in the Indo-West Pacific region, and the Pacific tropics tend to be slightly warmer than those in the Atlantic. In strong contrast to the terrestrial environment, where seasonal climatic variability tends to increase monotonically with latitude (Stevens 1989; Chown et al. 2004), seasonal variability in oceanic surface temperature is highest at intermediate temperate latitudes (fig. 10.2). The differences between the two major ocean basins, and also the asymmetry between the northern and southern hemispheres, are exemplified in the meridional data plotted in figure 10.2. These data show clearly how seasonal variability in surface temperatures is highest at intermediate latitudes. They also reveal marked differences between the northern and southern hemispheres in the spatial distribution of surface temperature.

These data reflect only the surface waters, which extend to the depth of the mixed layer, which is typically 100 m or less. Below the thermocline, modern oceanic temperatures are almost universally cold, and the mean temperature of all seawater is <4° C. These patterns have important implications for elucidation of any potential role for temperature in the physiology or ecology of marine organisms. The most important of these is that the

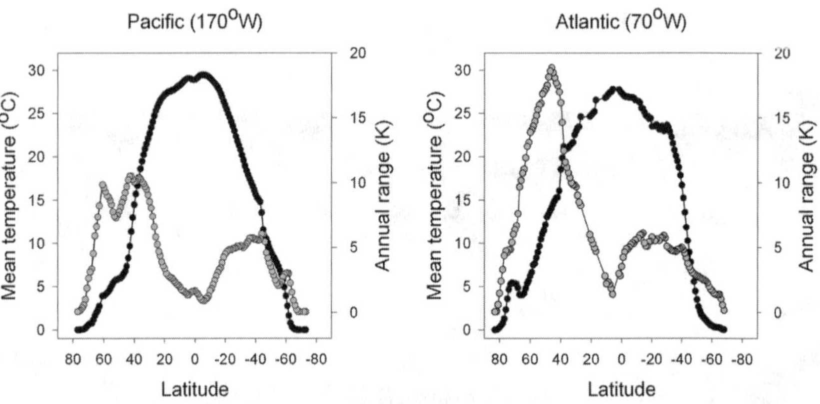

FIGURE 10.2 Mean and seasonal range of sea surface temperature along longitude 170°W in the Pacific Ocean (left panel) and latitude longitude 30°W in the Atlantic Ocean (right panel). Data were calculated for the period 1986–2001 [URL], and plots show the mean temperature (°C, black symbols, axis to left) and mean annual range (K, grey symbols, axis to right). Note the higher peak temperature in the tropical Pacific, and the very different range and latitudinal pattern of variability in the two ocean basins. Data are averages for the period 1983 to 2003. The AVHRR Pathfinder v5.0 SST data were obtained from the Physical Oceanography Distributed Active Archive Centre (PO.DAAC) at the NASA Jet Propulsion Laboratory, Pasadena, California (http://podaac.jpl.nasa.gov).

temperatures seen by satellites are experienced only by organisms living in surface waters or on shallow continental shelves. Marine organisms living below the thermocline experience a more-or-less uniform thermal environment with only very small spatial and temporal variations in temperature. If, as frequently has been proposed, temperature is a major factor in determining patterns of diversity in the sea, then we should expect these patterns to differ between the various thermal environments. In particular, we should expect to see very different patterns in shallow waters and in the deep sea. Before discussing the influence of temperature on marine diversity in the sea, it is important to establish the broad-scale (macroecological) patterns of marine species richness.

Patterns of Diversity in the Sea

Global patterns of diversity in the sea have been described by numerous authors, and hence only the salient points will be outlined here. For summaries of current knowledge see Angel (1997), Clarke, and Crame (1997), Gaston (2000), Gray (2002) and Stuart, Rex, and Etter (2003).

Shallow Waters

Although it had long been recognized that reef-building (hermatypic) corals and associated reef faunas were confined to warm, clear tropical waters, the presence of a global latitudinal diversity cline in the sea was probably first reported by Thorson (1957). Working in the North Atlantic, Thorson described a decrease in the diversity of the epifauna on shallow-water hard substrata with increasing latitude, but could detect no systematic difference in the diversity of soft-bottom fauna between polar, temperate, and tropical regions. Subsequent to Thorson's seminal work, the first clear description of broad-scale patterns of diversity in the sea was that of Stehli, McAlester, and Helsey (1967) who mapped the diversity of bivalve molluscs at the species, genus, and family level, pooled by 5° of latitude. This revealed a distinct cline in diversity from tropics to polar regions, with a clear center of diversity in the Indo-West Pacific. The data for the northern hemisphere were more complete than for the southern hemisphere, a situation that still obtains today, and the entire Antarctic was represented by only a single data point. Together with an earlier study of marine gastropods (Fischer 1960) and a slightly later one for foraminiferans (Stehli, Douglas, and Kefesceglíou 1972), these studies formed almost the entire basis until the early 1990s for the assumption of a latitudinal diversity cline in the shallow seas studied by most marine ecologists (Clarke 1992).

Since then, numerous studies have established the existence of a latitudinal cline in shallow-water diversity in a variety of taxa, including gastropods (Roy, Jablonski, and Valentine 1994, Roy et al. 1998; Valdovinos, Navarrete, and Marquet 2003), bivalves (Crame 2000), bryozoans (Clarke and Lidgard 2000) and decapod crustaceans (Boschi 2000; fig. 10.3). A notable feature of several of these studies is that diversity does not decline in a simple fashion from tropics to poles; rather, diversity is often uniformly high in the tropics, and starts to decline only around 20 to 30°N. In some cases, the area of highest diversity, although technically tropical, is not always centered on the equator but displaced slightly to higher latitudes.

These studies have typically considered the diversity of single taxa from which the existence of a general cline has often been inferred. In contrast, Witman, Etter, and Smith (2004) used a photographic technique to estimate the taxonomic richness of the fauna of subtidal rock walls, and detected a distinct latitudinal cline in assemblage richness over the range of 60°N to 60°S. For comparison, regional richness data were compiled from faunal lists, and were well fitted by a quadratic model with peak richness in a broad tropical zone of 20°N to 20°S (fig. 10.4, panel A). The assemblage data taken from photographic images could also be fitted by a quadratic model, but the explained variance was low and the data for the central tropics were strikingly low; here the highest observed assemblage diversities were around 20°N to 30°S (fig. 10.4, panel B).

Taken together, these studies demonstrate the existence of a number of clear global trends in the diversity of shallow-water taxa. The key features are a cline in regional diversity from tropics to poles, a marked asymmetry in this cline between northern and southern hemispheres (with high southern latitudes often markedly rich), and equally strong longitudinal clines, with centers of high diversity in the Indo-West Pacific and the Caribbean (Clarke and Crame 1997). There are thus two dominant macroecological patterns in the regional diversity of shallow-water marine diversity, one which broadly covaries with global patterns in shallow-water temperature and a second that does not.

At the assemblage level, patterns of richness are far more heterogeneous. Thorson (1957) first drew attention to the difference in patterns between the diversity of the epifauna of hard substrata and soft sediment assemblages in shallow waters. This pattern was confirmed by Kendall and Aschan (1993) and Kendall (1996), and Ellingsen and Gray (2002) were unable to detect any significant latitudinal trends in a detailed study of the soft sediment fauna of the Norwegian continental shelf. In contrast, a comparison of estuarine assemblages across a wide geographical and latitudinal extent indicated a

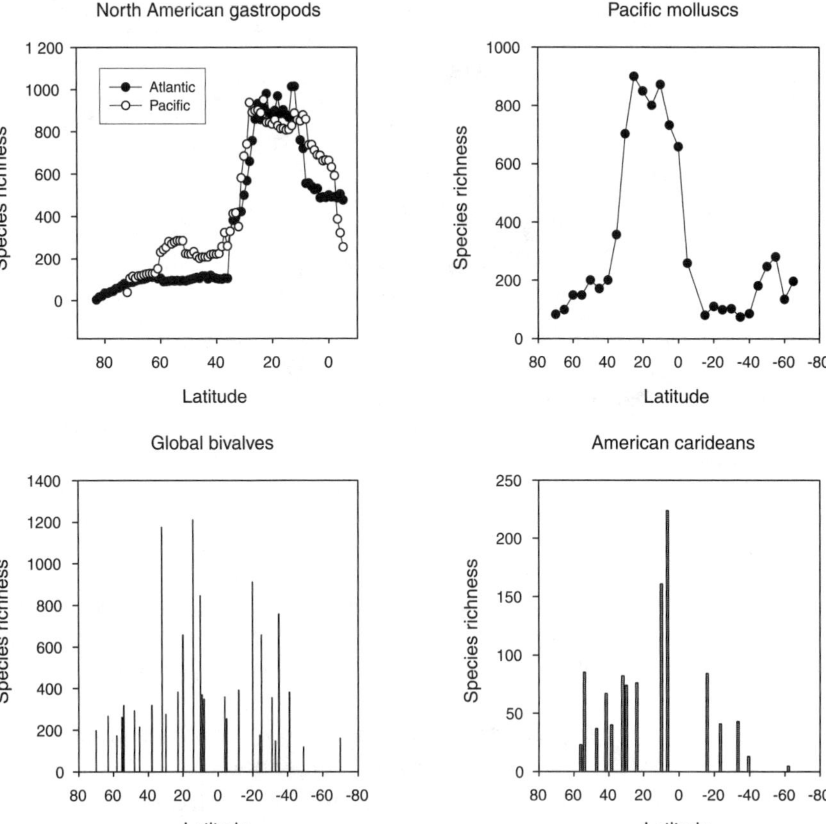

FIGURE 10.3 Latitudinal clines in the diversity of selected shallow-water taxa. By convention, southern latitudes are shown negative. (A) North American continental shelf gastropods (data from Roy et al. 1998); data pooled in bins of 1 degree of latitude, assuming species are found in all bins between range maximum and minimum. Note the striking decrease in diversity around 30°N on both Atlantic and Pacific continental shelves. (B) Molluscs along the Pacific coast of North and South America (data from Valdovinos, Navarrete, and Marquet 2003); data for 629 species (407 prosobranch gastropods, 166 bivalves, and 56 chitons) pooled in bins of 5 degrees of latitude, assuming taxa are found in all bins between range end points. Note the strong asymmetry of diversity about the equator. (C) The global pattern of regional bivalve diversity (data from Crame 2000); data pooled by biogeographic province, and plotted at mid-latitude for each province. Note the asymmetry about the equator, and the low values of some tropical provinces compared with some provinces with mid-latitudes in the range 20°–40° N and S. (D) Provinicial species richness of caridean decapods along the continental shelves of North and South America, plotted as a function of province mid-latitude (data from Boschi 2000). Note the overall decrease in diversity away from the tropics, and the asymmetry between northern and southern hemispheres.

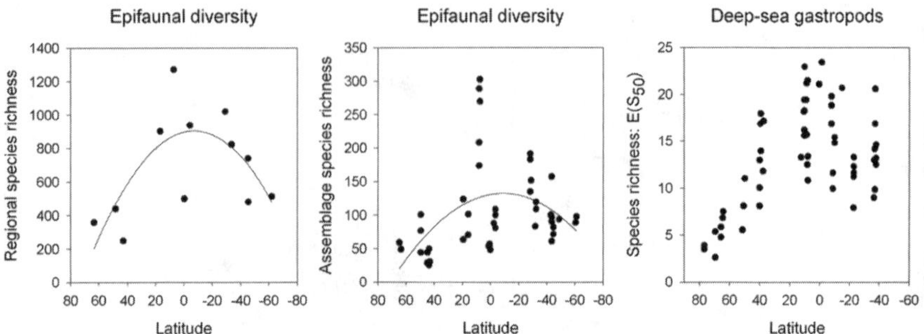

FIGURE 10.4 Latitudinal clines in epifauna and deep-sea benthos. By convention, southern latitudes are shown negative. (A) Regional diversity (species richness) for the epifauna of subtidal rock walls, estimated from published faunal lists and consultation with taxonomic experts (data from Witman, Etter, and Smith 2004). The line shows a quadratic model fitted to the entire data set; this model suggests a tropical peak in regional diversity, though it is asymmetric about the equator and predicts a higher regional diversity in the southern hemisphere compared with comparable latitudes in the northern hemisphere. (B) Assemblage species richness (Chao 2) for the epifauna of subtidal rock walls, estimated from photographic images of standard areas (data from Witman, Etter, and Smith 2004). The line shows a quadratic model fitted to the entire data set; note the low values in the central tropical sites. (C) Species richness, estimated by rarefaction, of gastropods in the deep sea (Rex et al. 1993). Diversity was estimated from rarefaction as the number of species expected for fifty individuals, $E(S_{50})$. Note the asymmetry of the patterns in the northern and southern hemisphere.

decrease in assemblage diversity from tropics to poles (Attrill, Stafford, and Rowden 2001), whereas Engle and Summers (1999) were unable to detect any such trends in a study of estuarine fauna along 17 degrees of latitude along the eastern North America coast. Data from shallow-water soft sediments thus reveal no consistent pattern that might reflect the influence of temperature on diversity.

The Deep Sea

These studies have all been undertaken in shallow waters or on continental shelves. Studies of depth gradients in diversity have suggested that diversity reaches a peak at intermediate depths, decreasing toward the deep sea (Grassle and Maciolek 1992; Rex et al. 1993; Gray 1997). The existence of a latitudinal diversity gradient in the deep sea was first suggested in a seminal paper by Sanders (1968). More recently, Rex et al. (1993; Rex, Stuart, and Coyne 2000; Rex et al. 2005) have provided evidence for a strong latitudinal gradient in species richness in gastropods, bivalves, and isopods in the North Atlantic; the gradients were far less steep in the southern hemisphere (fig. 10.4, panel C). These patterns have stimulated much debate, most notably concerning the extent to which the statistical patterns are determined by the low-diversity faunas from the Norwegian Sea, an area known to have suffered major disruption in the recent past (Gray 1994). Recent studies (Stuart, Rex, and Etter 2003) have tended to confirm the existence of a latitudinal cline in benthic macrofauna, and in a thorough review of all available data, Gray (2002) concluded that the deep sea tends to have fewer individuals per species than comparable shallow-water areas, that species density (number of species per unit area of sediment) is higher in the deep sea at local scales but similar to shallow waters at large scales, that diversity peaks at intermediate depths (1,500 to 2,500 m), and that species richness decreases from subtropical latitudes northward but not southward.

Latitudinal clines in deep-sea species diversity have also been reported for foraminifera (Culver and Buzas 2000) and nematodes (Lambshead et al. 2000, 2002). Benthic foraminifera were significantly more diverse in the southern hemisphere, and linear diversity/latitude relationships could be fitted to data from both hemispheres (Culver and Buzas 2000). The calcareous foraminifera are globally distributed and have an excellent fossil record; they thus provide a valuable group for investigating patterns of diversity in both time and space. Many species appear to be cosmopolitan, and hence global diversity may not be as high as local diversity might imply (Gooday et al. 1998).

Planktonic Diversity

Despite the enormous area of the oceans, and their great depth, planktonic communities are far less diverse than benthic assemblages, and the holozooplankton may contain as few as 5,000 species overall. The greatest diversity is attained in warm temperate and tropical latitudes, as illustrated by epipelagic calanoid copepods (Woodd-Walker, Ward, and Clarke 2002; fig. 10.5, panel A). The latitudinal distribution of taxonomic distinctness in these samples (fig. 10.5, panel B) does, however, point to important evolutionary patterns underpinning the deceptively simple pattern in richness. A pattern of higher diversity in the tropical regions is also seen in other groups including chaetognaths, pteropods, and euphausiids, although more detailed analyses indicates clearly that zooplankton biogeography is related most strongly to the dominant water masses that have long been known to oceanographers (Pierrot-Bults 1997). More recently Longhurst (1998) has used a variety of physical and biological criteria to divide the ocean into four primary biomes: westerlies, trades, coastal, and polar; these in turn are subdivided into fifty-one provinces. It seems likely that zooplankton biogeography and diversity will prove to be related more to these biomes than directly to single abiotic factors such as temperature (Woodd-Walker, Ward, and Clarke 2002).

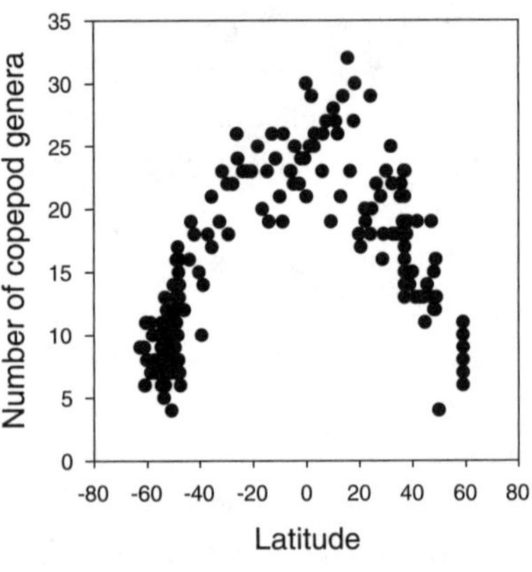

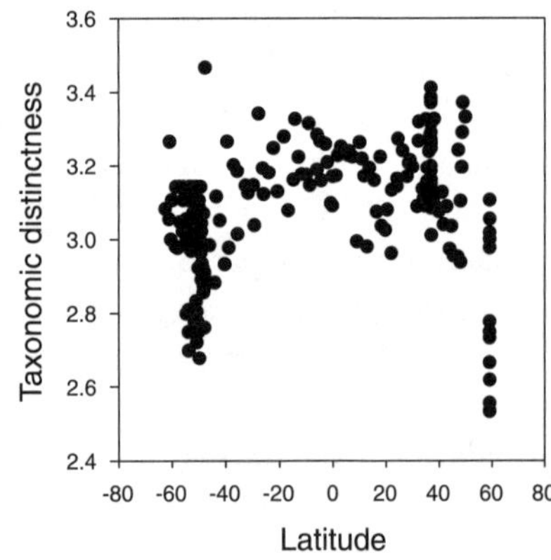

FIGURE 10.5 Diversity of epipelagic copepods from North and South Atlantic as a function of latitude (data from Woodd-Walker, Ward, and Clarke 2002). By convention, southern latitudes are shown negative. (A) Generic richness; note the asymmetry about the equator, with peak richness displaced to the north. (B) Taxonomic distinctness (Clarke and Warwick 1998); note the uniformly high values in the tropics and subtropics (roughly 30°N to 30°S), and the highly variable values at higher latitudes in both hemispheres.

Problems of Interpretation

The patterns of marine diversity on broad scales, summarized earlier, reveal strong biogeographical trends, the most important of which are latitudinal clines in diversity and marked longitudinal heterogeneity. Although these patterns have been reported in many studies, sufficient for meta-analyses to confirm the results as robust (Hillebrand 2004a, 2004b), there remain significant problems of interpretation. The most important of these problems concerns sampling scale. Concerns that comparative studies have mixed results from studies using very different sampling techniques have a long history (Abele and Walters 1979; Clarke 1992), and have been explored in detail by Gray (2000, 2002) and Gray et al. (1997). The core of the problem is that samples from different habitats and depths have been taken using different gears that collect over different spatial scales, processed using different screen sizes to retain the fauna, and analyzed with different statistical techniques.

Small samples from deep-sea sediments are too restricted to collect the entire local fauna, and samples therefore have to be pooled or statistical techniques used to estimate the species richness over areas comparable to those studies in shallow water. Marine ecologists have traditionally used rarefaction to estimate the species richness of a predetermined number of individuals, although Gray (2002) has argued that sample accumulation curves are preferable statistically. This is principally because the estimate of diversity produced by rarefaction is strongly dependent on the species abundance distribution. Other statistics that have been commonly used in marine ecology include Chao 2 and Fisher's alpha; while each has its particular strengths (Magurran 2004) the plethora of statistics makes comparison between studies very difficult.

Because marine sampling is typical blind and small scale, an inevitable result of the species/abundance relationship is that many of the less common organisms occurring within a given area are not collected. Diversity statistics such as Chao1 and 2 have been devised to minimize the impact of this problem on the estimate of local diversity (see, for example, Witman, Etter and Smith 2004) but they do not eliminate the problem entirely. For estimates of regional diversity, ecologists have typically assumed that a given species is found everywhere between the range endpoints and pooled species counts into bins of a size reflecting sampling intensity rather than any particular features of the environment or organism (Roy et al. 1994, 1998; Roy, Jablonski, and Valentine 2000: Valdovinos, Navarrete, and Marquet 2003).

Although the concept of local and regional scale introduced into studies of diversity by Whittaker (1972) has been utilized widely in terrestrial studies, it has made less of an impact in marine ecology. One reason for this is the difficulty of defining local and regional scales in the sea, since these depend both on the size of the organisms involved and the scale of dispersal; the definition of local scale is very different for sediment meiofauna, corals, and whales. Combined with the difficulties of much marine sampling, a pragmatic view has tended to prevail, whereby a sample or collection of samples from a single locality is taken to represent local diversity, and samples pooled over a larger area are regarded as representing regional diversity. Regional diversity in the sea should perhaps best be related to shallow-water marine provinces, in that these represent geographically defined regions that share a recent common evolutionary history, and that are typically separated from nearby such provinces by a geographical or oceanographic barrier. These provinces are typically quite large, and hence the technique of pooling samples into latitudinal bins of a particular size, although pragmatic, may thus give a measure of diversity that is neither local nor regional in an evolutionary sense. These distinctions are important because they affect the patterns observed, and hence the conclusions we may draw about the underlying mechanisms. A valuable step forward in our understanding of shallow-water diversity would be a global analysis of continental shelf biotic provinces to match that undertaken by Longhurst (1998) for the pelagic realm, and an important such analysis has recently been undertaken by Spalding et al. (2007).

Explaining the Macroecological Patterns of Marine Diversity

Bold macroevolutionary patterns such as the geographical distribution of diversity in the sea require explanation. We cannot, however, simply assume that these patterns have a biological explanation. A necessary first step is to eliminate the possibility that these patterns may be generated by purely random or neutral processes. Thus Colwell and Hurtt (1994) showed that placing species of randomly sized ranges at random on a globe would generate a latitudinal cline with no other mechanisms involved. The most influential of these null models has undoubtedly been the mid-domain model of Colwell and Lees (2000). While these models cannot reproduce the detail of the patterns we see in nature, they provide a valuable tool in highlighting the deviations from a random model that require a nonneutral explanation (Zapata, Gaston, and Chown 2003, 2005; Connolly, Bellwood, and Hughes 2003).

The boldest geographical patterns of marine diversity are the latitudinal clines, the longitudinal heterogeneity, and the variation with depth (Gray 2002). Food availability (productivity), history, and temperature have all been invoked as explanations of these patterns; here I will concentrate on the role of temperature.

Does Marine Diversity Correlate with Temperature?

Any detailed examination of a possible role for temperature in driving macroecological patterns of diversity must start with the question of how well diversity correlates with temperature. From a macroecological perspective, this means looking for large-scale correlations. A difficulty here is that the latitudinal cline in shallow-water marine diversity inevitably produces a correlation with a variety of climatic variables, including temperature; such a correlation may thus tell us nothing about the underlying ecological processes. Because the diversity and temperature in the sea are necessarily correlated at the broadest scale, clues as to the nature of the functional (ecological) relationship, if any, will be found in the fine detail.

Correlation of Diversity and Temperature in Space

An example of a tight spatial correlation between diversity and temperature is provided by epipelagic marine copepods (fig. 10.6, panel A). Here there is a strong linear relationship between temperature and generic richness in samples taken from the North and South Atlantic from 59°N to 63°S (Woodd-Walker, Ward, and Clarke 2002). In this study, genus was chosen

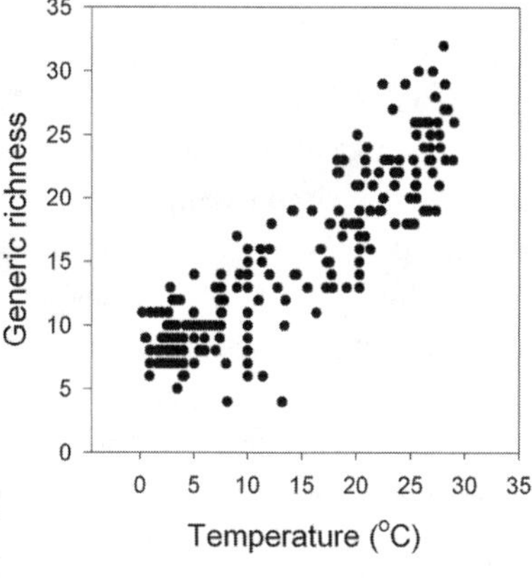

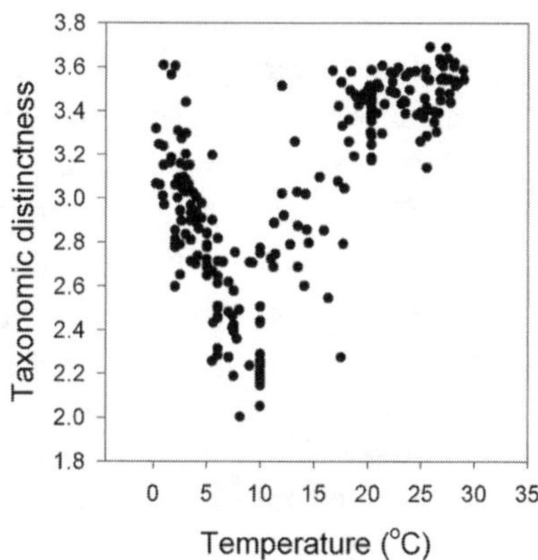

FIGURE 10.6 Diversity and temperature in epipelagic marine copepods (data from Woodd-Walker, Ward, and Clarke 2002). (A) Relationship between generic richness per sample and sea surface temperature at the time of collection. Richness was determined at the level of genus because of unresolved taxonomic difficulties at species level. (B) Taxonomic distinctness (sensu Clarke and Warwick 1998) and temperature for the same samples.

as the most appropriate level of taxonomic resolution for copepods examined over such a wide latitudinal range; identification to species level was not undertaken because of unreliable taxonomy within this diverse group, and the difficulties of identifying sub-adult stages to species. Use of higher taxonomic levels is not uncommon in diversity studies and has been shown to be tightly correlated with species richness for a wide range of organisms (Williams and Gaston 1994; Williams, Humphries, and Gaston 1994; Roy, Jablonski, and Valentine 1996). Moreover in some cases for analyses over large spatial scales such as here, trends are clearer at generic level than species level (Roy, Jablonski, and Valentine 1996).

The strong relationship between generic richness and temperature is not surprising given the symmetrical distribution of copepod diversity about the equator (fig. 10.5, panel A), and would seem intuitively to imply a significant role for temperature in regulating diversity. Furthermore, this symmetrical pattern is exhibited by a wide range of holozooplankton, including euphausiids, larvaceans, siphonophores, and fish, indicating that it is a general phenomenon in the pelagic marine environment (Angel 1997; Pierrot-Bults 1997). Similar patterns are also seen in the microbial planktonic fauna. Rutherford, D'Hondt, and Prell (1999) examined the diversity of planktonic foraminifera from sediment samples across a wide latitudinal range, and found that sea surface temperature explained nearly 90 percent of the variation in diversity. Diversity was actually highest in the mid-latitudes rather than the equatorial tropics, and Rutherford, D'Hondt, and Prell (1999) proposed that planktonic diversity was controlled by the physical structure of the mixed layer, for which they identified temperature as a major driver.

More detailed examination of epipelagic copepod assemblages, however, suggests that the drivers are more complex than simply temperature. Multivariate analyses showed that the assemblage composition mapped well onto the oceanic biomes proposed by Longhurst (1998) on the basis of oceanic physics and phytoplankton characteristics, though not to the finer-scale division of these biomes into provinces (Woodd-Walker, Ward, and Clarke 2002). Important underlying structure was also revealed using the taxonomic distinctness index of Clarke and Warwick (1998). This measure incorporates taxonomic structure into a univariate diversity index, and thereby provides insight into phylogenetic history. Both taxonomic distinctness (fig. 10.5, panel B) and evenness revealed a distinct pattern for high-latitude assemblages (those above ~40°N or S) to be far more variable than in the intervening tropical regions, which were characterized by uniformly high values of evenness and taxonomic distinctness (Woodd-Walker, Ward, and Clarke 2002). As a result, the relationship between taxonomic distinctness and temperature reveals not a simple pattern but rather separate relationships for warmer and cooler waters, with the boundary around 18°C (fig. 10.6, panel B). This underlying taxonomic structure implies an important role for evolutionary history within the oceanic biomes; this history does not, however, override a strong overall relationship between richness and temperature. It is possible that temperature could set an upper limit on diversity, while other factors affect which species comprise that diversity; our present knowledge of planktonic diversity do not allow us to separate these two different processes.

In contrast, data for benthic assemblages shows a far less coherent relationship between diversity and temperature. The most detailed data for shallow-water benthic diversity are for continental shelf gastropods and bivalves of North America (fig. 10.3). Although the patterns of diversity with latitude are remarkably similar along the Atlantic and Pacific continental shelves, the very different oceanographic regimes in these two basins mean that the relationship between diversity and temperature also differs. Along the Pacific continental shelf, the relationship is monotonic, but clearly nonlinear (fig. 10.7, panel A); the relationship is roughly exponential and is therefore linear in log/linear and log/log space. The relationship for the Atlantic continental shelf is quite different, and markedly biphasic. There is essentially no relationship between diversity and temperature over the range 5°C to 20°C, with diversity averaging ~100 species. There is a small decline from ~120 to

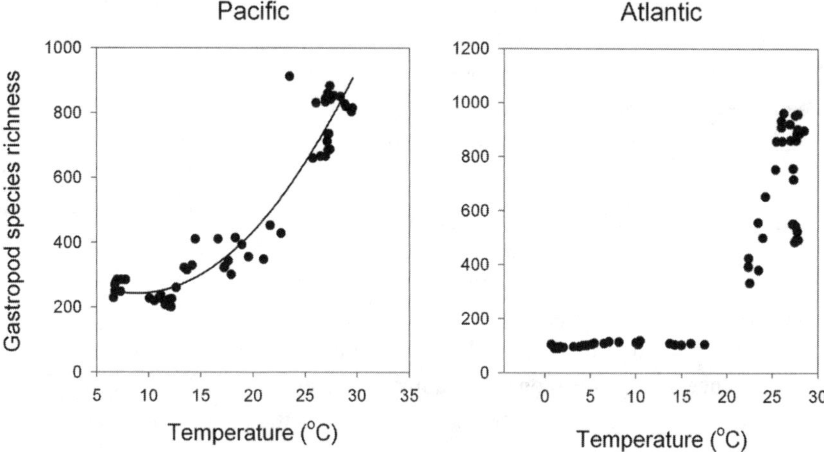

FIGURE 10.7 Relationship between diversity and temperature for shallow water gastropods from the Pacific and Atlantic continental shelves of North America (data from Roy et al. 1998). (A) Pacific continental shelves. Here the relationship can be fitted with a monotonic curve (a quadratic model is shown). (B) Atlantic continental shelves. Here the relationship is markedly biphasic, and cannot easily be modeled with a simple relationship. There is a small increase in species richness with temperature up to about 20°C, and a very steep relationship above about 22°C.

~90 species from 5°C to 0°C; however, in warmer waters there is a steep increase from ~400 species at 22°C to ~1,000 species at 28°C (fig. 10.7, panel B). This pattern remains after division of the gastropod fauna into broad trophic groups (carnivores/noncarnivores: Roy, Jablonski, and Valentine 2004).

The three examples—epipelagic copepods, and gastropods from the Pacific and Atlantic continental shelves, thus show three different relationships between temperature and diversity: linear, exponential, and complex. However, the most powerful test of the nature of any functional relationship between diversity and temperature comes from the deep sea. This is because, with the exception of a few areas such as the Red Sea, the temperature of the deep sea is fairly uniform over the globe, being dictated by the production of bottom water at high latitudes. Any theory that links diversity directly to temperature would necessarily predict a uniform distribution of diversity in the deep sea. This, however, is not what we see: despite this uniformity of thermal environment, deep-sea macrofaunal diversity shows strong regional differentiation, with a marked cline in the northern hemisphere (fig. 10.4, panel C). In the southern hemisphere, evidence for a strong latitudinal diversity cline is far more equivocal.

We must therefore conclude that the relationship between temperature and diversity in the sea is not a mechanistic one, for strong patterns in diversity exist where there is no variation in temperature, and in some areas strong gradients in temperature are accompanied by more-or-less uniform diversity. Since temperature is clearly not the major driver of diversity in the sea, we must look elsewhere for the factors that really drive marine diversity.

Correlation in Time

It has long been recognized that in both physical and biological processes in the sea, scales of space and time are intimately linked (Steele 1991; Steele and Henderson 1994). This relationship means that patterns over very large spatial scales tend to be determined by processes acting on long time scales, and the balance between ecological and historical (evolutionary) factors tends to shift with spatial scale (Clarke 2003, 2007). To understand the drivers of macroecological patterns at the global scales we therefore need to consider evolutionary processes acting over geological time spans.

During the Cenozoic, marine diversity has increased steadily (fig. 10.8, panel A) and the present diversity of life on earth is probably the highest it has ever been. The trajectory of both terrestrial and marine diversity since the K/T mass extinction event has been almost exclusively upward through the entire Cenozoic. There is no real indication that diversity has reached an asymptote, except possibly for the past five million years. Neither correction for sampling intensity (that is, volume of rock) nor taxonomic revision altered significantly the shape of the Cenozoic diversity curve (Signor 1985; Foote and Sepkoski 1999; Crampton et al. 2003; Jablonski et al. 2003; Bush, Markey, and Marshall 2004) (But see Alroy et al. 2008).

The Cenozoic increase in marine diversity has coincided with a steady decrease in global seawater temperature (fig. 10.8, panel B). This would suggest that at the global scale, temperature cannot be the primary driver of diversification. The relationship is, of course, made complex by factors such as tectonic effects on habitat area, and the synergistic effect of continental arrangement and oceanography on the degree of provincialization (Valentine 1968; Valentine, Foin, and Peart 1978). Bambach (1977) has shown that the Cenozoic marine diversification involved principally diversification within provinces, although there may also have been a simultaneous increase in the number of marine provinces as the latitudinal temperature gradient steepened and oceanographic patterns changed. Present diversity thus reflects a complex interaction between tectonics and climate, but the historical record shows clearly that there is no simple relationship between temperature and diversity. The trajectory of marine diversity during the Cenozoic would also suggest that the earth as a whole is not yet saturated with species (although of course specific habitats or areas may be).

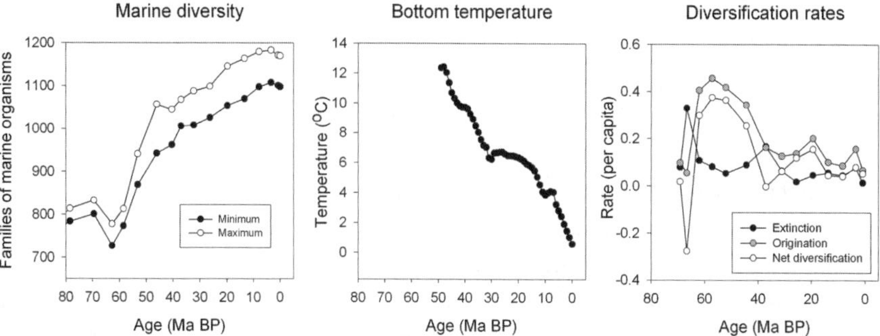

FIGURE 10.8 Marine diversity and oceanic temperature during the Cenozoic (redrawn from Clarke 2007). (A) Marine diversity during the Cenozoic; data from the Fossil II database (courtesy Mike Benton). Data are plotted at family level and both the maximum and minimum estimate of family diversity per stratigraphic interval are plotted. (B) Cenozoic oceanic temperatures determined from Mg/Ca ratios in the skeletons of foraminifera (Lear, Elderfield, and Wilson 2000). This palaeotemperature proxy avoids the difficulties associated with glaciation which beset the more widely used oxygen isotope proxy. The data are for deep-water benthic forams, and so may differ slightly from the temperatures experienced on the continental shelves (for which we currently have no reliable global proxy). Comparison with the diversity curve (A) reveals the inverse relationship through time of marine diversity and global bulk seawater temperature. (C) Origination and extinction rates for marine taxa since the Late Eocene (the period for which Mg/Ca palaeotemperatures are available; data from Foote 2005). Data are calculated as per capita rates averaged over stratigraphic intervals, and plotted as a function of the mean absolute date for that interval. Net diversification is the difference between the per capita rates of origination and extinction.

Although patterns of marine diversity at the macroecological scale suggest that seawater temperature is not the main driver for standing diversity, temperature may influence the rate of diversification. Foote (2005) has shown clearly that there was a peak in net diversification immediately following the K/T extinction, after which the rate slowed (fig. 10.8, panel C). Because the post-extinction pulse of diversification came at a time when the global ocean was significantly warmer than today, and the later reduced rates coincided with cooler temperatures, the per capita rate of origination of new marine taxa is significantly correlated with oceanic temperature. However, neither extinction rate nor net diversification rate correlate significantly with temperature for the period since the Lower Eocene, this being the time for which we have suitable palaeotemperature data (fig. 10.8, panel B). A recent analysis for the entire Phanerozoic by Mayhew, Jenkins, and Benton (2007) has shown that global marine benthic diversity (at least that preserved as fossils) was lower in greenhouse conditions than it was in icehouse climates, leading to an inverse relationship between estimated marine temperature and fossil diversity. This is precisely the opposite of the relationship predicted from a simple temperature control of marine diversity. The fossil data do, however, point to positive correlations between both diversification rate and extinction rate with global marine seawater temperature.

The steady increase in marine diversity through the Cenozoic has been accompanied by an intensification of the latitudinal diversity gradient. Careful detailed work on fossil bivalve faunas has shown that latitudinal diversity clines were present in the Late Jurassic, but were far less steep than observed today, and were underpinned by a tropical fauna that comprised a roughly equal number of infaunal and epifaunal taxa (Crame 2002). The reduced meridional temperature gradients that have characterised much of the past 100 Ma may also have allowed for more extensive exchange between faunas than is observed today (Crame 1993). During the Cenozoic there has been a spectacular bivalve radiation, which has involved predominantly the younger, infaunal clades of heteroconchs (Crame 2000b, 2002). These clades dominate modern tropical bivalve faunas, and in the case of bivalves point clearly to the development of the latitudinal (meridional) diversity cline through tropical diversification and the spread of these taxa to higher latitudes (Crame 2000a, 2000b, 2001, 2002; Crame and Rosen 2002). The diversification of marine clades in tropical regions, and particularly the Indo-West Pacific is a general feature (Briggs 2003, 2004), and it underpins both meridional and longitudinal clines in marine diversity we observe today (Clarke and Crame 1997, 2003). The latitudinal gradient in foraminiferal diversity is also a Cenozoic feature, having formed about 36 Ma BP (Culver and Buzas 2000).

Although modern macroecological patterns of marine diversity are dominated by Cenozoic tropical diversification, some clades have diversified at high southern latitudes (Clarke and Johnston 2003). Wilson (1998) has shown that the deep-sea isopod fauna comprises two distinct clades with different evolutionary origins. The asellotes are the most diverse, and include many families that are endemic to the deep sea. In contrast, the flabelliferans are a relatively recent clade that shows a significant cline in diversity from high values in the Southern Ocean (60°S) decreasing northward toward 50°N. Wilson (1998) interprets this pattern as indicating that flabelliferans originated and diversified on the high southern latitude continental shelves, probably during the Cenozoic, and have subsequently spread northward into deeper water. Other taxa that show evidence of diversification at high southern latitudes include some families of amphipods, pycnogonids, predatory gastropods, and some clades of cnidarians (Watling and Thurston 1989; Clarke and Johnston 2003). It is probable that a key feature common to the diversifications of marine fauna in high southern latitudes and the Indo-West Pacific was changes in habitat area driven by variations in the

Antarctic ice-sheet on Milankovitch frequencies (Clarke and Crame 1989, 1997, 2003). In the Indo-West Pacific changes in habitat area would have been driven by variations in sea level, whereas in the Southern Ocean the critical factor would have been periodic extension of the ice-sheet over the continental shelf.

The correlation between marine diversity and oceanic temperature is thus nonexistent when examined over very long temporal scales. A recent study of deep-sea sediment cores has, however, suggested the possibility of a tight coupling between foraminiferal richness and temperature over the past 130,000 years (Hunt, Cronin, and Roy 2005). Using data from ten cores (six Atlantic, four Pacific, 59°N to 48°S) diversity was estimated directly from microfossils in the core, and temperature from the nearest location for which suitable deep-sea palaeotemperature data were available. Although there was a positive relationship between diversity and temperature in the overall data set, the data for the different cores were widely divergent with individual relationships ranging from strongly positive to strongly negative.

Overall, the pattern of global marine diversity with time thus points to an important role for climate change on geological time-frames, coupled with associated tectonic and oceanographic changes. It does, however, also rule out a direct role for temperature in driving the patterns of marine diversity at the macroecological scale.

Marine Diversity and Temperature: What Might the Mechanisms Be?

We must conclude that although there is a broad spatial correlation between marine diversity and seawater temperature in some places, this correlation is not a universal one; it is frequently poor when examined in detail, and nonexistent in deep time. The spatial correlation is perhaps strongest in the northern hemisphere, where the most recent glacial maximum has led to a strong geographic covariation in diversity and climate. This covariation has led to a number of hypotheses that temperature controls diversity directly, the logical consequence of which is that diversity will necessarily be higher where temperatures are warmer. This prediction matches the intuitive feeling that warmer habitats are easier places to make a living than colder ones, and hence more species live there. There is as yet no compelling evidence that a polar marine ectotherm is any more physiologically stressed or disadvantaged than a tropical one, although evolutionary adaptation to a particular temperature does have powerful ecological consequences (Clarke 2003). We must also be wary of anthropocentric judgements of perceived stress; if some species of a group of organisms can evolve to live in apparently tough or demanding conditions, such as polar seas, why cannot another (a question posed succinctly by Hutchinson 1959)?

Neither is it clear precisely how temperature might regulate diversity directly. Allen, Brown, and Gillooly (2002) used the frequently observed correlation between diversity and temperature in a wide range of animal and plant taxa, to argue that diversity is linked directly and mechanistically to temperature through "the generally faster biological rates observed at higher temperatures" (Brown, Allen, and Gillooly 2003). The argument is based on the energy-equivalence rule, which states that the total energy flux of a population per unit area is invariant with respect to body size (Damuth 1987). Allen, Brown, and Gillooly (2002) extend the assumption of energy equivalence by incorporating the biochemical kinetics developed within the Metabolic Theory of Ecology (Brown et al. 2004), and predict a relationship between diversity and temperature, which they test with data for a variety of terrestrial, freshwater, and marine taxa. Several studies have suggested that the energy-equivalence rule does not apply to all communities (for example, Marquet, Navarrete, and Castilla 1995; Ackerman, Bellwood, and Brown 2004), and Storch (2003) has questioned the conceptual basis for the model. In particular, a model linking diversity to environmental temperature through body temperature should not apply to endotherms, and yet both birds and mammals exhibit strong global scale correlations between diversity and environmental temperature (Storch 2003). As discussed earlier, neither do the data for marine diversity support any direct link between diversity and temperature. Allen, Gillooly, and Brown (2007) have subsequently recast their hypothesis, abandoning the dependence of the hypothesis of energy equivalence and building instead a model linking temperature to diversity through its influence on evolutionary rate.

An alternative hypothesis is that the diversity of marine organisms is linked not to the mean temperature, but to its seasonality (Stevens 1989). Stevens suggested that organisms living at high latitudes required a more generalist physiology and ecology because of the strong seasonal variation in climate. Biogeographic ranges were thus wider at high latitudes and this thereby set a limit on the total diversity that could be supported. The idea that enhanced climate variability might reduce diversity was developed largely for terrestrial systems. Here, seasonality of climate (sensu lato) tends to correlate broadly with latitude, and hence the expected inverse correlation with diversity is often observed. The latitudinal pattern of temperature variability in the sea is, however, very different from that on land, with high latitudes typically having low but seasonally stable temperatures and the greatest seasonality being observed at intermediate temperate latitudes (fig. 10.2). The climate variability hypothesis would thus predict the highest marine diversities to be in the temperate latitudes, and this is not what we observe, either in the shallow continental shelves (fig. 10.3), in the deep sea (fig. 10.4), or pelagic zooplankton (fig. 10.5). We must conclude that there is no observational support for a link between climate variability and diversity in the sea.

An alternative class of hypothesis involves indirect effects, whereby temperature influences some factor that in turn regulates diversity. Particularly widely discussed in the marine realm have been the possible effects of temperature on speciation or extinction rates. If the rate of speciation were a positive function of temperature, then it would seem intuitive that the tropics would be more diverse than polar regions. This is a nonequilibrium explanation, for it implies that the process of diversification in colder regions has not proceeded as far as it has in tropical regions: given sufficient time, the diversity of polar regions would eventually match that of the tropical regions (Clarke 2007). It also assumes that there is no equivalent effect of temperature on extinction rate, since standing diversity reflects a dynamic balance between these two processes. Equilibrium explanations are also possible, whereby the balance between speciation and extinction is in some way dependent on temperature.

Why should speciation and/or extinction rates be faster at higher temperatures? Two possible mechanisms that have attracted considerable attention are temperature-related variations in mutation rate and generation time. Martin (1995) has suggested that the increased metabolic rate in ectotherms at higher temperatures may lead to an enhanced rate of mutation through increased free radical damage. Gillooly et al. (2005) have developed this idea within the formal scaling context established by the Metabolic Theory of Ecology, and Allen, Gillooly, and Brown (2007) link this approach with life history to develop a theory linking diversity to temperature.

Data to test this are hard to acquire; Held (2001) could find no evidence for a reduced rate of molecular substitutions in polar marine ectotherms, and Bromham and Cardillo (2003) could detect no latitudinal trends in the rates of molecular evolution of terrestrial vertebrates. Evidence from the fossil record is also equivocal. Crame and Clarke (1997) could find no significant difference in the diversification rate of tropical and polar families of gastropods, whereas Flessa and Jablonski (1996) examined the rate of turnover of whole bivalve faunas and showed that tropical faunas turned over more rapidly. Recently, Buzas, Collins, and Culver (2002) reported higher diversification rates in tropical than temperate foraminifera in the Cenozoic, though these data are not directly comparable with other studies because the authors used a different measure of diversification from the more traditional palaeoecological studies (specifically the proportional increase through time in diversity as measured by Fisher's alpha for the entire fossil community at a given location, rather than the more usual exponential diversification model).

Considerable attention has also been directed at the possibility that generation time might influence diversification rate (Marzluff and Dial 1991). Since generation time tends to be shorter in small organisms and among ectotherms at higher temperatures, this hypothesis leads to very specific predictions. Careful tests of these predictions have, however, yielded mixed results to date (Cardillo 1999; Bromham and Cardillo 2003) and none appear to have been performed for marine organisms.

In the terrestrial realm, considerable attention has been directed at the possible role of energy availability in driving diversity (Hawkins et al. 2003, 2007; Clarke and Gaston 2006). While local-scale studies have clearly shown relationships between resource availability and benthic diversity in the sea (Dayton and Oliver 1977), broad-scale studies in the sea are almost completely lacking (and in any event are complicated by the enormous distance between surface waters where most primary production takes place, and the seabed, together with the complicating factor of advection).

Other Macroecological Patterns

The field of macroecology is concerned with large-scale patterns in a number of ecological features, including abundance, size, and range (Brown 1995; Maurer 1999). Many of these are intimately related to diversity, though they frequently exhibit striking patterns of their own. Relatively few of these patterns, however, have been investigated in the marine realm, and hence our ability to compare these features in the marine and terrestrial realms is severely limited. Two areas where important results have been obtained recently are species abundance patterns and size.

Species Abundance Patterns

The description of species abundance patterns requires fully quantitative sampling, and this is unfortunately rare for the marine environment. Two studies suggest that there are striking differences in the abundance structure of gastropod assemblages from two contrasting locations. A recent detailed study by Bouchet et al. (2002) of a single location in New Caledonia involving 400 person-days of sampling sampled over 140,000 individuals comprising 2,738 species. These results confirm the local-scale species richness of tropical locations implied by regional taxonomic compilations, in that this site exceeds by several fold the highest molluscan species richness recorded for any location of comparable size anywhere in the world. The sample contained many rare species, with 20% being represented by single individuals. Furthermore, 28% of the taxa were recorded only as empty shells.

A comparable study from Antarctica emphasizes the very different species abundance structure found at higher latitudes. Picken (1980) studied epifaunal gastropods from a shallow-water site at Signy Island, Antarctica, and collected over 138,000 individuals, but only thirty-one taxa. Of these, two species were represented by only a single individual and one by two individuals; the next most abundant species had forty-one individuals. This indicates a gastropod assemblage dominated by fewer taxa and with relatively few rare species. The differences in regional diversity discussed earlier suggest that this is likely to be a general result, though more quantitative samples are badly needed from the marine environment to establish whether this is the case. If tropical assemblages are indeed characterized by much larger numbers of rare taxa, it poses the obvious question of to what extent (if any) the very different abundance structures mean that tropical and polar ecosystems operate differently.

Size and Temperature

A number of global trends in the size of marine organisms have long been recognized, though relatively few have been the subject of quantitative study. The tendency for cold-water molluscs to be small and thin-shelled was noted long ago (Nicol 1964, 1966, 1978), and Graus (1974) demonstrated a clear latitudinal cline in the degree of calcification in shallow-water gastropods. This is presumably related to the increasing energetic cost of calcification at lower temperatures (Clarke 1983). Although high latitude and deep-water molluscs tend to be small and thin-shelled, there are also striking examples of large species in polar regions, often referred to as polar gigantism. The general tendency for organisms to be larger at colder temperatures has been analyzed in terms of life-history theory (Atkinson and Sibly 1997). Recently, Chapelle and Peck (1999) analyzed the size frequency spectra of amphipod crustaceans from a range of marine and freshwater locations, and showed that whereas the size of the smallest species was largely invariant, the largest size achieved was related to the oxygen content of the water. Since the oxygen content of seawater is strongly temperature dependent, this means that temperature sets a constraint on the maximum size that species within a given clade or *bauplan* can achieve. What is currently unknown is whether this constraint affects just the maximum size, or whether there are also consequences for the overall size frequency distribution.

Concluding Remarks

The sea is the largest habitat on the face of the earth, and exhibits a wide range of temperatures. The diversity of marine organisms also show a range of patterns, dominant among which are variations with latitude, depth, and longitude. Although there is a strong tendency for diversity to be highest in the tropics and lowest toward the poles, marine diversity shows no consistent pattern with temperature, and we must conclude that other factors are more important, most notably history. It is likely that temperature does influence diversity through its effects on life history, but it is not clear whether habitat temperature constrains the maximum diversity that can be achieved. A full understanding of patterns of marine diversity will only be achieved through an integration of ecology and physiology with climatic, tectonic, and oceanographic changes.

ACKNOWLEDGMENTS

I thank Kaustuv Roy and Jon Witman for the invitation to write this article, and for the provision of data used in plotting some figures. I thank Carrie Lear for provision of palaeotemperature data, and Mike Benton for the Fossil 2 database data. Paul Mann helped with the extraction of the data for figures 10.1 and 10.2, and Alistair Crame has been a continual source of inspiration and insight into the historical aspects of marine diversity.

REFERENCES

Abele, L. G., and K. Walters. 1979. Marine benthic diversity: A critique and alternative explanation. *Journal of Biogeography* 6:115–26.

Ackerman, J. L., D. R. Bellwood, and J. H. Brown. 2004. The contribution of small individuals to density-body size relationships: Examination of energetic equivalence in reef fishes. *Oecologia* 139:568–71.

Allen, A. P., J. H. Brown, and J. F. Gillooly. 2002. Global biodiversity, biochemical kinetics, and the energy-equivalence rule. *Science* 297:1545–48.

Allen, A. P., J. F. Gillooly, and J. H. Brown. 2007. Recasting the species-energy hypothesis: The different roles of kinetic and potential energy in regulating biodiversity. In *Scaling biodiversity,* ed. D. Storch, P. A. Marquet, and J. H. Brown), 283–99. Cambridge: Cambridge University Press.

Alroy J., M. Aberhan, D. J. Bottjer, M. Foote, F. T. Fursich, P. J. Harries, A. J. W. Hendy, et al. 2008. Phanerozoic trends in the global diversity of marine invertebrates. *Science* 321:97–100.

Angel, M. V. 1997. Pelagic biodiversity. In *Marine biodiversity: Causes and consequences,* ed.

R. F. G. Ormond, J. D. Gage, and M. V. Angel, 35–68. Cambridge: Cambridge University Press.

Atkinson, D., and R. M. Sibly. 1997. Why are organisms usually bigger in colder environments? Making sense of a life history puzzle. *Trends in Ecology and Evolution* 12:235–39.

Attrill, M. J., R. Stafford, and A. A. Rowden. 2001. Latitudinal diversity patterns in estuarine tidal flats: Indications of a global cline. *Ecography* 24:318–24.

Bambach, R. K. 1977. Species richness in marine habitats through the Phanerozoic. *Paleobiology* 3:152–67.

Boschi, E. E. 2000. Species of decapod crustaceans and their distribution in the American marine zoogeographic provinces. *Revista de Investigación y Desarrollo Pesquero* 13:1–136.

Bouchet, P., P. Lozouet, P. Maestrati, and V. Heros. 2002. Assessing the magnitude of species richness in tropical marine environments: Exceptionally high numbers of molluscs at a New Caledonia site. *Biological Journal of the Linnaean Society* 75:421–36.

Briggs, J. C. 2003. Marine centres of origin as evolutionary engines. *Journal of Biogeography* 30:1–18.

———. 2004. Older species: A rejuvenation on coral reefs? *Journal of Biogeography* 31:5252–30.

Bromham, L., and M. Cardillo. 2003. Testing the link between the latitudinal gradient in species richness and rates of molecular evolution. *Journal of Evolutionary Biology* 16:200–7.

Brown, J. H. 1995. *Macroecology*. Chicago: University of Chicago Press.

Brown, J. H., A. P. Allen, and J. F. Gillooly. 2003. Response to heat and biodiversity (Huston). *Science* 299:512–13.

Brown, J. H., J. F. Gillooly, A. P. Allen, V. M. Savage, and G. B. West. 2004. Toward a metabolic theory of ecology. *Ecology* 85:1771–89.

Bush, A. M., M. J. Markey, and C. R. Marshall. 2004. Removing bias from diversity curves: The effects of spatially organized biodiversity on sampling-standardization. *Paleobiology* 30:666–86.

Buzas, M. A., L. S. Collins, and S. J. Culver. 2002. Latitudinal difference in biodiversity caused by higher tropical rate of diversification. *Proceedings of the National Academy of Sciences, USA* 99:7841–43.

Cardillo, M. 1999. Latitude and rates of diversification in birds and butterflies. *Proceedings of the Zoological Society of London, B* 266:1221–25.

Chapelle, G., and L. S. Peck. 1999. Polar gigantism dictated by oxygen availability. *Nature* 399:114–15.

Chown, S. L., B. J. Sinclair, H. P. Leinas, and K. J. Gaston. 2004. Hemispheric asymmetries in biodiversity—a serious matter for ecology. *PLoS Biology* 2:1701–7.

Clarke, A. 1983. Life in cold water: The physiological ecology of polar marine ectotherms. *Oceanography and Marine Biology: An Annual Review* 21:341–453.

———. (1992) Is there a latitudinal diversity cline in the sea? *Trends in Ecology & Evolution* 7:286–87.

———. 2003. Costs and consequences of evolutionary temperature adaptation. *Trends in Ecology & Evolution* 18:573–81.

———. 2007. Climate and diversity: The role of history. In *Scaling biodiversity*, ed. D. Storch, P. A. Marquet, and J. H. Brown, 225–45. Cambridge: Cambridge University Press.

Clarke, A., and J. A. Crame. 1989. The origin of the Southern Ocean marine fauna. In *Origins and evolution of the Antarctic biota*, ed. J. A. Crame, 253–68. London: The Geological Society. Geological Society Special Publications, 47.

———. 1997. Diversity, latitude and time: Patterns in the shallow sea. In *Marine biodiversity:*

Causes and consequences, ed. R. F. G. Ormond, J. D. Gage, and M. V. Angel, 122–47. Cambridge: Cambridge University Press.

———. 2003. The importance of historical processes in global patterns of diversity. In *Macroecology: Concepts and consequences,* ed. T. M. Blackburn and K. J. Gaston, vol. 43, 130–51. Oxford: Blackwell.

Clarke, A., and K. J. Gaston. 2006. Temperature, energy and diversity. *Proceedings of the Royal Society of London B* 273:2257–66.

Clarke, A., and N. M. Johnston. 2003. Antarctic marine benthic diversity. *Oceanography and Marine Biology: An Annual Review* 41:47–114.

Clarke, A., and S. Lidgard. 2000. Spatial patterns of diversity in the sea: Bryozoan species richness in the North Atlantic. *Journal of Animal Ecology* 69:799–814.

Clarke, K. R., and R. M. Warwick. 1998. A taxonomic distinctness index and its statistical properties. *Journal of Applied Ecology* 35:523–31.

Colwell, R. K., and G. C. Hurtt. 1994. Nonbiological gradients in species richness and a spurious Rapoport effect. *American Naturalist* 144:570–95.

Colwell, R. K., and D. C. Lees. 2000. The mid-domain effect: Geometric constraints on the geography of species richness. *Trends in Ecology & Evolution* 15:70–76.

Connell, J. H., and E. Orias. 1964. The ecological regulation of species diversity. *American Naturalist* 48:399–412.

Connolly, S. R., D. R. Bellwood, and T. P. Hughes. 2003. Indo-Pacific biodiversity of coral reefs: Deviations from a mid-domain model. *Ecology* 84:2178–90.

Crame, J. A. 1993. Latitudinal range fluctuations in the marine realm through geological time. *Trends in Ecology and Evolution* 8:162–66.

———. 2000a. Evolution of taxonomic diversity gradients in the marine realm: Evidence from the composition of Recent bivalve faunas. *Paleobiology* 26:188–214.

———. 2000b. The nature and origin of taxonomic diversity gradients in marine bivalves. In *The evolutionary biology of the Bivalvia,* ed. E. M. Harper, J. D. Taylor, and J. A. Crame, vol. 177, 347–60. London: The Geological Society.

———. 2001. Taxonomic diversity gradients through geological time. *Diversity and Distributions* 7:175–89.

———. 2002. Evolution of taxonomic diversity gradients in the marine realm: A comparison of Late Jurassic and Recent bivalve faunas. *Paleobiology* 28:184–207.

Crame, J. A., and B. R. Rosen. 2002. Cenozoic palaeogeography and the rise of modern biodiversity patterns. In *Palaeobiogeography and biodiversity change: The Ordovician and Mesozoic-Cenozoic radiations,* ed. J. A. Crame and A. W. Owen, vol. 194, 153–68. London: Geological Society of London.

Crampton, J. S., A. G. Beu, R. A. Cooper, C. M. Jones, B. Marshall, and P. A. Maxwell. 2003. Estimating the rock volume bias in paleobiodiversity studies. *Science* 301:358–60.

Culver, S. J., and M. A. Buzas. 2000. Global latitudinal species diversity gradient in deep-sea benthic foraminifera. *Deep-Sea Research, Part I* 47:259–75.

Damuth, J. 1987. Interspecific allometry of population density in mammals and other animals: The independence of body mass and population energy use. *Biological Journal of the Linnaean Society* 31:93–246.

Dayton, P. K., and J. S. Oliver. 1977. Antarctic soft-bottom communities in oligotrophic and eutrophic environments. *Science* 197:55–58.

Ellingsen, K. E., and J. S. Gray. 2002. Spatial patterns of benthic diversity: Is there a latitudinal gradient along the Norwegian continental shelf? *Journal of Animal Ecology* 71:373–89.

Engle, V. D., and J. K. Summers. 1999. Latitudinal gradients in benthic community composition in Western Atlantic estuaries. *Journal of Biogeography* 26:1007–23.

Fischer, A. G. 1960. Latitudinal variation in organic diversity. *Evolution* 14:64–81.

Flessa, K. W., and D. Jablonski. 1996. The geography of evolutionary turnover: A global analysis of extant bivalves. In *Evolutionary paleobiology*, ed. D. Jablonski, D. H. Erwin, and J. H. Lipps, 376–97. Chicago: University of Chicago Press.

Foote, M. 2005. Pulsed origination and extinction in the marine realm. *Paleobiology* 31:6–20.

Foote, M., and J. J. Sepkoski. 1999. Absolute measures of the completeness of the fossil record. *Nature* 398:415–17.

Gaston, K. J. 2000. Global patterns in biodiversity. *Nature* 405:220–27.

Gillooly, J. F., A. P. Allen, G. B. West, and J. H. Brown. 2005. The rate of DNA evolution: Effects of body size and temperature on the molecular clock. *Proceedings of the National Academy of Sciences, USA* 102:140–45.

Gooday, A. J., B. J. Bett, R. Shires, and P. J. D. Lambshead. 1998. Deep-sea foraminiferal species diversity in the NE Atlantic and NW Arabian sea: A synthesis. *Deep-Sea Research, Part II* 45:165–201.

Grassle, J. F., and N. J. Maciolek. 1992. Deep sea species richness: Regional and local diversity estimates from quantitative bottom samples. *American Naturalist* 139:313–41.

Graus, R. R. 1974. Latitudinal trends in the shell characteristics of marine gastropods. *Lethaia* 7:303–14.

Gray, J. S. 1994. Is deep-sea species diversity really so high? Species diversity of the Norwegian coastal shelf. *Marine Ecology Progress Series* 112:205–9.

———. 1997a. Gradients of marine biodiversity. In *Marine biodiversity: Causes and consequences*, ed. R. F. G. Ormond, J. D. Gage, and M. V. Angel, 18–34. Cambridge: Cambridge University Press.

———. 1997b. Marine biodiversity: Patterns, threats and conservation needs. *Biodiversity and Conservation* 6:153–75.

———. 2000. The measurement of marine species diversity, with an application to the benthic fauna of the Norwegian continental shelf. *Journal of Experimental Marine Biology and Ecology* 250:23–49.

———. 2002. Species richness of marine soft sediments. *Marine Ecology Progress Series* 244:285–97.

Gray, J. S., G. C. B. Poore, K. I. Ugland, F. Olsgard, and Ø. Johannessen. 1997. Coastal and deep-sea benthic diversities compared. *Marine Ecology Progress Series* 159:97–103.

Hawkins, B. A., F. S. Albuquerque, M. B. Araújo, J. Beck, L. M. Bini, F. J. Cabrero-Sañudo, I. Castro-Parga, et al. 2007. A global evaluation of metabolic theory as an explanation for terrestrial species richness gradients. *Ecology* 88:1877–88.

Hawkins, B. A., R. Field, H. V. Cornell, D. J. Currie, J.-F. Guégan., D. M. Kaufman, J. T. Kerr, et al. 2003. Energy, water, and broad-scale geographic patterns of species richness. *Ecology* 84:3105–17.

Held, C. 2001. No evidence for slow-down of molecular substitution rates at subzero temperatures in Antarctic serolid isopods (Crustacea, Isopoda, Serolidae). *Polar Biology* 24:497–501.

Hillebrand, H. 2004a. Strength, slope and variability of marine latitudinal gradients. *Marine Ecology Progress Series* 273.

———. 2004b. On the generality of the latitudinal diversity gradient. *American Naturalist* 163:192–211.

Hunt, G., T. M. Cronin, and K. Roy. 2005. Species-energy relationship in the deep sea: A test using the Quaternary fossil record. *Ecology Letters* 8:739–47.

Hutchinson, G. E. 1959. Homage to Santa Rosalia, or why are there so many kinds of animals? *American Naturalist* 93:145–59.

———. 1961. The paradox of the plankton. *American Naturalist* 95:137–45.

Jablonski, D., K. Roy, J. W. Valentine, R. M. Price, and P. S. Anderson. 2003. The impact of the pull of the recent on the history of marine diversity. *Science* 300:1133–35.

Kendall, M. A. 1996. Are Arctic soft sediment macrobenthic communities impoverished? *Polar Biology* 16:393–99.

Kendall, M. A., and M. Aschan. 1993. Latitudinal gradients in the structure of macrobenthic communities: A comparison of Arctic, temperate and tropical sites. *Journal of Experimental Marine Biology and Ecology* 171:57–169.

Lambshead, P. J. D., C. J. Brown, T. J. Ferrero, N. J. Mitchell, C. R. Smith, L. E. Hawkins, and J. Tietjen. 2002. Latitudinal diversity gradients of deep-sea marine nematodes and organic fluxes: A test from the central equatorial Pacific. *Marine Ecology Progress Series* 236:129–35.

Lambshead, P. J. D., J. Tietjen, T. J. Ferrero, and P. Jensen. 2000. Latitudinal diversity gradients in the deep sea with special reference to North Atlantic nematodes. *Marine Ecology Progress Series* 194:159–67.

Lear, C. H., H. Elderfield, and P. H. Wilson. 2000. Cenozoic deep-sea temperatures and global ice volumes from Mg/Ca in benthic foraminiferal calcite. *Science* 287:269–72.

Longhurst, A. 1998. *Ecological geography of the sea.* San Diego: Academic Press.

Magurran, A. E. 2004. *Measuring ecological diversity.* Oxford: Blackwell Science.

Marquet, P. A., S. A. Navarrete, and J. C. Castilla. 1995. Body size, population density and the energy-equivalence rule. *Journal of Applied Ecology* 64:325–32.

Martin, A. P. 1995. Metabolic rate and directional nucleotide substitution in animal mitochondrial DNA. *Molecular Biology and Evolution* 12:1124–31.

Marzluff, J. M., and K. P. Dial. 1991. Life history correlates of taxonomic diversity. *Ecology* 72:428–39.

Maurer, B. A. 1999. *Untangling ecological complexity: The macroscopic perspective* Chicago: The University of Chicago Press.

Mayhew, P. J., G. B. Jenkins, and T. G. Benton. 2007. A long-term association between global temperature and biodiversity, origination and extinction in the fossil record. *Trends in Ecology and Evolution* 22:516–21.

Mittelbach, G. G., D. W. Schemske, H. V. Cornell, A. P. Allen, J. M. Brown, M. B. Bush, S. P. Harrison, et al. 2007. Evolution and the latitudinal diversity gradient: Speciation, extinction and biogeography. *Ecology Letters* 10:315–31.

Nicol, D. 1964. An essay on the size of marine pelecypods. *Journal of Paleontology* 38:968–74.

———. 1966. Size of pelecypods in recent marine faunas. *Nautilus* 79:109–13.

———. 1978. Size trends in living pelecypods and gastropods with calcareous shells. *Nautilus* 92:70–79.

Picken, G. B. 1980. The nearshore prosobranch gastropod epifauna of Signy Island, South Orkney Islands. PhD diss., University of Aberdeen, Aberdeen, Scotland.

Pierrot-Bults, A. C. 1997. Biological diversity in oceanic macrozooplankton: More than counting species. In *Marine biodiversity: Causes and consequences,* ed. R. F. G. Ormond, J. D. Gage, and M. V. Angel, 69–93. Cambridge: Cambridge University Press.

Rex, M. A., J. A. Crame, C. T. Stuart, and A. Clarke. 2005. Large scale biogeographic patterns in marine mollusks: A confluence of history and productivity. *Ecology* 86:2288–97.

Rex, M. A., C. T. Stuart, and G. Coyne. 2000. Latitudinal gradients of species richness in the deep-sea benthos of the North Atlantic. *Proceedings of the National Academy of Sciences, USA* 8.

Rex, M. A., C. T. Stuart, R. R. Hessler, J. A. Allen, H. L. Sanders, and G. D. F. Wilson. 1993. Global scale latitudinal patterns of species diversity in the deep-sea benthos. *Nature* 365:636–39.

Rosenzweig, M. L. 1995. *Species diversity in space and time* Cambridge: Cambridge University Press.

Roy, K., D. Jablonski, and J. W. Valentine. 1994. Eastern Pacific molluscan provinces and latitudinal diversity gradient: No evidence for "Rapoport's rule." *Ecology* 91:8871–74.

———. 1996. Higher taxa in biodiversity studies: Patterns from eastern Pacific marine molluscs. *Philosophical Transactions of the Royal Society of London B* 351:1605–13.

———. 2000. Dissecting latitudinal diversity gradients: Functional groups and clades of marine bivalves. *Proceedings of the Royal Society of London B* 267.

———. 2004. Beyond species richness: Biogeographic patterns and biodiversity dynamics using other metrics of diversity. In *Frontiers of biogeography: New directions in the geography of nature,* ed. M. V. Lomolino and L. R. Heaney, 151–170. Sunderland, MA: Sinauer.

Roy, K., D. Jablonski, J. W. Valentine, and G. Rosenberg. 1998. Marine latitudinal diversity gradients: Tests of causal hypotheses. *Proceedings of the National Academy of Sciences, USA* 95:3699–3702.

Rutherford, S., S. D'Hondt, and W. Prell. 1999. Environmental controls on the geographic distribution of zooplankton diversity. *Nature* 400:749–53.

Sanders, H. L. 1968. Marine benthic diversity: A comparative study. *American Naturalist* 102:243–82.

Signor, P. W. 1985. Real and apparent trends in species richness through time. In *Phanerozoic diversity patterns: Profiles in macroevolution,* ed J. W. Valentine, 129–50. Princeton, NJ: Princeton University Press.

Spalding, M. D., H. E. Fox, G. R. Allen, N. Davidson, Z. A. Ferdaña, M. Finlayson, B. S. Halpern, et al. 2007. Marine ecoregions of the world: A bioregionalization of coastal and shelf areas. *Bioscience* 57:573–83.

Steele, J. H. 1991. Marine ecosystem dynamics: Comparison of scales. *Ecological Research* 6:175–83.

Steele, J. H., and E. W. Henderson. 1994. Coupling between physical and biological scales. *Philosophical Transactions of the Royal Society of London B* 343:5–9.

Stehli, F. G., R. Douglas, and I. Kefescegliou. 1972. Models for the evolution of planktonic foraminifera. In *Models in paleobiology,* ed. T. J. M. Schopf, 116–28. San Francisco: Freeman Cooper.

Stehli, F. G., A. L. McAlester, and C. E. Helsey. 1967. Taxonomic diversity of recent bivalves and some implications for geology. *Bulletin of the Geological Society of America* 78:455–66.

Stevens, G. C. 1989. The latitudinal gradient in geographical range: How so many species coexist in the tropics. *American Naturalist* 133:240–56.

Storch, D. 2003. Comment on "global biodiversity, biochemical kinetics, and the energy-equivalence rule." *Science* 299:346b.

Stuart, C. T., M. A. Rex, and R. J. Etter. 2003. Large-scale spatial and temporal patterns of deep-sea benthic species diversity. In *Ecosystems of the deep oceans,* ed. P. A. Tyler, vol. 28, 295–311. Amsterdam: Elsevier.

Thorson, G. 1952. Zur jetzeigen Lage der marinen Bodentier-Ökologie. *Zoologischer Anzeiger (Supplement)* 16:276–327.

———. 1957. Bottom communities (sublittoral or shallow shelf). In *Treatise on marine ecology and paleoecology,* ed. J. W. Hedgpeth, 461–534. New York: Geological Society of America.

Valdovinos, C., S. A. Navarrete, and P. A. Marquet. 2003. Mollusk species diversity in the Southeastern Pacific: Why are there more species towards the pole? *Ecography* 26:139–44.

Valentine, J. W. 1968. Climatic regulation of species diversification and extinction. *Geological Society of America Bulletin* 79:273–75.

Valentine, J. W., T. C. Foin, and D. Peart. 1978. A provincial model of Phanerozoic marine diversity. *Paleobiology* 4:55–66.

Vogel, S. 1981. *Life in moving fluids: The physical biology of flow.* Princeton, NJ: Princeton University Press.

Watling, L., and M. H. Thurston. 1989. Antarctica as an evolutionary incubator: Evidence from the cladistic biogeography of the amphipod family Iphimediidae. In *Origins and evolution of the Antarctica biota,* ed. J. A. Crame, vol. 47, 297–309. Bath, UK: The Geological Society.

Whittaker, R. H. 1972. Evolution of measurements of species diversity. *Taxon* 21:213–51.

Williams, P. H., and K. J. Gaston. 1994. Measuring more of biodiversity: Can higher-taxon richness predict wholesale species richness? *Biological Conservation* 67:211–17.

Williams, P. H., C. J. Humphries, and K. J. Gaston. 1994. Centers of seed plant diversity: The family way. *Proceedings of the Zoological Society of London B* 256:67–70.

Wilson, G. D. F. 1998. Historical influences on deep-sea isopod diversity in the Atlantic Ocean. *Deep-Sea Research, Part II* 45:279–301.

Witman, J. D., R. J. Etter, and F. Smith. 2004. The relationship between regional and local species diversity in marine benthic communities: A global perspective. *Proceedings of the National Academy of Sciences, USA* 101:15664–69.

Woodd-Walker, R. S., P. Ward, and A. Clarke. 2002. Large scale patterns in diversity and community structure of surface water copepods from the Atlantic Ocean. *Marine Ecology Progress Series* 56:309–16.

Zapata, F. A., K. J. Gaston, and S. L. Chown. 2003. Mid-domain models of speceis richness gradients: Assumptions, methods and evidence. *Journal of Animal Ecology* 72:677–90.

———. 2005. The mid-domain effect revisited. *American Naturalist* 166:E144–E148.

CHAPTER ELEVEN

MACROECOLOGICAL THEORY AND THE ANALYSIS OF SPECIES RICHNESS GRADIENTS

SEAN R. CONNOLLY

Introduction

Explaining regional and global scale patterns of biodiversity has occupied naturalists since before the dawn of modern ecology (Hawkins 2001). This endeavor has accelerated recently, sparked in part by the emergence of macroecology as a discipline (Ricklefs and Schluter 1993; Brown 1995); by substantial increases in the availability of tools for manipulating, analyzing, and graphically displaying data on species' geographical distributions (e.g., Geographical Information Systems software); and by the increasingly urgent need to prioritize regions for conservation on a global scale (see Roberts et al. 2002; Hughes, Bellwood, and Connolly 2002; Worm et al. 2005 for marine examples). This recent work has confirmed that many very different taxa exhibit similar species richness gradients. In the marine realm, for instance, species richness frequently exhibits a hump-shaped pattern, with species richness decreasing with latitudinal and longitudinal distance from the Indo-Australian Archipelago (e.g., fig. 11.1; also see Stehli and Wells 1971; Crame 2000; Ellison 2002; Roberts et al. 2002), although many taxa also exhibit a secondary hotspot in the Caribbean (Stehli and Wells 1971; Duke, Lo, and Sun 2002; Roberts et al. 2002). Bathymetric gradients often are also hump-shaped, peaking at intermediate depths (Piñeda and Caswell 1998; Rex et al. 2005; but see Gray 2001). Exceptions to these general rules have

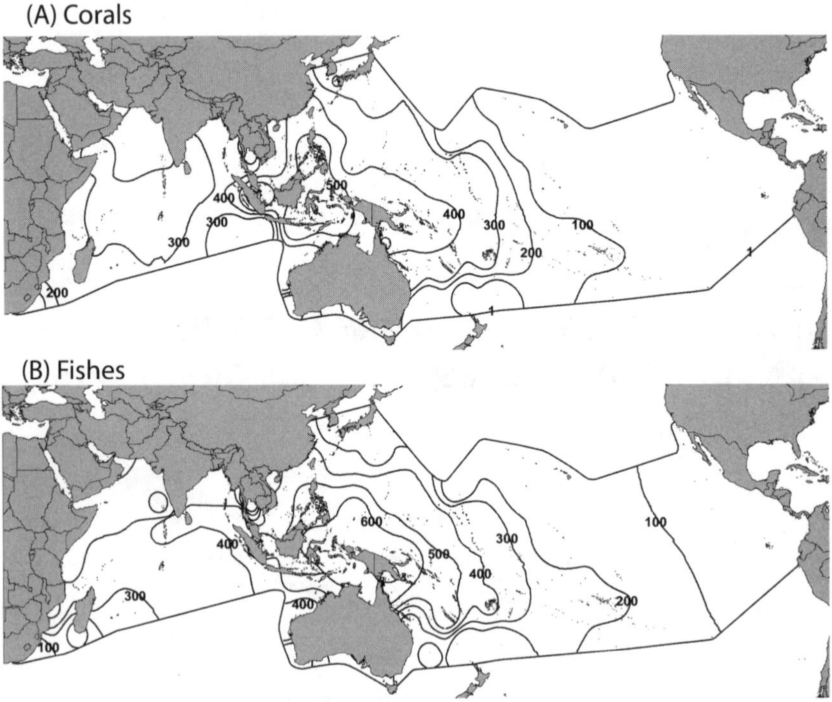

(A) Corals

(B) Fishes

FIGURE 11.1 Species richness contours for Indo-Pacific (A) corals and (B) reef fishes, illustrating the Indo-Pacific biodiversity hotspot. Modified after Bellwood et al. (2005).

also been identified. For instance, marine counterexamples to the latitudinal diversity gradient include macroalgae (Kerswell 2006), fish parasites (Rohde 1998), seals, and seabirds (Proches 2001).

Species richness gradients arise from geographical patterns in origination, range expansion, and persistence; many debates about biodiversity revolve around identifying which of these proximate mechanisms is the principal cause of particular species richness gradients (Chown and Gaston 2000). For instance, in the marine realm, there has been substantial attention to the question of whether global biodiversity hotspots are centers of origination of new species, accumulation of species that have become extinct elsewhere, or overlap of species ranges from neighboring biogeographical provinces (Veron 1995; Bellwood and Wainwright 2002; Mora et al. 2003; Goldberg et al. 2005). Ultimately, however, geographical patterns in origination, range expansion, and speciation themselves have environmental causes. It is the study of these environmental causes of species richness that are the focus of this chapter.

Latitudinal, longitudinal, and bathymetric diversity gradients, and the current evidence for and against hypothesized causes of those gradients, have been reviewed extensively of late (Gray 2001; Bellwood and Wainwright 2002; Willig, Kaufman, and Stevens 2003; Rex et al. 2005). Therefore, I focus instead on the way in which we presently use ecological data to assess hypotheses about effects of environmental factors on species richness. I argue that the statistical models we usually fit to species richness data often correspond only loosely to the conceptual models that motivate our statistical hypotheses, particularly with respect to provincial-scale species richness gradients. I then discuss, at some length, an increasingly prevalent way that macroecologists are seeking to move beyond these traditional, correlative approaches: examining the patterns in the sizes and locations of species ranges that give rise to species richness gradients. To date, these models have focused overwhelmingly on "null models" that seek to omit effects of environmental factors. However, fresh insights also have emerged from predicting the patterns of range sizes and locations implied by particular hypotheses about environmental gradients (Stevens 1989; Chown and Gaston 2000; Connolly, Bellwood, and Hughes 2003). This suggests that a body of rigorous macroecological theory relating environmental variables to species' geographical distributions has the potential to shed important new light on the environmental causes of species richness patterns. I conclude by discussing a few promising avenues for the development of such a theory.

Environmental Correlates of Biodiversity

The practice of attributing geographical variation in species richness to coincident variation in environmental conditions has a long history. For instance, the latitudinal diversity gradient was attributed to latitudinal differences in climatic variability as far back as 1807 (Hawkins 2001), and hypotheses invoking other mechanisms have accrued at an increasing rate since then, particularly in the last half century (Willig, Kaufman, and Stevens 2003). In the marine realm, species richness gradients have been attributed to some of the same variables that have been invoked on land, such as average temperatures (Rohde 1992; Fraser and Currie 1996; Worm et al. 2005), seasonal climatic variability (Stevens 1989), habitat availability (Bellwood and Hughes 2001; Bellwood, Hughes, and Connolly 2005), and productivity (Rex et al. 2005). Explanations for particular patterns in the marine realm (e.g., longitudinal gradients in the Indo-Pacific) have also invoked mechanisms specific to those contexts, such as ocean currents (Jokiel and Martinelli 1992; Connolly, Bellwood, and Hughes 2003), and vicariance events due to Plio-Pleistocene sea-level fluctuations (McManus 1985).

Although initial assessments of many of these hypotheses were qualitative, more recent attempts to understand the relationship between environmental variables and species richness have been based on multiple regression analyses. Early approaches utilized prevailing ordinary least-squares (OLS) regression techniques and selection of best-fit models by stepwise algorithms (e.g., Fraser and Currie 1996; Roy et al. 1998; Bellwood and Hughes 2001). Increasingly, however, workers have adopted more contemporary techniques for incorporating spatial autocorrelation and selecting best-fit models (Bellwood et al. 2005; Worm et al. 2005). Such regression-based approaches are designed to assess a model's predictive, rather than explanatory, power. In other words, they choose variables (or models) that are expected to best predict species richness at unobserved locations within the region analyzed (Cressie 1993). Thus, they are ideal when a study's principal aim is prediction (e.g., estimating the number of species protected under different reserve design systems). In contrast, most macroecological analyses of species richness aim to explain: to understand how much an environmental factor has influenced how the number of species present at a location became established and has been maintained. Superficially, it seems reasonable to expect that important environmental variables (i.e., variables whose geographical distributions have had a substantial influence on the origination, colonization, or persistence of species) probably will correlate strongly with species richness levels. For instance, consider a comparison of species pools in different biogeographical provinces (fig. 11.2, panel A): if the environment in one province has tended to promote speciation, or inhibit extinction, more than the environment in another province, then species richness should be higher in the first province. A similar intuition applies for nearby locations that access a common regional species pool (fig. 11.2, panel B). In this case, if one local environment promotes colonization of species, or inhibits local extinction, more than another local environment, then, again, species richness should be higher at the first location. On the other hand, consider a comparison of species richness across an entire biogeographical region (fig. 11.2, panel C). In this case, the number of species present at a location, or whose ranges encompass a particular location, depends not only on the environmental conditions at that location (which influence colonization and local extinction rates), but also on the broad-scale distribution (present-day and historical) of environmental conditions that have influenced where species originated, and how readily they have expanded their ranges toward the location in question. For this reason, the strength of statistical concordance between an environmental variable and species richness may not be as indicative of causal importance as we expect at the smaller and larger scales.

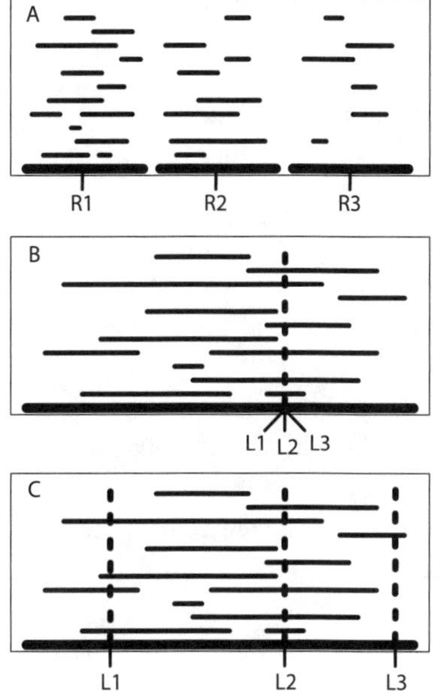

FIGURE 11.2 Schematic illustrating three scales at which species richness gradients may be analyzed. In each panel, the thick line at the bottom represents a hypothetical biogeographical domain, and the thinner horizontal lines represent the range extents of individual species. (A) Comparison of the size of three distinct regional species pools (labeled R1, R2, and R3). (B) Comparison of local species richness for three nearby locations (L1, L2, and L3) that access a common regional species pool. (C) Comparison of species richness (local or regional pool size) at three locations (L1, L2, and L3) dispersed widely within a biogeographical province.

Species Ranges and Biodiversity

Rapoport's Rule

The geographical distribution of environmental conditions influences species richness through its effects on the origination, colonization, and persistence of species. These dynamic processes influence not only variation in species richness, but also patterns in the sizes and locations of species' ranges. Therefore, one way to move beyond statistical concordance between environmental factors and species richness is to predict the geographical distribution of species' ranges that a particular hypothesized environmental mechanism implies. One of the first such attempts led to the development of Rapoport's rule, according to which species in tropical locations have smaller ranges, on average, than those outside the tropics (Stevens 1989). The mechanism hypothesized to underlie this pattern is based on a climatic stability hypothesis: greater seasonal climatic variability outside the tropics should favor species with broad environmental tolerances, whereas the more stable climes of the tropics should more readily support species with narrow tolerances. If species with narrow climatic tolerances tend to have smaller range sizes, such small-range species should be disproportionately concentrated in tropical regions. Consequently, range sizes should, on average, be smaller in the tropics (Stevens 1989).

Early attempts to test Rapoport's rule quantified whether mean range size was an increasing function of distance from the equator, an approach known as the "Stevens method" (Stevens 1989). This approach has been criticized: mean range sizes are not statistically independent because the same species contribute to mean range size at multiple latitudes (Rohde, Heap, and Heap 1993). A common alternative approach, known as the "midpoint method," also calculates mean range size at different latitudes, but species only contribute to mean range size for the latitude closest to their range midpoint. This approach resolves the problem of statistical nonindependence among latitudes, but it is not, strictly speaking, a test of Rapoport's rule. Although this problem has been noted previously, the midpoint method is still commonly used, perhaps because many studies that apply both methods obtain similar results (Gaston, Blackburn, and Spicer 1998). However, the apparent tendency for these two methods to agree is likely to be quite fragile, as can be illustrated with a simple idealized example. Consider a single tropical region, bordered by two extratropical regions of equal size (fig. 11.3, panel A). There are twenty pandemics (species whose ranges encompass all three regions), and twenty-four small-range endemics (species whose ranges are confined to one region). Twenty endemics are located in the tropics, and four are located outside the tropics (two in each extratropical region). In this case, there is a latitudinal species richness gradient (forty species in the tropics versus twenty-two outside the tropics), and it is due entirely to Rapoport's rule (higher numbers of small-range endemics in the tropical region). The Stevens method correctly identifies the larger range sizes outside the tropics, but the midpoint method does not, instead indicating (erroneously) that range sizes are larger in the tropics (fig. 11.3, panel B). Indeed, Rohde, Heap, and Heap (1993) were aware of this potential problem, and assessed its potential to bias their analyses. However, for other data sets, particularly where there are many species with very large ranges, this bias can be quite severe (see Fortes and Absalão 2004 for discussion of a marine example).

Although the overwhelming majority of studies investigating Rapoport's rule utilize either the Stevens method or the midpoint method (see Gaston, Blackburn, and Spicer 1998 for review), there have been recent attempts to devise more robust tests of Rapoport's rule. One approach uses randomization methods to incorporate spatial nonindependence in the expected (i.e., null) distribution of regression slopes. These studies yield mixed results: support for a Rapoport effect for New World bats and marsupials (Lyons and Willig 1997), but not for Indo-Pacific corals and fishes (Hughes, Bellwood, and Connolly 2002). An alternative approach directly quantifies the latitudinal distribution of small-range species and identifies regions where small-

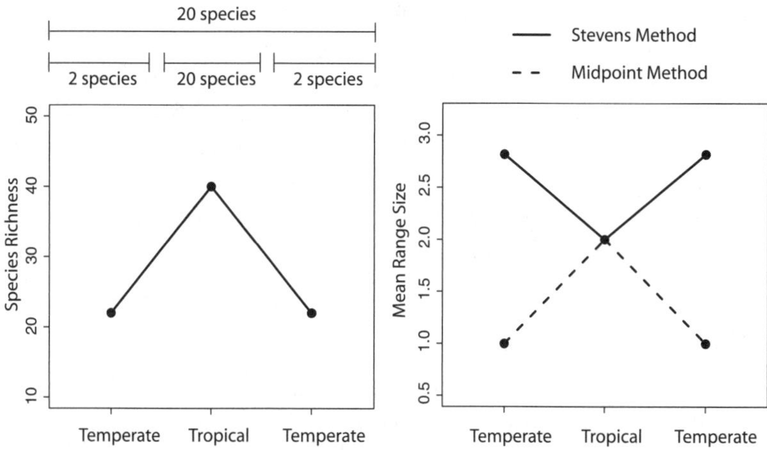

FIGURE 11.3 Idealized example of a Rapoport effect. (A) A hypothetical latitudinal species richness gradient (across three regions), with twenty pandemics (species whose ranges encompass all regions), and twenty-four endemics (species whose ranges are individually confined to a single region). (B) Analysis of hypothetical data, according to the Stevens and midpoint methods.

range species are concentrated (the "cross-species method;" e.g., Blackburn and Gaston 1996; McCain 2003). This latter approach corresponds more directly with the hypothesized mechanism underlying Rapoport rule than tests based on the central tendency of range size (e.g., mean, median), which may not provide a representative picture of where the small-range endemics are concentrated (Roy, Jablonski, and Valentine 1994). Although initially suggested by Rohde, Heap, and Heap (1993), the cross-species method has not been widely used, probably because it has been based on a qualitative, visual inspection of the data. However, randomization methods have recently been used to quantify more rigorously where small-range species are more or less concentrated than expected by chance. For instance, Indo-Pacific corals and reef fishes show concentrations of small-range species *away* from the equator, near the tropical margins, contrary to the expectation under the climatic stability hypothesis (Connolly, Bellwood, and Hughes 2003; see Jetz, Rahbek, and Colwell 2004 for a terrestrial example using a similar approach).

A recent review argued against the generality of Rapoport's Rule, given the apparently considerable variability among regions and taxa in how range size varies with latitude (Gaston, Blackburn, and Spicer 1998). The most recent marine studies also yield mixed results, with most studies failing to find support for it (Hughes, Bellwood, and Connolly 2002; Connolly, Bellwood, and Hughes 2003; Smith and Gaines 2003; Hernández, Moreno, and Rozbaczylo

2005; but see MacPherson 2003; Fortes and Absalão 2004). More fundamentally, an assumption that underlies the hypothesized cause of Rapoport's rule—that range size is correlated with climatic tolerance—is generally untested. In the one marine example of an explicit test of this prediction, no relationship was found between range size and climatic tolerance (Parmesan et al. 2005). Clearly, more such tests are needed if we are to determine whether the relatively few marine instances of concordance with Rapoport's rule do, in fact, provide evidence for the climatic stability hypothesis proposed to explain it.

The Mid-Domain Effect

As indicated earlier, one way to conduct quantitative tests of hypotheses about the geographical distribution of species is by constructing null models: statistical expectations designed to approximate the pattern expected in the absence of environmental gradients (cf. Gotelli and Graves 1996). The first of these null models were generating by randomizing, in one dimension, the sizes and locations of species' geographical ranges, subject only to the "geometric constraints" imposed by the outer limits of the biogeographical domain itself (Colwell and Hurtt 1994). These constraints are typically depicted by a triangular region on a graph of range size versus location along a geographical axis (fig. 11.4, panel A). For instance, a pandemic species, which (by definition) occupies the entire domain, must be centered in the middle of the domain, at the triangle's apex. By contrast, small ranges (ranges sizes at the bottom of the triangle) can be centered almost anywhere within the domain. When these geometric constraints were applied to construct a null model, an important and unexpected result emerged. Specifically, quasi-parabolic (i.e., hump-shaped) gradients in species richness are generated if species' range sizes and range locations are chosen randomly (i.e., all geometrically feasible range sizes/locations are equally likely; fig. 11.4, panel B). Based on the discovery of these "mid-domain effects" (MDEs), Colwell and Hurtt (1994) proposed that hump-shaped species richness gradients could arise in nature, even in the absence of gradients in environmental conditions.

The null models described previously generate range size and location by means of random draws from an invariant (generally uniform) distribution; that is, the same prediction is generated, regardless of which taxon's species richness gradient is under investigation (Willig and Lyons 1998; Bokma, Bokma, and Mönkkönen 2001; Grytnes 2003). This approach makes the implicit assumption that the frequency distribution of range size exhibited by a group of species depends only on the domain's size and shape, and

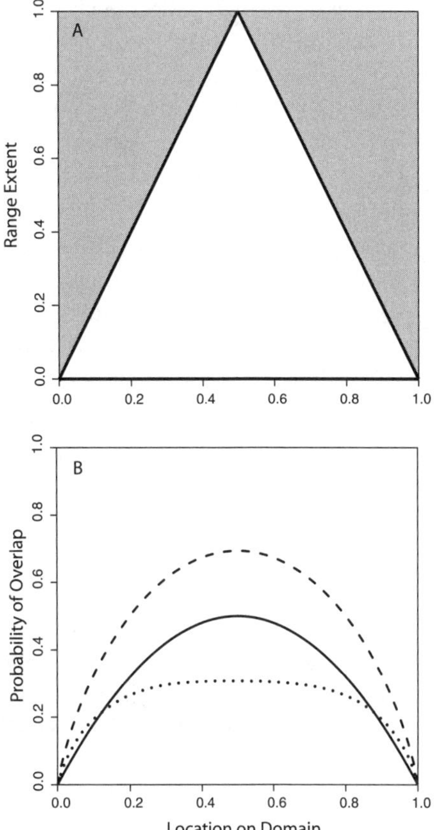

FIGURE 11.4 The Mid-Domain Effect. (A) Triangle illustrating the geometric constraints on species' geographical distributions. (B) Predictions of alternative "fully randomized" models. Solid line: all geometrically feasible ranges (i.e., all points within the triangle of panel A) are equally likely. Dashed line: range extent is chosen from a uniform distribution, then range location is chosen (with uniform probability) from among the geometrically feasible values, conditional on range extent. Dotted line: location of range midpoint is chosen at random, then range extent is chosen (with uniform probability) from among geometrically feasible values, conditional on location of range midpoint.

thus is independent of characteristics of the taxon itself (Connolly 2005). In nature, however, the species belonging to some higher taxa may have greater dispersal ability or broader environmental tolerances, on average, than species belonging to other taxa. In other words, rates of origination, colonization, or local extinction might well differ among taxa, even in the absence of geographical gradients of those rates. As a result, frequency distributions of range size (RSFDs) may well differ among taxa, independent of any effects of environmental gradients.

To move beyond the invariant RSFDs of the first generation of null models, workers have increasingly utilized observed RSFDs and randomized only their locations (e.g., Piñeda and Caswell 1998; Lees, Kremen, and Andriamampianina 1999; Sanders 2002; Jetz and Rahbek 2001; Connolly, Bellwood, and Hughes 2003). Most such studies apply a "range shuffling" algorithm: each species' range extent in one dimension (e.g., latitude, longitude, altitude, depth) is randomly reallocated to a location on the domain, with all geometrically feasible locations (i.e., range midpoints within the triangle shown in fig. 11.4, panel A) considered equally likely. For this model, the probability that a species' range will overlap some location on a biogeographical domain depends on its range size (fig. 11.5, panel A); thus, the predicted species richness gradient depends on the range sizes of the species being analyzed, with MDEs being most pronounced for intermediate range sizes.

An alternative to the range-shuffling algorithm is the "spreading dye model" (Jetz and Rahbek 2001). In this model, a point of origin is chosen at random, and the species' range spreads out in all directions, until the entire range has been allocated. In practice, this takes place on gridded maps, with ranges spreading into adjacent grid cells with equal probability. Consequently, stochastic binomial variation creates some asymmetry in the range's outward spread from its point of origin. However, this asymmetry is an artifact of how coarsely the domain is divided into grid cells, not a characteristic of the underlying biology. In the limit as these cells become small, the range's randomized location becomes determined completely by its point of origin, and, at least for the one-dimensional case, an analytical solution can be obtained (Connolly 2005). This analytical solution reveals some striking differences with the range-shuffling model (fig. 11.5, panel B). First, the probability that a species' range abuts a domain boundary is not zero, so species richness does not decrease to zero at domain boundaries. Second, probability of overlap jumps up or down abruptly partway into the domain, depending on whether range size is greater or less than 50 percent of the domain size. Although there is, as yet, no analytical solution to the two-dimensional spreading dye model, numerical analysis suggests that it exhibits qualitatively similar behavior (fig. 11.5 panels C, D); however, for large ranges, the steepening of the species richness gradient partway toward the mid-domain occurs much less abruptly than in the one-dimensional case (compare fig. 11.5, panel D with the dotted line in fig. 11.5, panel B). Although randomizations of entire RSFDs typically do not exhibit such pronounced dips or jumps in species richness, because they average over a large number of differently sized ranges, it is important to recognize that even these predictions have, as their basis, an underlying probability model that does possess such discontinuities.

One limitation of mid-domain models that use observed range sizes has been that they make an implicit assumption that range size is an intrinsic property of species, and thus is independent of geographical gradients in environmental conditions (Connolly 2005). Because preservation of range sizes is not universally agreed to entail this assumption (e.g., Colwell, Rah-

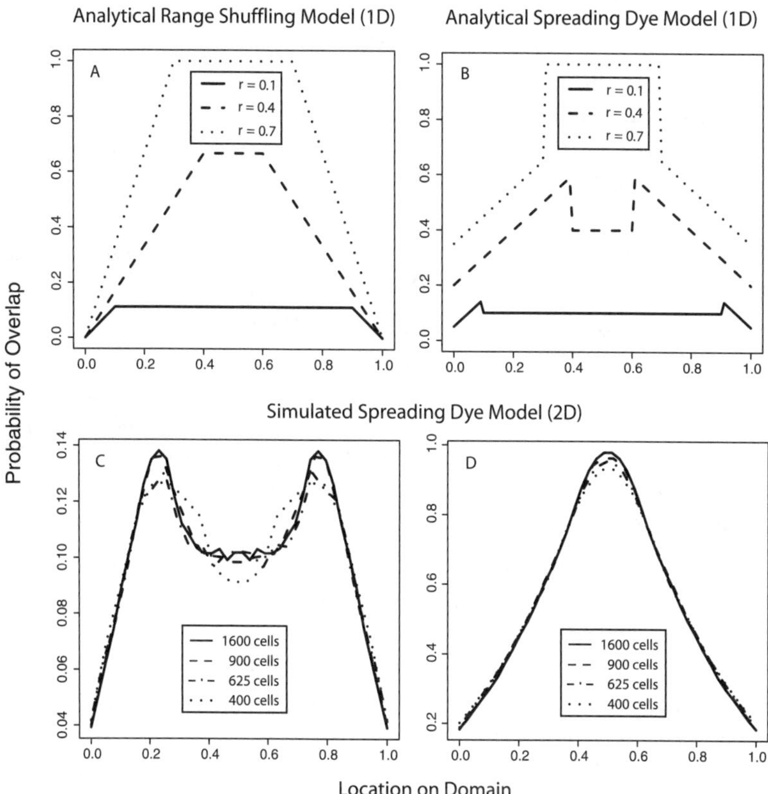

FIGURE 11.5 Probability of range overlap for different range extents, as a function of location on the domain for randomization models that use observed range sizes. (A) Range shuffling model (analytical solution). (B) One-dimensional spreading dye model (analytical solution). (C, D) Simulation-based approximations of the two-dimensional spreading-dye model prediction, illustrating convergence of the prediction as the grid-cell subdivision of the domain increases. (C) Range area = 10 percent of the domain area. (D) Range Area = 50 percent of domain area.

bek, and Gotelli 2005), it is worth developing the rationale more fully here. Range size, in nature, arises from an interaction between species' intrinsic properties (e.g., vagility, breadth of environmental tolerances) and characteristics of the domain, including environmental gradients: these will jointly determine the geographical distribution of origination, colonization, and extinction rates that ultimately determine the distribution of range locations and range sizes that produce the species richness pattern. When one generates a null expectation by randomizing the locations of observed range sizes, then one implicitly assumes that, had environmental gradients played no role in determining the species richness patterns exhibited by this particular group of species, the range size frequency distribution would still have resembled the observed one. If not, then the randomization could not be assumed to be a reasonable approximation of the relevant null expectation. Moreover, qualitative features of null model predictions can be highly sensitive to this use of empirical range sizes. For example, if one modifies the spreading-dye framework to allow the encounter of range limits to be determined by a stochastic process (rather than constraining range size to be equal to that observed), then the discontinuous jumps and dips of the spreading dye algorithm disappear (Connolly 2005: see next section for further discussion of such a model).

A second limitation of randomization-based null models is that different, but seemingly equally plausible, randomization approaches generate markedly different predictions. As a result, the interpretation of null model analyses can be very sensitive to which null model is used. A comparison of range shuffling and (one-dimensional) spreading dye predictions for the longitudinal gradients of Indo-Pacific corals and coral reef fishes serves to illustrate this (fig. 11.6). There are some points of agreement: compared with both models, the data show higher-than-predicted richness in the Indo-Australian Archipelago (IAA) and along the African coast, and a depauperate mid-Pacific. However, there are also differences. Most notably, the American coastline is depauperate compared to the spreading-dye model, but not the range-shuffling model. In addition, compared with the range-shuffling model, the IAA and African margin are comparably enriched, relative to expectation; however, compared with the spreading-dye model, the IAA is the much larger anomaly. These differences highlight a risk associated with generating null expectations by randomization: when there are multiple, seemingly equally plausible ways to randomize a data set, it can be impossible to determine which one, if any, is the most appropriate null expectation for any given biological hypothesis. Indeed, many criticisms of null model approaches in macroecology are based on the lack of a clear link between statistical and biological expectations (Roughgarden 1983; Hawkins and Diniz-Filho 2002; Zapata, Gaston, and Chown 2003).

Recently, attempts have been made to model the interaction between effects of domain boundaries and of environmental gradients, by means of weighted randomizations in which probabilities associated with different points of origin, or probabilities of range spread, depend on the values of environmental variables at the relevant locations (Storch et al. 2006; Rahbek et al. 2007). This approach has yielded fits to empirical richness patterns that can be surprisingly good, given their simplicity (e.g., a spreading-dye

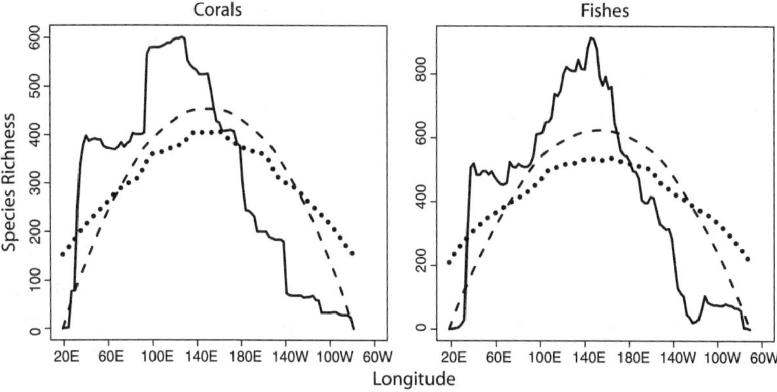

FIGURE 11.6 Longitudinal species richness of (A) Indo-Pacific corals and (B) Indo-Pacific reef fishes, comparing observed gradients (solid lines) with predictions from the range shuffling (dashed line) and spreading dye (dotted line) null models. Modified from Connolly, Bellwood, and Hughes (2003).

model weighted according to evapotranspiration alone explains 60% of the variation in global avian richness: Storch et al. 2006). Such algorithms will, of course, inherit any limitations associated with the randomization algorithms on which they are based (e.g., biases associated with the preservation of observed range sizes). For instance, in nature, evolution of species ranges in areas with high topographic relief may well lead to a disproportionate number of small-range endemics, as a consequence of the steep environmental gradients associated with such regions. However, in a weighted randomization, a species randomly chosen to originate in such an area will have a range size that is drawn from the same frequency distribution as a species randomly chosen to originate in a relatively environmentally homogeneous part of the domain. Indeed, it is striking that both of the studies conducted to date have specifically noted that best-fit models tend to perform most poorly in high-altitude, tropical locations.

Whatever their limitations, however, weighted randomizations represent the first attempts to explain environmental effects on species richness by means of their influences on species' geographic ranges. In particular, the use of a fixed, empirical distribution of range sizes means that they are unlikely to adequately capture patterns that are driven by effects of environmental gradients on range size (e.g., concentrations of small-range species in areas of high topographic relief). Nevertheless, such models do generate predictions about the geographical distribution of species ranges, and thus can be tested against a richer set of predictions than conventional regression approaches. Moreover, as we become increasingly familiar with these approaches, our ability to identify those kinds of model predictions that are most likely to be robust, or, conversely, most prone to bias, will undoubtedly enhance our capacity to accurately assess and interpret model performance.

Process-Based Models of Species Distributions

Attempts to move beyond the controversies about randomization algorithms described earlier typically involve models of species' geographical distributions that are more process-oriented than are randomization methods. Indeed, one of the few points of agreement among proponents and critics of MDEs has been the need for such approaches (Zapata, Gaston, and Chown 2003; Colwell, Rahbek, and Gotelli 2004). Most of this work has focused on testing the logic underlying the MDE hypothesis, aiming to generate a distribution of range sizes and locations with a model whose assumptions are consistent with the absence of geographical gradients in origination, colonization, or persistence. The models to date reveal two very different interpretations of what the "null expectation" for species richness patterns is, even though they all might be considered null models because they aim to generate the pattern expected in the absence of gradients in origination, extinction, or persistence (Gotelli and Graves 1996). Under one interpretation, the environment is homogeneous; that is, individual species' demographic and evolutionary rates are constant throughout a biogeographical domain (Bokma, Bokma, and Mönkkönen 2001; Hawkins and Diniz-Filho 2002; Davies, Grenyer, and Gittleman 2005; Rangel and Diniz-Filho 2005). Under the second interpretation, the environment is assumed to vary, so that individual species tolerate only a subset of the available range of conditions; however, in the aggregate (i.e., across all species), no one region of the domain contains conditions more likely (or unlikely) to be suitable for a species than any other (Willig and Lyons 1998; Colwell and Lees 2000; Connolly, Bellwood, and Hughes 2003).

When the domain is environmentally homogeneous, a reasonable null model assumes that each species' demographic and evolutionary rates are the same everywhere on the domain (Bokma, Bokma, and Mönkkönen 2001; Davies, Grenyers, and Gittleman 2005; Rangel and Diniz-Filho 2005). In the models developed to date, the domain is divided into an arbitrary number of discrete grid cells. Species' ranges expand as they colonize previously unoccupied grid cells, and contract as they go locally extinct from grid cells. Details of these models vary. Bokma, Bokma, and Mönkkönen (2001) and Davies, Grenyers, and Gittleman (2005) model colonization and extinction probabilities at the scale of subpopulations occupying grid cells, whereas Rangel and Diniz-Filho (2005) analyze a neutral model in which birth, death, and dispersal probabilities are identical for all individuals, regardless of species (Bell 2001; Hubbell 2001). Depending on the model, new species occur by differentiation of subpopulations within a species' range (Bokma, Bokma, and Mönkkönen 2001); by vicariance of existing species ranges (Davies, Grenyers, and Gittleman 2005); or not at all (Rangel and Diniz-Filho 2005). Analysis of these models, based on stochastic simulation, has prompted markedly different conclusions. Bokma, Bokma, and Mönkkönen (2001) and Rangel and Diniz-Filho (2005) found MDEs, but they were shallower than predicted by randomization models. However, Davies, Grenyers, and Gittleman (2005) found that stochastic variation in the location of peak species richness was so great that no tendency toward an MDE was apparent at all.

To better understand the differences between these studies, I have constructed one-dimensional analytical models whose assumptions are akin to those of the models summarized earlier. Within this model, I define species' distributions by the locations of their lower and upper range endpoints (x_L,x_U). The domain is divided into an arbitrary number (H) of discrete cells, and the dynamics of species' ranges are modeled in discrete time steps, as in previous approaches. In any time step, a species' range endpoint expands outward according to a colonization probability, p_C, contracts inward according to an extinction probability, p_E, or remains unchanged. If a species occupies only one habitat patch (i.e., $x_L = x_U$), the species becomes globally extinct if the remaining population goes extinct. Species' ranges cannot extend beyond domain boundaries (i.e., $p_C = 0$ where $x_L = 1$ or $x_U = H$). Existing species give rise to a new species (initially occupying only a single cell) with probability p_S for each cell that it occupies (I will refer to this as the "founder speciation" model). From these assumptions, we can generate speciation and range-state transition matrices that are analogous to the fecundity and transition matrices used in standard density-independent stage-structured population models (Caswell 2001), and from which we can derive the species richness gradient to which the system tends in the long run (Appendix A).

This model generates a quasi-parabolic species richness gradient, and that gradient is most pronounced when colonization and extinction rates are approximately equal (fig. 11.7, panel A). This is consistent with analyses of simulation models (Bokma, Bokma, and Mönkkönen 2001). When extinction exceeds colonization, the species richness peak decreases, but a gradient is still apparent. By contrast, as colonization increasingly exceeds extinction, the species richness gradient becomes flat-topped, with a decline toward domain boundaries confined to a narrow region near those boundaries (fig. 11.7, panel A). These results are largely insensitive to the speciation mode assumed: analysis of a vicariance model (Appendix A) yields similar results (fig. 11.7, panel B). These results help to explain why Davies, Grenyer, and Gittleman (2005) did not find strong evidence for an MDE, in contrast to Bokma, Bokma, and Mönkkönen (2001). In their simulations, probabilities of range expansion substantially exceed those of range contraction (by a minimum factor of 2). In such cases, the expected MDE is quite weak (fig. 11.7). Most species occupy the majority of the domain, except for those that have recently undergone vicariance. Consequently, the location of peak richness is driven largely by stochastic variation in where these most recent vicariance events occurred, and a strong tendency towards an MDE should not be apparent (cf. Davies, Grenyer, and Gittleman 2005).

The analytical model presented implicitly assumes range contiguity. That is, when a peripheral population goes extinct, occupancy of the adjacent patch is assumed, so the range endpoint contracts inward by only one cell. This assumption can be relaxed by modeling occupancy of all patches of habitat in between the upper and lower range endpoints. However, if a species' state is defined as a vector of ones and zeroes indicating presence or absence in each patch, there are 2^H possible states. Solving explicitly for the limiting distribution of species richness is computationally prohibitive (e.g., a domain with twenty cells entails only 210 states for a range endpoint model, but over a million states for an occupancy model). Nevertheless, to

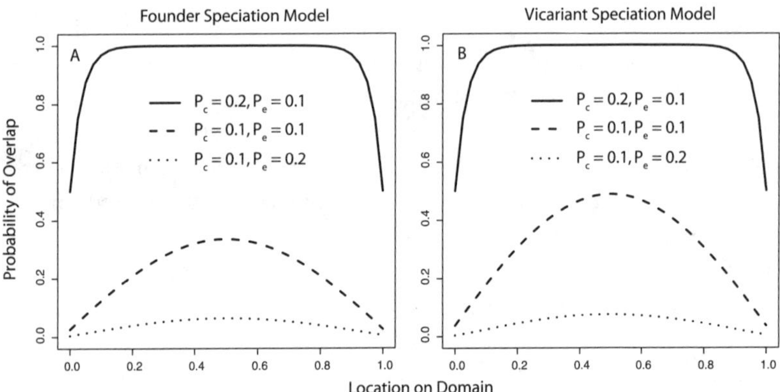

Figure 11.7 Species richness gradients generated by the analytical model of dynamic, contiguous ranges in a homogeneous environment (Appendix A). (A) Founder speciation model. (B) Vicariant speciation model. For each set of colonization and extinction parameters, the speciation probability was chosen so that the matrix's leading eigenvalue was approximately unity (i.e., the species pool would remain approximately constant over time).

illustrate the potential effects of the contiguity assumption, I have analyzed a patch-occupancy model by simulation (Appendix B). This analysis indicates that gradients in local occupancy can be much less pronounced (specifically, more flat-topped) than corresponding gradients in range overlap (fig. 11.8, panels A, B). This model also highlights how simulation results can depend on how the domain is divided into grid cells. For instance, a domain divided into a relatively small number of cells may produce a much more quasi-parabolic species richness gradients than a domain divided into a large number, even for the same parameter values (fig. 11.8, panels C, D). Clearly, while the work to date has highlighted some important potential

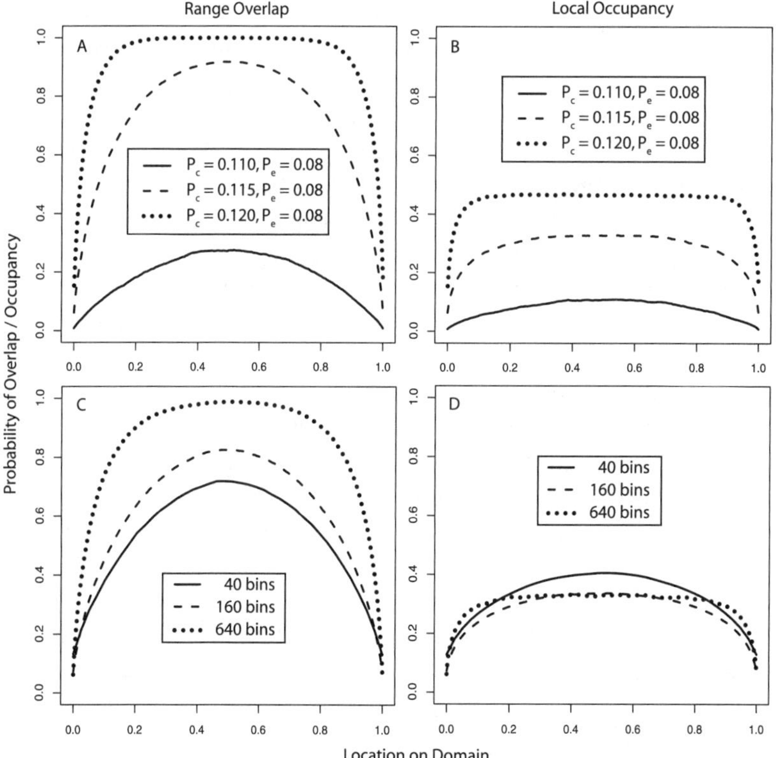

Figure 11.8 Species richness gradients for the patch occupancy model (Appendix B), approximated by stochastic simulation. (A, B) Corresponding gradients in "regional" richness (i.e., range overlap), and (B) "local" richness (i.e., cell occupancy), for multiple parameter values. The domain has been gridded into 320 cells. (C, D) Illustration of sensitivity of predicted species richness to the number of cells into which the domain is subdivided. For panels (C) and (D), $p_C = 0.115$ and $p_E = 0.08$.

effects of peripheral colonization-extinction dynamics on species richness gradients, a formal theory for the dynamics of range endpoints in continuous space is needed to more comprehensively and rigorously assess how predicted species richness patterns may depend on the patch structure of the domain.

An alternative to modeling colonization and extinction at the patch scale is to explicitly model individual birth, death, and migration events. Indeed, given the recent attention to "neutral" models of biodiversity (i.e., models assuming that individuals have the same demographic rates, regardless of location, time, or species: Bell 2001; Hubbell 2001), it is somewhat surprising that there have not been more attempts to predict the species richness gradients generated by such models. The only study conducted to date suggests that MDEs occur, but only when dispersal distances are very short (Rangel and Diniz-Filho 2005). Because this model omits speciation, the species richness gradients are transient, on a trajectory toward monodominance (i.e., all species extinct except one). Nevertheless, the results do show the potential for neutral models to exhibit MDEs. Whether they do so consistently (e.g., under a broad range of initial conditions, speciation assumptions, or ways of subdividing the landscape into local communities) will have to await further work.

Although the assumption of environmental homogeneity allows speciation, colonization, and local extinction to be modeled explicitly, it reflects a more restrictive interpretation of the null expectation for species richness than is implicit in many verbal justifications for mid-domain models (e.g., Willig and Lyons 1998; Colwell and Lees 1999). Specifically, if the environment varies geographically, with species adapted to only a subset of the available environmental conditions, then individual species' colonization and extinction rates also will vary geographically. In the aggregate, however, different locations on the domain may still be equally likely to be suitable or unsuitable for a species chosen at random from any given species pool. It has been a longstanding conjecture that this might lead to locations of species ranges being distributed approximately randomly (i.e., uniformly) among geometrically feasible locations (Willig and Lyons 1998; Colwell and Lees 1999; Connolly, Bellwood, and Hughes 2003).

Assessing this conjecture requires formally specifying a set of assumptions that are consistent with this conceptual model, and then explicitly deriving the species richness pattern that those assumptions imply. For instance: (a) the location of species' optima are uniformly distributed within the domain (i.e., any one location is as likely to be optimal as any other, as in the spreading dye randomization model); and (b) a species is as likely to encounter a range limit in any one location as any other (Connolly 2005). In contrast to the demographic approaches outlined previously, this framework sacrifices the explicit characterization of demographic stochasticity (colonization and extinction of peripheral populations) in order to relax the assumption of environmental homogeneity in a tractable way. In effect, we assume that the locations of range endpoints are determined by the distribution of suitable environmental conditions, with any fluctuations in the locations of range endpoints due to peripheral colonization and extinction confined to negligibly small regions near species' tolerance limits, where colonization and extinction rates are approximately balanced.

Under these assumptions, there is always a quasi-parabolic MDE, regardless of the particular stochastic process that determines the (geographical) rate at which range limits are encountered (see Connolly 2005 for a formal proof). However, this MDE is also relatively constrained, allowing a maximum twofold difference in species richness between domain boundaries and the mid-domain. Although this has been shown for two specific models (Poisson and Weibull models of Connolly 2005), some first-year university calculus can be employed to prove that this result is general (Appendix C). Interestingly, relaxation of the range contiguity assumption increases the decoupling between regional richness (i.e., range overlap) and local richness (i.e., habitat occupancy), with MDEs in local richness becoming increasingly diminished as habitat patchiness increases (Connolly 2005). This result is strikingly analogous to the effect of relaxing the range contiguity assumption for the colonization-extinction models described above (e.g., fig. 11.8, panels A, B).

Although process-based theory for species distributions and species richness is at an early stage of development, I believe that the work conducted to date is sufficient to propose some generalizations. Firstly, hump-shaped species richness gradients do appear to be a reasonable null expectation for species richness in bounded biogeographical domains. These MDEs can be substantial in magnitude (e.g., on the order of twofold). In homogeneous environments, MDEs are induced by the tendency for peripheral extinction to keep species range endpoints away from domain boundaries (Bokma, Bokma, and Mönkkönen 2001; Rangel and Diniz-Filho 2005). This effect appears to be confined to a small region near the edge of the domain when colonization rates exceed extinction rates (Davies, Grenyer, and Gittleman 2005), or when recolonization occurs regularly through long-distance dispersal (Rangel and Diniz-Filho 2005). In contrast, when colonization and extinction rates are comparable, these processes may induce a gradient throughout much of the domain (figs. 11.7, 11.8; Bokma, Bokma, and Mönkkönen 2001; Rangel and Diniz-Filho 2005). It is presently unclear whether these more pronounced MDEs are artifacts of the way in which the domain is divided into a particular number of grid cells, but the models analyzed in this chapter indicate that this possibility warrants further investigation (fig. 11.8, panels C, D). In heterogeneous environments without environmental gradients, pronounced MDEs occur when ranges of intermediate size dominate the RSFD (i.e., species encounter distributional limits at an intermediate geographical rate [Connolly 2005]). These models cannot produce more than a two-fold gradient in species richness between the edge and middle of the domain (Appendix C), and, often, the gradients produced are much smaller than twofold (Connolly 2005). However, results of the colonization-extinction models suggest that these MDEs could be amplified somewhat by stochastic fluctuations in the locations of species' range endpoints, where colonization and extinction rates are comparable. Both kinds of process-based null models suggest that relaxing the range contiguity assumption diminishes MDEs, at least when local species richness (i.e., habitat occupancy rather than range overlap) is considered.

A Future for Analysis of Species Richness

Most macroecological analyses of geographical variation in species richness have involved one of two approaches: a regression of species richness against environmental variables, or an assessment of the statistical concordance between an empirical species richness pattern and one produced by a randomization of species' ranges in geographical space (or, more recently, weighted randomizations). A limitation of both approaches is that the statistical expectation used to analyze empirical data is not derived explicitly from a set of assumptions consistent with the underlying biological expectation. The need to overcome this limitation has been widely recognized for null models, as outlined earlier in this chapter. Inexplicably, calls for a similar movement away from regression-based approaches are largely absent from the literature. Consequently, there is, as yet, no formal theoretical framework for deriving expected species richness patterns from the geographical distribution of environmental factors, based on assumptions about how those factors mediate origination, colonization, or persistence. However, some recent work suggests a few avenues along which such a theory might develop.

One approach is to build on the foundation laid by process-based null models. Specifically, species' ecological optima, rather than being uniformly distributed within the domain, as in some null models (e.g., Jetz and Rahbek 2001; Grytnes 2003; Connolly 2005), may be more or less likely to occur at different points, depending on the environmental conditions there. Similarly, the probability that a species encounters species-specific distributional limits in a particular region of the domain need not be independent of where that region is located within the domain. Instead, the probability can depend upon environmental conditions (e.g., the absolute value of an environmental variable, how different it is from its value at the species' optimum, or how steep the gradient in the environmental variable is at that point [sensu Connolly 2005]). Such models would, in many ways, be process-based analogues of the weighted randomization approaches described previously (Storch et al. 2006; Rahbek et al. 2007); however (at least as envisioned here), they would predict range size as well as species richness. Indeed, for many environmental variables, probability distributions of optima and thresholds for species may be calibrated using data already available for many biogeographical databases. Indeed, recent work suggests that predictive modeling of species distributions is enhanced by models that assume species' environmental tolerances are drawn from common distributions (Elith et al. 2006). In principle, such predictions are testable by a combination of experimental and observational studies at finer scales (e.g., Parmesan et al. 2005), although logistical considerations are likely to limit such tests to a few species and locations that are well suited to such tests.

A disadvantage of the framework just described is the difficulty of incorporating explicitly an historical dynamic. Historical effects might be incorporated phenomenologically (e.g., effects of environmental variables on speciation rates might be incorporated via the probability distribution of species optima). However, a more explicitly mechanistic interpretation of the assumptions implicit in this approach is that species are found where they are able to persist, and not found where they cannot persist. Thus, the theory is likely to be most applicable where the sizes and locations of species ranges have been determined substantially by a particular geographical distribution of environmental conditions (e.g., because the distribution has changed very little over time, or because present-day species ranges have been overwhelmingly determined by present-day environmental conditions).

As a general rule, data on present-day biogeographical distributions are much more extensive than historical data on changes in those distributions through evolutionary time. However, when historical data are available, there are clearly advantages to a theory that incorporates the effects of environmental variables on the dynamics of diversification. Hypotheses that link environmental factors to biodiversity via different proximate mechanisms (e.g., increasing speciation rates versus facilitating range expansion or persistence) may entail different predictions about the dynamics of species ranges through time. Similarly, if the geographical distribution of environmental conditions has changed markedly through time, the potential effects of such changes on present-day biodiversity would require an explicitly historical approach. Moreover, historical, evolutionary models can make a richer set of testable predictions than ahistorical approaches, such as age distributions (i.e., time since origination) of taxa (Goldberg et al. 2005), or phylogenetic patterns in geographical location, range size, or environmental tolerances (Sugihara et al. 2003; Webb and Gaston 2003; Hunt, Roy, and Jablonski 2005).

Between these extremes of strictly ahistorical or explicitly historical models are approaches that use present-day distributions of environmental conditions, but explicitly model evolutionary diversification. Rangel et al. (2007) pioneered this approach, modeling the diversification of South American birds on a map of present-day environmental conditions. Such hybrid frameworks will inevitably require some compromises. For instance, in the model of Rangel et al. (2007), the dynamics of speciation depend on the distribution of environmental conditions (and thus are potentially sensitive to the assumption that those conditions have not changed over the history of diversification), and speciation is driven by range fragmentation induced by hypothetical sinusoidal fluctuations in environmental conditions. Nevertheless, such approaches can potentially capture important real features of diversification (for instance, heritability of environmental tolerances, and the tendency for descendant species to arise in geographical proximity to their parents), and thus may still approximate the effects of such processes on species richness patterns. They can also generate many of the same kinds of testable predictions about the phylogenetic patterning of biodiversity as fully historical approaches. Moreover, given the limited availability of high-resolution data on changes in the geographical distribution of environmental conditions through geological time, such hybrid approaches are likely to be the only means available to incorporate evolutionary dynamics into process-based models of species distributions and species richness gradients.

An additional challenge for biodiversity studies is the incorporation of effects of species interactions on geographical distributions. Both of the approaches outlined previously assume, at least implicitly, that species' distributional limits are determined independently of one another (at least within the group of species under investigation in a particular study). Presently, we do not have any theory that can be applied (at least in a tractable way) to examine effects of species interactions on range limits for entire floras or faunas. Nevertheless, theoretical work (mainly on two-species models) shows that interactions among closely related species can have important effects on the locations of range limits, and empirical evidence suggests that these effects may be important in at least some cases (Case et al. 2005). One major effect of species interactions is to sharpen species' distributional limits along environmental gradients (Case and Taper 2000; Case et al. 2005), and such effects may not create substantial problems for models that assume individualistic responses to environmental gradients. In other cases, however, species interactions can produce range limits in the absence of environmental gradients, for a variety of reasons (Case et al. 2005). Although the empirical evidence for such effects is presently sparse, it would be premature to conclude that they are not widespread. As Turchin (2003) has noted, evidence for apparent competition did not begin to accumulate until well after Holt (1977) first identified the phenomenon in simple community models.

Tests of ecological theory are most convincing when they move beyond curve-fitting (i.e., parameter estimation and goodness-of-fit testing: McGill 2003). In a very few cases, ecological theory has been used to derive, a priori, predictions about the values of particular parameters (e.g., Harte, Blackburn, and Ostling 2001; Savage et al. 2004). In the context of species richness gradients, this would require a theory about effects of environmental variables on species' geographical range dynamics that specifies the functional forms and range of parameter values that define the relationship between environmental variables and species' origination, colonization, or persistence. However, the fundamentally contextual nature of ecological and macroevolutionary dynamics suggests that the estimation of parameters from data will be common in macroecology in general, and the analysis of species richness in particular, for some time to come. Consequently, it is important, where possible, to use parameters estimated from fits to one ecological pattern to quantitatively predict patterns in other systems or at other scales (McGill 2003). This approach has proved powerful in other areas of macroecology, such as the analysis of species relative abundance (Adler et al. 2005; Dornelas et al. 2006). A theory that seeks to explain variation in species richness at the provincial scale should, in principle, also explain the sizes and locations of species ranges that give rise to those patterns. Moreover, where there is an implicit or explicit historical dynamic to the hypothesis, the theory might also imply historical elements of these patterns, such as times of origination of species, or historical changes in the size and location of species' ranges. Other predictions, such as patterns of abundance within a species range, or phylogenetic patterns in the sizes and locations of species ranges, might also be implicit in particular hypotheses about the causes of species richness patterns. A more rigorous and extensive body of theory about species richness would help to draw out such predictions. Indeed, this theory need not all be mathematical: rigorous conceptual models can also be used to derive novel, qualitative predictions about species ranges based on hypothesized environmental mechanisms (e.g., Chown and Gaston 2000).

Ultimately, scientific inquiry aims to determine how well our conceptual models of how nature works (our theory, in other words) allow us to understand nature as it actually is, and we do this by confronting expectations about nature (i.e., hypotheses) with empirical data. Few ecologists would dispute the importance of rigorous methodological protocols for empirical observations, protocols designed to ensure that our data provide as representative a picture of nature as possible. However, it is just as important that our expectations—the statistical models that we fit to empirical data—actually follow from our conceptual models of how nature works. In the study of species richness gradients, the statistical models we use are rarely straightforward mathematical translations of our conceptual models. In such cases, tools that can assist our thinking (e.g., formalizing our conceptual models, and explicitly deriving predictions from those models) become particularly important. Indeed, other areas of macroecology, such as the study of patterns of relative abundance, have been reinvigorated by the development of statistical expectations derived explicitly from process-based models (Engen and Lande 1996; Hubbell 2001; Sugihara et al. 2003). I opened this chapter by noting that there have been substantial, almost revolutionary improvements in recent years in our ability to characterize species' geographical distributions and species richness patterns in ingenious and rigorous ways at large scales. Unfortunately, our macroecological theory—the lens through which we examine these large-scale patterns—lags far behind, and similarly revolutionary developments will be required in this area if our understanding of the origin and maintenance of species richness gradients is to continue to progress.

APPENDIX A
ANALYTICAL COLONIZATION-EXTINCTION MODELS

I illustrate the structure of the matrix model for a very simple case where the domain consists only of three patches of habitat. First, consider a "founder speciation" model: each species gives rise to a new species of range size 1 with a fixed probability for each cell that its range encompasses. Thus, we have the matrices

$$\mathbf{S}=\begin{array}{c} \\ 11 \\ 12 \\ 22 \\ 13 \\ 23 \\ 33 \end{array}\!\!\begin{array}{c} \begin{array}{cccccc} 11 & 12 & 22 & 13 & 23 & 33 \end{array} \\ \begin{pmatrix} P_s & P_s & 0 & P_s & 0 & 0 \\ 0 & 0 & 0 & 0 & 0 & 0 \\ 0 & P_s & P_s & P_s & P_s & 0 \\ 0 & 0 & 0 & 0 & 0 & 0 \\ 0 & 0 & 0 & 0 & 0 & 0 \\ 0 & 0 & 0 & P_s & P_s & P_s \end{pmatrix} \end{array} \qquad \text{(A.1)}$$

for speciation, and

$$\mathbf{T}=\begin{array}{c} \\ 11\\12\\22\\13\\23\\33\end{array}\begin{array}{c} \begin{array}{cccccc}11 & 12 & 22 & 13 & 23 & 33\end{array}\\ \begin{pmatrix} 1-P_c-P_e & (1-P_e)P_e & 0 & 0 & 0 & 0\\ P_c & (1-P_e)(1-P_c-P_e) & P_c/2 & (1-P_e)P_e & P_cP_e & 0\\ 0 & P_e(1-P_c-P_e) & 1-P_c-P_e & P_e^2 & (1-P_c-P_e)P_e & 0\\ 0 & (1-P_e)P_c & 0 & (1-P_e)^2 & P_c(1-P_e) & 0\\ 0 & P_eP_c & P_c/2 & P_e(1-P_e) & (1-P_c-P_e)(1-P_e) & P_c\\ 0 & 0 & 0 & 0 & P_e(1-P_e) & 1-P_c-P_e \end{pmatrix}\end{array} \tag{A.2}$$

for range-state transitions. For example, the second column of matrix **S** indicates that a species whose range extends from cell 1 to cell 2 (i.e., in range state "12") can give rise to a new species in either of the patches it occupies, with probability p_S in each case (eq. A.11.1). The same species transitions from its current state ("12") to state "22" (i.e., upper and lower limits both in cell 2) if the population in patch 1 goes extinct (this occurs with probability p_E), and the upper-range endpoint neither expands nor contracts (this occurs with probability $1-p_E-p_C$) (row 3, column 2 in eq. A. 11.2). If, instead, the upper-range endpoint expands (with probability p_C), the species would transition to state "23" (row 5, column 2); if it contracts (with probability p_E), the species goes extinct. The sum of the speciation and extinction matrices, $\mathbf{A}=\mathbf{S}+\mathbf{T}$, gives us a projection matrix for species' range endpoints.

Alternatively, we can consider a "vicariant speciation" model. In this case, at each time step, each species with range extent >1 cell splits into two species with probability p_S, or undergoes a range-state transition with probability $1-p_S$. If all points inside the range are equally likely points of origination, then our projection model becomes:

$$\mathbf{A}=p_S\,\mathbf{S}+\left(1-p_S\right)\mathbf{T} \tag{A.3}$$

where **T** is as indicated in Eq. (A.11.2), and

$$\mathbf{S}=\begin{array}{c} \\ 11\\12\\22\\13\\23\\33\end{array}\begin{array}{c} \begin{array}{cccccc}11 & 12 & 22 & 13 & 23 & 33\end{array}\\ \begin{pmatrix} 0 & 1 & 0 & 0.5 & 0 & 0\\ 0 & 0 & 0 & 0.5 & 0 & 0\\ 0 & 1 & 0 & 0 & 1 & 0\\ 0 & 0 & 0 & 0 & 0 & 0\\ 0 & 0 & 0 & 0.5 & 0 & 0\\ 0 & 0 & 0 & 0.5 & 1 & 0 \end{pmatrix}\end{array} \tag{A.4}$$

Thus, for instance, for a species in range state "12," only one vicariant event is possible: a split into species with range states "11" and "22." Conversely, a species in range state "13" splits between cells 1 and 2 with probability 0.5 (giving rise to species with range states "11" and "23"), and between cells 2 and 3 with probability 0.5 as well.

By normalizing the eigenvector, **v**, associated with the leading eigenvalue of **A**, we can find the long-run fraction of species in each state, and thus we can calculate the long-run proportion of species whose ranges include cell h by summing up the eigenvector elements for those range locations that overlap h (Caswell 2001).

$$S(h)=\sum_{(x_L,x_U)}\begin{cases}\mathbf{v}_{x_L,x_U} & \text{if } x_L\le h \text{ and } x_U\ge h\\ 0 & \text{otherwise}\end{cases}$$

APPENDIX B
ALGORITHM FOR PATCH-OCCUPANCY SIMULATION

To illustrate the effect of relaxing the range contiguity assumption on predicted species richness gradients for homogeneous domains, I analyzed simple, one-dimensional simulation models. For each occupied patch, a species has a probability p_C of colonizing any adjacent, unoccupied cells in each time step. It also has a probability p_E of going locally extinct in any occupied cells. To prevent an explosion in memory demands and computer time, species richness was held constant by adding a new species each time an existing species became globally extinct. The new species was given a range size of one cell, and the location of the new species' origination was determined by local species richness levels: the probability that a new species originated in cell h was linearly proportional to the number of species occupying that cell. This speciation assumption should amplify any emerging gradients in species richness, and thus, if anything, is likely to produce stronger MDEs than an alternative speciation algorithm.

This model was analyzed numerically, by stochastic simulation. Local richness was quantified over time as the number of species occupying a particular patch. Regional richness was quantified as the number of species whose ranges encompassed a particular patch (whether or not the patch was occupied). Plots of the species richness gradients over time were inspected to determine convergence. Convergence occurred in all cases except the 640-cell case, for which the species richness gradient had not yet equilibrated after approximately 20,000 iterations. Visual inspection of the range overlap and local occupancy graphs for this case indicated that the both gradients were still becoming more flat-topped over time.

APPENDIX C
TWO-FOLD CONSTRAINT ON MDE FOR A GENERAL CLASS OF NULL MODELS

For the general class of null models analyzed in Connolly (2005), the probability that a species' range encompasses location x is given by:

$$p(x) = \int_{y=0}^{x} S(x-y)\,dy + \int_{y=x}^{1} S(y-x)\,dy \tag{C.1}$$

$S(x\text{-}y)$ and $S(y\text{-}x)$ are the probabilities that a species extends from optima at y up to x, or down to x, respectively. With changes of variable ($z = x\text{-}y$ in the first integral and $y\text{-}x$ in the second integral), we have:

$$p(x) = \int_{z=0}^{x} S(z)dz + \int_{z=0}^{1-x} S(z)dz \tag{C.2}$$

Thus, the ratio of species richness at the mid-domain ($x = 0.5$) relative to a domain boundary ($x = 0$) is:

$$\frac{p_{Mid}}{p_{Edge}} = \frac{2\int_{z=0}^{0.5} S(z)dz}{\int_{z=0}^{1} S(z)dz} = \frac{2\int_{z=0}^{0.5} S(z)dz}{\int_{z=0}^{0.5} S(z)dz + \int_{z=0.5}^{1} S(z)dz} \tag{C.3}$$

$S(z)$ is the probability that no distributional limits are encountered over a distance x within the domain. Consequently, it must be a nonincreasing function of x (i.e., it is either flat or decreasing), and it is bounded between zero and unity. This means that the second term in the denominator must be equal to or less than the first term, but it must also be greater than or equal to zero:

$$0 \leq \int_{z=0.5}^{1} S(z)dz \leq \int_{z=0}^{0.5} S(z)dz \tag{C.4}$$

The inequality on the left of (C.4) means that the minimum possible value for the richness ratio is unity (this would only be attained if there was literally a zero probability of encountering a distributional limit anywhere within the domain—such that both numerator and denominator of eq. C.3 integrated to unity). Conversely, the maximum possible value for the richness ratio is 2 (when the second term in the denominator of eq. C.3 is zero).

ACKNOWLEDGMENTS

I wish to thank D. Bellwood and T. Hughes; our collaborative work on coral reef biogeography has sparked many of the ideas that are expressed in this chapter. I am also grateful to K. Roy and J. Witman for encouragement and advice during the process of writing this chapter. I thank R. Colwell, N. Gotelli, A. Helfgott, M. Hisano, M. Hoogenboom, and M. Roughan for helpful comments and assistance at various stages of this work. I also thank colleagues in the working group "Modelling Species Diversity" at the National Center for Ecological Analysis and Synthesis for stimulating discussion on many of the topics covered in this chapter. The Australian Research Council is acknowledged for financial support.

REFERENCES

Adler, P. B., E. P. White, W. K. Lauenroth, D. M. Kaufman, A. Rassweiler, and J. A. Rusak. 2005. Evidence for a general species-time-area relationship. *Ecology* 86:2032–39.

Bell, G. 2001. Neutral macroecology. *Science* 293:2413–18.

Bellwood, D. R., and T. P. Hughes. 2001. Regional-scale assembly rules and biodiversity of coral reefs. *Science* 292:1532–34.

Bellwood, D. R., T. P. Hughes, S. R. Connolly, and J. Tanner. 2005. Environmental and geometric constraints on Indo-Pacific coral reef biodiversity. *Ecology Letters* 8:643–51.

Bellwood, D. R., and P. C. Wainwright. 2002. The history and biogeography of fishes on coral reefs. In *Ecology of fishes on coral reefs*, ed. P. F. Sale, 5–32. Amsterdam: Elsevier.

Blackburn, T. M., and K. J. Gaston. 1996. Spatial patterns in the geographic range sizes of bird species in the New World. *Philosophical Transactions of the Royal Society of London, B* 351:897–912.

Bokma, F., J. Bokma, and M. Mönkkönen. 2001. Random processes and geographic species richness patterns: Why so few species in the north? *Ecography* 24:43–49.

Brown, J. H. 1995. *Macroecology.* Chicago: University of Chicago Press.

Case, T. J., R. D. Holt, M. A. McPeek, and T. H. Keitt. 2005. The community context of species' borders: Ecological and evolutionary perspectives. *Oikos* 108:28–46.

Case, T. J., and M. L. Taper. 2000. Interspecific competition, gene flow, and the coevolution of species borders. *American Naturalist* 155:583–605.

Caswell, H. 2001. Matrix population models, 2nd ed. Sunderland, MA: Sinauer.

Chown, S. L., and K. J. Gaston. 2000. Areas, cradles and museums: The latitudinal gradient in species richness. *Trends in Ecology & Evolution* 15:311–15.

Colwell, R. K., and G. C. Hurtt. 1994. Nonbiological gradients in species richness and a spurious Rapoport effect. *American Naturalist* 144:570–95.

Colwell, R. K., and D. C. Lees. 2000. The mid-domain effect: Geometric constraints on the geography of species richness. *Trends in Ecology & Evolution* 15:70–76.

Colwell, R. K., C. Rahbek, and N. J. Gotelli. 2004. The mid-domain effect and species richness patterns: What have we learned so far? *American Naturalist* 163:E1–E23.

———. 2005. The mid-domain effect: There's a baby in the bathwater. *American Naturalist* 166:E149–E154.

Connolly, S. R. 2005. Process-based models of species distributions and the mid-domain effect. *American Naturalist* 166:1–11.

Connolly, S. R., D. R. Bellwood, and T. P. Hughes. 2003. Indo-Pacific biodiversity of coral reefs: Deviations from a mid-domain model. *Ecology* 84:2178–90.

Crame, J. A. 2000. Evolution of taxonomic diversity gradients in the marine realm: Evidence from the composition of Recent bivalve faunas. *Paleobiology* 28:184–207.

Cressie, N. A. C. 1993. *Statistics for spatial data.* New York: Wiley.

Davies, T. J., R. Grenyer, and J. L. Gittleman. 2005. Phylogeny can make the mid-domain effect an inappropriate null model. *Biology Letters* 1:143–46.

Dornelas, M., S. R. Connolly, and T. P. Hughes. 2006. Coral reef diversity refutes the neutral theory of biodiversity. *Nature* 440:80-82.

Duke, N. C., E. Y. Y. Lo, and M. Sun. 2002. Global distribution and genetic discontinuities of mangroves: Emerging patterns in the evolution of *Rhizophora*. *Trees* 16:65–79.

Elith, J., C. H. Graham, R. P. Anderson, M. Dudik, S. Ferrier, A. Guisan, R. J. Hijmans, et al. Novel methods improve prediction of species' distributions from occurrence data. *Ecography* 29:129–51.

Ellison, A. M. 2002. Macroecology of mangroves: Large-scale patterns and processes in tropical coastal forests. *Trees* 16:181–94.

Engen, S., and R. Lande. 1996. Population dynamic models generating the lognormal species abundance distribution. *Mathematical Biosciences* 132:69–183.

Fortes, R. R., and R. S. Absalao. 2004. The applicability of Rapoport's rule to the marine molluscs of the Americas. *Journal of Biogeography* 31:1909–16.

Fraser, R. H., and D. J. Currie. 1996. The species richness-energy hypothesis in a system where historical factors are thought to prevail: Coral reefs. *American Naturalist* 148:138–59.

Gaston, K. J., T. M. Blackburn, and J. I. Spicer. 1998. Rapoport's rule: Time for an epitaph? *Trends in Ecology & Evolution* 13:70–74.

Gaston, K. J., and F. He. 2002. The distribution of species range size: A stochastic process. *Proceedings of the Royal Society of London, B* 269:1079–86.

Goldberg, E. E., K. Roy, R. Lande, and D. Jablonski. 2005. Diversity, endemism, and age distributions in macroevolutionary sources and sinks. *American Naturalist* 165:623–33.

Gotelli, N. J., and G. R. Graves. 1996. *Null models in ecology.* Washington, DC: Smithsonian Institution Press.

Gray, J. S. 2001. Marine diversity: the paradigms in patterns of species richness examined. *Scientia Marina* 65 (Suppl 2): 41–56.

Grytnes, J. A. 2003. Ecological interpretations of the mid-domain effect. *Ecology Letters* 6:883–88.

Harte, J., T. Blackburn, and A. Ostling. 2001. Self-similarity and the relationship between abundance and range size. *American Naturalist* 157:374–86.

Hawkins, B. A. 2001. Ecology's oldest pattern? *Trends in Ecology and Evolution* 16:470.

Hawkins, B. A., and J. A. F. Diniz-Filho. 2002. The mid-domain effect cannot explain the diversity gradient of Nearctic birds. *Global Ecology and Biogeography* 11:419–26.

Hernandez, C. E., R. A. Moreno, and N. Rozbaczylo. 2005. Biogeographical patterns and Rapoport's rule in southeastern Pacific benthic polychaetes of the Chilean coast. *Ecography* 28:363–73.

Holt, R. D. 1977. Predation, apparent competition, and the structure of prey communities. *Theoretical Population Biology* 12:197–229.

Hubbell, S. P. 2001. *The unified neutral theory of biodiversity and biogeography.* Princeton, NJ: Princeton University Press.

Hughes, T. P., D. R. Bellwood, and S. R. Connolly. 2002. Biodiversity hotspots, centers of endemicity, and the conservation of coral reefs. *Ecology Letters* 5:775–84.

Hunt, G., K. Roy, and D. Jablonski. 2005. Species-level heritability reaffirmed: A comment on "On the heritability of geographic range sizes." *American Naturalist* 166:129–35.

Huntley, B., R. E. Green, Y. C. Collingham, J. K. Hill, S. G. Willis, P. J. Barlein, W. Cramer, W. J. M. Hagemeijer, and C. J. Thomas. 2004. The performance of models relating species geographical distributions to climate is independent of trophic status. *Ecology Letters* 7:417–26.

Jetz, W., and C. Rahbek. 2001. Geometric constraints explain much of the species richness pattern in African birds. *Proceedings of the National Academy of Sciences, USA* 98:5661–66.

Jetz, W., C. Rahbek, and R. K. Colwell. 2004. The coincidence of rarity and richness and the potential signature of history in centres of endemism. *Ecology Letters* 7:1180–91.

Jokiel, P., and F. J. Martinelli. 1992. The vortex model of coral reef biogeography. *Journal of Biogeography* 19:449–58.

Kerswell, A. 2006. Global biodiversity patterns of benthic marine algae. *Ecology* 87:2479–88.

Lees, D. C., C. Kremen, and L. Andriamampianina. 1999. A null model of species richness gradients: Bounded range overlap of butterflies and other rainforest endemics in Madagascar. *Biological Journal of the Linnean Society* 67:529–94.

Lyons, S. K., and M. R. Willig. 1997. Latitudinal patterns of range size: Methodological concerns and empirical evaluations for New World bats and marsupials. *Oikos* 79:568–80.

MacPherson, E. 2003. Species range size distributions for some marine taxa in the Atlantic Ocean. Effect of latitude and depth. *Biological Journal of the Linnean Society* 80:437–55.

McCain, C. 2003. North American desert rodents: A test of the mid-domain effect in species richness. *Journal of Mammalogy* 84:967–80.

McGill, B. J. 2003. Strong and weak tests of macroecological theory. *Oikos* 102:679–85.

McManus, J. W. 1985. Marine speciation, tectonics, and sea-level changes in southeast Asia. In *Proceedings of the Fifth International Coral Reef Congress (Tahiti, 1985). vol. 4,* ed. C. Gabrie and M. Harmelin Vivien, 133–38. French Polynesia: Antenne Museum-Ephe, Moorea.

Mora, C., P. M. Chittaro, P. F. Sale, J. P. Kritzer, and S. A. Ludsin. 2003. Pattern and process in reef fish diversity. *Nature* 421:933–36.

Parmesan, C., S. Gaines, L. Gonzalez, D. M. Kaufman, J. Kingsolver, A. T. Peterson, and R. Sagarin. 2005. Empirical perspectives on species borders: From traditional biogeography to global change. *Oikos* 108:58–75.

Piñeda, J., and H. Caswell. 1998. Bathymetric species-diversity patterns and boundary constraints on vertical range distributions. *Deep-Sea Research* 45:81–101.

Proches, S. 2001. Back to the sea: Secondary marine organisms from a biogeographical perspective. *Biological Journal of the Linnean Society* 74:197–203.

Rahbek, C., N. J. Gotelli, R. K. Colwell, G. L. Entsminger, T. F. L. V. B. Rangel, and G. R. Graves.

2007. Predicting continental-scale patterns of bird species richness with spatially explicit models. *Proceedings of the Royal Society, B* 274:165–74.

Rangel, T. F. L. V. B., and J. A. F. Diniz-Filho. 2005. Neutral community dynamics, the mid-domain effect, and spatial patterns in species richness. *Ecology Letters* 8:783–390.

Rangel, T. F. L. V. B., J. A. F. Diniz-Filho, and R. K. Colwell. 2007. Species richness and evolutionary niche dynamics: A spatial pattern-oriented simulation experiment. *American Naturalist* 170:602–16.

Rex, M. A., J. A. Crame, C. T. Stuart, and A. Clarke. 2005. Large-scale biogeographic patterns in marine mollusks: A confluence of history and productivity? *Ecology* 86:2288–97.

Ricklefs, R. E., and D. Schluter, eds. 1993. *Species diversity in ecological communities.* Chicago: University of Chicago Press.

Roberts, C. M., C. J. McClean, J. E. N. Veron, J. P. Hawkins, G. R. Allen, D. E. McAllister, C. G. Mittermeier, et al. 2002. Marine biodiversity hotspots and conservation priorities for tropical reefs. *Science* 295:1280–84.

Rohde, K. 1992. Latitudinal gradients in species diversity: the search for the primary cause. *Oikos* 65:514–27.

———. 1998. Latitudinal gradients in species diversity: Area matters, but how much? *Oikos* 82:184–90.

Rohde, K., M. Heap, and D. Heap. 1993. Rapoport's Rule does not apply to marine teleosts and cannot explain latitudinal gradients in species richness. *American Naturalist* 142:1–16.

Roughgarden, J. 1983. Competition and theory in community ecology. *American Naturalist* 122:583–601.

———. 1996. *Theory of population genetics and evolutionary ecology: An introduction.* Upper Saddle River, NJ: Prentice Hall.

Roy, K., D. Jablonksi, and J. W. Valentine. 1994. Eastern Pacific molluscan provinces and latitudinal diversity gradient: No evidence for "Rapoport's rule." *Proceedings of the National Academy of Sciences, USA* 91:8871–74.

Roy, K., D. Jablonski, J. W. Valentine, and G. Rosenberg. 1998. Marine latitudinal diversity gradients: Tests of causal hypotheses. *Proceedings of the National Academy of Sciences, USA* 95:3699–3702.

Sanders, N. J. 2002. Elevational gradients in ant species richness: Area, geometry, and Rapoport's Rule. *Ecography* 25:25–32.

Savage, V. M., J. F. Gillooly, J. H. Brown, G. B. West, and E. L. Charnov. 2004. Effects of body size and temperature on population growth. *American Naturalist* 163:429–41.

Smith, K. F., and S. D. Gaines. 2003. Rapoport's bathymetric rule and the latitudinal species diversity gradient for Northeast Pacific fishes and Northwest Atlantic gastropods: Evidence against a causal link. *Journal of Biogeography* 30:1153–59.

Stehli, F. G., and J. W. Wells. 1971. Diversity and age patterns in hermatypic corals. *Systematic Zoology* 20:115–26.

Stevens, G. C. 1989. The latitudinal gradient in geographical range: How so many species coexist in the tropics. *American Naturalist* 133:240–56.

Storch, D., R. G. Davies, S. Zajicek, C. D. L. Orme, V. Olson, G. H. Thomas, T.-S. Ding, et al. 2006. Energy, range dynamics, and global species richness patterns: Reconciling mid-domain effects and environmental determinants of avian diversity. *Ecology Letters* 9:1308–20.

Sugihara, G., L. Bersier, T. R. E. Southwood, S. L. Pimm, and R. M. May. 2003. Predicted correspondence between species abundances and dendrograms of niche similarities. *Proceedings of the National Academy of Sciences, USA* 100:5246–51.

Turchin, P. 2003. *Complex population dynamics: A theoretical/empirical synthesis*. Princeton, NJ: Princeton University Press.

Veron, J. E. N. 1995. *Corals in space and time: The biogeography and evolution of the sclearactinia*. Sydney: UNSW Press.

Webb, T. J., and K. J. Gaston. 2003. On the heritability of geographic range sizes. *American Naturalist* 161:553–66.

Willig, M. R., D. M. Kaufman, and R. D. Stevens. 2003. Latitudinal gradients of biodiversity: Pattern, process, scale, and synthesis. *Annual Review of Ecology and Systematics* 34:273–309.

Willig, M. R., and S. K. Lyons. 1998. An analytical model of latitudinal gradients of species richness with an empirical test for marsupials and bats in the New World. *Oikos* 81:93–98.

Worm, B., M. Sandow, A. Oschlies, H. K. Lotze, and R. A. Myers. 2005. Global patterns of predator diversity in the open oceans. *Science* 309:1365–69.

Zapata, F. A., K. J. Gaston, and S. L. Chown. 2003. Mid-domain models of species richness gradients: Assumptions, methods, and evidence. *Journal of Animal Ecology* 72:677–90.

CHAPTER TWELVE

MACROECOLOGICAL CHANGES IN EXPLOITED MARINE SYSTEMS

DEREK P. TITTENSOR, BORIS WORM, AND RANSOM A. MYERS

Introduction

The oceans have been used as an important source of food and materials for much of human history and prehistory, and until relatively recently were viewed as inexhaustible. In 1883, T. H. Huxley, addressing the International Fisheries Exhibition in London, famously declared that "Any tendency to over-fishing will meet with its natural check in the diminution of the supply . . . this check will always come into operation long before anything like permanent exhaustion has occurred." Yet with the advent of industrialized fishing such predictions have been comprehensively invalidated, with fish stocks crashing and catches decreasing drastically in many parts of the world (e.g., Myers, Hutchings, and Barrowman 1997; Pauly et al. 2002), bycatch adversely affecting non-targeted species (e.g., Lewison et al. 2004) and marine resource management conflicts being placed squarely on the agenda of many governments. Indeed, exploitation may be the major driver of recent extinctions in the oceans, having an effect greater than that of habitat loss, climate change, pollution, disease, and species invasions (Dulvy, Sadovy, and Reynolds 2003).

As an example, in the Canadian waters around eastern Labrador and Newfoundland, the cod (*Gadus morhua*) fishery underwent heavy exploitation, resulting in the species becoming commercially extinct and a fishery moratorium being announced in 1992 (Hutchings and Myers 1994) in what was once the largest cod fishery in the world (McGrath 1911; Thompson 1943). Virtual population analysis suggests that spawner biomass underwent a 99 percent decline between its maximum and the year in which the moratorium was imposed. Similar patterns have been observed in many other heavily fished regions (Myers, Hutchings, and Barrowman 1997; Christensen et al. 2003; Myers and Worm 2005). The magnitude and rapidity of these declines in fish populations can be severe. Figure 12.1 shows community biomass in thirteen open ocean and shelf ecosystems under exploitation pressure (Myers and Worm 2003). Industrialized fisheries typically reduced community biomass by 80 percent within fifteen years, and large predatory fish biomass was estimated to be around 10 percent of preindustrial levels (Myers and Worm 2003).

Our perception of what is the "natural" abundance of a species or community may have changed over the course of the past half-century of heavy fishing—the "shifting baseline" effect (Pauly 1995; Dayton et al. 1998; Baum and Myers 2004). In addition, there can be a time lag between the onset of fishing and the detection of its effects on populations and communities, such as the period between the last reported sighting of a species and its extinction (Dulvy, Sadovy, and Reynolds 2003). Analysis of historical data suggests that the lag between overfishing and subsequent ecosystem changes could range from decades to centuries (Jackson et al. 2001; Springer et al. 2003). Significant fishing effort and large impacts on species and communities may also have occurred far earlier than originally thought (e.g., Barrett, Locker, and Roberts 2004).

Fishing has an effect at many levels in addition to that of the population and the species. Life-history changes (e.g., Hutchings and Baum 2005), loss of genetic diversity (e.g., Hauser et al. 2002), habitat alteration (e.g., Cranfield, Michael, and Doonan 1999) and changes in community structure (e.g., Witman and Sebens 1992; Tegner and Dayton 1999; Dulvy, Freckleton, and Polunin 2004; Worm, Lotze, and Myers 2005) may also result, and these can all lead to detectable changes in the macroecology of the marine environment. In this chapter we take a broad view of macroecology as being the effect of local and small-scale processes upon large-scale patterns in the marine environment, and the analysis and utilization of these large-scale statistical patterns to infer ecological change from local to global scales. The new tools and analytical processes that a macroecological viewpoint provide enable us to view the interactions and synergies between biological and ecological processes in multiple dimensions. While we consider four major macroecological patterns, namely species richness, range size, abundance,

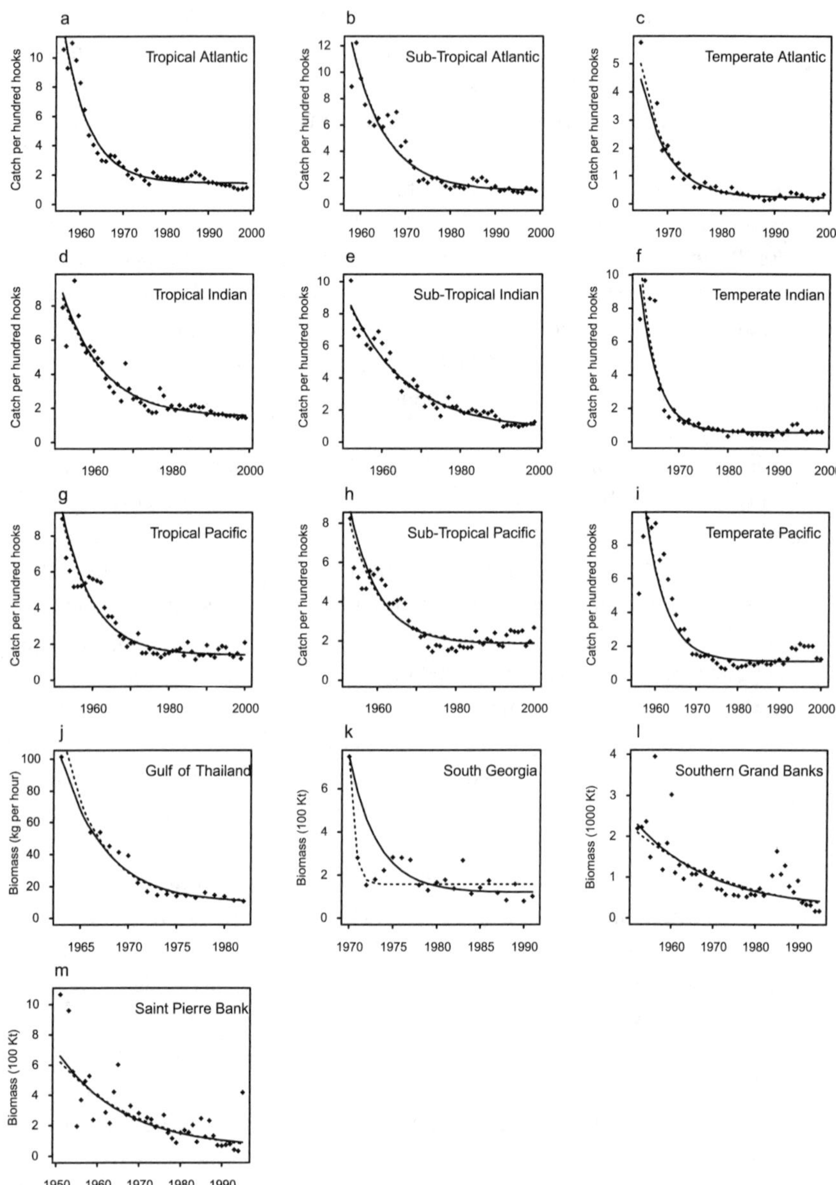

FIGURE 12.1 Trends in community biomass of large predatory fishes in open ocean (a–i) and shelf (j–m) ecosystems. Points denote relative biomass estimates from the beginning of industrialized fishing. Solid curves are individual maximum-likelihood fits. Dashed curves are empirical Bayes predictions derived from fitting a mixed-model. Redrawn with permission from Myers and Worm (2003).

and body size (Gaston and Blackburn 2000), industrialized fishing affects many of the parameters and processes of piscatorial communities and habitats discussed in this chapter. We structure this chapter according to the scale of these various effects, moving from individuals to species to whole ecosystems.

We first consider changes at the level of individual life-histories, then discuss the loss of stocks, populations, and species. Changes in ecosystem structure, biodiversity, and habitat are covered as the next level of our approach. Finally, we conclude by considering future research needs for further understanding and mitigation of the macroecological effects of fishing. This chapter is not intended to be an exhaustive survey of the effects of fishing on marine environments and communities, but rather an examination of some of the commonly encountered macroecological perturbations that are expressed under severe exploitation pressure. We also limit our discussion to the effects of exploitation through fishing, as opposed to other human impacts on the oceans.

Changes in Life-History Parameters

The life-history traits of many marine species, such as slow growth rates and aggregating behavior, leave them vulnerable to exploitation, and even highly fecund species—exactly those predicted by Lamarck during the first half of the nineteenth century to be safe from threat—are at risk (Sadovy and Cheung 2003). Exploited fish populations often manifest changes in life history traits such as age and size at maturity, and growth rate (Hutchings and Baum 2005). These changes can have cascading effects on survival until maturity, reproductive lifespan, and subsequently lifetime fecundity (Ernande, Dieckmann, and Heino 2004), and are the subject of this section.

Figure 12.2 shows changes in age and length at maturity of individuals for a number of different fish stocks and species under exploitation pressure; the general trend toward a reduction in both age and length, across a number of regions and species, is clear (Hutchings and Baum 2005). The decline in age at maturity is substantial, averaging 21%, with an average decline in length of 13% across all species (Hutchings and Baum 2005). The earlier maturation time and smaller size at maturity is echoed in a reduction of the mean age and weight of spawning individuals in most of these populations. Such changes in the life history of individuals could have consequences for the vulnerability of the species to extinction (Dulvy, Sadovy, and Reynolds 2003). Smaller individuals may be at greater risk of predation (Munday and Jones 1998), and experimental work suggests that genetic changes in life his-

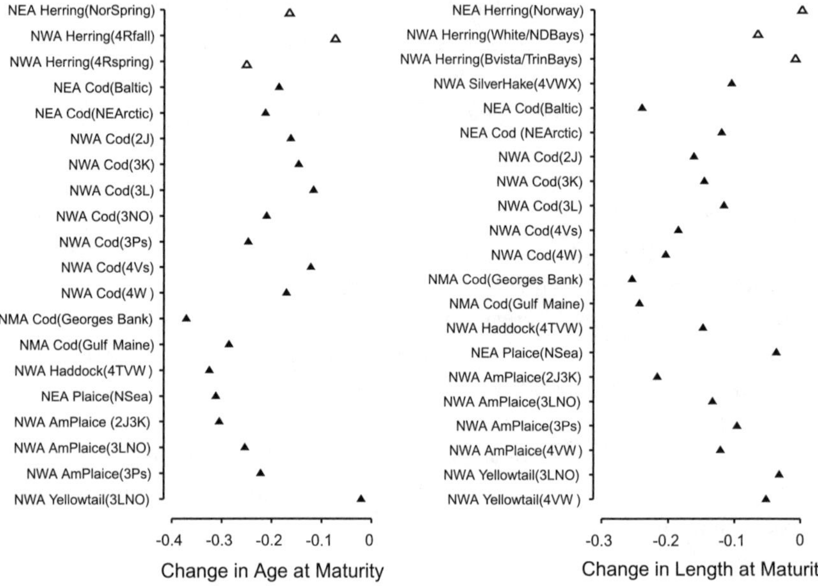

FIGURE 12.2 Comparative changes under exploitation in mean age and length at maturity for marine fishes from the north-temperate regions of the Atlantic and Pacific Oceans. The period of time each point represents is different among populations. Open triangles represent pelagic species, closed triangles demersal species. Redrawn with permission from Hutchings and Baum (2005).

tory may reduce the capacity for population recovery after overharvesting (Walsh et al. 2006).

Much of the debate on the life-history effects of fishing focuses on whether the observed changes are genetic or phenotypic (Hutchings 2004). The case for genetic change (e.g., Conover and Munch 2002; Grift et al. 2003; Olsen et al. 2004) depends on the fact that individuals genetically determined for late maturation have a much higher chance of being harvested before reproduction. This leads to more favorable conditions for those individuals that mature early (and hence have a smaller body mass) and a subsequent rapid evolutionary effect. Rapid evolution in response to human-induced selection pressures has been observed in many natural systems (Thompson 1998; Palumbi 2001), and conservation efforts should take the implications of this "contemporary evolution" into account (Stockwell, Hendry, and Kinnison 2003). An analysis of maturation patterns for cod (*Gadus morhua*) stocks from Georges Bank and the Gulf of Maine suggested that evolutionary forces are at least partially responsible for shifts in life-history parameters (Barot et al. 2004).

The converse of this argument is that phenotypic plasticity can result in a higher growth rate and earlier maturity in response to declines in population density caused by fishing (e.g., Roff 2002). Such changes seem to be largely responsible for life-history changes in Norwegian spring-spawning herring (*Clupea harengus*), with evolutionary responses playing at most a minor role (Engelhard and Heino 2004). Disentangling the relative roles of these processes is tricky, and some combination of both is likely in effect in many instances (Law 2000; Stokes and Law 2000).

Although there has been some exploration of the relationship between body size (another life-history trait) and other frequently examined macroecological parameters (e.g., range size, abundance patterns, species richness) in marine fishes (e.g., Munday and Jones 1998; Goodwin, Dulvy, and Reynolds 2005); in general, these patterns have received limited study (Macpherson et al., chapter 5, this volume). The observed relationships can be weak or inconclusive, such as the correlation between body size and abundance (Macpherson et al., chapter 5, this volume), or variable between taxonomic groups, such as the patterns of maximum body size and mean depth range in pelagic fishes (Smith and Brown 2002). It is likely that fisheries exploitation has a smearing effect on observed relationships, especially given that individuals are often targeted by body size (Bianchi et al. 2000); this deserves further study. To understand underlying macroecological processes, the effects of exploitation should be removed by examining pristine or near-pristine systems (Macpherson et al., chapter 5, this volume), although this is rarely, if ever, possible. Instead, macroecological methods can be used to reconstruct the historical state of a system (e.g., Jennings and Blanchard 2004).

Changes at the scale of the individual can influence the macroecological structure of entire populations, species, and communities, the interplay of processes at multiple scales becoming progressively more transparent as greater numbers of systems are combined in meta-analyses. We consider examples from the population and species level in the next section.

Loss of Populations and Species

The rapid and dramatic decline of marine fish populations and species in areas of intensive fishing activity has been well documented in a number of recent studies (e.g., Jackson et al. 2001; Baum et al. 2003; Myers and Worm 2003; Jennings and Blanchard 2004; Myers and Worm 2005; Worm et al. 2005). Fisheries affect both species that are directly targeted, and those that are unintentionally caught as bycatch or affected through indirect means. We consider all of these various effects.

Directed Fisheries

It is the extent of the decline in economically important, actively targeted food fish such as cod (*Gadus morhua*) that commands much of our attention and leads to some of the most intense debate on effective management strategies (e.g., Hutchings and Myers 1994). In many targeted species, less than 10 percent of most populations are still extant (e.g., Myers, Hutchings, and Barrowman 1997). As is evident from figure 12.1, such declines appear in populations throughout the world's oceans, and are often present irrespective of the methodology used for calculations (Myers and Worm 2005). The loss of populations can be so rapid that fisheries lose their viability shortly after opening. For instance, the marbled rockcod (*Notothenia rossii*) off South Georgia Island in the South Atlantic was fished for three years with a 99 percent decline in catch, and then abandoned in the fourth year (Kock 1992). Once species suffer such drastic declines, the time scale for recovery may be very long. A meta-analysis of many heavily fished species showed that for nonclupeid fishes there is little evidence of recovery since overfishing occurred (Hutchings 2000).

Species need not disappear entirely (global extinction) to be dramatically impacted by overexploitation; local extinctions can result in severe disruption of community and habitat structure (e.g., Estes, Duggins, and Rathbun 1989; see also Community Structure and Ecosystem Dynamics section), and species can also be reduced to such low densities that they are functionally extinct in the marine environment (ecological extinction). As shown in the influential work of Ricker (1958), it is also possible for subpopulations to go extinct in an economically sustainable fishery, if one stock has a greater catchability than the other.

Bycatch Effects

Losses are not restricted to targeted fish; exploitation can also affect non-target species through direct effects such as gear-induced mortality, and indirectly such as through the alteration of food supplies. Bycatch (the capture of species other than those that are directly targeted) can affect such diverse groups as dolphins (Tudela et al. 2005), leatherback turtles (Lewison, Freeman, and Crowder 2004; James, Ottensmeyer, and Myers 2005), seabirds (Tasker et al. 2000), sharks (Baum et al. 2003), and whales (Springer et al. 2003; Roman and Palumbi 2003). Tudela et al. (2005) studied the impact of fishing on non-target species by the driftnet fleet operating in the Alboran Sea at the western end of the Mediterranean, targeting swordfish (*Xiphias gladius*). Annual catch rates for the dolphin species *Delphinus delphis* and *Stenella coeruleoalba* were estimated to be greater than 10 percent of their population sizes in the Alboran Sea. Shark species suffered the heaviest bycatch rates, with blue shark (*Prionace glauca*), shortfin mako (*Isurus oxyrinchus*), and thresher shark (*Alopias vulpinus*) combined reaching half the total numerical captures of the target species *Xiphias gladius* (Tudela et al. 2005).

The Pacific leatherback turtle is another species at risk as bycatch. During the 1990s, it was estimated that 1,500 female leatherback turtles were caught per year in the Pacific Ocean by trawling, longlining, and gillnet fisheries directed for other species (Spotila et al. 2000). This corresponds to an annual mortality rate of between 23 and 33 percent, an astonishingly high figure that clearly has critical consequences for the leatherback, now IUCN-listed as critically endangered and facing extinction in the Pacific (Spotila et al. 2000; James, Ottensmeyer, and Myers 2005). Such deleterious effects of bycatch pressure may be especially prominent in long-lived, low-fecundity species such as sharks and whales (Musick 1999).

Range Contraction

A positive abundance-range size (abundance-distribution) relationship is an almost universal macroecological pattern in animal assemblages (Gaston and Blackburn 2000, but see Gaines et al., chapter 9, this volume). Reductions in population and species abundance are accompanied by concomitant range contraction for some exploited marine species. We consider single species here; interspecific changes in the abundance-distribution relationship are discussed below (Changes in Community Structure and Ecosystem Dynamics section).

Even large changes in range size can go undetected without analysis; an example is the barndoor skate (*Raja levis*), which disappeared from most of its range—likely due to being caught as bycatch in other fisheries—without notice until nearly a half-century of trawl data were analyzed (Casey and Myers 1998). Density-dependent differential changes in habitat use between marginal and prime habitats have also been predicted in North Sea cod (Myers and Stokes 1989); such changes are obviously of note, given the pressures and abundance declines caused by exploitation.

Range contractions are not limited to fish populations; reef modification and disturbance from 130 years of fishing in the Foveaux Strait, New Zealand, has led to the reduction of oyster density to such low values in some areas that fishers have abandoned them (Cranfield, Michael, and Doonan 1999). Meta-analysis of oyster fisheries on three continents showed that this pattern of fishery collapse is commonplace (Kirby 2004).

Such effects are not confined to the recent past. An example of the effects of preindustrialized exploitation on species ranges concerns pinnipeds in New Zealand (Smith 2005). The New Zealand fur seal (*Arctocephalus for-*

steri), one of eight species of southern fur seals, was hunted as a means of subsistence during the prehistoric period by the Maori, who colonized the region at around 750 years B.P. It was then killed for its skin during the historic period when Europeans colonized, beginning in A.D. 1769. During these time periods, seal colonies underwent a substantial range reduction. Figure 12.3 shows the estimated northern limits for the fur seal breeding ranges, and their successive retreat with time, from the beginnings of Maori settlement up to early colonization by Europeans. Historically, exploitation is known to be the major contributing factor responsible for reductions in fur seal abundance and distribution, given the large number of individuals taken (Smith 2005). Prehistorically, however, it is more difficult to determine whether exploitation, or another factor such as habitat degradation, climate change,

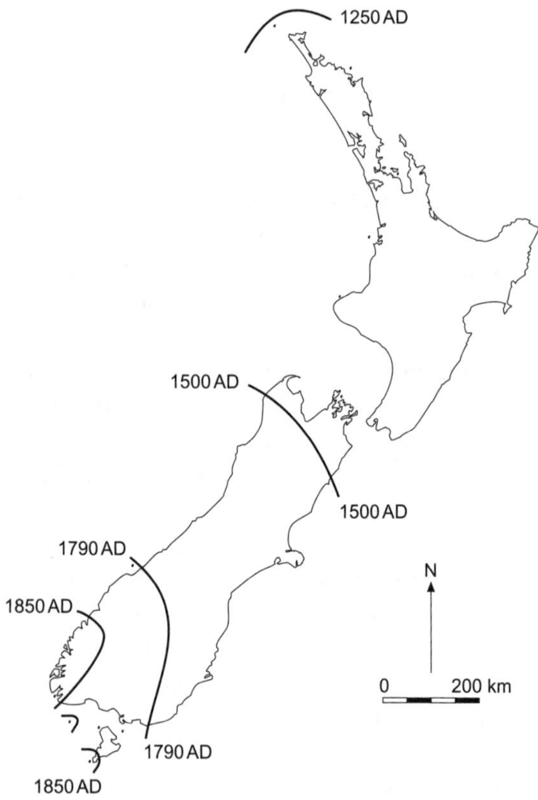

FIGURE 12.3 Changes in the northern limits of fur seal breeding range in New Zealand. Estimates are for the beginning (1250), middle (1500), and end (1790) of local prehistory, and after historic sealing (1850). Redrawn with permission from Smith (2005).

or introduced predators, was the underlying cause. The analysis by Smith (2005) suggested that everything but exploitation could be ruled out as having a significant effect on range reduction. Thus human-induced impacts upon the range of this species have been occurring for centuries. The effects of exploitation changed according to the time period and the intensity of effort, but the combined long-term impact was that the fur seal's range was reduced to a fraction of its original span in New Zealand.

The Allee Effect

In the wake of population declines and range contractions, the Allee effect, also known as depensation, has been hypothesized to be a contributing factor to the slow rate of recovery of many overexploited marine species (Frank and Brickman 2000; Hutchings and Reynolds 2004), although evidence for this is not always readily apparent (Myers et al. 1995). When population sizes fall below a certain threshold level of abundance, the lower density of individuals may result in reduced population growth per capita. There are several mechanisms (for example, reduced probability of encountering potential spawning partners) that may play a role in this effect. Such complications need to be considered when modeling observed macroecological patterns in exploited systems. If the Allee effect is not incorporated in population dynamical models, it can be an oversight that ignores what may be an effective reduction of biomass (Liermann and Hilborn 1997).

Loss of Genetic Diversity

There are further losses under exploitation, namely those of genetic variability in subpopulations, which should be of great concern due to the vast time scale on which such variability has evolved, and the correspondingly large time scale necessary for it to be recovered. Though studies on the loss of genetic diversity in exploited marine species have only recently begun to be conducted and are few in number, in at least one case, namely the New Zealand snapper (*Pagrus auratus*), there appears to have been a reduction in genetic variability due to exploitation (Hauser et al. 2002); figure 12.4 shows the change in genetic structure of two snapper stocks. The Hauraki Gulf stock was first exploited in the 1800s and showed signs of overfishing by the mid-1980s, with a corresponding decline in standing stock biomass during this period. The Tasman Bay stock followed a different exploitation pattern, with the fishery only commencing in the middle of the twentieth century. In figure 12.4, therefore, the Tasman Bay time series starts with the stocks essentially at a natural level, while the Hauraki Gulf time series depicts a stock that has already been exploited. Two measures of genetic diversity (number

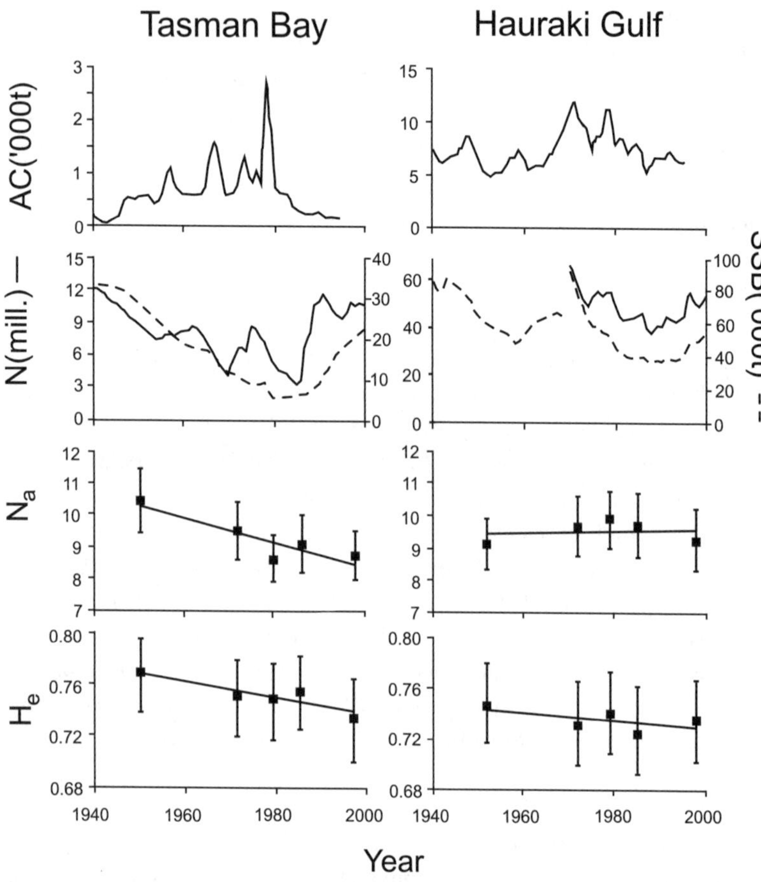

FIGURE 12.4 Temporal changes in New Zealand snapper (*Pagrus auratus*) populations in two locations. From top: Annual catch (*AC*), spawning stock numbers (*N* : thin line) and biomass (*SSB* : studded line), genetic diversity (mean number of alleles per locus (N_a), and mean expected heterozygosity (H_e); both means ± 95 percent confidence limits of 30 individuals). Redrawn with permission from Hauser et al. (2002).

of alleles per locus, and mean expected heterozygosity) showed significant decreases in the Tasman Bay region during the fifty-year period, but only random fluctuations in the Hauraki Gulf. It is possible that much of the genetic diversity had already been lost from the Hauraki Gulf stock by 1950. Such losses can drastically reduce the effective population size relative to the censused population size; should this be commonplace it may have a large-scale, long-term effect on population persistence and adaptability (Hauser et al. 2002). If loss of diversity commonly occurs during the early phases of exploitation, then studies initiated after the commencement of fisheries will likely detect no significant changes in diversity.

Such patterns, however, are not consistent. A study of North Sea cod (*Gadus morhua;* Hutchinson et al. 2003) showed an apparent decrease, then recovery, of genetic diversity under heavy exploitation; another, in contrast, showed no apparent loss of genetic variability (Poulson et al. 2006); while a study of Newfoundland cod (*Gadus morhua*) showed no loss in genetic diversity despite drastic changes in population abundances (Ruzzante et al. 2001). The diverse range of results within such a slim set of studies, coupled with the potential importance of loss of genetic diversity, marks this out as an important area for future research.

Historical and Prehistorical Exploitation

As noted previously, the impacts of exploitation on populations and species are not limited to modern-day industrialized fishing. Sea otters (*Enhydra lutris*), once active across the Pacific Rim, were driven to numerous local extinctions by the fur trade prior to the twentieth century (Estes, Duggins, and Rathbun 1989). Even before this, the hunting of otters by aboriginal Aleuts caused substantial changes in otter densities. These prehistoric changes in otter populations have been implicated as the probable cause of a shift between two alternate nearshore community stable-states in the Aleutian islands (Simenstad, Estes, and Kenyon 1978). The impacts of single-species exploitation can thus propagate and become visible at macroecological scales involving multiple species. Transformations such as these, of whole communities and ecosystems, are the subject of the next section.

Changes in Community and Ecosystem Structure, Biodiversity, and Habitat

The changes effected by fisheries at the level of the ecosystem are extensive and challenging to interpret. Multiple anthropogenic impacts on the marine environment produce complex yet recognizable symptoms of degradation (Tegner and Dayton 1999; Jackson et al. 2001; Lotze and Milewski 2004), and in some cases appear to have been affecting entire ecosystems for centuries (e.g., Pandolfi et al. 2003). The macroecological makeup of large regions, visible in the community structure, as well as the diversity, range size, abundance, life history, and evenness of constituent species, can be substantially altered by industrial fishing practices. Such effects are difficult to disentangle from those caused by a broad spectrum of other processes—for example, climate change (Hughes et al. 2003; Worm et al. 2005). Yet reports of habitat and ecosystem-scale shifts due to overfishing are surfacing with increasing regularity for a variety of marine ecosystems.

Changes in Community Structure and Ecosystem Dynamics

The worldwide phenomenon known as "fishing down the food web" (Pauly et al. 1998; Pauly et al. 2001) involves targeting progressively lower trophic levels for fishing as catches begin to stagnate or decline. Initially, this results in an ephemeral increase in catch, but ultimately it leads to cascading changes in relative species abundance and community structure, and as a result a reduction in the mean trophic level of fisheries landings. A different, but related, shift was observed by Bianchi et al. (2000) in a meta-analysis of fishing data from many regions. The size composition of exploited demersal communities appeared to change in high-latitude regions, with a relative decline of larger-sized fish.

It is very difficult to assess change where baseline details of the preimpacted system are not available. Recently, macroecological theory has been applied in order to estimate original, unexploited stock levels and to provide a theoretical comparison to current conditions (Jennings and Blanchard 2004). The biomass and abundance of fishes that would be expected in an unexploited North Sea community were calculated using a theoretical abundance-body mass relationship (size spectrum) based on measurements of primary production and predator-prey body mass ratios. These estimates were then compared to 2001 species-size-abundance data from trawl surveys. The biomass of the fish community as a whole in the 2001 surveys was less than half that expected for an unexploited ecosystem. Large fish were particularly severely affected, showing a reduction of approximately one hundred-fold in biomass; see figure 12.5, panel A. This approach highlighted the differential vulnerability of large fish to exploitation. (Jennings and Blanchard 2004).

The abundance-distribution relationship, a correlation between local abundance and geographical distribution (e.g., Brown 1984; also see section 3.3), is another macroecological pattern for which changes have been observed within an exploited marine community. The abundance-distribution relationships for twenty-four common marine fishes on the Scotian Shelf and in the Bay of Fundy, Canada, were examined by Fisher and Frank (2004) over a period of thirty-one years. In addition to intraspecific temporal trends for some stocks (i.e., correlated changes in abundance and distribution within a species), a trend was observed in the slope of the interspecific abundance-distribution relationship (i.e., that of the multispecies system). The value of the slope almost doubled between 1970 and 2001 (Fisher and

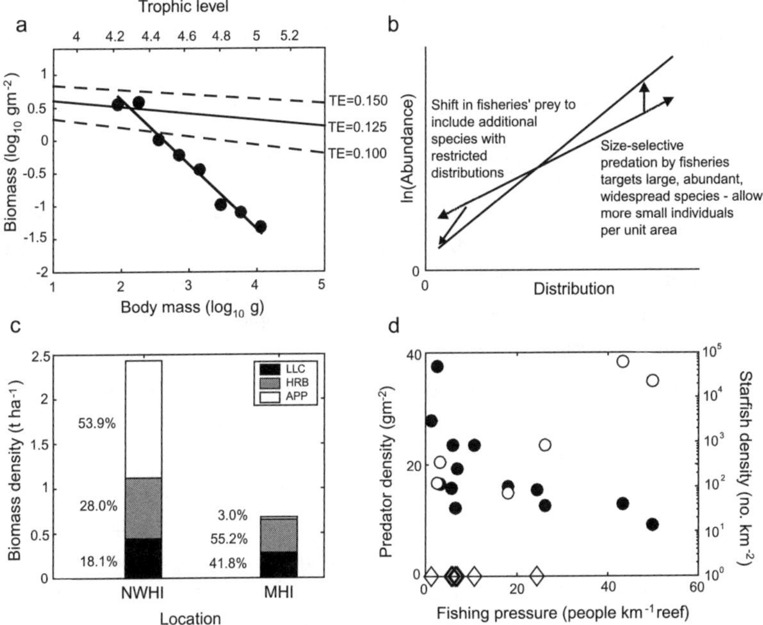

FIGURE 12.5 (A) Fish biomass plotted against body mass (circles) for the fish community in the North Sea in 2001 and the fitted size spectrum (steep bold line). Remaining lines indicate predicted slopes and locations of unexploited size spectra. Three size spectra are presented, corresponding to transfer efficiencies (TE) of 0.100, 0.125 and 0.150, with the spectrum for TE=0.125 in bold. Redrawn with permission from Jennings and Blanchard (2004). (B) Schematic representation of two forces interacting to produce different interspecific abundance-distribution relationships based on the same assemblage of species at different points in time. Redrawn with permission from Fisher and Frank (2004). (C) Fish biomass density and community structure in the northwestern Hawaiian islands (NWHI) and the main Hawaiian islands (MHI). White bars represent apex predators (APP), grey bars herbivores (HRB), and black bars lower level carnivores (LLC). Redrawn from with permission Friedlander and DeMartini (2002). (D) Fishing intensity and average density of predatory fishes (filled circles), and crown-of-thorns starfish (empty circles) for thirteen Fijian islands. Islands without starfish show an empty diamond for starfish density. Redrawn with permission from Dulvy et al. (2004).

Frank 2004). The authors proposed that direct fishery effects, or fishery-driven second-order trophic effects, probably drove this change. Among the factors suggested as being potentially responsible were increased abundance due to decreased body size (and therefore resource requirements) of targeted species, commercial fisheries switching exploited species, and density compensation through increases in prey and competitor species. A schematic of the interspecific abundance-distribution shift is presented in figure

12.5, panel B. This is a remarkable example of an observed temporal change in a macroecological pattern; such manipulations cannot easily be intentionally (experimentally) implemented at the large scales at which these processes operate. Thus marine systems that have a long history of exploitation and observation provide us with a valuable opportunity to examine the causal factors that underlie macroecological processes.

Another macroecological pattern observed to change under exploitation in the marine environment is the species-area relationship (SAR). An analysis of SARs for reef fish assemblages in the Atlantic, Indian, Pacific, and Mediterranean found that in all regions fishing pressure consistently resulted in a lower slope for the SAR, with the effect being proportional to the intensity of the fishing (Tittensor et al. 2007). This represents an alteration of the fundamental rate of scaling of diversity with area, one of the oldest known general laws in ecology.

It may prove feasible to use changes in observed macroecological patterns as assessment tools and ecological indicators of the effects of exploitation. For example, species-area relationships could potentially be used to quantify the impact of fishing on reef fish diversity, as an indicator of community-level changes (Tittensor et al. 2007).

Another means of assessing impacts is to compare geographically similar regions with differing exploitation histories. Using this approach, Friedlander and DeMartini (2002) found striking differences in the community structure between lightly and heavily exploited reef systems of the Hawaiian islands. The grand mean standing stock of the northwestern Hawaiian islands, a remote and lightly fished area, was 260% that of the main Hawaiian islands, an urbanized and heavily fished area. The community composition was also remarkably different, as depicted in figure 12.5, panel C, with the fish biomass of the northwestern Hawaiian islands being dominated by apex predators (54 percent), whereas in the main Hawaiian islands herbivores accounted for 55 percent, small-bodied lower-level carnivores 42 percent, and apex predators merely 3 percent. Grand mean weight per individual was also higher in the northwestern Hawaiian islands (Friedlander and DeMartini 2002). Using a comparable approach, changes to a suite of reef fish and benthic community variables were found along a population and fishing gradient in the northern Line Islands by Sandin et al. (2008).

Clearly, the Hawaiian example shows the dramatic effects of heavy fishing. Yet even relatively small takes can result in cascading effects on community structure. In their 1996 study of the effects of fishing on the structure of Fijian reef fish communities, Jennings and Polunin found that a 5 percent annual removal of fish biomass may have been responsible for drastic community changes (Jennings and Polunin 1996). An example of indirect effects apparently induced by artisanal exploitation of large predatory fishes, once more from Fiji, is the study of Dulvy, Freckleton, and Polunin (2004), shown in figure 5, panel D, in which loss of predators coincided with an increase in crown-of-thorns starfish density. The increased starfish population resulted in higher predation on reef-building corals and coralline algae, and their replacement by nonreef building taxa—a habitat shift that may have been induced, albeit indirectly, by exploitation. The scale and consequences of this effect, in a region of nonindustrialized fishing, should perhaps give cause for concern.

Given the potential consequences of ecosystem change, it is important to understand the forces that interact to effect these disturbances. A meta-analytic study of the coupled predator-prey pair Atlantic cod (*Gadus morhua*) and northern shrimp (*Pandalus borealis*) attempted to uncover some of these interactions (Worm and Myers 2003). Biomass time-series trends were used to determine whether control mechanisms were "top-down" or "bottom-up." Eight of nine regions showed negative correlations between cod and shrimp, a pattern consistent with top-down control (see figure 12.6). This finding implies that overexploitation of a predator can result in cascading effects at lower trophic levels, confirming an earlier study of the community-level effects of overfishing in the northwest Atlantic (Witman and Sebens 1992). The highly diverse communities of cold-water kelp ecosystems can show a similar dynamic, in which overexploitation of sea urchin predators facilitates destructive grazing by this species. This leads to cascading effects for other species in the habitat (Tegner and Dayton 2000; Steneck et al. 2002). Frank et al. (2005) provide evidence for a trophic cascade in the formerly cod-dominated Scotian Shelf ecosystem off eastern Canada. Systems such as these, with top-down control, should be managed under a multi-species framework, with species interactions taken into consideration (Worm and Myers 2003).

Large-scale changes in community composition caused by exploitation are also thought to have occurred in the Bering Sea and the Grand Banks food webs, driven by whaling and cod fishing, respectively (Worm et al. 2005). Similarly, shifts at this scale have been observed in the North Pacific Ocean, where there has been a sequential collapse in the populations of seals, sea lions, and sea otters (Springer et al. 2003). Both studies present evidence that top-down forcing mechanisms are responsible for these community phase shifts, with industrial whaling inducing a series of successive community changes over time.

Attempts have also been made to document and understand the driving forces behind community change on a global scale. In an analysis of Japa-

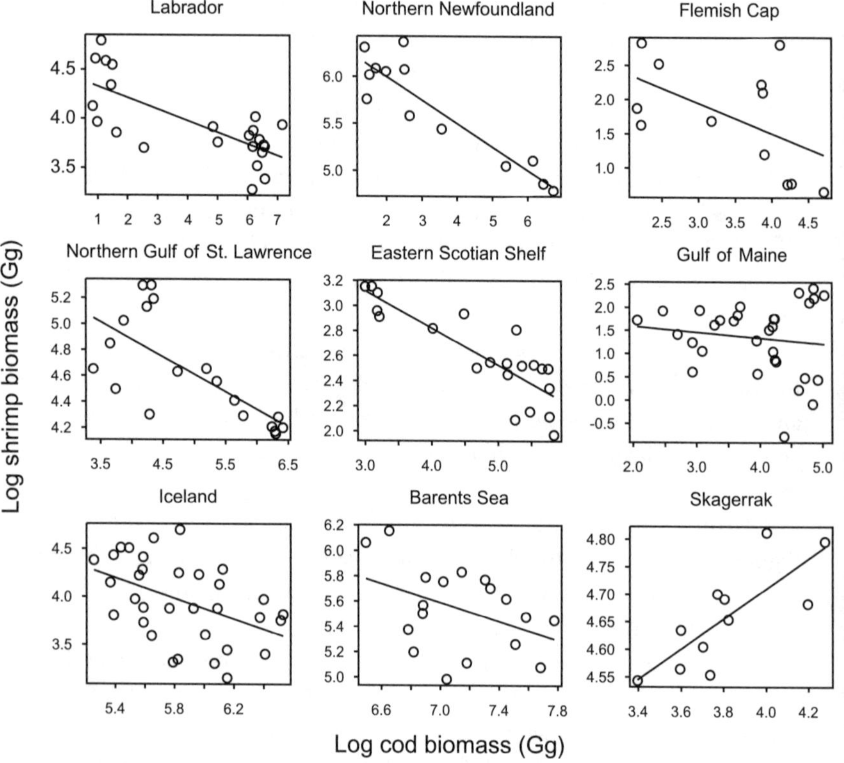

FIGURE 12.6 Linear correlations between cod and shrimp biomass time series from nine North Atlantic locations are suggestive of top-down control. Gg = gigagram, equivalent to 1,000 metric tons. Redrawn with permission from Worm and Myers (2003).

nese longlining data from 1952 to 2000, Worm et al. (2005) uncovered a decline of tuna and billfish diversity of between 10 and 50 percent in all oceans studied, a pattern linked to both climate and fishing. Their analysis suggests that the long-term, low-frequency variation in community composition was driven by the effects of exploitation, whereas the higher-frequency year-to-year variation was climate induced, modifying decadal trends only when lasting regime shifts such as the Pacific Decadal Oscillation occurred (Worm et al. 2005).

New attempts are being made to document the impacts of fishing at many structural levels and hence reach a more complete understanding of the dynamics involved. In this way, simultaneous changes in community composition and individual-level traits have been observed under fishing pressure. Ward and Myers (2005) used fisheries observer and scientific sur-

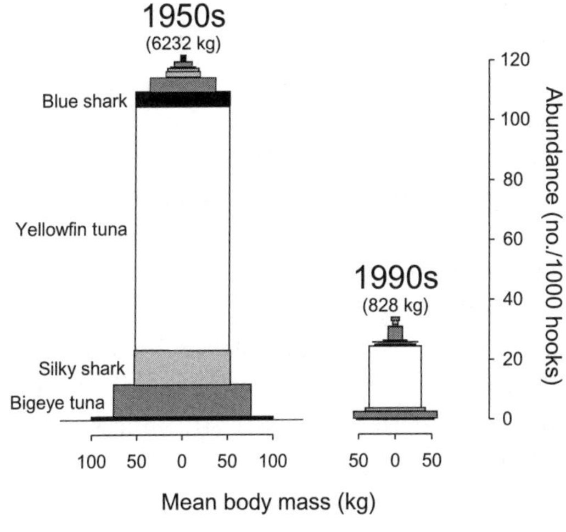

FIGURE 12.7 Changes in species abundance, mean body mass, and community composition in the tropical Pacific between the 1950s and the 1990s. The four most abundant species in the 1950s are labeled. Redrawn with permission from Ward and Myers (2005).

vey data from the 1950s and 1990s to detail changes in size composition, species abundance, and biomass of communities in the tropical Pacific targeted by longlining operations. Figure 12.7 shows simultaneous changes in structure across three biological dimensions that represent three hierarchical levels: body mass, species abundance, and community structure. The summed overall abundance of species captured per 1,000 hooks was drastically reduced, with large predators showing the greatest declines in abundance, along with sizeable decreases in mean body mass. There are concomitant changes in the community structure and relative abundance of species. The available evidence implicates fishing as the most likely cause of these changes (Ward and Myers 2005). This example shows the linkages between macroecological processes at multiple biological scales in exploited marine systems, as described in the introduction to this chapter.

While we have largely focused thus far in this section on the impacts of modern fishing, community shifts are not an entirely recent phenomenon. Historical overexploitation of sea otters (*Enhydra lutris*) as a food source has been indicted as a probable cause for the shift between two alternate stable-state community structures in the Aleutian islands (Simenstad, Estes, and Kenyon 1978), one dominated by macroalgae (and sea otters), the other by epibenthic herbivores such as sea urchins. When a substantial sea otter population exists, it keeps the sea urchins and other herbivorous epibenthic macroinvertebrates in check through predation, allowing the macroalgae to persist. When the otter populations are at meager levels, herbivorous invertebrates are present in much greater densities, and subsequently overgraze and virtually exclude macroalgae. Overexploitation of sea otters by aboriginal Aleuts may well have disrupted the system into this second state (Simenstad, Estes, and Kenyon 1978).

Indeed, we are starting to discover that large-scale removals and losses occurred throughout the coastal oceans prior to the twentieth century, resulting in vast anthropogenically induced changes (Jackson 2001). Analyses of historic and prehistoric patterns are important and useful tools, not only to piece together changes in community structure that occurred before the functional patterns visible in contemporary ecosystems, but also to predict the shifts that may occur with further exploitation. It is likely that further examples of trophic cascades caused by fisheries exploitation, both past and present, will be unearthed as increasing numbers of systems are targeted for study (Pinnegar et al. 2000). The importance of these structural shifts in communities is gaining international recognition, as evidenced by the fact that the trophic integrity of ecosystems is under consideration as one of a suite of global biodiversity indicators by the United Nations Environmental Programme (UNEP 2004). More information on these transitions is needed in order to gain a greater understanding of when and where they are likely to occur in the future.

Habitat Modification

Habitat modification is another side effect of exploitation that can indirectly affect the macroecology of target and non-target species. Although habitat damage can occur through destructive fishing practices such as dynamiting and cyanide fishing, here we concentrate on the effects of trawling and dredging. For instance, in Foveaux Strait, New Zealand, estimates derived from von-Bertalanffy growth models suggest that a reduction in habitat complexity and an increase in disturbance due to dredging from the oyster fishery have impeded the growth of juvenile blue cod (*Parapercis colias*) in the region (Carbines, Jiang, and Beentjes 2004). A comprehensive survey of the multifarious effects of the many different types of fishing on the marine habitat is beyond the scope of this chapter, but as an example, bottom-trawling or dredging-induced habitat degradation has been observed in regions including Ireland and Norway (Hall-Spencer, Allain, and Foss 2002), Georges Bank (Auster, et al. 1996; Collie, Escanero, and Valentine 1997), the Gulf of Maine (Auster et al. 1996), offshore California (Friedlander et al. 1999), and the North Sea (Jennings et al. 2001; Schratzberger and Jennings 2002; Schratzberger, Dinmore, and Jennings 2002). Benthic trawling is well known to scour the seafloor, weighty steel trawl doors gouging bottom habitat and leaving marks that are readily detectable with sidescan-sonar (Friedlander et al. 1999). Such damage, and its resultant effects on habitat complexity (Auster et al. 1996) and disturbance (Schratzberger, Dinmore, and Jennings 2002), often reduces species richness and biomass and can change community structure and species composition (e.g., Collie, Escanero, and Valentine 1997). Long-term study of bottom trawling in the North Sea (Jennings et al. 2001) has indicated that infaunal and epifaunal biomass showed significant decreases from trawling disturbance, but there appeared to be a minimal effect on the trophic structure of the community.

Experimental studies on the effects of trawling provide greatly differing estimates of mortality and destruction (Moran and Stephenson 2000), a disparity that may be attributable to differences in the nature of the gear that is used. Such results give hope that enforcement of fishing practices with reduced effects on habitat and non-target species can be used to mitigate the effects of exploitation. Monitoring and enforcement of fishing practices, however, remains problematic (Moran and Stephenson 2000; Tudela et al. 2005), particularly as the fishing practices that are most effective at maximizing catch of target species are often the most destructive.

Emerging Fisheries

Given the deleterious effects of exploitation on ecosystems and communities, there are urgent concerns about emerging deep-sea fisheries (Roberts 2002), especially on seamounts (Clark et al. 2006), some of which may be areas of high biodiversity and endemism (Richer de Forges, Koslow, and Poore 2000). Fisheries are often displaced onto these fragile habitats after shelf and slope fisheries have collapsed or had moratoriums imposed. Deepwater fish species are often long lived, late maturing, slow growing, and have low fecundity, characteristics that render them extremely vulnerable to the effects of overexploitation, especially as many species aggregate on seamounts (Koslow et al. 2000). These life-history features also result in long-lasting changes in community structure after fishing, with a correspondingly slow recovery time. There is a common "boom-and-bust" cycle for seamount fisheries, in which a highly productive resource is exploited, yielding high catches, and then collapses rapidly, at which point another seamount is targeted with the same effect (Clark et al. 2006). Such concerns have caused several governments to prepare legislation and carry out scientific surveys of these features within their waters.

Destabilization of Ecosystems

Changes in the biodiversity of ecosystems through the loss of species may also have additional unexpected and deleterious effects. Theoretical work has shown that complex food webs may maintain stability through many weak trophic interactions (McCann, Hastings, and Huxel 1998), or the fluctuating selection of trophic links (Kondoh 2003). Worm et al. (2006), using meta-analyses across a broad range of scales, found that the loss of biodiversity in marine systems had a destabilizing effect, resulting in increased resource collapse and decreased recovery potential, together with negative impacts on ocean ecosystem services.

Taken together, the range of studies outlined in this section indicate that multiple concurrent shifts in many aspects of biological communities can occur under many levels of exploitation pressure across a variety of marine environments. The evidence that is being accumulated could be incorporated into future management strategies, while more research and elucidation of these large-scale changes is also vital. These subjects are discussed in the final section.

Conclusions

The new tools of macroecology allow us to synthesize data, sift for patterns, and hypothesize processes. Recent studies have provided evidence for changes of breathtaking extent under exploitation (Baum et al. 2003; Worm and Myers 2003; Jennings and Blanchard 2004; Ward and Myers 2005; Worm 2005). The macroecological patterns of marine species, communities, and ecosystems are changing in fundamental ways. Although not all such changes may be immediately apparent, their long-term consequences can be profound. Among the patterns discussed in this chapter, shifts in life history, population ranges, species abundances and densities, and community composition have all been observed. The array of modifications and impacts that humans have caused in the oceans is impressively—or perhaps depressingly—large (Peterson and Estes 2001).

Achieving fisheries that are sustainable in the long term is an outcome that is typically sought by all stakeholders—fishermen, their communities, scientists, fisheries managers, and politicians. Yet the work that remains in order to inform attempts to mitigate or reverse our impacts is vast. There is very little research addressing how exploitation changes the relationships *between* macroecological variables in marine systems. We believe that more studies are needed to further interpret the linkages between commonly examined macroecological parameters (such as species richness, species range, body size, and abundance), and the effect of exploitation on these couplings (Fisher and Frank 2004; Jennings and Blanchard 2004). Historical studies are necessary to quantify pre-exploitation abundances, and meta-analyses are needed now more than ever to synthesize the vast quantities of data at regional, basin, and global scales (Christensen et al. 2003). The vulnerability of marine species to extinction is an area of current research and concern (Dulvy, Sadovy, and Reynolds 2003; Hutchings and Reynolds 2004; Myers and Ottensmeyer 2005), as it is now recognized that species could be at greater risk of extinction than previously thought (e.g., Roberts and Hawkins 1999). Predicting the spectrum of future scenarios for marine species is a research area in which work is needed to guide and optimize management strategies for sustainability (Myers and Worm 2005), as is the use of macroecological parameters as tools to assess the extent of changes wrought upon marine ecosystems (Fisher and Frank, 2004; Jennings and Blanchard 2004; Tittensor et al. 2007). Incomplete species inventories can also affect the description of macroecological patterns (Mora, Tittensor and Myers 2008), a particularly important issue for many marine habitats, given the sampling challenges.

Effective management can mitigate impacts and lead to recovery, while lack of action can lead to full-scale population collapse and extinction (Myers and Worm 2005). As discussed within this chapter, future management strategies need to take into consideration all of the following: potential effects of both genotypic and phenotypic changes in the life history of fishes (Conover and Munch 2002), multispecies interactions, community- and ecosystem-level dynamical shifts, the impacts of habitat alteration on the macroecology of marine assemblages, and the optimal utilization of tools such as marine reserves (Roberts, Hawkins, and Gell 2005). Superimposed on this challenge is the difficulty that each new generation of fisheries scientists experiences an altered set of initial conditions in the systems they observe. This can result in the "shifting baseline" effect, in which the new abundances that serve as the baseline for the system are radically different from those of the natural environment, having suffered from another generation's worth of anthropogenic disturbance and exploitation (Pauly 1995). Due to the enormous declines and shifts that have been shown to occur across a wide range of species, scales, and systems, fundamental changes in our approach to the management of fisheries are necessary if we are to implement sustainability.

Fishing appears to be both the earliest and greatest form of anthropogenic disturbance in many marine ecosystems (Jackson et al. 2001; Dulvy, Sadovy, and Reynolds 2003). Macroecological studies have shown that these disturbances often result in simultaneous shifts in dynamics at multiple scales. Given the rapidity of species collapse and ecosystem disturbance commonly observed under exploitation, it is important we recognize the time constraints under which we operate, in order to help guide us in our attempts to sustain fisheries and alleviate our long-term impacts on the macroecology of these vast environments.

ACKNOWLEDGMENTS

We wish to thank J. K. Baum, C. A. Ottensmeyer, R. M. Tittensor, and J. Witman for helpful and constructive comments. We acknowledge funding from the Sloan Census of Marine Life, the Pew Charitable Trust, and the Natural Sciences and Engineering Research Council of Canada.

REFERENCES

Auster, P. J., R. J. Malatesta, R. W. Langton, L. Watling, P. C. Valentine, C. L. S. Donaldson, E. W. Langton, A. N. Shepard, and I. G. Babb. 1996. The impacts of mobile fishing gear on seafloor habitats in the Gulf of Maine (Northwest Atlantic): Implications for conservation of fish populations. *Reviews in Fisheries Science* 4:185–202.

Barot, S., M. Heino, L. O'Brien, and U. Dieckmann. 2004. Long-term trend in the maturation reaction norm of two cod stocks. *Journal of Applied Ecology* 14:1257–71.

Barrett, J. H., A. M. Locker, and C. M. Roberts. 2004. The origins of intensive marine fishing in medieval Europe: The English evidence. *Proceedings of the Royal Society of London, B* 271:2417–21.

Baum, J. K., and R. A. Myers. 2004. Shifting baselines and the decline of pelagic sharks in the Gulf of Mexico. *Ecology Letters* 7:135–45.

Baum, J. K., R. A. Myers, D. G. Kehler, B. Worm, S. J. Harley, and P. A. Doherty. 2003. Collapse and conservation of shark populations in the Northwest Atlantic. *Science* 299:389–92.

Bianchi, G., H. Gislason, K. Graham, L. Hill, X. Jin, K. Koranteng, S. Manickchand-Heileman, I. Payá, K. Sainsbury, F. Sanchez, and K. Zwanenburg. 2000. Impact of fishing on size composition and diversity of demersal fish communities. *ICES Journal of Marine Science* 57:558–71.

Brown, J. H. 1984. On the relationship between abundance and distribution of species. *American Naturalist* 124:255–79.

Carbines, G., W. M. Jiang, and M. P. Beentjes. 2004. The impact of oyster dredging on the growth of blue cod, *Parapercis colias*, in Foveaux Strait, New Zealand. *Aquatic Conservation: Marine and Freshwater Ecosystems* 14:491–504.

Casey, J. M., and R. A. Myers. 1998. Near extinction of a large, widely distributed fish. *Science* 281:690–92.

Christensen, V., S. Guenette, J. J. Heymans, C. J. Walters, R. Watson, D. Zeller, and D. Pauly. 2003. Hundred-year decline of North Atlantic predatory fishes. *Fish and Fisheries.* 4:1–24.

Clark, M. R., D. Tittensor, A. D. Rogers, P. Brewin, T. Schlacher, A. Rowden, K. Stocks, and M. Consalvey. 2006. *Seamounts, deep-sea corals and fisheries: Vulnerability of deep-sea corals to fishing on seamounts beyond areas of national jurisdiction.* Cambridge, UK: UNEP-WCMC.

Collie, J. S., G. A. Escanero, and P. C. Valentine. 1997. Effects of bottom fishing on the benthic megafauna of Georges Bank. *Marine Ecology Progress Series* 155:159–72.

Conover, D. O., and S. B. Munch. 2002. Sustaining fisheries yields over evolutionary time scales. *Science* 297:94–96.

Cranfield, H. J., K. P. Michael, and I. J. Doonan. 1999. Changes in the distribution of epifaunal reefs and oysters during 130 years of dredging for oysters in Foveaux Strait, southern New Zealand. *Aquatic Conservation: Marine and Freshwater Ecosystems* 9:461–83.

Dayton, P. K., M. J. Tegner, P. B. Edwards, and K. L. Riser. 1998. Sliding baselines, ghosts, and reduced expectations in kelp forest communities. *Journal of Applied Ecology* 8:309–22.

Dulvy, N. K., R. P. Freckleton, and N. V. C. Polunin. 2004. Coral reef cascades and the indirect effects of predator removal by exploitation. *Ecology Letters* 7:410–16.

Dulvy, N. K., Y. Sadovy, and J. D. Reynolds. 2003. Extinction vulnerability in marine populations. *Fish and Fisheries* 4:25–64.

Engelhard, G. H., and M. Heino. 2004. Maturity changes in Norwegian spring-spawning herring *Clupea harengus*: Compensatory or evolutionary responses? *Marine Ecology Progress Series* 272:245–56.

Ernande, B., U. Dieckmann, and M. Heino. 2004. Adaptive changes in harvested populations: Plasticity and evolution of age and size at maturation. *Proceedings of the Royal Society of London, B* 271:415–23.

Estes, J. A., D. O. Duggins, and G. B. Rathbun. 1989. The ecology of extinctions in kelp forest communities. *Conservation Biology* 3:252–64.

Fisher, J. A. D., and K. T. Frank. 2004. Abundance-distribution relationships and conservation of exploited marine fishes. *Marine Ecology Progress Series* 279:201–13.

Frank, K. T., and D. Brickman. 2000. Allee effects and compensatory population dynamics within a stock complex. *Canadian Journal of Fisheries and Aquatic Sciences* 57:513–17.

Frank, K. T., B. Petrie, J. S. Choi, and W. C. Leggett. 2005. Trophic cascades in a formerly cod-dominated ecosystem. *Science* 308:1621–23.

Friedlander, A. M., G. W. Boehlert, M. E. Field, J. E. Mason, J. V. Gardner, and P. Dartnell. 1999. Sidescan-sonar mapping of benthic trawl marks on the shelf and slope off Eureka, California. *Fish Bulletin* 97:786–801.

Friedlander, A. M., and E. E. DeMartini. 2002. Contrasts in density, size, and biomass of reef fishes between the northwestern and the main Hawaiian islands: The effects of fishing down apex predators. *Marine Ecology Progress Series* 230:253–64.

Gaston, K. J., and T. M. Blackburn. 2000. *Pattern and process in macroecology*. Oxford, UK: Blackwell Science.

Goodwin, N. B., N. K. Dulvy, and J. D. Reynolds. 2005. Macroecology of live-bearing in fishes: Latitudinal and depth range comparisons with egg-laying relatives. *Oikos* 110:209–18.

Grift, R. E., A. D. Rijnsdorp, S. Barot, M. Heino, and U. Dieckmann. 2003. Fisheries-induced trends in reaction norms for maturation in North Sea plaice. *Marine Ecology Progress Series* 257:247–57.

Hall-Spencer, J., V. Allain, and J. H. Fosså. 2002. Trawling damage to Northeast Atlantic ancient coral reefs. *Proceedings of the Royal Society of London, B* 269:507–11.

Hauser, L., G. J. Adcock, P. J. Smith, J. H. B. Ramirez, and G. R. Carvalho. 2002. Loss of microsatellite diversity and low effective population size in an overexploited population of New Zealand snapper (*Pagrus auratus*). *Proceedings of the National Academy of Sciences* 99:11742–47.

Hughes, T. P., A. H. Baird, D. R. Bellwood, M. Card, S. R. Connolly, C. Folke, R. Grosberg, et al. 2003. Climate change, human impacts, and the resilience of coral reefs. *Science* 301:929–33.

Hutchings, J. A. 2000. Collapse and recovery of marine fishes. *Nature* 406:882–85.

———. 2004. Evolutionary biology: The cod that got away. *Nature* 428:899–900.

Hutchings, J. A., and J. K. Baum. 2005. Measuring marine fish biodiversity: Temporal changes in abundance, life history, and demography. *Philosophical Transactions of the Royal Society, B* 360:315–38.

Hutchings, J. A., and R. A. Myers. 1994. What can be learned from the collapse of a renewable resource? Atlantic cod, *Gadus morhua*, of Newfoundland and Labrador. *Canadian Journal of Fisheries and Aquatic Sciences* 51:2126–46.

Hutchings, J. A., and J. D. Reynolds. 2004. Marine fish population collapses: Consequences for recovery and extinction risk. *Bioscience* 54:297–309.

Hutchinson, W. F., C. van Oosterhout, S. I. Rogers, and G. R. Carvalho. 2003. Temporal analysis of archived samples indicates marked genetic changes in declining North Sea cod (*Gadus morhua*). *Proceedings of the Royal Society of London, B* 270:2125–32.

Jackson, J. B. C. 2001. What was natural in the coastal oceans? *Proceedings of the National Academy of Sciences, USA* 98:5411–18.

Jackson, J. B. C., M. X. Kirby, W. H. Berger, K. A. Bjorndal, L. W. Botsford, B. J. Bourque, R. H. Bradbury, et al. 2001. Historical overfishing and the recent collapse of coastal ecosystems. *Science* 293:629–38.

James, M. C., C. A. Ottensmeyer, and R. A. Myers. 2005. Identification of high-use habitat and threats to leatherback sea turtles in northern waters: New directions for conservation. *Ecology Letters* 8:195–201.

Jennings, S., and J. L. Blanchard. 2004. Fish abundance with no fishing: Predictions based on macroecological theory. *Journal of Animal Ecology* 73:632–42.

Jennings, S., J. K. Pinnegar, N. V. C. Polunin, and K. J. Warr. 2001. Impacts of trawling disturbance on the trophic structure of benthic invertebrate communities. *Marine Ecology Progress Series* 213:127–42.

Jennings, S., and N. V. C. Polunin. 1996. Effects of fishing effort and catch rate upon the structure and biomass of Fijian reef fish communities. *Journal of Applied Ecology* 33:400–12.

Kirby, M. X. 2004. Fishing down the coast: Historical expansion and collapse of oyster fisheries along continental margins. *Proceedings of the National Academy of Sciences, USA* 101:13096–99.

Kock, K.-H. 1992. *Antarctic fish and fisheries*. Cambridge: Cambridge University Press.

Kondoh, M. 2003. Foraging adaptation and the relationship between food-web complexity and stability. *Science* 299:1388–91.

Koslow, J. A., G. W. Boehlert, J. D. M. Gordon, R. L. Haedrich, P. Lorance, and N. Parin. 2000. Continental slope and deep-sea fisheries: Implications for a fragile ecosystem. *ICES Journal of Marine Science* 57:548–57.

Law, R. 2000. Fishing, selection, and phenotypic evolution. *ICES Journal of Marine Science* 57:659–68.

Lewison, R. L., L. B. Crowder, A. J. Read, and S. A. Freeman. 2004. Understanding impacts of fisheries bycatch on marine megafauna. *Trends in Ecology and Evolution* 19:598–604.

Lewison, R. L., S. A. Freeman, and L. B. Crowder. 2004. Quantifying the effects of fisheries on threatened species: The impact of pelagic longlines on loggerhead and leatherback sea turtles. *Ecology Letters* 7:221–31.

Liermann, M., and R. Hilborn. 1997. Depensation in fish stocks: A hierarchic Bayesian meta-analysis. *Canadian Journal of Fisheries and Aquatic Sciences* 54:1976–84.

Lotze, H. K., and I. Milewski. 2004. Two centuries of multiple human impacts and successive changes in a North Atlantic food web. *Journal of Applied Ecology* 14:1428–47.

McCann, K., A. Hastings, and G. R. Huxel. 1998. Weak trophic interactions and the balance of nature. *Nature* 395:794–98.

McGrath, P. T. 1911. *Newfoundland in 1911.* London: Whitehead, Morris, and Co.

Mora, C., Tittensor, D. P., and Myers, R. A. 2008. The completeness of taxonomic inventories for describing the global diversity and distribution of marine fishes. *Proceedings of the Royal Society of London, B* 275:149–55.

Moran, M. J., and P. C. Stephenson. 2000. Effects of otter trawling on macrobenthos and management of demersal scalefish fisheries on the continental shelf of north-western Australia. *ICES Journal of Marine Science* 57:510–16.

Munday, P. L., and G. P. Jones. 1998. The ecological implications of small body size among coral reef fishes. *Oceanography and Marine Biology* 36:373–411.

Musick, J. A. 1999. *Life in the slow lane: Ecology and conservation of long-lived marine animals.* Bethesda, MD: American Fisheries Society.

Myers, R. A., N. J. Barrowman, J. A. Hutchings, and A. A. Rosenberg. 1995. Population dynamics of exploited fish stocks at low population levels. *Science* 269:1106–8.

Myers, R. A., J. A. Hutchings, and N. J. Barrowman. 1997. Why do fish stocks collapse? The example of cod in Atlantic Canada. *Journal of Applied Ecology* 7:91–106.

Myers, R. A., and Ottensmeyer, C. A. 2005. Extinction risk in marine species. In *Marine conservation biology: the science of maintaining the sea's biodiversity*, ed. E. A. Norse and L. B. Crowder, 58–79. Washington, DC: Island Press.

Myers, R. A., and K. Stokes. 1989. Density dependent habitat utilization of groundfish and the improvement of research surveys. *International Council for the Exploration of the Sea Committee Meeting 1989/D* 15:1–11.

Myers, R. A., and B. Worm. 2003. Rapid worldwide depletion of predatory fish communities. *Nature* 423:280–83.

———. 2005. Extinction, survival or recovery of large predatory fishes. *Proceedings of the Royal Society of London, B* 360:13–20.

Olsen, E. M., M. Heino, G. R. Lilly, M. J. Morgan, J. Brattey, B. Ernande, and U. Dieckmann. 2004. Maturation trends indicative of rapid evolution preceded the collapse of northern cod. *Nature* 428:932–35.

Palumbi, S. R. 2001. Humans as the world's greatest evolutionary force. *Science* 293:1786–90.

Pandolfi, J. M., R. H. Bradbury, E. Sala, T. P. Hughes, K. A. Bjorndal, R. G. Cooke, D. McArdle, et al. 2003. Global trajectories of the long-term decline of coral reef ecosystems. *Science* 301:955–58.

Pauly, D. 1995. Anecdotes and the shifting baseline syndrome of fisheries. *Trends in Ecology and Evolution* 10:430.

Pauly, D., V. Christensen, J. Dalsgaard, R. Froese, and F. Torres. 1998. Fishing down marine food webs. *Science* 279:860–63.

Pauly, D., V. Christensen, S. Guenette, T. J. Pitcher, U. R. Sumaila, C. J. Walters, R. Watson, and D. Zeller. 2002. Towards sustainability in world fisheries. *Nature* 418:689–95.

Pauly, D., M. L. Palomares, R. Froese, P. Sa-a, M. Vakily, D. Preikshot, and S. Wallace. 2001. Fishing down Canadian aquatic food webs. *Canadian Journal of Fisheries and Aquatic Sciences* 58:51–62.

Peterson, C. H., and J. A. Estes. 2001. Conservation and management of marine communities. In *Marine community ecology*, ed. M. D. Bertness, S. D. Gaines, and M. E. Hay, 469–507. Sunderland, MA: Sinauer.

Pinnegar, J. K., N. V. C. Polunin, P. Francour, F. Badalamenti, R. Chemello, M. L. Harmelin-Vivien, B. Hereu, et al. 2000. Trophic cascades in benthic marine ecosystems: Lessons for fisheries and protected-area management. *Environmental Conservation* 27:179–200.

Poulson, N. A., E. E. Nielsen, M. H. Schierup, V. Loeschcke, and P. Grønkjær. 2006. Long-term stability and effective population size in North Sea and Baltic Sea cod (*Gadus morhua*). *Molecular Ecology* 15:321–31.

Richer de Forges, B., J. A. Koslow, and G. C. B. Poore. 2000. Diversity and endemism of the benthic seamount fauna in the southwest Pacific. *Nature* 405:944–47.

Ricker, W. E. 1958. Maximum sustained yields from fluctuating environments and mixed stocks. *Journal of the Fisheries Research Board of Canada* 15:991–1006.

Roberts, C. M. 2002. Deep impact: The rising toll of fishing in the deep sea. *Trends in Ecology and Evolution* 17:242–45.

Roberts, C. M., and J. P. Hawkins. 1999. Extinction risk in the sea. *Trends in Ecology and Evolution* 14:241–46.

Roberts, C. M., J. P. Hawkins, and F. R. Gell. 2005. The role of marine reserves in achieving sustainable fisheries. *Philosophical Transactions of the Royal Society, B* 360:123–32.

Roff, D. A. 2002. *Life history evolution*. Sunderland, MA: Sinauer.

Roman, J., and S. R. Palumbi. 2003. Whales before whaling in the North Atlantic. *Science* 301:508–10.

Ruzzante, D. E., C. T. Taggart, R. W. Doyle, and D. Cook. 2001. Stability in the historical pattern of genetic structure of Newfoundland cod (*Gadus morhua*) despite the catastrophic decline in population size from 1964 to 1994. *Conservation Genetics* 2:257–69.

Sadovy, Y., and W. L. Cheung. 2003. Near extinction of a highly fecund fish: The one that nearly got away. *Fish and Fisheries* 4:86–99.

Sandin, S. A., J. E. Smith, E. E. DeMartini, E. A. Dinsdale, S. D. Donner, A. M. Friedlander, T. Konotchick, et al. 2008. Baselines and degredation of coral reefs in the northern Line Islands. *PLoS One*. 3:e1548.

Schratzberger, M., T. A. Dinmore, and S. Jennings. 2002. Impacts of trawling on the diversity, biomass and structure of meiofauna assemblages. *Marine Biology* 140:83–93.

Schratzberger, M., and S. Jennings. 2002. Impacts of chronic trawling disturbance on meiofaunal communities. *Marine Biology* 141:991–1000.

Simenstad, C. A., J. A. Estes, and K. W. Kenyon. 1978. Aleuts, sea otters, and alternate stable-state communities. *Science* 200:403–11.

Smith, I. 2005. Retreat and resilience: Fur seals and human settlement in New Zealand. In *The exploitation and cultural importance of sea mammal*, ed. G. G. Monks, 6–18. Oxford: Oxbow Books.

Smith, K. F., and J. H. Brown. 2002. Patterns of diversity, depth range and body size among pelagic fishes along a gradient of depth. *Global Ecology and Biogeography* 11:313–22.

Spotila, J. R., R. D. Reina, A. C. Steyermark, P. T. Plotkin, and F. V. Paladino. 2000. Pacific leatherback turtles face extinction. *Nature* 405:529–30.

Springer, A. M., J. A. Estes, G. B. van Vliet, T. M. Williams, D. F. Doak, E. M. Danner, K. A. Forney, and B. Pfister. 2003. Sequential megafaunal collapse in the North Pacific Ocean: An ongoing legacy of industrial whaling? *Proceedings of the National Academy of Sciences, USA* 100:12223–28.

Steneck, R. S., M. H. Graham, B. J. Bourque, D. Corbett, J. M. Erlandson, J. A. Estes, and M. J. Tegner. 2002. Kelp forest ecosystems: Biodiversity, stability, resilience and future. *Environmental Conservation* 29:436–59.

Stockwell, C. A., A. P. Hendry, and M. T. Kinnison. 2003. Contemporary evolution meets conservation biology. *Trends in Ecology and Evolution* 18:94–101.

Stokes, K., and R. Law. 2000. Fishing as an evolutionary force. *Marine Ecology Progress Series* 208:307–9.

Tasker, M. L., M. C. J. (Kees)Camphuysen, J. Cooper, S. Garthe, W. A. Montevecchi, and S. M. Blaber. 2000. The impacts of fishing on marine birds. *ICES Journal of Marine Science* 57:531–47.

Tegner, M. J., and P. K. Dayton. 1999. Ecosystem effects of fishing. *Trends in Ecology and Evolution* 14:261–62.

———. 2000. Ecosystem effects of fishing in kelp forest communities. *ICES Journal of Marine Science* 57:579–89.

Thompson, H. 1943. A biological and economic study of cod (*Gadus callarias, L.*) in the Newfoundland area, including Labrador. *Newfoundland Department of Natural Resources Fisheries Bull.* 14:1–160.

Thompson, J. N. 1998. Rapid evolution as an ecological process. *Trends in Ecology and Evolution* 13:329–32.

Tittensor, D. P., Micheli, F., Nyström, M., and Worm, B. 2007. Human impacts on the species-area relationship in reef fish assemblages. *Ecology Letters* 10:760–72.

Tudela, S., A. Kai Kai, F. Maynou, M. El Andalossi, and P. Guglielmi. 2005. Driftnet fishing and biodiversity conservation: The case study of the large-scale Moroccan driftnet fleet operating in the Alboran Sea (SW Mediterranean). *Biological Conservation* 121:65–78.

UNEP (United Nations Environmental Programme). 2004. Provisional global indicators for assessing progress towards the 2010 biodiversity target. Nairobi, Kenya: CBD COP Document UNEP/CBD/COP/7/INF/33.

Walsh, M. R., S. B. Munch, S. Chiba, and D. O. Conover. 2006. Maladaptive changes in multiple traits caused by fishing: Impediments to population recovery. *Ecology Letters* 9:142–48.

Ward, P., and R. A. Myers. 2005. Shifts in open-ocean fish communities coinciding with the commencement of commercial fishing. *Ecology* 86:835–847.

Witman, J. D., and K. P. Sebens. 1992. Regional variation in fish predation intensity: A historical perspective in the Gulf of Maine. *Oecologia* 90:305–15.

Worm, B., H. K. Lotze, and R. A. Myers. 2005. Ecosystem effects of fishing and whaling in the North Pacific and Atlantic Ocean. In *Whales, whaling and ocean ecosystems,* ed. J. A. Estes, R. L. Brownell, D. P. DeMaster, D. F. Doak, and T. M. Williams, 333–41. Berkeley, CA: University of California Press.

Worm, B., M. Sandow, A. Oschlies, H. K. Lotze, and R. A. Myers. 2005. Global patterns of predator diversity in the open oceans. *Science* 309:1365–69.

Worm, B., E. B. Barber, N. Beaumont, J. E. Duffy, C. Folke, B. S. Halpern, J. B. C. Jackson, et al. 2006. Impacts of biodiversity loss on ocean ecosystem services. *Science* 314:787–90.

Worm, B., and R. A. Myers. 2003. Meta-analysis of cod-shrimp interactions reveals top-down control in oceanic food webs. *Ecology* 84:162–73.

PART THREE
EXPERIMENTAL APPROACHES TO MARINE MACROECOLOGY

CHAPTER THIRTEEN

EXPERIMENTAL MARINE MACROECOLOGY: PROGRESS AND PROSPECTS

JON D. WITMAN AND KAUSTUV ROY

As originally defined, the goal of macroecology is to gain insight into large-scale ecological phenomena mainly through rigorous statistical analyses of information on species abundance, diversity, body size, and range size (Brown and Maurer 1989; Brown 1995). Over the last decade or so, such analyses have revealed a number of large-scale trends that hold across taxa and across the land-sea boundary. For example, the allometric scaling relationship between body size and abundance has been quantified in a wide variety of groups, from higher plants to marine invertebrates and phytoplankton, although debate continues over whether there is a universal exponent (Marquet, Naverrete, and Castilla 1990; Enquist and Niklas 2001; Belgrano et al. 2002; Ackerman and Bellwood 2003; Ackerman, Bellwood, and Brown 2004; Russo, Robinson, and Terborgh 2003; Li, this volume). Similarly, species diversity of a variety of taxa, from plants and terrestrial vertebrates to corals and marine mollusks, has been shown to scale positively with available energy (Roy et al. 1998; Hawkins et al. 2003; Currie et al. 2004; Witman et al. 2008). Such statistical trends are being increasingly documented for both marine and terrestrial organisms. While the trends themselves are robust and some of them may even be universal, the processes that generate such trends are still very poorly understood and the subject of considerable debate. For example, mechanistic explanations of the species-energy relationship are seldom borne out by empirical evidence (Srivastava and Lawton 1998; Roy et al. 1998; Gaston 2000; Currie et al. 2004; Hawkins et al. 2003, 2007) and whether the observed scaling relationships between body size and abundance can be explained by the energetic equivalence rule remains a matter of debate (e.g., Blackburn et al. 1993; Marquet et al. 2003; Russo, Robinson, and Terborgh 2003). Part of the problem here is that we still lack process-based models for macroecological patterns, although recent advances in metabolic theory of ecology (Brown et al. 2004) and the development of other models (e.g., Connolly, this volume) are starting to fill this void. But even when process-based models are available, testing those models can be challenging, especially over large spatial and temporal scales. Where such models make specific predictions regarding scaling relationships, they are generally tested using statistical analyses of relevant empirical data (Allen, Brown, and Gillooly 2002; Hunt, Cronin, and Roy 2005). While such tests are the only option in some cases (e.g., for historical data, Hunt, Cronin, and Roy 2005), they tend to lack controls and as such are confirmatory tests. An alternative would be to use an experimental approach that would improve our ability to identify the mechanisms underlying macroecological patterns. It is certainly true that the spatial and temporal scale of macroecological analyses would preclude complete experimental tests in many cases, and large-scale manipulative experiments are not feasible in many marine habitats (e.g., the deep sea) due to logistical difficulties. However, with technological advances, it is now possible to replicate experiments over considerable spatial scales and we suggest that experimental macroecology would not only complement the statistical approach, but experimental tests in conjunction with statistical analyses of observational data are the best way to resolve some of the long-standing debates about the processes underlying macroecological relationships.

Scaling

Scaling has long been recognized as a central issue in ecological investigations (Steele 1978; Dayton and Tegner 1984; Wiens 1986; Levin 1992). For example, Wiens (1989) wrote "If we study a system at an inappropriate scale, we may not detect its actual dynamics but may instead identify patterns that are artifacts of scale" (390). The problem in experimental ecology is, of course, that ecological variables are manipulated mostly in small areas for short periods of time, while the major issues challenging the predictive abilities of ecologists, such as climate change and anthropogenic modification of nutrient cycling and food webs, impact large geographic regions. Moreover, environmental problems are caused by the propagation of effects across scales

(Ricklefs 1990; Schneider 2001). The need to understand how to extrapolate or "scale-up" the results of small-scale studies to larger scales is still vitally important (Lawton 2000) but remains unresolved despite substantial work on the topic of ecological scaling in the last twenty years (Schneider 2001), which is reflected in several volumes (Schneider 1994; Petersen and Parker 1998; Gardner et al. 2001), workshops (Thrush et al. 1997) and numerous papers. Central to this body of work is the concept that scaling represents minimally two components; grain and extent (Wiens 1989; Schneider 1994, 2003). The grain of a study refers to the minimum resolvable area (e.g., the area of a plot) or time period. Changing the spatial grain would extend the size of the area over which a phenomenon is studied. The extent of a study or phenomenon refers to the spatial or temporal range. Increasing the spatial extent of a study would extend the generality of the results to different places (Wiens 1989; Thrush et al. 1997). Macroecologists are concerned with both aspects of scale, although most of the emphasis on scaling in descriptive macroecology has been placed on increasing the extent of the results (Brown 1995).

Descriptive versus Experimental Marine Ecology: A Temporal Perspective

Marine ecologists have been interested in large-scale ecological patterns long before the term macroecology was coined and have used both observational and experimental approaches to study spatial variation in species diversity and community structure. For example, Gunnar Thorson was among the first to use the comparative observational approach in the ocean to advance ideas about geographic variation in reproductive mode of marine invertebrates, noting that direct developing invertebrates were more common at high latitudes than those with planktonic dispersal (Thorson 1950). He also made observations about geographic variation in the community structure of tropical versus temperate "parallel" bottom communities (Thorson 1957). It was also during the 1950s that some of the first quantitative studies were conducted at regional spatial scales, on barnacle distribution patterns along rocky intertidal shores of southern England (Southward et al. 2005). On larger spatial scales, studies such as Stephenson and Stephenson's (1972) comprehensive, qualitative comparison of global patterns of rocky intertidal zonation led to a better understanding of geographic variation in marine communities. What we now call marine macroecology has also been an element of fisheries biology from the outset (Leichter and Witman, this volume), as it involved establishing and interpreting patterns of recruitment, body size, and condition, and their correlation with ocean currents and productivity on geographic scales. All of this early marine research was fundamentally important to marine macroecology, as it established major patterns of geographic variation, but it wasn't as focused on testing and constructing theory as recent efforts.

In contrast, for almost half a century now, marine ecologists have tested process-based hypotheses by adding experimental manipulations to the comparative observational approach (Connell 1961; Paine 1966; Dayton 1971; Menge 1976; Underwood 1978; Sanford and Bertness, this volume). This observational-experimental approach, so popular in marine ecology today, is obviously an extremely powerful tool, but the demanding logistical constraints of conducting manipulative experiments initially restricted their application to local spatial scales. By 1981, the comparative observational experimental approach was being applied across large geographic distances to examine tropical-temperate differences in predation and intertidal community structure (e.g., Bertness, Garrity, and Levings 1981; Menge and Lubchenco 1981), although as we discuss in the following, the spatial extent and grain of most such experiments still lag far behind those of descriptive studies. Another major constraint here is the difficulty of working in deeper parts of the world ocean; the vast majority of manipulative experiments are still concentrated in the easily accessible intertidal habitats, with a few studies in the subtidal and even fewer in deeper waters.

The emphasis on larval dispersal and recruitment as a key process driving local patterns of intertidal population structure (Gaines and Roughgarden 1985; Underwood and Denley 1984; Gaines et al., this volume) was a turning point in the development of marine macroecology, as it forced marine ecologists to think beyond the local spatial scale of their immediate study site and increase the extent of research to encompass the influence of regional processes (Witman and Dayton 2001). Evaluating the degree to which local marine populations are open to extrinsic propagule (larvae, asexual fragments, seeds) supply versus "closed," or dominated by small-scale, within-site supply remains an active area of study today (e.g., Swearer et al. 1999; Hellberg et al. 2002). Similarly, Dayton and Tegner (1984) drew attention to the overriding impact that oceanographic processes have on local marine assemblages and to the importance of scale. Severe ENSO events (Glynn 1988) also prompted the development of regional-scale ecological investigations. The rise of regional scale marine ecology was also spurred on by efforts to test Terborgh and Faaborg's (1980) ideas about the influence of the regional species pool on local species richness. Rigorous tests of this theory based on standardized sampling have recently shown a large influence of the regional species pool on local diversity (Karlson, Cornell, and Hughes

2004; Witman, Etter, and Smith 2004), strongly suggesting that local marine diversity is influenced by the interplay of regional and local processes (Russell et al. 2006).

To trace the historical development of what can now be called marine macroecology, we surveyed papers on marine population and community ecology published in *Ecology* and *Ecological Monographs* from 1960 to 2004. This enabled us to assess temporal trends in the spatial scale (extent) and the degree of replication of marine ecological research. We classified each paper as descriptive or experimental. Descriptive papers were those that solely quantified patterns. Some papers contained a small amount of descriptive data to set the stage for experimental manipulations. In these cases, less than 25 percent of the results (figures and tables) were descriptive, so these papers were classified as experimental.

Figure 13.1 shows trends in the spatial scale of descriptive and experimental investigations of marine benthic communities from 1960 to 2004. The spatial extent of descriptive papers ranged from 0.5 to nearly 25,000 km maximum distance between study sites. Using Mittlebach et al.'s (2001) definitions of spatial scale with "local" as 0 to 20 km, "landscape" from 20 to 200 km, "regional" from 200 to 4,000 km, and "continental-global" as greater than 4,000 km, it is clear that descriptive studies were conducted at landscape and even larger regional spatial scales right from the start of the survey period in the early 1960s (fig. 13.1, panel A). They were performed on continental-global scales by the mid 1970s, with the density of studies performed at the largest spatial extent peaking during the 1970s. Subsequently, there has been a general decline in the frequency and spatial scale of marine studies based solely on descriptive data in these two journals. Two exceptions are Findley and Findley's (2001) study of the relation between regional and local species richness of butterfly fish, which sampled eighteen regions around the globe, and Estes and Duggins' (1995) test of the generality of a sea otter-dominated trophic cascade along approximately 3,700 km of Pacific Northwest coast. The declining trend in the frequency of large scale-descriptive studies from 1980 to 2004, shown in figure 13.1, panel A is likely an artifact of the journals we chose to focus on. Our choice of *Ecology* and *Ecological Monographs* was based on the fact that most of the early work in marine ecology appeared in these journals.

A comparison of figure 13.1, panels A and B makes an important distinction; that while descriptive studies started out at large scales, experimental studies began at small local spatial scales (extent) and have shown a dramatic increase in spatial scale ever since. By 1976 to 1980, experimental studies of predation, herbivory, and competition in New England rocky

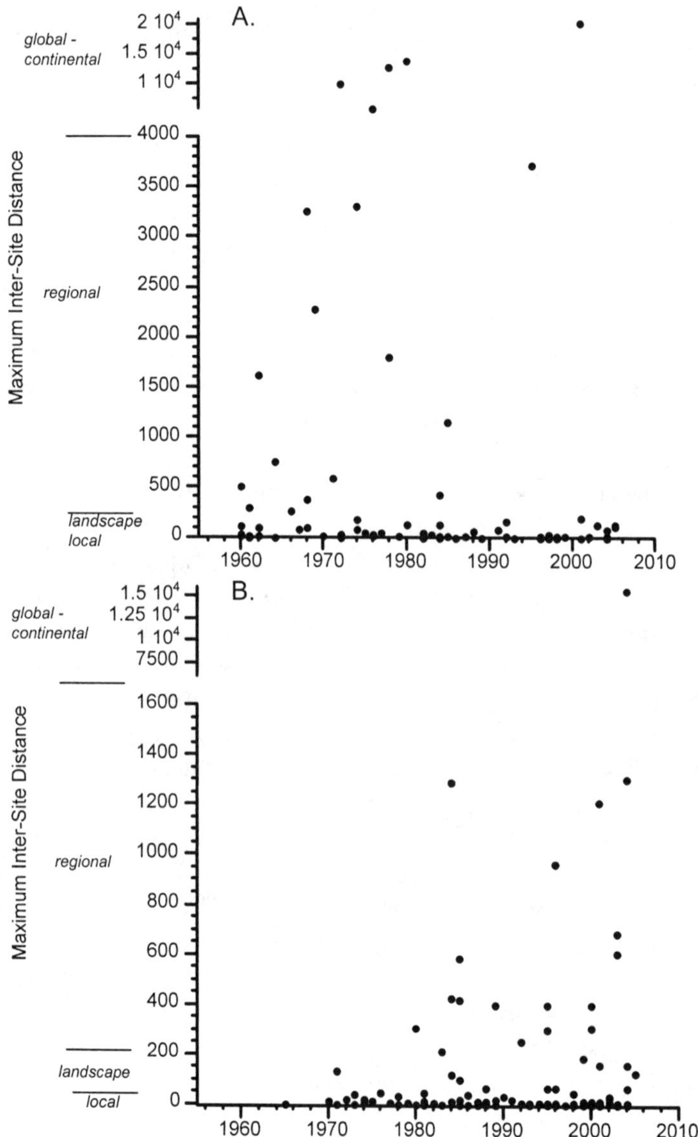

FIGURE 13.1 Trends in spatial scale, represented by the maximum distance between study sites, in descriptive (A) and experimental (B) studies published in *Ecology and Ecological Monographs* from 1960 to 2004, which is a measure of spatial extent (Wiens 1989). Scale classifications are: local (0 to 20 km), landscape (from 20 to 200 km), regional (200 to 4,000 km), and continental–global (>4,000 km). Note that descriptive studies have been conducted at large spatial scales of the landscape and region since 1960 to 1970 (A), while investigations employing experimental manipulations began at local scales and expanded (B).

intertidal communities were conducted by replicating small experimental units at sites across a regional spatial scale (Menge 1976; Lubchenco 1980). Regional scale experimentation became relatively common from the mid-1980s on (figure 13.1, panel B). Thus, the field of experimental marine macroecology arguably dates back to the mid-1980s. The "macro" scale of these experiments relates to their large spatial extent, not to the grain of investigation as the size of the areas manipulated by ecologists remains small (generally less than 2.0 m^2; Karieva and Andersen 1988) although there are a few exceptions (e.g., Thrush et al. 1997), particularly in the case of mobile predators (Englund 1997; Ellis et al. 2007).

Experimental Macroecology: Prospects

Experimental macroecology could represent a complementary approach to the traditional descriptive macroecology outlined by Brown and colleagues (Brown 1995; Brown and Maurer 1989). It considers simple manipulative experiments replicated over large spatial scales as the units of analysis. Both descriptive and experimental macroecology achieve the same goal of fostering a better understanding of processes governing the distribution, abundance, and diversity of species on large spatial scales, albeit by different means. For example, a recent large-scale study of intertidal limpet grazing from the British Isles to southern Portugal demonstrates the potential of experimental macroecology to reveal latitudinal variation in biological processes (Coleman et al. 2006). As mentioned previously, experimental macroecology can potentially test hypotheses about the processes underlying large-scale patterns more directly than descriptive macroecology, but it is more limited in application, since experimental manipulations on regional or global scales are either logistically difficult or even impossible for many systems. Also, historical processes that are increasingly being recognized as important determinants of macroecological patterns (Webb et al. 2002; Currie et al. 2004; Jablonski, Roy, and Valentine 2006; Mittlebach et al. 2007; Clarke, this volume; Roy and Witman, this volume) are obviously beyond the scope of experiments.

A conceptual model of macroecology illustrates the two main approaches to understanding the influence of large-scale processes (fig. 13.2). One approach is to describe patterns at large spatial scales, and since the ability to generalize increases with spatial scale, that is, grain or extent, (solid diagonal line increasing from left to right in figure 13.2), this is the attraction of descriptive macroecology. The limitation of the descriptive approach is that one has to infer causal processes driving the patterns. The other approach

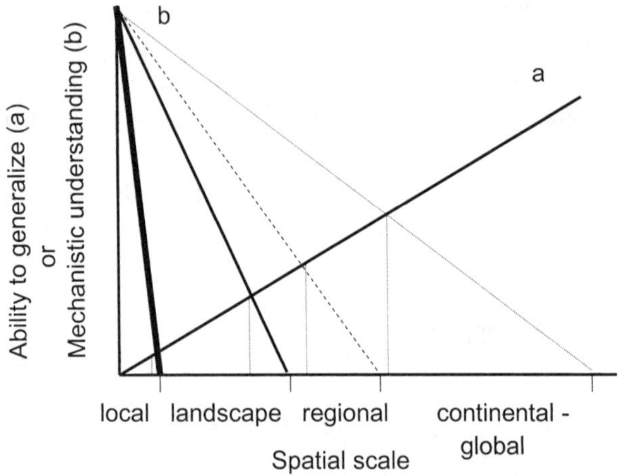

FIGURE 13.2 Conceptual model of macroecology showing how the ability to generalize and the level of mechanistic understanding varies with spatial scale. The ability to generalize (A) increases with spatial scale as more populations and communities are sampled. This is the rationale for, and attraction of, descriptive macroecology. In contrast, the level of mechanistic understanding (B) gleaned from experimentation is a decreasing function of spatial scale, due to the logistic difficulty of conducting manipulative experiments at larger scales. Line thickness is proportional to the amount of experimentation conducted at local-global spatial scales. The majority of mechanistic understanding in ecology is based on experiments performed on local spatial scales, as indicated by the thick black line. Progress is being made in conducting experiments at increasingly larger spatial scales, but there aren't many manipulative efforts at regional-global scales. The intersection of the generalization line with the mechanistic understanding line (marked by vertical dashed lines) represents process-based generalization about pattern. A major objective of macroecology is to make process-based generalizations about large-scale patterns. Performing experimental manipulations at spatial scales larger than local ones is the domain of experimental macroecology.

is to replicate experiments at large spatial scales (extent). However, the level of mechanistic insight gleaned from experimentation necessarily decreases with spatial extent simply due to the logistical difficulty of performing experimental manipulations at the largest scales. But as shown by the dashed lines of figure 13.2 and in figure 13.1, panel B, progress in increasing the spatial scale of experimentation in marine ecology is occurring. The intersection of the solid and dashed lines in figure 13.2 represents process-based generalization about pattern, one of the strongest inferences in ecology (Platt 1964), and a major goal of marine macroecology is to make these generalizations across large scales (Brown 1995; Gaston and Blackburn 2000; Lawton 2000). Clearly, there is a role for both descriptive and experimental approaches in macroecology, because while experimentation may provide a faster route to causation, only a few factors can be manipulated at one time and there are multivariate causes of ecological patterns. Also, restricting experimental manipulations to small areas may not be the most realistic simulation of ecological processes (Levin 1992; Lawton 2000). One way to overcome some of these constraints is to increase the spatial extent by replicating relatively small experimental units across regional scales (Dethier and Duggins 1988; Witman and Sebens 1992; Menge et al. 1999; Jenkins et al. 2001; Sanford et al. 2003; Coleman et al. 2006) or across multiple regions (Menge et al. 2002, Connell and Irving, this volume). Increasing the grain size of experiments to big areas is also needed to simulate and capture the full range of direct and indirect trophic interactions that might be altered by climatic change and/or human intervention. Although both approaches are needed, progress in macroecology—in terms of making process-based generalizations across large scales—is probably more readily achieved by increasing the extent of experiments rather than by increasing their grain.

Figure 13.3 shows trends in site replication of descriptive and experimental marine benthic papers published in *Ecology* and *Ecological Monographs* from 1960 to 2004. Although 230 experimentally based papers qualified for the analysis shown in figure 13.3, panel B, nearly half (47.4 percent) of them were conducted at one site only and were omitted from the graph. This underscores an important fact: that despite the trend of increasing regional-scale experimentation among the studies that *are* replicated (fig. 13.1, panel A) surprisingly large numbers of manipulative studies are still not replicated at more than one site. Thus, we have a long way to go before process-based ecological generalizations are available across large geographic distances and for large areas of the marine environment. The site replication of descriptive studies was higher than experimental ones, with only 25.2 percent of 103 descriptive papers (versus 47.4 percent experimental) lacking site replication (not shown in fig. 13.3). Descriptive studies showed little overall trend in the number of study sites utilized over time. Site replication was high (> ten sites) in the early 1960s, reaching a peak of 400 sites in Stephenson, Williams, and Lance's (1970) study at a landscape spatial scale in Moreton Bay, Australia (Stephenson, Williams, and Lance 1970). Site replication has remained high, with many descriptive papers based on the analysis of patterns at 10 to twenty sites after 2000. In stark contrast to the lack of temporal variation in site replication in descriptive studies, the amount of replication in experimental investigations has increased since the mid-1960s, with three marine studies based on > ten sites by the late 1970s, expanding to the maximum of eight papers in 2000–2004, where research was conducted at more than ten sites (fig. 13.3, panel B).

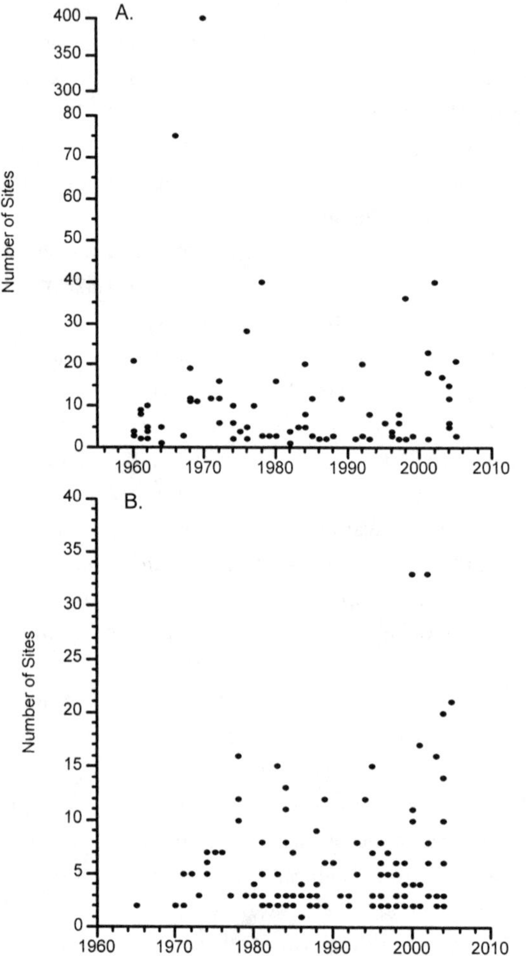

FIGURE 13.3 Variation in site replication in descriptive (A) and experimental (B) investigations published in *Ecology and Ecological Monographs* from 1960 to 2004. There is little temporal trend in site replication of descriptive studies (B). Almost half (47.4 percent) of the papers published that used experimental manipulations didn't have any site replication, and so they weren't plotted. Of those that were replicated, the amount of replication has increased over time.

Progress in identifying patterns and elucidating the mechanisms driving them could be made in marine macroecology by designing more scale-sensitive experiments, where the grain or extent of the experiment bracket some of the scales of the process under investigation (Englund 1997; Thrush et al. 1997; Smith and Witman 1999). This obviously requires prior information and is not always practical, as the operational scale of ecological processes is often unknown (Thrush et al. 1997). However, technological advances in the measurement of environmental variables has produced extensive data for choosing the relevant scales for the experimental analyses of physical forcing of biological interactions, distribution, and diversity across the largest spatial scales. Explicitly including a hierarchy of scales in experimental manipulations would formally identify the scale dependence of pattern and process. It would also increase the chance of identifying nonlinearities, such as thresholds, where the form of the relationship between ecological variables and spatial scale changes, such as in relations between patch size and soil nutrient concentration (Ludwig, Wiens, and Tongway 2000), productivity and diversity at different scales of investigation (Whittaker, Willis, and Field 2001), and in self organizing systems (Van de Koppel et al. 2006). Little is known about thresholds in scaling relationships, especially in the marine environment, yet they are fundamental to understanding how to extrapolate results from smaller to larger spatial scales, as thresholds represent areas where predictability breaks down (Wiens 2001). Spatial autocorrelation is another method to detect scale dependence in ecological data (Legendre 1993; Tomita, Hirabuki, and Seiwa 2002) that is generally underutilized in macroecology. Finally, since more than one environmental or ecological factor drives local pattern across large spatial scales, macroecologists in particular should be attentive to confounding factors in large-scale correlative studies such as the covarying effects of productivity and salinity on local species richness (Witman et al. 2008) and in experimental macroecology, where the reductionist nature of manipulations constrains the number of treatments possible.

Taken together, all of this suggests that manipulative experiments can play an important, yet somewhat restricted role in macroecological studies. For logistical and ethical reasons the spatial scale (extent and grain) over which such experiments can be conducted will always be much smaller than that of descriptive studies. For example, even if experimental manipulation of temperature or productivity levels over whole regions were possible, it would undoubtedly be unethical. And as discussed earlier, historical processes, a critical component of macroecology, are beyond the scope of experimental manipulations. On the other hand, experimental manipulations remain the only direct means of testing process-based hypotheses, and may be the only way to successfully resolve some long-standing debates about processes driving macroecological patterns (e.g., the role of productivity in regulating species richness). We think that small-scale manipulative experiments replicated over large spatial scales (e.g., Connell and Irving, this volume) are particularly useful in this context and the results of such studies should provide insights into macroecological processes that complement the inferences derived from statistical analyses. Ultimately, a combination of such large-scale experiments and statistical analyses of observational data is likely to be the key to understanding the processes driving the trends macroecologists love to study.

REFERENCES

Ackerman, J. L., and D. R. Bellwood. 2003. The contribution of small individuals to density-body size relationships. *Oecologia* 136:137–40.

Ackerman, J. L., D. R. Bellwood, and J. H. Brown. 2004. The contribution of small individuals to density-body size relationships: Examination of energetic equivalence in reef fishes. *Oecologia* 139:568–71.

Allen, A. P., J. H. Brown, and J. F. Gillooly. 2002. Global biodiversity, biochemical kinetics, and the energetic-equivalence rule. *Science* 297:1545–48.

Belgrano, A., A. P. Allen, B. J. Enquist, and J. F. Gillooly. 2002. Allometric scaling of maximum population density: A common rule for marine phytoplankton and terrestrial plants. *Ecology Letters* 5:611–13.

Bertness, M. D., S. D. Garrity, and S. C. Levings. 1981. Predation pressure and gastropod foraging: A tropical-temperate comparison. *Evolution* 35:995–1007.

Blackburn, T. M., V. K. Brown, B. M. Doube, J. J. D. Greenwood, J. H. Lawton, and N. E. Stork. 1993. The relationship between abundance and body size in natural animal assemblages. *Journal of Animal Ecology* 62:519–28.

Blackburn, T. M., and K. J. Gaston. 1998. Some methodological issues in macroecology. *American Naturalist* 151:68–83.

Brown, J. H. 1995. *Macroecology.* Chicago: University of Chicago Press.

Brown, J. H., J. F. Gillooly, A. P. Allen, V. M. Savage, and G. B. West. 2004. Toward a metabolic theory of ecology. *Ecology* 85:1771–89.

Brown, J. H., and B. A. Maurer. 1989. Macroecology: The division of food and space among species on continents. *Science* 243:1145–50.

Carpenter, S. R., S. W. Chisholm, C. J. Krebs, D. W. Schindler, and R. F. Wright. 1995. Ecosystem experiments. *Science* 269:324–27.

Coleman, R. A., A. J. Underwood, L. Benedetti-Cecchi, P. Aberg, F. Arenas, J. Arrontes, J. Castro, et al. 2006. A continental scale evaluation of the role of limpet grazing on rocky shores. *Oecologia* 147:556–64.

Connell, J. H. 1961. The influence of interspecific competition and other factors on the distribution of the barnacle *Chthamalus stellatus. Ecology* 42:710–23.

Currie, D. J., G. G. Mittelbach, H. V. Cornell, R. Field, J.-F. Guégan, B. A. Hawkins, D. M. Kaufman, et al. 2004. Predictions and tests of climate-based hypotheses of broad-scale variation in taxonomic richness. *Ecology Letters* 7:1121–34.

Dayton, P. K. 1971. Competition, disturbance, and community organization: The provision and subsequent utilization of space in a rocky intertidal community. *Ecological Monographs* 41:351–89.

Dayton, P. K., and M. J. Tegner. 1984. The importance of scale in community ecology: A kelp forest example with terrestrial analogs. In *A new ecology: Novel approaches to interactive systems,* ed. P. W. Price, C. N. Slobodchikoff, and W. S. Gaud, 457–81. New York: Wiley.

Dethier, M. N., and D. O. Duggins. 1988. Variation in strong interactions in the intertidal zone along a geographic gradient: A Washington-Alaska comparison. *Marine Ecology Progress Series* 50:97–105.

Ellis, J .C., M. J. Shulman, M. Wood, J. D. Witman, and S. Lozyniak. 2007. Regulation of intertidal and shallow subtidal food webs by avian predators on New England rocky shores. *Ecology* 88:853–63.

Englund, G. 1997. The importance of spatial scale and prey movements in predator caging experiments. *Ecology* 78:2316–25.

Enquist, B. J., and K. J. Niklas. 2001. Invariant scaling relations across tree-dominated communities. *Nature* 410:655–60.

Estes, J. A., and D. O. Duggins. 1995. Sea otters and kelp forests in Alaska: Generality and variation in a community ecological paradigm. *Ecological Monographs* 65:75–100.

Findley, J. T., and M. T. Findley. 2001. Global, regional and local patterns in species richness and abundance of butterflyfishes. *Ecological Monographs* 71:69–91.

Gaines, S. D., and J. Roughgarden. 1985. Larval settlement rate: A leading determinant in the structure of an ecological community of the rocky intertidal zone. *Proceedings of the National Academy of Sciences USA* 82:3707–11.

Gardner, R. H., W. M. Kemp, V. S. Kennedy, and J. E. Petersen, eds. 2001 *Scaling relations in experimental ecology.* New York: Columbia University Press.

Gaston, K. J. 2000. Global patterns in biodiversity. *Nature* 405:220–27.

Gaston, K. J., and T. M. Blackburn. 2000. Pattern and process in macroecology. Malden, MA: Blackwell Science.

Glynn, P. W. 1988. El Nino-Southern Oscillation 1982–1983: Nearshore population, community, and ecosystem responses. *Ann. Rev. Ecol. Syst.* 19:309–45.

Hawkins, B. A., R. Field, H. V. Cornell, D. J. Currie, J.-F. Guégan, D. M. Kaufman, J. T. Kerr, et al. 2003. Energy, water and broad-scale geographic patterns of species richness. *Ecology* 84:3105–17.

Hawkins, B. A., F. S. Albuquerque, M. B. Araújo, J. Beck, L. M. Bini, F. J. Cabrero-Sañudo, I. Castro-Parga, et al. 2007. A global evaluation of metabolic theory as an explanation for terrestrial species richness gradients. *Ecology* 88:1877–88.

Hellberg, M. E., R. S. Burton, J. E. Neigel, and S. R. Palumbi. 2002. Genetic assessment of connectivity among marine populations. *Bulletin of Marine Science* 70 (Suppl): 273–90.

Hunt, G., T. M. Cronin, and K. Roy. 2005. Species-energy relationship in the deep sea: A test using the Quaternary fossil record. *Ecology Letters* 8:739–47.

Huston, M. A. 1999. Local processes and regional patterns: Appropriate scales for understanding variation in the diversity of plants and animals. *Oikos* 86:393–401.

Jablonski, D., K. Roy, and J. W. Valentine. 2006. Out of the tropics: Evolutionary dynamics of the latitudinal diversity gradient. *Science.* 314:102–6.

Jenkins, S. R., F. Arenas, J. Arrontes, J. Bussell, J. Castro, R. A. Coleman, S. J. Hawkins, et al. 2001. European-scale analysis of seasonal variability in limpet grazing activity and microalgal abundance. *Marine Ecology Progress Series* 211:193–203.

Karieva, P., and M. Andersen. 1988. Spatial aspects of species interactions: The wedding of models and experiments. In *Community ecology,* ed. A. Hastings, 35–50. New York: Springer-Verlag.

Karlson, R. H., H. V. Cornell, and T. P. Hughes. 2004. Coral communities are regionally enriched along an oceanic biodiversity gradient. *Nature* 429:867–70.

Lawton, J. H. 2000. Community ecology in a changing world. 21385 Oldendorf/Luhe, Germany: Ecology Institute.

Legendre, P. 1993. Spatial autocorrelation: Trouble or new paradigm? *Ecology* 74:1659–73.

Levin, S. A. 1992. The problem of pattern and scale in ecology. *Ecology* 73:1943–83.

Lubchenco, J. 1980. Algal zonation in the New England rocky intertidal community: An experimental analysis. *Ecology* 333–44.

Ludwig, J. A., J. A. Wiens, and D. J. Tonggway. 2000. A scaling rule for landscape patches and how it applies to conserving soil resources in savannas. *Ecosystems* 3:84–97.

Marquet, P. A., S. A. Navarrete, and J. C. Castilla. 1990. Scaling population density to body size in rocky intertidal communities. *Science* 250:1125–27.

Marquet, P. A., R. A. Quinones, S. Abades, F. Labra, M. Tognelli, M. Arim and M. Rivadeneria. 2003. Scaling and power laws in ecological systems. *The Journal of Experimental Biology* 208:1749–69.

Menge, B. A. 1976. Organization of the New England rocky intertidal community: Role of predation, competition and environmental heterogeneity. *Ecological Monographs* 46:355–93.

Menge, B. A., B. A. Daley, J. Lubchenco, E. Sanford, E. Dahlhoff, P. M. Halpin, G. Hudson, and J. L. Burnaford. 1999. Top-down and bottom-up regulation of New Zealand rocky intertidal communities. *Ecological Monographs* 69:297–330.

Menge, B. A., and J. Lubchenco. 1981. Community organization in temperate and tropical rocky intertidal habitats: Prey refuges in relation to consumer pressure gradients. *Ecological Monographs* 51:429–50.

Menge, B. A., E. Sanford, B. A. Daley, T. L. Freidenburg, G. Hudson, and J. Lubchenco. 2002. An inter-hemispheric comparison of bottom-up effects on community structure: Insights revealed using the comparative-experimental approach. *Ecological Research* 17:1–16.

Mittlebach, G. G., C. F. Steiner, S. M. Scheiner, K. L. Gross, H. L. Reynolds, R. B. Waide, M. R. Willig, S. I. Dodson, and L. Gough. 2001. What is the observed relationship between species richness and productivity? *Ecology* 82:2381–96.

Mittelbach, G. G., D. W. Schemske, H. V. Cornell, A. P. Allen, J. M. Brown, M. Bush, S. P. Harrison, et al. 2007. Evolution and the Latitudinal Diversity Gradient: Speciation, extinction, and biogeography. *Ecology Letters* 10:315–31.

Paine, R. T. 1966. Food web complexity and species diversity. *American Naturalist* 100:65–75.

Petersen, D. L., and V. T. Parker, eds. 1998. *Ecological scale: Theory and applications.* New York: Columbia University Press.

Platt, J. R. 1964. Strong inference. *Science* 146:347–53.

Ricklefs, R. E. 1990. Scaling pattern and process in marine ecosystems. In *Large marine ecosystems,* ed. K. Sherman, L. M. Alexander, and B. D. Gold, 169–78. Washington, DC: American Association for the Advancement of Science.

Roy, K., D. Jablonski, J. W. Valentine, and G. Rosenberg. 1998. Marine latitudinal diversity gradients: Tests of causal hypotheses. *Proceedings of the National Academy of Sciences, USA* 95:3699–3702.

Russell, R., S. A. Wood, G. Allison, and B. A. Menge. 2006. Scale, environment, and trophic status: The context dependency of community saturation in rocky intertidal communities. *American Naturalist* 167:E-158–E-170.

Russo, S. E., S. K. Robinson, and J. Terborgh. 2003. Size-abundance relationships in an Amazonian bird community: Implications for the energetic equivalence rule. *American Naturalist* 161:267–83.

Sanford, E., M. S. Roth, G. C. Johns, J. P. Wares, and G. N. Somero. 2003. Local selection and latitudinal variation in a marine predator-prey interaction. *Science* 300:1135–37.

Schneider, D. C. 1994. Quantitative ecology: Spatial and temporal scaling. Orlando, FL: Academic Press.

———. 2001. The rise of the concept of scale in ecology. *BioScience* 51:545–54.

Smith, F., and J. D. Witman. 1999. Species diversity in subtidal landscapes: Maintenance by physical processes and larval recruitment. *Ecology* 80:51–69.

Srivastava, D. S., and J. H. Lawton. 1998. Why more productive sites have more species: An experimental test of theory using tree hole communities. *American Naturalist* 152:510–29.

Steele, J. H. 1978. Some comments on plankton patches. In *Spatial pattern in plankton communities,* ed. J. H. Steele, 1–20. New York: Plenum.

Stephenson, T. A., and A. Stephenson. 1972. Life between tidemarks on rocky shores. San Francisco: W. H. Freeman.

Stephenson, W., W. T. Williams, and G. N. Lance. 1970. The macrobenthos of Moreton Bay. *Ecological Monographs* 40:459–94.

Southward, A. J., et al. 2005. Long term oceanographic and ecological research in the western English Channel. *Advances in Marine Biology* 47:1–105.

Swearer, S. E., J. E. Caselle, D. W. Lea, and R. R. Warner 1999. Larval retention and recruitment in an island population of a coral-reef fish. *Nature* 402:799–802.

Terborgh, J., and J. Faaborg. 1980. Saturation of bird communities in the West Indies. *American Naturalist* 116:178–95.

Thorson, G. 1950. Reproductive and larval ecology of marine bottom invertebrates. *Biological Reviews of the Cambridge Philosophical Society* 50:1–45.

———. 1957. Bottom communities (sublittoral or shallow shelf). In *Treatise on marine ecology and paleoecology,* ed. J. W. Hedgepeth. Vol. 1, Geological Society of America Memoirs. 67:461–534.

Thrush, S. E., R. D. Pridmore, R. G. Bell, V. J. Cummings, P. K. Dayton, et al.1997. The sandflat habitat: Scaling from experiments to conclusions. *Journal of Experimental Marine Biology and Ecology* 216:1–9.

Tomita, M., Y. Hirabuki, and K. Seiwa. 2002. Post-dispersal changes in the spatial distribution of *Fagus crenata* seeds. *Ecology* 83:1560–65.

Underwood, A. J. 1978. An experimental evaluation of competition between three species of intertidal prosobranch gastropods. *Oecologia* 33:185–202.

Underwood, A. J., and E. J. Denley. 1984. Paradigms, explanations, and generalizations in models for the structure of intertidal communities on rocky shores. In *Ecological communities: Conceptual issues and the evidence,* ed. D. R. Strong, D. Simberloff, L. G. Abele, and A. B. Thistle, 151–80. Princeton, NJ: Princeton University Press.

Van de Koppel, J., M. Rietkerk, N. Dankers, and P. M. J. Herman. 2005. Scale-dependent feedback and regular spatial patterns in young mussel beds. *American Naturalist* 165:E66–E77.

Webb, C. O., D. D. Ackerly, M. A. McPeek, and M. J. Donoghue. 2002. Phylogenies and community ecology. *Annual Review of Ecology and Systematics* 33:475–505.

Wiens, J. A. 1986. Spatial scale and temporal variation in studies of shrubsteppe birds. In *Community ecology,* ed. J. Diamond and T. J. Case, 154–72. New York: Harper and Row.

———. 1989. Spatial scaling in ecology. *Functional Ecology* 3:385–97.

———. 2001. Understanding the problem of scale in experimental ecology. In *Scaling relations in experimental ecology,* ed. R. H. Gardner, W. M. Kemp, V. S. Kennedy, and J. E. Petersen, 61–88. New York: Columbia University Press.

Whittaker, R. J., K. J. Willis, and R. Field. 2001. Scale and species richness: Towards a general, hierarchical theory of species diversity. *Journal of Biogeography* 28:453–70.

Witman, J. D., and P. K. Dayton. 2001. Rocky subtidal communities. In *Marine community ecology,* ed. M. D. Bertness, S. D. Gaines, and M. Hay, 339–66. Sunderland, MA: Sinauer.

Witman, J. D., R. J. Etter, and F. Smith. 2004. The relationship between regional and local species diversity in marine benthic communities: A global perspective. *Proceedings of the National Academy of Sciences USA* 101:15664–69.

Witman, J. D., and K. P. Sebens. 1992. Regional variation in fish predation intensity: A historical perspective in the Gulf of Maine. *Oecologia* 90:305–15.

Witman, J. D., M. Cusson, P. Archambault, A. J. Pershing, and N. Mieszkowska. 2008. The relation between productivity and species diversity in temperate–arctic marine ecosystems. *Ecology* 89 (11) Supplement: S66–S80.

CHAPTER FOURTEEN

LATITUDINAL GRADIENTS IN SPECIES INTERACTIONS

ERIC SANFORD AND MARK D. BERTNESS

Introduction

In recent decades, ecologists have focused increasingly on understanding patterns and processes that operate over broad spatial scales, ranging from hundreds to thousands of kilometers. Given that ecological studies have traditionally been conducted at one or a few local field sites, this new emphasis represents a significant shift in research focus. This change has been driven in part by a desire to address pressing large-scale environmental issues such as the impacts of climate change, eutrophication, habitat fragmentation, and invasive species (Lubchenco et al. 1991; Vitousek 1994). But the impetus to expand the scale of ecological studies has also come from the realization that studying local processes alone has seldom been able to account for all of the important variation among natural communities (Dayton and Tegner 1984; Menge 1992; Brown 1995).

An interest in large-scale patterns is not new to the field of ecology. Latitudinal patterns of species diversity have long fascinated ecologists working both in terrestrial and marine systems (Pianka 1966; Rosenzweig 1995). Geographic gradients, such as the striking increase in species diversity at lower latitudes, have been attributed to numerous physical and biological processes (Pianka 1966; MacArthur 1972; Rhode 1992). In the 1960s, biological interactions were central in ecologists' minds because field experiments had recently demonstrated that competition and predation could directly regulate local patterns of species diversity (Paine 1966). Not surprisingly, species interactions were also invoked as a potential explanation for latitudinal patterns of diversity. For example, one hypothesis suggested that intense predation in the tropics reduced competition among prey species and favored greater coexistence and speciation (Paine 1966; Pianka 1966). Higher prey diversity, in turn, was argued to support a greater diversity of predators, thus completing a positive feedback loop. However, this and other hypotheses regarding the role of biological interactions in generating geographic variation in species diversity were criticized for being circular (Rhode 1992). Although predation can clearly contribute to the maintenance of diversity, this hypothesis did not adequately address the initial origin of more predator species in the tropics.

A more recent perspective places species interactions within a set of nested factors that operate at different spatial scales (Ricklefs 1987; Menge and Olson 1990; Huston 1999). In this view, the diversity and structure of a local community are influenced by a combination of geographic, regional, and local processes. The regional species pool is a product of rates of speciation, extinction, and invasion, and these processes are shaped by area, temperature, environmental stability, geological history, and other factors (Mora et al. 2003; and Witman and Roy, this volume). Although local communities are assembled from this regional species pool, ecological interactions can modify local patterns of species richness and community structure (Huston 1999). Moreover, if species interactions vary predictably along environmental gradients, then a mechanistic understanding of this variation should increase our understanding of how communities vary across geographic scales (Gaines and Lubchenco 1982).

This chapter focuses on the nature and regulation of latitudinal variation in species interactions. During the past forty years, community ecologists have had enormous success using field experiments to demonstrate that biological interactions such as predation, herbivory, competition, and facilitation play a central role in determining the diversity and structure of communities. Most of our knowledge regarding community structure and dynamics comes from these experiments done at the local level. But the extent to which results from local studies can be extrapolated to larger spatial scales has remained an open, and sometimes contentious, question (Foster 1990; Paine 1991). Until recently, we have known little about the processes that might alter the intensity and outcome of species interactions across latitudinal spatial scales.

Significant advances have come from expanding the spatial and temporal scale of ecology to reach beyond that of the local field experiment (Witman and Roy, this volume). Marine ecologists embraced this challenge in the 1980s, and the last two decades have seen a growing number of studies aimed at understanding how biological interactions vary across broad scales. These studies have generated exciting results that point toward two general conclusions: (a) Species interactions can vary profoundly across large spatial scales with the abiotic and biotic context in which they occur, and (b) species interactions can be modified via both phenotypic and genotypic mechanisms, acting over a variety of spatial and temporal scales. Exploring these links is the goal of this chapter.

Recent calls for a broader, macroscopic perspective in ecology have inspired a variety of different research approaches. Most studies now categorized as "macroecology" emphasize the statistical analysis of large-scale patterns rather than experimental manipulations. This reliance on statistical approaches has been motivated by a desire to address questions that are simply not amenable to manipulation. For other macroscopic questions, including many of those addressed in this chapter, experimental approaches remain a powerful tool, particularly when combined with nonexperimental data. As pointed out by James Brown (1995), the relative emphasis on nonexperimental versus experimental approaches is often a matter of practicality, rather than philosophy.

In this chapter, we focus largely on studies that have used a comparative-experimental approach (Menge et al. 2002). We hope to demonstrate that these experimental studies can make important contributions to our understanding of marine macroecology. We begin the chapter by tracing the history of macroscopic approaches in marine systems. Through a review of recent studies, we then investigate a set of potential sources of geographic variation in species interactions. We further highlight these sources of variation and their potential community consequences by presenting two case studies from our work in rocky intertidal habitats along the Pacific coast and salt marshes along the Atlantic coast. We conclude by identifying gaps in our knowledge, with recommendations for promising areas of research.

A Brief History of Macroscopic Studies in Marine Ecology

In his book *Macroecology*, Brown (1995) reviews the transformation of ecology from a descriptive natural history into a rigorous modern science, beginning in the 1960s. He notes that the key development in this process was the adoption of experimental methods to test models and hypotheses. Brown argues that this new experimental approach was so powerful that many ecologists became convinced that controlled experiments were the only legitimate form of empirical ecology.

Nowhere was this enthusiastic acceptance of experimental ecology more apparent than in the field of marine ecology. This was not surprising, given that many of the pioneering experiments in modern ecology were conducted in marine systems (e.g., Connell 1961; Paine 1966). Moreover, these early experiments often produced dramatic results demonstrating that field experiments could help tease apart the complexities of natural communities. Thus, throughout the 1970s and early 1980s, marine ecologists used field manipulations to reveal the importance of biological interactions in determining the structure of local communities (e.g., Dayton 1971; Sutherland 1974; Menge 1976; Lubchenco 1978; Sousa 1979; Hixon and Brostoff 1983; Bertness 1984; Castilla and Duran 1985).

Brown suggests that in their embrace of the experimental approach, ecologists largely overlooked the limitations of field manipulations. In his words, experimental ecology became the study of "small places for short periods" (Brown 1995). This realization concerned a number of ecologists who recognized during the 1980s that communities were often influenced by processes operating at scales much larger than those of their experimental manipulations (Dayton and Tegner 1984; Wiens et al. 1986; Ricklefs 1987; Roughgarden, Gaines, and Possingham 1988; Brown and Maurer 1989; Menge and Olson 1990). These authors urged the use of new approaches to address large-scale processes that impinged upon communities. These authors, and subsequent discussions by Brown (1995) and Maurer (1999), did not dismiss the tremendous power of the experimental approach. Rather, they pointed out that nonexperimental, macroscopic data could complement small-scale experiments by placing them in a larger perspective that explicitly considered (a) spatial and temporal variation, (b) openness to exchange, and (c) the influence of history (Brown 1995).

A successful manifestation of these goals has been the "comparative-experimental" approach that establishes identical, replicated experiments at multiple sites along an environmental gradient (Menge et al. 2002). This design explores how physical conditions that are not amenable to manipulation may influence species interactions to generate large-scale patterns of community variation (Menge et al. 2002). In addition, since experiments are often replicated in communities that also differ in recruitment supply or evolutionary history, biotic and historical factors are inevitably addressed, as well. The comparative-experimental approach has been used infrequently in terrestrial systems (e.g., Jeanne 1979; Louda 1982; Brown et al., 1986), but has a history of more than three decades in marine systems. This research strategy was pioneered by Paul Dayton, working in rocky intertidal habitats in North America (Dayton 1971). Dayton conducted identical field experiments at sites ranging from the wave-beaten outer coast of the Olympic Peninsula, Washington, to the more sheltered San Juan Islands. His results suggested that variation in wave forces and desiccation altered the relative importance of disturbance and competition along this gradient stretching >100 km (Dayton 1971).

Following Dayton's work, a comparative-experimental approach was soon adopted by other marine ecologists, who replicated experiments across distances that ranged from several hundred kilometers along the New England coast (Menge 1976; Lubchenco and Menge 1978) to much broader experimental comparisons among the temperate rocky shores of North America, New Zealand, and Chile (Paine, Castillo, and Cancino 1985). Experiments were also undertaken to quantify and compare the role of predation in temperate versus tropical regions (Bertness, Garrity, and Levings 1981; Menge and Lubchenco 1981). Evolutionary biologists had given ecologists good reason to believe that predation was more intense in tropical regions. Most notably, Geerat Vermeij (1974, 1978, 1980) demonstrated that the shells of tropical marine gastropods had evolved a variety of predator-resistant adaptations including strong sculpture, low spires, thickened walls, and narrow apertures with teeth. Similarly, Vermeij (1977) analyzed the morphology of crab claws to show that the size and crushing strength of tropical crabs exceeded that of temperate crabs. Field experiments comparing consumer effects in Panama versus New England confirmed that predation and herbivory were indeed more intense in the tropical intertidal habitats studied (Menge and Lubchenco 1981; Bertness, Garrity, and Levings 1981; Gaines and Lubchenco 1982; see also Heck and Wilson 1987).

Although these and other studies began to consider how species interactions might vary over large spatial scales, it was not until the mid-1980s that marine ecologists began to explicitly incorporate the influence of large-scale processes into their work. At this time, several key papers stressed that larval recruitment was a neglected, and potentially important, source of variation in marine communities (Dayton and Tegner 1984 Underwood and Denley 1984). These authors emphasized the need to integrate the results of local experiments with an understanding of large-scale oceanographic processes responsible for the delivery of larvae and nutrients. Subsequently, the late 1980s saw an explosion of studies that examined the role of recruitment variation in marine communities (e.g., Gaines, Brown, and Roughgarden 1985; Keough 1988 Roughgarden, Gaines, and Possingham 1988; Jones 1990). This body of work led to a greater awareness of the "open" nature of marine systems and focused the attention of ecologists on large-scale processes that could not be easily manipulated, including larval dispersal, currents, climate, and fishing pressure (e.g., Witman and Sebens 1992).

In the last two decades, marine ecologists have actively pursued the goal of placing local experiments within a broader spatial and temporal context. Experiments are now typically replicated at a number of study sites, frequently spanning tens to hundreds of kilometers (Witman and Roy, this volume). Nonexperimental data are also considered a vital part of most studies. Monitoring levels of larval recruitment, nutrients, phytoplankton, and salinity has become routine. The use of temperature data-loggers, wave force dynamometers, current meters, and other instruments is also commonplace. Satellite images and oceanographic data are used to place study sites within a broader spatial context that recognizes currents, gyres, and other oceanographic features (e.g., Menge et al. 1997; Menge et al. 2003; Nielsen and Navarrete 2004). Population genetics and paleontology have also been integrated into some ecological studies as tools to learn more about gene flow, population connectivity, and the historical geography of species assemblages (Hellberg, Balch, and Roy 2001; Wares and Cunningham 2001; Sanford et al. 2003; Sotka, Wares, and Hay 2003). A growing subset of recent studies has focused on understanding how and why abiotic and biotic changes generate latitudinal variation in species interactions. In the next sections we review these results.

Species Interactions along Environmental Gradients

Working primarily at local scales, ecologists have long recognized that changes in environmental conditions can modify the strength, or even the outcome, of a given biological interaction (Park 1954; Menge and Sutherland 1987; Dunson and Travis 1991; Bertness and Callaway 1994; Travis 1996; Thompson 1998). A logical extension of this concept is that species interactions may be regulated along geographic gradients by variation in temperature, light, desiccation, salinity, nutrients, rainfall, and other environmental factors (Travis 1996). For example, the growth and calcification of corals is sharply reduced by cooler water temperatures. Thus, at higher latitudes in eastern Australia, corals may be less able to compete for space with temperate biota such as polychaete tube worms (Harriott and Banks 2002).

Similarly, Wethey's experimental studies in the northwest Atlantic have shown that the outcome of competition between barnacle species varies predictably along a latitudinal gradient of environmental stress (Wethey 1983, 1984, 2002). Rocky shores of southern New England are a region of geographic overlap for a southern barnacle (*Chthamalus fragilis*) and a more northern, boreo-arctic barnacle (*Semibalanus balanoides*). *Semibalanus* can typically overgrow and outcompete *Chthamalus* by virtue of its larger size and superior growth rate. This is particularly true north of Cape Cod, where *Semibalanus* rarely experiences thermal stress and thus can competitively exclude *Chthamalus* from the region (Wethey 2002). In contrast, south of Cape Cod, *Semibalanus* is often weakened and killed by heat and desiccation in the high intertidal zone. Here, *Chthamalus* is able to persist in a refuge above the upper limit of *Semibalanus*. Thus, competition between these two barnacle species is mediated by heat and desiccation, factors that vary with latitude in New England (Wethey 1983, 1984, 2002). Competition among intertidal barnacles may also be influenced by water temperature, and long-term observations in Europe suggest that the competitive balance between two barnacle species has varied over many decades in association with shifts in ocean temperature (Southward and Crisp 1956; Southward, Hawkins, and Burrows 1995).

In addition to modifying the direction or outcome of species interactions (e.g., Wethey 1983, 1984, 2002; Leonard 2000), latitudinal gradients in temperature may regulate rates of species interactions. Most marine fishes and invertebrates are ectotherms and thus their body temperatures are greatly modified by environmental temperature. For these animals, rates of metabolism, respiration, locomotion, and consumption are strongly influenced by thermal variation (Cossins and Bowler 1987; Hochachka and Somero 2002). For example, a decrease in temperature of 5°C is expected to decrease rates of biological processes by ~30 percent to 43 percent (given a typical Q_{10} of 2 to 3). Although these effects are large and well known to physiologists, marine ecologists have seldom considered small changes in temperature to be an important source of variation in species interactions (Sanford 2002b).

Under natural field conditions, many ectotherms show varying degrees of acclimatization to temperature change (Clarke 1987). For example, prolonged exposure to colder temperatures may trigger partial or total compensation, such that initial reductions in biological rates gradually diminish or disappear. Compensation may be brought about via a number of phenotypic mechanisms including changes in enzyme concentrations, production of enzyme variants, or modifications of the intracellular environment (Clarke 1987). However, ectotherms differ in their potential for acclimatization and it is difficult to predict a priori how a particular species will respond to temperature change. Moreover, marine physiologists have typically studied acclimation under constant conditions in the laboratory, and considerably less is known about patterns of compensation among geographically separated, natural populations. Clearly, individuals belonging to a single species will often experience very different thermal regimes in different parts of their geographic range.

Sanford (1999, 2002b) examined the effects of temporal variation in water temperature on rates of predation by the intertidal sea star, *Pisaster ochraceus*. Along the Pacific coast of the United States, *Pisaster* can play a keystone role by controlling the lower distribution of mussels (*Mytilus californianus*), the dominant competitor for primary space in this community (Paine 1966, 1974; Menge et al. 1994). Wind-driven upwelling along the Pacific coast causes water temperatures to drop 3 to 5°C during episodic events lasting from a few days to several weeks. Sanford (1999) assessed sea star predation rates in the field by following the fate of mussels transplanted into the low intertidal zone on the Oregon coast. In a series of fourteen-day experiments, per capita rates of sea star predation were roughly halved during periods of upwelling when water temperatures declined by ~4°C. Similar results were observed when *Pisaster* were held under controlled temperature treatments in the laboratory (Sanford 2002a). These studies indicate that temporal variation in water temperature can directly regulate rates of predation by this sea star.

Sanford (1999) hypothesized that water temperature may also modify *Pisaster* predation rates across large spatial scales. Both *Pisaster* and *Mytilus californianus* have broad geographic distributions, ranging from Baja California, Mexico, to Alaska. If water temperature regulates sea star feeding, per capita impacts on prey may increase with decreasing latitude. In contrast, if sea stars are acclimatized to local conditions, then *Pisaster*'s per capita effects could remain relatively constant regardless of thermal variation across its geographic range. A recent study by Menge and collaborators provides a partial test of these hypotheses (Menge et al. 2004). Sea star predation was quantified using mussels experimentally transplanted into the low intertidal zone at thirteen study sites in Oregon, Central California, and Southern California. Per capita predation by *Pisaster* did not differ significantly between Oregon and Central California (fig. 14.1). However, the temperature difference between these regions is slight; water temperatures in Central California are typically only 1 to 2°C warmer than in Oregon during the summer months (Sanford, unpublished data). *Pisaster* densities were very low in Southern California and thus per capita predation was only estimated at one location (site 3). However, these data did not suggest greater per capita effects south of Point Conception, where water temperatures are often 5 to 8°C warmer than in Oregon. This may indicate that predicted increases in per capita predation by this sea star are modified by acclimatization.

In contrast, limited evidence from other studies suggests that water tem-

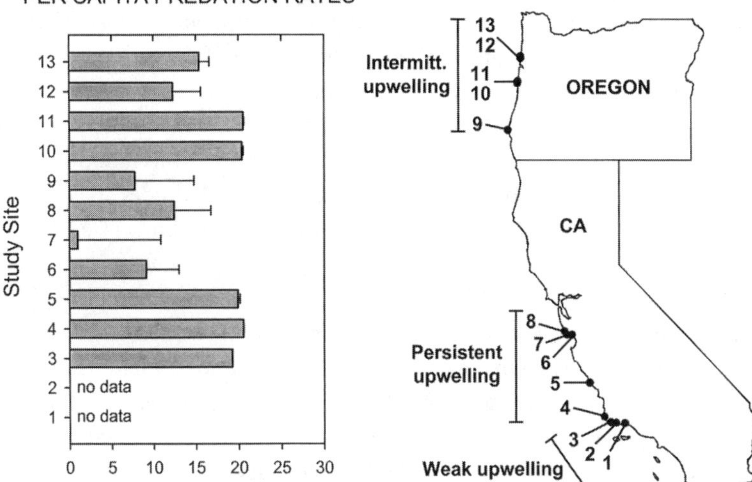

FIGURE 14.1 Latitudinal variation in per capita predation rates of sea stars feeding on mussels (modified from Menge et al. 2004). Map shows the thirteen study sites located in regions where upwelling was persistent (sites 4 to 8), intermittent (sites 9 to 13), or weak (sites 1 to 3). At each site, mussels (*Mytilus californianus*) were translocated to the low intertidal zone. Predation rates were assessed by comparing rates of mussel mortality in plots where sea stars (*Pisaster ochraceus*) were present at natural densities versus plots where sea stars were excluded. Per capita predation rates were calculated using average sea star density in each plot. To make values positive, ordinate values (ln mussels/day/sea star) were coded by adding 21 to each value. Data are means (+ 1 SE) from experiments conducted in summer 1999. See Menge et al. (2004) for a full description of experimental methods.

perature may sometimes play an important role in regulating geographic variation in predation. The crown-of-thorns sea star *Acanthaster planci* can be a voracious predator on corals. During periodic outbreaks, masses of sea stars may devastate large areas of coral reefs in the Pacific. However, the effects of this predator appear to vary greatly with geographic region. In Guam, an outbreak of *Acanthaster* killed 90 percent of corals inhabiting 38 km of coastline in less than two years (Birkeland and Lucas 1990). In contrast, an outbreak of *Acanthaster* in Hawaii moved little and had minimal impacts on the reef. Further studies showed that individual sea stars in Guam consumed about twice as much coral tissue and had rates of movement that were five times greater than individuals in Hawaii. Although not tested directly, differences in water temperature were one likely explanation for the observed geographic variation in sea star impacts (Birkeland and Lucas 1990). In Guam, temperatures are typically 28 to 29°C whereas temperatures in Hawaii are often below 25°C. Although further studies are needed, it appears that acclimatization does not modify the effects of temperature on *Acanthaster* predation. Rather, relatively small differences in water temperature appear to play a significant role in regulating the geographic impact of this predator.

Similar considerations may apply to consumption rates of herbivores (Jenkins et al. 2001; Pennings and Silliman 2005). In the northeast Atlantic, large fucoid algae dominate the mid-intertidal zone in northern Europe, but are increasingly restricted to sheltered shores at lower latitudes (Ballantine 1961). This pattern appears to be related to the effects of limpets. At lower latitudes, limpet density is greater and per capita grazing rates may also be higher (Hawkins et al. 1992; Jenkins et al. 2001). This latitudinal trend in grazing activity may be driven in part by temperature, although other factors may also play a role (Jenkins et al. 2001). At higher latitudes, where grazing is less intense and algal propagules are abundant, the early life stages of fucoids may stand a greater chance of escaping grazing and becoming established on wave-exposed shores (Jenkins et al. 2005; Coleman et al. 2006).

In addition to temperature and desiccation, large-scale variations in oceanographic processes including coastal upwelling and nutrient fluxes have been explored for their potential role in regulating species interactions (Bustamante et al. 1995; Menge et al. 1997, 2003; Broitman et al. 2001; Witman and Smith 2003; Nielsen and Navarrete 2004). Along the Pacific coast of North America, alongshore winds and upwelling are stronger and more persistent in California than they are further north in Oregon and Washington State. As a result, the advection of larvae offshore may be greater in California (Roughgarden, Gaines, and Possingham 1988), and the onshore recruitment of barnacles and mussels is one to two orders of magnitude less than along the Oregon coast (Connolly, Menge, and Roughgarden 2001). Similar geographic gradients of recruitment are present in other marine systems (e.g., Chile and Australia; Broitman et al. 2001; Hughes et al. 2002), although underlying causes may differ.

Based on this latitudinal gradient in California and Oregon, Connolly and Roughgarden (1999) constructed models to examine how variation in larval input might influence both competition for space and predator-prey dynamics in rocky intertidal communities. Their models estimated both per capita interaction strength (the effect of an individual of one species on the per capita population growth rate of another species) and population interaction strength (the net effect of a population on the per capita growth rate of another species). In contrast to traditional closed models, most marine ecosystems are essentially open because predator and prey (e.g., sea stars, mussels, and barnacles) often have planktonic larvae (Gaines and Lafferty 1995). As a result, the Connolly and Roughgarden models predict that the per capita effects of predators on prey should be independent of upwelling intensity. This occurs because predators do not directly influence the production or settlement of their prey (produced by distant upstream sources). In contrast, the population or total effect of predators on prey should decrease with upwelling intensity. This occurs because increased upwelling reduces the recruitment and abundance of predators.

Experiments by Menge et al. (2004) tested the predictions of this predator-prey model. As described earlier, identical experiments were conducted to quantify *Pisaster* predation on mussels in three regions that differed greatly in upwelling intensity: Central California (persistent upwelling), the Oregon coast (intermittent upwelling), and Southern California (weak upwelling). Consistent with model predictions, per capita predation rates did not differ between oceanographic regions (fig. 14.1). Contrary to theory, the total effect of sea stars on mussels (population interaction strength, fig. 14.2) did not differ between Central California (sites 4 through 8) and Oregon (sites 9 through 13). Why did total predation fail to vary consistently along the gradient of upwelling intensity? The model predicts that latitudinal variation in

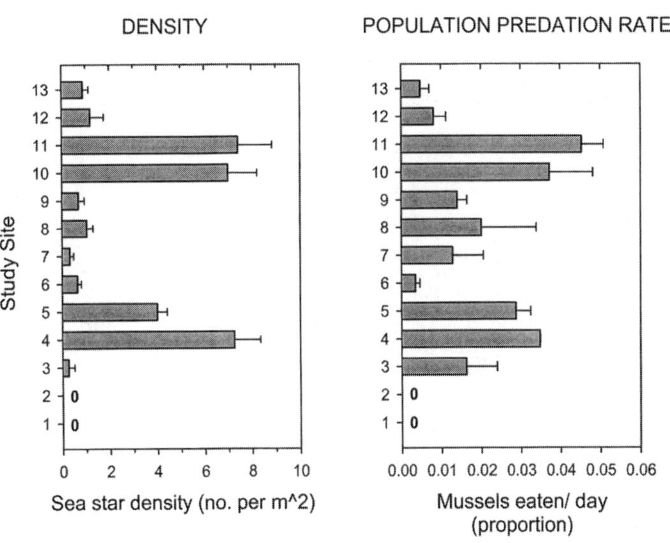

FIGURE 14.2 Geographic variation in sea star density and predation effects (modified from Menge et al. 2004). Left panel shows density of sea stars (*Pisaster ochraceus*) in the low intertidal zone at the thirteen study sites. See figure 14.1 map for site locations. Data are mean number of sea stars per m^2 (+ 1 SE) recorded during surveys conducted in June/July 1999. Right panel shows per population predation rates (mean + 1 SE) at each site during summer 1999. Data are mussel mortality rates corrected for background mortality (i.e., the difference between rates of mussel mortality in plots with sea stars present minus mortality rates in plots where sea stars were excluded). See Menge et al. (2004) for a full description of experimental methods.

upwelling should determine predator recruitment and abundance and thus produce consistent variation in the strength of predation. The results of the Menge et al. study confirmed that total predation at each site was indeed a function of sea star abundance. However, rather than varying consistently with oceanographic regime, sea star abundance (and hence predation) varied strongly among sites within each region (fig. 14.2). In other words, *Pisaster* abundance was quite patchy even among sites separated by <100 km (see also Sagarin and Gaines 2002). Menge et al. (2004) conclude that broad generalizations about upwelling in California versus Oregon may overlook significant variation in upwelling that occurs over mesoscales (tens of kilometers). This variation, driven by shelf bathymetry, headlands, gyres, fronts, and other near-shore features, may influence the advection and transport of sea star larvae. These same processes may also regulate the local strength of bottom-up forces, including the supply of nutrients, phytoplankton, and mussel recruits (Menge et al. 1997). Thus, *Pisaster* abundance and the strength of the sea star-mussel interaction may ultimately reflect oceanographic processes operating over mesoscales (Menge et al. 2004).

Biological and Evolutionary Context of Species Interactions

A key point illustrated by the Menge et al. (2004) study is that geographic variation in abundance can be an important source of variation in species interactions. This and other studies indicate that the biological context in which a species interaction occurs will often vary with latitude in ways that modify the interaction. For example, the relative abundance of species may frequently vary among sites, even when the resident communities consist of the same set of species. Alternatively, the species composition of a community will often vary with latitude (Brown and Kurzius 1987; Bustamante and Branch 1996; Broitman et al. 2001), in association with changes in the regional species pool. These changes can trigger both phenotypic and evolutionary modifications of species and their interactions (Gaines and Lubchenco 1982).

Species abundance seldom varies in a uniform manner with latitudinal gradients in temperature or other abiotic factors (Broitman et al. 2001; Sagarin and Gaines 2002). Rather, many marine species have a patchy geographic distribution, with strong variation in abundance occurring over a spatial scale of 100 km or less (Sagarin and Gaines 2002; Gilman 2005). A similar pattern has been noted for terrestrial species, with many species having adjacent "hot spots" and "cool spots" of high and low local abundance, respectively (Brown 1995). Significantly, the geographic location of these "hot spots" is often consistent through time in both terrestrial and marine systems (Brown 1995; Menge et al. 1994). This suggests that the local population abundance may be a function of relatively stable physical and/or biotic factors. Since variation in population density can clearly modify the strength and importance of a species interaction, increased attention should be devoted to understanding when and why local abundance is patchy at the landscape level.

If the relative abundance of key species varies with latitude, then even communities with essentially identical species composition may have fundamentally different dynamics. Paine's experiments in Washington State demonstrated that the sea star *Pisaster* could play a keystone role by controlling the distribution of the mussel *Mytilus californianus* (Paine 1966; 1974). However, Paine (1980) speculated that *Pisaster* was "just another starfish" in Torch Bay, Alaska, because *M. californianus* is uncommon in the northern part of its geographic range. With the reduced presence of this competitively dominant mussel, the community importance of *Pisaster* may be fundamentally changed at Torch Bay. A similar study compared the role of an herbivorous chiton, *Katharina tunicata*, in Washington State versus Torch Bay (Dethier and Duggins 1988). Identical experiments were conducted in the two regions by removing chitons from large plots in the mid to low intertidal zone for >3 years. In Washington, the experimental removal of this grazer resulted in the rapid transformation of the low intertidal zone into a dense bed of kelps. In contrast, removal of this chiton in Alaska failed to produce a discernible impact on the community (Dethier and Duggins 1988). These very different responses were attributed to the relative abundance of a single kelp species. The perennial kelp *Hedophyllum sessile* was rare in Alaska, but was common in Washington and rapidly colonized once chitons were excluded. These and other studies (e.g., Estes and Duggins 1995; Jenkins et al. 2005) suggest that the community-level response that arises following the addition or removal of a consumer is often dependent on geographic variation in the recruitment of foundation species and other key players in the community. The overall lesson is clear: the biological context in which an interaction occurs can be just as significant as its environmental setting.

All of the studies we have presented thus far consider how an interaction between the same two species varies with latitude. Understanding patterns of latitudinal variation becomes more complex when changes in species composition are considered. For example, Paine, Castillo, and Cancino (1985) tested how intertidal communities in Chile, New Zealand, and North America (Washington State) responded to the long-term, press removal of important invertebrate consumers. In all three systems, exclusions of sea stars resulted in a remarkably similar response: mussels increased in abundance and extended their distribution downshore into the low intertidal zone. The exclusions were eventually abandoned and sea stars were allowed to return to experimental areas. In North America and New Zealand, the mussels that had spread into the low zone subsequently coexisted with their predators for >14 years, apparently because these mussels had attained a size refuge and were too large to be consumed. In contrast, when consumers were allowed to return to the community in Chile, they rapidly eliminated the low zone mussels and the assemblage returned to its control state. The striking difference in these recovery patterns was attributed to evolutionary differences in the mussel genera that occupy these regions. In North America and New Zealand, *Mytilus* and *Perna* can grow to large sizes that are essentially immune to predation. In contrast, the Chilean mussel *Perumytilus purpuratus* is a small-bodied species that rarely exceeds 40 mm in length and is thus easily consumed. Although these communities had all evolved similar sea star-mussel interactions, evolutionary differences in mussel body size led to important differences in how these communities recover from disturbance. This may be a common result in independently evolved species assemblages (Moreno and Sutherland 1982). Species in different geographic regions have evolved along different paths and the resulting differences in morphology, body size, physiology, or behavior can generate significant, sometimes idiosyncratic, differences in how species interact.

There is a growing awareness that the importance of evolution in marine communities is not restricted to effects dividing distantly related assemblages of species. Rather, spatial variation in selection may act on distinct populations of a single species to generate important variation in species interactions. The potential contribution of genetic variation among populations to geographic differences in species interactions has only recently been considered in marine systems (e.g., Stachowicz and Hay 2000; Sotka and Hay 2002; Sanford et al. 2003). Many marine species have planktonic dispersal and thus gene flow among populations is expected to be high, diminishing the potential for local adaptation in marine species (Grosberg and Cunningham 2001). However, an increasing number of studies have discovered high levels of genetic differentiation among populations of marine fishes and invertebrates (Palumbi 1995; Barber et al. 2000; Taylor and Hellberg 2003; Sotka et al. 2004). These unexpected results suggest limited dispersal, perhaps resulting from larval behavior or local retention of larvae in fronts, gyres, and other oceanographic features.

Biologists working in terrestrial systems have led the way in considering the role of evolution in generating geographic variation in species interactions. In many terrestrial species, dispersal is limited, populations show a high level of genetic differentiation, and the potential for local adaptation is high. Since abiotic and biotic conditions vary geographically, natural selection may favor important evolutionary differences among populations (Endler 1977; Thompson and Cunningham 1999). The strength of selection may be particularly strong for pairs of tightly coevolved species, linked by mutualistic, parasitic, or predator-prey interactions. Theoretical and empirical work suggests that a geographic gradient or mosaic of varying abiotic and biotic conditions may create a landscape of coevolutionary hotspots and coldspots (Thompson and Cunningham 1999; Benkman, Holimon, and Smith et al. 2001).

Only a handful of studies have addressed the potential for similar evolutionary processes to shape interactions in marine systems (Sotka 2005). In most cases, these studies have considered how latitudinal variation in selection may generate differences in behavior. For example, geographic variation in defensive behavior has been documented in subtidal decorator crabs (*Libinia dubia*) living along the Atlantic coast of North America (Stachowicz and Hay 2000). Juvenile *Libinia* defend themselves against predation by camouflaging their carapace with algae, sessile invertebrates, and debris gathered from their environment. Crabs in North Carolina were found to selectively decorate their carapace with a chemically noxious macroalga (*Dictyota menstrualis*), a behavior that significantly reduced predation. Although *Libinia* ranges north into Southern New England, *Dictyota* is a southern alga that is not present north of Virginia (fig. 14.3). At northern field sites beyond the range of *Dictyota*, crabs did not exhibit strong camouflage preferences. To further explore this pattern, crabs from northern and southern sources were simultaneously offered equal amounts of eight magroalgal species in laboratory choice assays. *Dictyota* represented over 80 percent of the algae selected and used for camouflage by the southern crabs. In contrast, northern crabs were generalists, selecting more evenly from three or four algal species, with *Dictyota* typically being <20 percent of the total (fig. 14.3). Stachowicz and Hay (2000) suggest that these behavioral differences may have a genetic basis. Significantly, crabs from Alabama showed a strong preference for this alga even though they did not have contact with *Dictyota* in the field. Previous field studies indicate that *Libinia* are vulnerable to fish consumption in the south, whereas these predators are rare or absent at northern sites. This suggests that selection on crabs has been stronger in the south, perhaps leading to geographic variation in the evolution of defensive behaviors (Stachowicz and Hay 2000; see also Fawcett 1984 for an additional example).

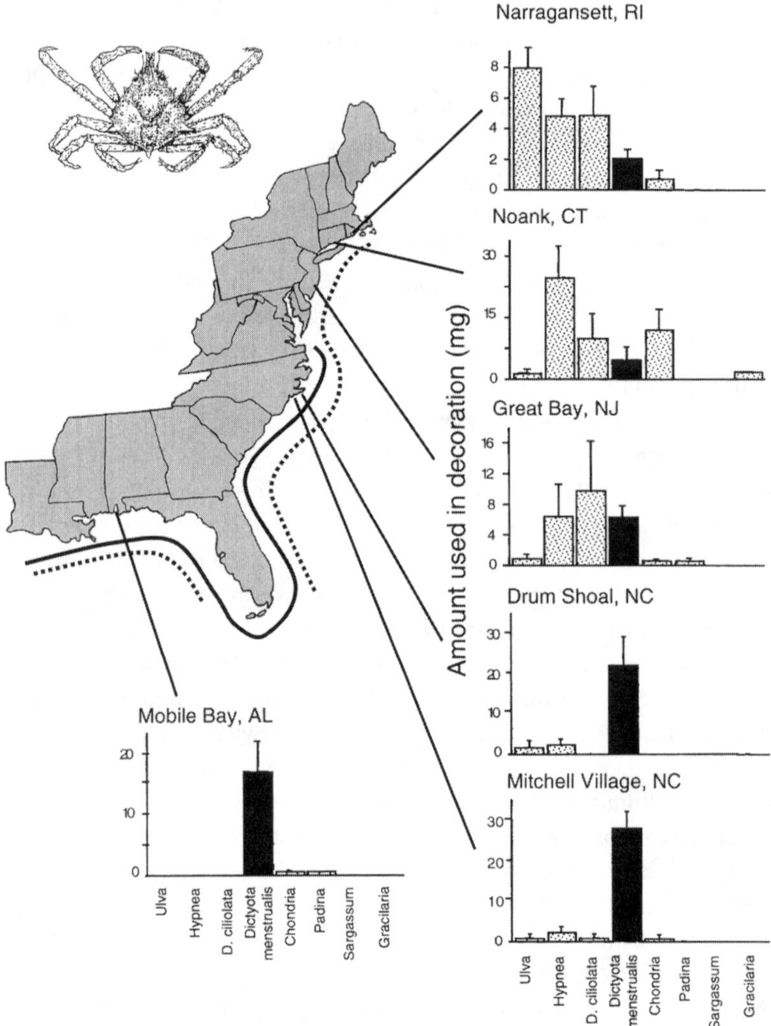

FIGURE 14.3 Camouflage preferences of decorator crabs (*Libinia dubia*) from six populations along the Atlantic coast of the United States (from Stachowicz and Hay 2000). The range of occurrence of *L. dubia* (dashed line) and *Dictyota menstrualis* (solid line) are shown on the map. Results are from laboratory assays in which crabs from each site were offered a simultaneous choice of the same eight algal species. Accumulated algae (mean + 1 SE) was removed from crabs after ~40 hours and weighed. See Stachowicz and Hay 2000 for additional details. Note that southern crabs used the noxious *Dictyota menstrualis* (dark bars) preferentially, whereas northern crabs selected a broader mix of algal species.

A similar pattern of geographic variation has been documented in the tolerance of an herbivorous amphipod (*Ampithoe longimana*) for the chemically rich alga *Dictyota* (Sotka and Hay 2002; Sotka, Wares, and Hay 2003). As with decorator crabs, the geographic distribution of this amphipod extends into New England. Thus, amphipod populations from Virginia to Maine live beyond the northern range of *Dictyota*. Laboratory feeding assays demonstrated that amphipods from southern populations fed more readily on *Dictyota* and were more tolerant of its defensive compounds than northern amphipods. Moreover, differences among these herbivore populations persisted through five generations in the laboratory, suggesting that this variation has a genetic basis. Sotka and Hay (2002) hypothesized that these differences reflect geographic variation in selection. Analysis of mitochondrial DNA and nuclear sequences indicated a strong historical separation between northern and southern *Ampithoe* populations, increasing the potential for adaptive differences to arise (Sotka, Wares, and Hay 2003). An evolved tolerance for *Dictyota*'s secondary metabolites may allow southern amphipods to avoid fish predation by living in association with this noxious alga. In contrast, selection may be diminished at northern sites where fish predation is less severe and *Dictyota* is absent. This study provides convincing evidence that herbivore-algal interactions may be modified over latitudinal scales by spatially varying selection.

Research Challenges and Approaches: Two Case Studies

The experimental studies we have summarized in the previous sections suggest some general conclusions regarding latitudinal variation in species interactions. First, it is clear that there is considerable geographic variation in the strength and outcome of species interactions. This variation is often a function of the abiotic and biotic context in which an interaction occurs. To the extent that factors modifying species interactions vary predictably with latitude, community dynamics and structure may exhibit regular patterns of geographic variation. However, many of the studies we have reviewed suggest that important abiotic and biotic drivers, such as upwelling intensity or the abundance of a key species, can show patchy variation over meso-spatial scales (<100 km). In some cases, even seemingly predictable variables like temperature may show complex spatial variation. For example, a latitudinal gradient in the body temperature of intertidal mussels is absent along the Pacific coast of North America, obscured by spatial variation in the timing of low tide (Helmuth et al. 2002). Such patterns of variation represent a challenge for ecologists interested in predicting how communities will vary across latitudinal scales. Moreover, there is a growing realization that evolution in marine species may operate over finer spatial and temporal scales than has generally been appreciated. Natural selection will shape and modify species interactions and may complicate attempts to extrapolate local experimental results to broader spatial scales (Pennings et al. 2003; Sanford et al. 2003). In the following sections, we provide two case studies from our research to further highlight some of these issues.

Local Selection and Latitudinal Variation in Whelk Predation

An important interaction between a carnivorous snail and mussel changes dramatically along the Pacific coast of North America, and this pattern appears to reflect variation in local selection (Sanford et al. 2003). The channeled dogwhelk, *Nucella canaliculata*, is common on wave-exposed rocky shores from just north of Point Conception, California, to Alaska. Among its prey is *Mytilus californianus*, a mussel that forms mid-intertidal beds along much of this coastline. Whelks feed on mussels and barnacles by using acid secretions and their radula to drill a borehole through the protective shell of their prey. Sanford et al. (2003) conducted field surveys of mussel beds at sixteen sites spanning 1,500 km of the coasts of California, Oregon, and Washington (fig. 14.4). The frequency of drilled *M. californianus* varied markedly with latitude. Whelk predation on *M. californianus* was intense at most sites in California. In contrast, drilling appeared weak at northern sites, particularly on the central Oregon coast where drilled shells were nearly absent at some sites. This pattern was even more surprising given that densities of *Nucella canaliculata* were at least twice as high at the Oregon sites than in California.

Geographic variation in drilling could reflect prey choice. Along the Oregon coast, many whelks were observed feeding intensely on their preferred prey, the thin-shelled blue mussel (*Mytilus trossulus*). Blue mussels were rare on the California coast, but were locally abundant in Oregon. Therefore, *M. californianus* may have escaped being drilled in Oregon, simply because whelks had access to more preferred prey. To test this hypothesis, we held adult *Nucella* from each of the sixteen sites in replicate laboratory containers with only *M. californianus* available as prey (Sanford et al. 2003). Drilling rates were monitored for one year, producing results that matched those observed in the field (fig. 14.4). Whereas most whelks from California sites drilled mussels intensely, drilling rates were low among Oregon whelks. Remarkably, many Oregon whelks did not drill a single mussel during the one-year study (apparently surviving on diatoms and debris in the container).

These results suggested that the ability to drill *M. californianus* could be a

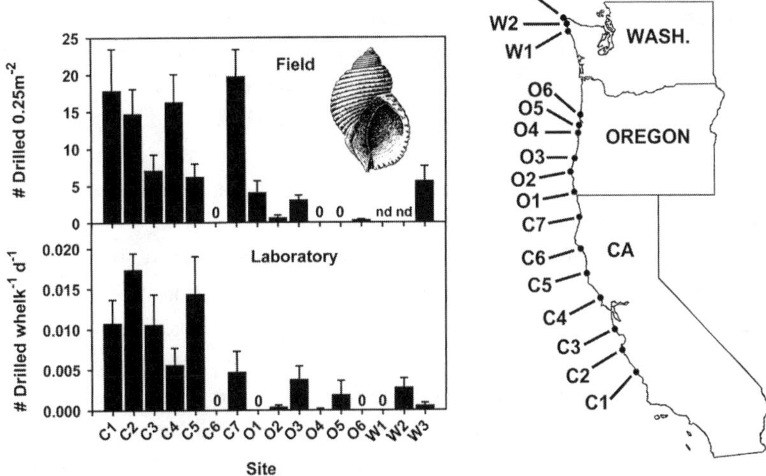

FIGURE 14.4 Latitudinal variation in predation by the channeled dogwhelk (*Nucella canaliculata*) along the Pacific coast of the United States (from Sanford et al. 2003). Map shows the location of study sites in California (C1–C7), Oregon (O1–O6), and Washington (W1–W3). See Sanford et al. 2003 for site names and coordinates. Field data (top panel) are mean density (+1 SE) of drilled mussels (*Mytilus californianus*) recorded in surveys of wave-exposed mussel beds at each site. Bars in lower panel are mean drilling rates (+ 1 SE) for field-collected whelks held in laboratory containers for one year with only *M. californianus* available as prey. In both panels, whelks from California show a greater tendency to drill *M. californianus* than snails from more northern populations. Nd = no data.

fixed trait. Unlike many marine invertebrates, *Nucella canaliculata* are direct developers with juvenile offspring that hatch and crawl away from intertidal egg capsules. Thus, *Nucella* lacks planktonic larvae and dispersal is limited. This life history favors low gene flow among populations and may increase the potential for adaptation to local conditions (Grosberg and Cunningham 2001). Geographic variation in predation could thus reflect evolutionary differences in drilling capacity. Alternatively, the drilling variation observed among field-collected whelks may have represented the persistent influence of prior feeding history.

To explore these hypotheses, we conducted a "common garden" experiment with snails reared from egg capsules under common conditions (Sanford et al. 2003). Egg capsules were collected from eight of the sixteen sites and returned to the laboratory where development was completed. Newly hatched whelks were then reared to adult size (in ~10 months) on a common diet of blue mussels. These lab-reared snails were then placed in independent containers with *Mytilus californianus*, a species they had never encountered. Each individual was scored as a "driller" or "nondriller" during a four-month feeding trial. The percentage of drillers varied sharply with the source of egg capsules: 75.9 percent of California whelks drilled *M. californianus*, compared to only 9.2 percent of Oregon whelks, and 4.1 percent of Washington whelks (fig. 14.5). Because whelks were raised under identical laboratory conditions and diet, these results strongly suggest that latitudinal differences in drilling capacity have a genetic basis.

We used mitochondrial DNA sequences to test whether there had been a historical separation between southern and northern populations of *Nucella canaliculata* (Sanford et al. 2003). Whelks were collected from thirteen of the sixteen sites, and 870 bp of the mitochondrial gene-encoding cytochrome c oxidase subunit I (COI) were sequenced. Analysis of the COI sequences did not support a distinct phyloegeographic split between southern and northern populations. Nor was there evidence of a second cryptic species among these populations. Rather, there was a clear signal of isolation by distance, consistent with a species having limited dispersal and low gene flow, distributed in a continuous, linear array of populations (Grosberg and Cunningham 2001).

Low gene flow among these populations increases the potential for spatial variation in selection to shape drilling ability. We hypothesized that the

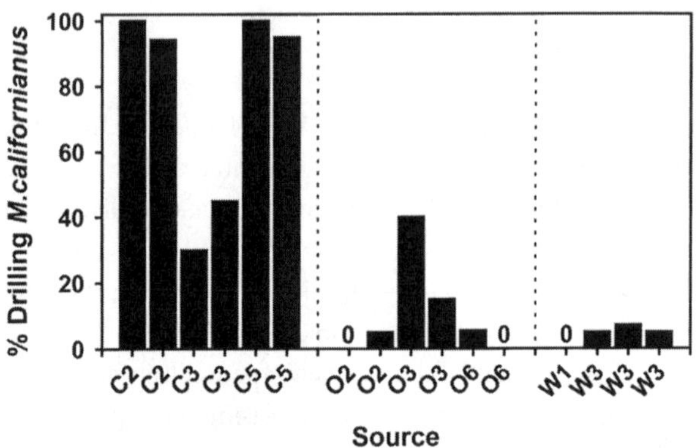

FIGURE 14.5 Variation in drilling predation among whelks (*Nucella canaliculata*) raised under common laboratory conditions (from Sanford et al. 2003). Egg capsules were collected from eight source populations (coded on the x-axis, as in fig. 14.4) and hatchlings were raised to adult size on a common diet. These naive whelks were then enclosed with *M. californianus* (shell length = 45 to 60 mm) and scored for their ability to drill this species. Bars are the percentage of lab-reared whelks (n = 10 to 20) from each set of capsules (n = 1 to 3 sets per site) that drilled *M. californianus* during a four-month trial.

feeding performance of whelks may be modified by latitudinal variation in upwelling intensity and prey availability. As noted earlier, the recruitment of barnacles and mussels is one to two orders of magnitude higher on the Oregon coast than in California, perhaps as a result of regional differences in upwelling patterns. These striking differences may alter the relative profitability of different prey species. In Oregon, thin-shelled prey such as blue mussels and acorn barnacles are abundant, and therefore the ability to drill *M. californianus* may have little effect on fitness. In contrast, whelks living at wave-exposed sites in central California are often dependent on *M. californianus* because preferred prey are found in low abundance here. This suggests that there may be a strong fitness advantage for whelks in California that are able to prey on *M. californianus*. Whether these differences in drilling performance are associated with differences in chemosensory detection, the structure of the radula, or some other mechanism, is under investigation.

Although whelk predation clearly varied among regions, preliminary evidence also suggested that finer scale variation may exist within regions. Thus, although *M. californianus* drillers were uncommon in most Oregon whelk populations, at least one population in Southern Oregon (Cape Arago: Site O3, fig. 14.4) contradicted this pattern. Significantly, this was a site marked by persistent upwelling and low recruitment, similar to most sites in California. Similar variation was observed among some sites in California (separated by ~100 km). This suggests that the strength of selection for *M. californianus* drilling could be modified over scales ≤100 km by local patterns of prey recruitment. Evidence points increasingly to the importance of shelf bathymetry, headlands, gyres, and fronts in determining patterns of larval transport and recruitment. Since many of these influences are spatially persistent features, sites may show consistent differences in recruitment. Thus, rather than wholesale variation between geographic regions, the drilling capacity of whelks may instead be shaped by a more subtle selection mosaic, imposed by local oceanographic processes. Experiments are underway in California and Oregon to test this hypothesis.

Geographic variation in whelk predation has important consequences for the dynamics of these intertidal communities. *Mytilus californianus* is a dominant competitor for space (Paine 1974) and an important habitat-forming species (Suchanek 1986). Because sea stars are larger and more efficient predators than whelks (Navarrete and Menge 1996), *Pisaster* probably plays a much greater role in regulating the lower distribution of mussels. However, how would these communities respond if a disturbance, such as disease (e.g., Dungan, Miller, and Thomson 1982; Leighton et al. 1991), reduced *Pisaster* abundances along the Pacific coast? We hypothesized that

Nucella would be able to partially fill *Pisaster*'s role, and that this response would vary with latitude (Sanford et al. 2003). To test this idea, *M. californianus* were transplanted to the low intertidal zone at two sites in central California and two sites in Oregon. Mussels were covered by plastic mesh cages that protected them from sea star predation. However, the mesh contained openings that were large enough to allow whelks to move freely in and out of the cages. After fourteen months, *Nucella canaliculata* had drilled 37 percent of the experimental mussels at the California sites, whereas not a single mussel was drilled in Oregon (Sanford, unpublished data). Remarkably, the density of *Nucella* in cages was seven times higher in Oregon than in California. Given the low recruitment and slow growth of mussels in California, these results suggest that following a loss of *Pisaster*, whelks might partially fill this species' functional role in California. These effects would be restricted to the very wave-exposed habitats that *Nucella canaliculata* occupies (*Pisaster* occurs over a much broader range of wave exposure). In contrast, it appears that whelks in Oregon would have little impact on *Mytilus californianus*. This study thus suggests that the functional role and community importance of a species may vary strongly with latitude. In this case, this result appears to be driven by restricted gene flow and spatial variation in selection, factors that deserve greater attention in marine systems.

Latitudinal Variation in Plant Interactions in Atlantic Coast Salt Marshes

Salt marshes are relatively simple systems and one of the most common shoreline communities on the east coast of North America from the Canadian Maritimes to Central Florida. Their extensive latitudinal distribution and simplicity make them an ideal model system to experimentally examine geographic variation in community structure and organization. Analogous to the formation of tropical reefs by corals, salt marshes are biogenic communities that are built and maintained by salt marsh organisms (Pennings and Bertness 2001). On the Atlantic coast of North America, salt marshes develop when the cordgrass, *Spartina alterniflora*, colonizes a shoreline. Cordgrass facilitates the development of salt marshes by increasing sedimentation and substrate stability and producing the belowground roots, rhizomes, and debris that form peat (Bruno 2000).

Salt marshes are built on a foundation of cordgrass peat and sediments that have accreted as sea level has increased over the past several hundred years (Donnelly and Bertness 2001), and so are the product of facilitated succession, where initially colonizing species pave the way for later colonizing species. The striking zonation of halophytic plants across Atlantic coast marshes, however, is largely driven by interspecific plant competition and the sharp elevation gradient in physical stress across marsh landscapes. In New England, high marsh elevations are dominated by competitively superior marsh plants that have a dense turflike clonal morphology that allow them to monopolize nutrient resources and space. These competitively dominant turfs displace competitively subordinate marsh plants to lower elevations that are too anoxic and stressful for the competitive dominants to survive. This simple assembly rule (Weiher and Keddy 1999), that competitively dominate plants displace subordinates to lower elevations, leads to a striking zonation of plants across salt marsh landscapes (Bertness 1999; Pennings et al. 2003).

Whereas the zonation of plants across New England is driven by interspecific plant competition, positive interspecific plant interactions (or facilitations) are also an integral process in salt marsh communities. These facilitations are typically the result of plant neighbors shading the substrate and limiting the evaporation of pore water in marsh sediments and the development of potentially lethal hypersaline soil conditions. For example, secondary succession in these systems occurs when competitively subordinate plants initially colonize disturbance-generated bare space and then facilitate the invasion of the competitive dominants (Bertness 1990). The mechanism of this facilitation has been shown to be salt stress amelioration (Bertness and Shumway 1993). When disturbances in marshes result in the creation of unvegetated bare space, the evaporation of surface pore water leads to bare patches having elevated soil salinities often over 100 ppt. These high soil salinities generally prevent successful seedling development and limit the colonization of hypersaline bare patches to salt tolerant clonal plants that move into patches assisted by the clonal integration of invading rhizome runners with neighbors in the dense vegetation. Once the initial clonal invaders colonize a bare patch, they shade the substrate, reducing surface evaporation and reducing the soil salinity. This leads to the invasion and displacement by competitively dominate turfs that have low salt tolerances. The amelioration of potentially lethal hypersaline conditions on southern New England salt marshes is also responsible for the coexistence of most of the forbs (i.e., broad-leaved herbaceous plants) that live in the high marsh clonal turf matrix (Hacker and Bertness 1999). These forbs are responsible for most of the species richness in these habitats and are dependent on their clonal neighbors to ameliorate hypersaline soil conditions. Without the high marsh clonal turf, the forbs die due to salt stress.

Since the development of hypersaline conditions in southern New England salt marshes is driven by solar radiation evaporating surface pore water from marsh soil, Bertness and Ewanchuk (2002) hypothesized that the relationship between plant neighbors was a function of climate. Specifically, we hypothesized that at lower latitudes with higher solar radiation, plant neighbor interactions in high salt marsh habitats would be largely facilitative due to neighbors ameliorating potentially lethal hypersaline soil conditions. In contrast, at higher latitudes, with cooler temperatures and less solar radiation, plant neighbor relationships were hypothesized to be largely competitive. We tested this latitudinal hypothesis by quantifying neighbor relationships in salt marshes north and south of Cape Cod, a major biogeographic and thermal stress boundary on the east coast of North America. Annually for four years (1996 to 1999), common New England marsh plants were transplanted into natural dense high marsh vegetation and into high marsh habitats where neighboring vegetation was removed.

The results revealed strong latitudinal as well as interannual variation in the strength and nature of plant neighbor interactions (fig. 14.6). South of Cape Cod in Narragansett Bay salt marshes, plants transplanted into the high marsh matrix of clonal plants generally had higher survival and growth than plants transplanted into high marsh habitats where neighbors had been

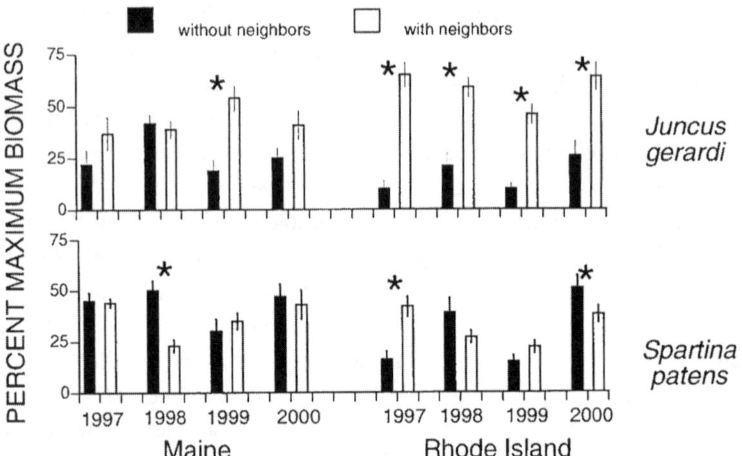

FIGURE 14.6 Field results demonstrating variation in the relative strength of facilitation among salt marsh plants in Rhode Island versus Maine (from Bertness and Ewanchuk 2002). Marsh hay (*Spartina patens*) and Black rush (*Juncus gerardi*) were transplanted into plots with and without neighbors during 1997 to 2000. Bars show the percent maximum biomass (mean ± 1 SE) attained in Maine and Rhode Island, based on pooled data from two sites in each region. An asterisk indicates a significant difference ($p < 0.05$, Scheffe test) between the with- and without-neighbor treatments. Note that plants in Rhode Island benefited from neighbors whereas interactions in Maine were generally neutral or competitive.

experimentally removed, revealing that south of Cape Cod, positive neighbor interactions prevailed. This general result, however, varied significantly among years. In cooler years, neighbor interactions south of Cape Cod were only weakly positive or slightly negative, while in warmer years neighbor interactions south of Cape Cod were strongly positive. This interannual variation in the nature of plant interactions revealed a very tight coupling between climate and the strength and nature of marsh plant interactions. North of Cape Cod in Southern Maine marshes, plants transplanted into the high marsh clonal plant matrix generally had lower survival and growth than plants transplanted into identical habitats where plant neighbors had been removed. Thus, north of Cape Cod, high marsh plant interactions were largely competitive. As was found south of Cape Cod, however, interannual variation in climate influenced this result. In warmer years plant interactions north of Cape Cod revealed less intense interspecific competition and neutral interspecific plant interactions.

Results from north and south of Cape Cod showing a tight relationship between climate and the nature and strength of salt marsh plant species interactions suggest that in the expansive marshes of southeastern North America (where summer temperatures are far higher than in New England and persist for six months of the year), plant species interactions would be strongly positive (fig. 14.7). In the salt marshes of Georgia, the Carolinas, and Florida, the evaporation of surface water from salt marshes is so severe that salt pans are common at intermediate elevations of many marshes. Salt pans are areas of marshes where the accumulation of salt is so extreme that most marsh plants cannot live in the hypersaline soils (Pennings and Bertness 2001).

Because of the increased hypersalinity of southern marshes, we hypothesized that neighbor relations among the plants in these marshes would be strongly positive, because neighbor salt stress amelioration would be more important than in New England marshes experiencing less intense solar radiation (Pennings et al. 2003). We tested this hypothesis using the same basic experimental design that we used in New England. We transplanted high marsh plants in salt marshes in Georgia and Alabama into dense natural vegetation and into plots where dense vegetation had been removed. The results were striking and unambiguous. Although we found that removing vegetation in southern marshes resulted in elevated salinities as expected, all of the southern marsh plants survived and grew better in unvegetated bare space than they did in natural dense vegetation. In contrast to our prediction that facilitation would be stronger in southern marshes than in New England marshes, we found no evidence for salt ameliorating facilitations

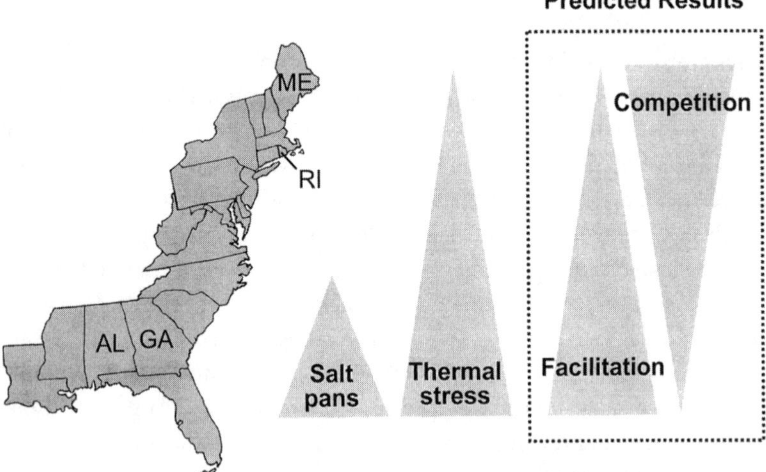

FIGURE 14.7 Latitudinal variation in salt marsh processes along the Atlantic coast of the United States (modified from Pennings and Bertness 2001). In southern marshes, greater heat and desiccation produce salt pans. Extrapolating from experimental results in New England, facilitation (i.e., soil shading by neighbors) was predicted to play an increased role under these stressful southern conditions. However, experiments conducted in Georgia and Alabama suggested that local marsh species were well adapted to salt stress, reducing the expected importance of facilitation (Pennings et al. 2003).

in southern salt marshes, and strong evidence that interspecific plant interactions in southern salt marshes were strongly competitive (Pennings et al. 2003).

What was responsible for this unexpected result? When we extrapolated from experimental work in New England to predict that the nature and intensity of plant species interactions would shift as a function of climate north and south of Cape Cod, our results strongly supported the hypothesis that interspecific plant species interactions among salt marsh plants were tightly linked to climate. When we tried to extend this result to a larger spatial scale, the predictions made by extrapolating results from local experiments to large continental spatial scales failed. In retrospect, the reason for this breakdown was clear. When working in New England salt marshes, the plant species composition of the marshes north and south of Cape Cod were all relatively similar—the plants were largely northern salt marsh plants. In our work in Georgia and Alabama, only one of the plants we worked with was also found at our northern sites, the common cordgrass, *Spartina alterniflora* (Pennings and Silliman 2005). The salt marshes of Georgia and

Alabama are instead dominated by a southern guild of plants that are much more salt tolerant than their northern marsh counterparts. Consequently, the prediction that positive plant neighbor interactions would be stronger in southern marshes than in New England marshes broke down because we moved into a different biogeographic region dominated by salt-tolerant plants. Thus, as was found on the Pacific coast with whelk drilling patterns, experimental results accurately predicted process and pattern at local spatial scales where high gene flow limited adaptation, but evolutionary processes prevented simple extrapolation from local predictions to latitudinal spatial scales.

Conclusions and Recommendations for Future Research

Using small-scale field experiments, ecologists have demonstrated convincingly that species interactions often influence local community structure. Thus, it is not surprising that species interactions also play a role in determining how communities vary across geographic scales. Rigorously examining such large-scale processes, however, has created logistical and conceptual challenges. The developing discipline of macroecology has encouraged ecologists to address these challenges by recognizing the value of a broader spatial and temporal perspective. Among the most powerful tools to emerge has been the comparative-experimental approach. This approach uses identical experiments replicated along a broad spatial gradient to test the potential influence of changing abiotic and biotic conditions that are not amenable to manipulation. In recent years, novel insights have been generated by combining these experimental results with data from satellite images, physical instrumentation, surveys of community composition, population genetics, physiology, and paleontology.

As more studies have quantified how interactions vary across large spatial scales, it appears that latitudinal gradients in species interactions may be more complex than originally imagined. Nevertheless, progress has been made in identifying the potential sources of geographic variation in interactions. More work will be needed to develop a predictive framework for how species interactions, and ultimately community dynamics, vary over large spatial scales. In pursuit of this goal, we have identified some critical areas for future research:

1. Geographic variation in interactions can be driven by the physiological response of organisms to their abiotic surroundings. However, direct extrapolations from laboratory studies or local field experiments may not

accurately predict how environmental variation will alter the strength of an interaction. To what extent are geographically separate populations influenced by physiological acclimatization and/or adaptation to local conditions? How does the potential for acclimatization vary among species and how is this related to other parameters (e.g., a species' vertical distribution, the size of its geographic range)?

2. A host of abiotic and biotic factors vary along any latitudinal gradient. How do we tease apart the relative influence of these factors on the strength of a particular species interaction? Does the same factor regulate other species interactions within the community?
3. Do the environmental factors that modify species interactions vary predictably with latitude? For example, in marine systems, coastal upwelling, nutrient availability, water temperature, and intertidal substrate temperatures are all influenced by local and regional features that can prevent a clear latitudinal gradient from developing. Improving our understanding of these processes will require closer collaborations between ecologists, oceanographers, and other physical scientists.
4. The strengths of species interactions are influenced by the density of the players involved. Many species have a patchy geographic distribution consisting of local "hot spots" and "cool spots." What factors determine variation in local abundance across broad spatial scales? How is the strength of a focal species interaction influenced by changes in community composition (i.e., the diversity and abundance of other "background" species in the community)?
5. To what extent are the strengths of local interactions shaped by evolutionary processes? Marine populations were viewed historically as being open and homogeneous, but recent molecular evidence suggests that many marine species have restricted gene flow among their populations. How common is local adaptation in marine species? To what extent is this influenced by the life history and phylogeographic history of the species involved?
6. How does variation in species interactions contribute to geographic variation in community dynamics and structure? How important are species interactions to patterns of local diversity relative to changes in the regional species pool?

With anthropogenic effects and habitat loss accelerating, we may be running out of time to answer these very basic questions about the organization of natural communities. Our best opportunity to examine these issues may be in the remaining natural intertidal systems such as exposed rocky shores and undisturbed salt marshes. Both of these systems are relatively easy to work with and occur in a reasonably intact condition across a large latitudinal gradient in both the northern and southern hemisphere. Regardless of the study system, we are optimistic that progress in macroecology will continue to be made by blending experimental and nonexperimental approaches within an interdisciplinary framework.

ACKNOWLEDGMENTS

We thank Jon Witman and Kaustuv Roy for inviting us to contribute to this volume. This chapter benefited from discussions with Bruce Menge, Bob Paine, and Brian Silliman, and was improved by constructive suggestions from Jon Witman and several anonymous reviewers. Our work was supported by National Science Foundation grant OCE-06-22924 (E.S.), the Partnership for Interdisciplinary Studies of Coastal Oceans (PISCO), funded by the David and Lucile Packard Foundation (E.S.), and the Ecology and Biological Oceanography Programs of the National Science Foundation (M.D.B).

REFERENCES

Ballantine, W. J. 1961. A biologically defined exposure scale for the comparative description of rocky shores. *Field Studies* 1:1–19.

Barber, P. H., S. R. Palumbi, M. V. Erdmann, and M. K. Moosa. 2000. Biogeography—A marine Wallace's line? *Nature* 406:692–93.

Benkman, C. W., W. C. Holimon, and J. W. Smith. 2001. The influence of a competitor on the geographic mosaic of coevolution between crossbills and lodgepole pine. *Evolution* 55:282–94.

Bertness, M. D. 1984. Habitat and community modification by an introduced herbivorous snail. *Ecology* 65:370–81.

———. 1990. Interspecific interactions among high marsh perennials. *Ecology* 72:125–37.

———. 1999. *The ecology of Atlantic shorelines.* Sunderland, MA: Sinauer.

Bertness, M. D., and R. Callaway. 1994. Positive interactions in communities. *Trends in Ecology and Evolution* 9:191–93.

Bertness, M. D., and P. J. Ewanchuk. 2002. Latitudinal and climate-driven variation in the strength and nature of biological interactions in New England marshes. *Oecologia* 132:392–401.

Bertness, M. D., S. D. Garrity, and S. C. Levings. 1981. Predation pressure and gastropod foraging: A tropical-temperate comparison. *Evolution* 35:995–1007.

Bertness, M. D., and S. W. Shumway. 1993. Competition and facilitation in marsh plants. *American Naturalist* 142:718–24.

Birkeland, C., and J. S. Lucas. 1990. *Acanthaster planci: Major management problem of coral reefs.* Boca Raton, FL: CRC Press.

Broitman, B .R., S. A. Navarrete, F. Smith, and S. D. Gaines. 2001. Geographic variation of southeastern Pacific intertidal communities. *Marine Ecology Progress Series* 224:21–34.

Brown, J. H. 1995. *Macroecology.* Chicago: University of Chicago Press.

Brown, J. H., D. W. Davidson, J. C. Munger, and R. S. Inouye. 1986. Experimental community ecology: The desert granivore system. In *Community ecology,* ed. J. Diamond and T. J. Case, 41–61. New York: Harper and Row.

Brown, J. H., and M. A. Kurzius. 1987. Composition of desert rodent faunas: Combinations of coexisting species. *Annales Zoologici Fennici* 24:227–37.

Brown, J. H., and B. A. Maurer. 1989. Macroecology: The division of food and space among species on continents. *Science* 243:1145–50.

Bruno, J. F. 2000. Facilitation of cobble beach plant communities through habitat modification by *Spartina alterniflora*. *Ecology* 81:1179–92.

Bustamante, R. H., and G. M. Branch. 1996. Large scale patterns and trophic structure of South African rocky shores: The roles of geographic variation and wave exposure. *Journal of Biogeography* 23:339–51.

Bustamante, R. H., G. M. Branch, S. Eekhout, B. Robertson, P. Zoutendyk, M. Schleyer, A. Dye, et al. 1995. Gradients of intertidal productivity around the coast of South Africa and their relationships with consumer biomass. *Oecologia* 102:189–201.

Castilla, J. C., and L. R. Duran. 1985. Human exclusion from the rocky intertidal zone of central Chile: The effects on *Concholepas concholepas* (Gastropoda). *Oikos* 45:391–99.

Clarke, A. 1987. The adaptation of aquatic animals to low temperatures. In *The effects of low temperatures on biological systems,* ed. B. W. W. Grout and G. J. Morris, 315–48. London: Edward Arnold.

Coleman, R. A., A. J. Underwood, L. Benedetti-Cecchi, P. Åberg, F. Arenas, J. Arrontes, J. Castro, et al. 2006. A continental scale evaluation of the role of limpet grazing on rocky shores. *Oecologia* 147:556–64.

Connell, J. H. 1961. The influence of interspecific competition and other factors on the distribution of the barnacle *Chthamalus stellatus*. *Ecology* 42 (4): 710–23.

Connolly, S. R., and J. Roughgarden. 1999. Theory of marine communities: Competition, predation, and recruitment-dependent interaction strength. *Ecological Monographs* 69:277–96.

Connolly, S. R., B. A. Menge, and J. Roughgarden. 2001. A latitudinal gradient in recruitment of intertidal invertebrates in the northeast Pacific ocean. *Ecology* 82:1799–1813.

Cossins, A. R., and K. Bowler. 1987. *Temperature biology of animals.* New York: Chapman and Hall.

Dayton, P. K. 1971. Competition, disturbance, and community organization: The provision and subsequent utilization of space in a rocky intertidal community. *Ecological Monographs* 41:351–89.

Dayton, P. K., and M. J. Tegner. 1984. The importance of scale in community ecology: A kelp forest example with terrestrial analogs. In *A new ecology: Novel approaches to interactive systems,* ed. P. W. Price, C. N. Slobodchikoff, and W.S. Gaud, 47–81. New York: Wiley.

Dethier, M. N., and D. O. Duggins. 1988. Variation in strong interactions in the intertidal zone along a geographic gradient: A Washington-Alaska comparison. *Marine Ecology Progress Series* 50:97–105.

Donnelly, J. P., and M. D. Bertness. 2001. Rapid shoreward encroachment of salt marsh cordgrass in response to accelerated sea-level rise. *Proceedings of the National Academy of Sciences USA* 98:14218–223.

Dungan, M. L., T. E. Miller, and D. A. Thomson. 1982. Catastrophic decline of a top carnivore in the Gulf of California rocky intertidal zone. *Science* 216:989–91.

Dunson, W. A., and J. Travis. 1991. The role of abiotic factors in community organization. *American Naturalist* 138:1067–91.

Endler, J. A. 1977. *Geographic variation, speciation, and clines.* Princeton, NJ: Princeton University Press.

Estes, J. A., and D. O. Duggins. 1995. Sea otters and kelp forests in Alaska: Generality and variation in a community ecological paradigm. *Ecological Monographs* 65:75–100.

Fawcett, M. H. 1984. Local and latitudinal variation in predation on an herbivorous marine snail. *Ecology* 65:1214–30.

Foster, M. S. 1990. Organization of macroalgal assemblages in the Northeast Pacific: The assumption of homogeneity and the illusion of generality. *Hydrobiologia* 192:21–33.

Gaines, S., S. Brown, and J. Roughgarden. 1985. Spatial variation in larval concentrations as a cause of spatial variation in settlement for the barnacle, *Balanus glandula. Oecologia* 67:267–72.

Gaines, S. D., and K. D. Lafferty. 1995. Modeling the dynamics of marine species: The importance of incorporating larval dispersal. In *Ecology of marine invertebrate larvae,* ed. L. McEdward, 389–412. Boca Raton, FL: CRC Press.

Gaines, S. D., and J. Lubchenco. 1982. A unified approach to marine plant-herbivore interactions. II. Biogeography. *Annual Review of Ecology and Systematics* 13:111–38.

Gilman, S. 2005. A test of Brown's principle in the intertidal limpet *Collisella scabra* (Gould, 1846). *Journal of Biogeography* 32:1583–89.

Grosberg, R. K., and C. W. Cunningham. 2001. Genetic structure in the sea: From populations to communities. In *Marine community ecology,* ed. M. D. Bertness, S. D. Gaines, and M. E. Hay, 61–84. Sunderland, MA: Sinauer.

Hacker, S. D., and M. D. Bertness. 1999. The role of positive interactions in the plant species diversity of salt marsh plant communities. *Ecology* 80:2064–73.

Harriott, V. J., and S. A. Banks. 2002. Latitudinal variation in coral communities in eastern Australia: A qualitative biophysical model of factors regulating coral reefs. *Coral Reefs* 21:83–94.

Hawkins, S. J., R. G. Hartnoll, J. M. Kain, and T. A. Norton. 1992. Plant animal interactions on hard substrata in the north-east Atlantic. In *Plant-animal interactions in the marine benthos,* ed. D. M. John, S. J. Hawkins, and J. H. Price, 1–32. New York: Oxford University Press.

Heck, K. L., and K. A. Wilson. 1987. Predation rates on decapod crustaceans in latitudinally separated seagrass communities: A study of spatial and temporal variation using tethering techniques. *Journal of Experimental Marine Biology and Ecology* 107:87–100.

Hellberg, M. E., D. P. Balch, and K. Roy. 2001. Climate-driven range expansion and morphological evolution in a marine gastropod. *Science* 292:1707–10.

Helmuth, B., C. D. G. Harley, P. Halpin, M. O'Donnell, G. E. Hofmann, and C. Blanchette. 2002. Climate change and latitudinal patterns of intertidal thermal stress. *Science* 298:1015–17.

Hixon, M. A., and W. N. Brostoff. 1983. Damselfish as keystone species in reverse: intermediate disturbance and diversity of reef algae. *Science* 220:511–13.

Hochachka, P. W., and G. N. Somero. 2002. *Biochemical adaptation: Mechanism and process in physiological evolution.* New York: Oxford University Press.

Hughes, T. P., A. H. Baird, E. A. Dinsdale, V. J. Harriott, N. A. Moltschaniwskyj, M. S. Pratchett, J. E. Tanner, and B. L. Willis. 2002. Detecting regional variation using meta analysis and large scale sampling: Latitudinal patterns of recruitment. *Ecology* 83:436–51.

Huston, M. A. 1999. Local processes and regional patterns: Appropriate scales for understanding variation in the diversity of plants and animals. *Oikos* 86:393–401.

Jeanne, R. L. 1979. A latitudinal gradient in rates of ant predation. *Ecology* 60:1211–24.

Jenkins, S. R., F. Arenas, J. Arrontes, J. Bussell, J. Castro, R. A. Coleman, S. J. Hawkins, et al. 2001. European-scale analysis of seasonal variability in limpet grazing activity and microalgal abundance. *Marine Ecology Progress Series* 211:193–203.

Jenkins, S. R., R. A. Coleman, P. Della Santina, S. J. Hawkins, M. T. Burrows, and R. G. Hartnoll. 2005. Regional scale differences in the determinism of grazing effects in the rocky intertidal. *Marine Ecology Progress Series* 287:77–86.

Jones, G. P. 1990. The importance of recruitment to the dynamics of a coral reef fish population. *Ecology* 71:1691–98.

Keough, M. J. 1988. Benthic populations: Is recruitment limiting or just fashionable? Townsville, AUS: *Proceedings of the Sixth International Coral Reef Symposium* 1:141–48.

Leighton, B. J., J. D. G. Boom, C. Bouland, E. B. Hartwick, and M. J. Smith. 1991. Castration and mortality in *Pisaster ochraceus* parasitized by *Orchitophyra stellarum* (Ciliophora). *Diseases of Aquatic Organisms* 10:71–73.

Leonard, G. H. 2000. Latitudinal variation in species interactions: A test in the New England rocky intertidal zone. *Ecology* 81:1015–30.

Louda, S. A. 1982. Distribution ecology: Variation in plant recruitment over a gradient in relation to insect seed predation. *Ecological Monographs* 52:25–41.

Lubchenco, J. 1978. Plant species diversity in a marine intertidal community: Importance of herbivore food preference and algal competitive abilities. *American Naturalist* 112:23–39.

Lubchenco, J., and B. A. Menge 1978. Community development and persistence in a low rocky intertidal zone. *Ecological Monographs* 48:67–94.

Lubchenco, J., A. Olson, L. Brubaker, S. Carpenter, M. Holland, S. Hubbell, S. Levin, et al. 1991. The sustainable biosphere initiative: An ecological research agenda. *Ecology* 72:371–412.

MacArthur, R. H. 1972. *Geographical ecology: Patterns in the distribution of species.* Princeton, NJ: Princeton University Press.

Maurer, B. A. 1999. *Untangling ecological complexity: The macroscopic perspective.* Chicago: University of Chicago Press.

Menge, B. A. 1976. Organization of the New England rocky intertidal community: Role of predation, competition and environmental heterogeneity. *Ecological Monographs* 46:355–93.

———. 1992. Community regulation: Under what conditions are bottom-up factors important on rocky shores? *Ecology* 73:755–65.

Menge, B. A., E. L. Berlow, C. A. Blanchette, S. A. Navarrete, and S. B. Yamada. 1994. The keystone species concept: Variation in interaction strength in a rocky intertidal habitat. *Ecological Monographs* 64:249–86.

Menge, B. A., C. Blanchette, P. Raimondi, T. Freidenburg, S. Gaines, J. Lubchenco, D. Lohse, et al. 2004. Species interaction strength: Testing model predictions along an upwelling gradient. *Ecological Monographs* 74:663–84.

Menge, B. A., B. Daley, P. A. Wheeler, E. Dahlhoff, E. Sanford, and P. T. Strub. 1997. Benthic-pelagic links in rocky intertidal communities: Evidence for bottom-up effects on top-down control. *Proceedings of the National Academy of Sciences USA* 94:14530–35.

Menge, B. A., B. A. Daley, J. Lubchenco, E. Sanford, E. Dahlhoff, P. M. Halpin, G. Hudson, and J. L. Burnaford. 1999. Top-down and bottom-up regulation of New Zealand rocky intertidal communities. *Ecological Monographs* 69:297–330.

Menge, B. A., and J. Lubchenco. 1981. Community organization in temperate and tropical rocky intertidal habitats: Prey refuges in relation to consumer pressure gradients. *Ecological Monographs* 51:429–50.

Menge B. A., J. Lubchenco, M. E. S. Bracken, F. Chan, M. M. Foley, T. L. Freidenburg, S. D. Gaines, et al. 2003. Coastal oceanography sets the pace of rocky intertidal community dynamics. *Proceedings of the National Academy of Sciences USA* 100:12229–34.

Menge, B. A., and A. M. Olson. 1990. Role of scale and environmental factors in regulation of community structure. *Trends in Ecology and Evolution* 5:52–57.

Menge, B. A., E. Sanford, B. A. Daley, T. L. Freidenburg, G. Hudson, and J. Lubchenco. 2002. An inter-hemispheric comparison of bottom-up effects on community structure: Insights revealed using the comparative-experimental approach. *Ecological Research* 17:1–16.

Menge, B. A., and J. P. Sutherland. 1987. Community regulation: Variation in disturbance, competition, and predation in relation to environmental stress and recruitment. *American Naturalist* 130:730–57.

Mora, C., P. M. Chittaro, P. F. Sale, J. P. Kritzer, and S. A. Ludsin. 2003. Patterns and processes in reef fish diversity. *Nature* 421:933–36.

Moreno, C. A., and J. P. Sutherland. 1982. Physical and biological processes in a *Macrocystis pyrifera* community near Valdivia, Chile. *Oecologia* 55:1–6.

Navarrete, S. A., and B. A. Menge. 1996. Keystone predation and interaction strength: Interactive effects of predators on their main prey. *Ecological Monographs* 66:409–29.

Nielsen, K. J., and S. A. Navarrete. 2004. Mesoscale regulation comes from the bottom-up: Intertidal interactions between consumers and upwelling. *Ecology Letters* 7:31–41.

Paine, R. T. 1966. Food web complexity and species diversity. *American Naturalist* 100:65–76.

———. 1974. Intertidal community structure: Experimental studies on the relationship between a dominant competitor and its principal predator. *Oecologia* 15:93–120.

———. 1980. Food webs: Linkage, interaction strength and community infrastructure. *Journal of Animal Ecology* 49:667–85.

———. 1991. Between Scylla and Charybdis—Do some kinds of criticism merit a response? *Oikos* 62:90–92.

Paine, R. T., J. C. Castillo, and J. Cancino. 1985. Perturbation and recovery patterns of starfish-dominated intertidal assemblages in Chile, New Zealand, and Washington State. *American Naturalist* 125:679–91.

Palumbi, S. R. 1995. Using genetics as an indirect estimator of larval dispersal. In *Ecology of marine invertebrate larvae,* ed. L. McEdward, 389–412. Boca Raton, FL: CRC Press.

Park, T. 1954. Experimental studies of interspecies competition. II. Temperature, humidity and competition in two species of *Tribolium*. *Physiological Zoology* 27:177–238.

Pennings, S. C., and M. D. Bertness. 2001. Salt marsh communities. In *Marine community ecology,* ed. M. D. Bertness, S. D. Gaines, and M. E. Hay, 289–316. Sunderland, MA: Sinauer.

Pennings, S. C., E. Selig, L. T. Houser, and M. D. Bertness. 2003. Geographic variation in positive and negative interactions among salt marsh plants. *Ecology* 84:1527–38.

Pennings, S. C. and B. R. Silliman. 2005. Linking biogeography and community ecology: Latitudinal variation in plant-herbivore interaction strength. *Ecology* 86:2310–19.

Pianka, E. R. 1966. Latitudinal gradients in species diversity: A review of concepts. *American Naturalist* 100:33–46.

Rhode, K. 1992. Latitudinal gradients in species diversity: The search for the primary cause. *Oikos* 65:514–27.

Ricklefs, R. E. 1987. Community diversity: Relative roles of local and regional processes. *Science* 235:167–71.

Rosenzweig, M. L. 1995. *Species diversity in space and time.* New York: Cambridge University Press.

Roughgarden, J., S. Gaines, and H. Possingham. 1988. Recruitment dynamics in complex life cycles. *Science* 241:1460–66.

Sagarin, R. D., and S. D. Gaines. 2002. Geographical abundance distributions of coastal invertebrates: Using one-dimensional ranges to test biogeographic hypotheses. *Journal of Biogeography* 29:985–97.

Sanford, E. 1999. Regulation of keystone predation by small changes in ocean temperature. *Science* 283:2095–97.

———. 2002a. The feeding, growth and energetics of two rocky intertidal predators (*Pisaster ochraceus* and *Nucella canaliculata*) under water temperatures simulating episodic upwelling. *Journal of Experimental Marine Biology and Ecology* 273:199–218.

———. 2002b. Water temperature, predation, and the neglected role of physiological rate effects in rocky intertidal communities. *Integrative and Comparative Biology* 42:881–91.

Sanford, E., M. S. Roth, G. C. Johns, J. P. Wares, and G. N. Somero. 2003. Local selection and latitudinal variation in a marine predator-prey interaction. *Science* 300:1135–37.

Sotka, E. E. 2005. Local adaptation in host use among marine invertebrates. *Ecology Letters* 8:448–59.

Sotka, E. E. and M. E. Hay. 2002. Geographic variation among herbivore populations in tolerance for a chemically rich seaweed. *Ecology* 83:2721–35.

Sotka, E. E., J. P. Wares, J. A. Barth, R. K. Grosberg, and S. R. Palumbi. 2004. Strong genetic clines and geographic variation in gene flow in the rocky intertidal barnacle *Balanus glandula*. *Molecular Ecology* 13:2143–56.

Sotka, E. E., J. P. Wares, and M. E. Hay. 2003. Geographic and genetic variation in feeding preferences for chemically defended seaweeds. *Evolution* 57:2262–76.

Southward, A. J., and D. J. Crisp. 1956. Fluctuations in the distribution and abundance of intertidal barnacles. *Journal of the Marine Biological Association UK* 35:211–29.

Southward, A. J., S. J. Hawkins, and M. T. Burrows. 1995. Seventy years' observations of changes in distribution and abundance of zooplankton and intertidal organisms in the western English Channel in relation to rising sea temperature. *Journal of Thermal Biology* 20:127–55.

Sousa, W. P. 1979. Disturbance in marine intertidal boulder fields: The nonequilibrium maintenance of species diversity. *Ecology* 60:1225–39.

Stachowicz, J. J., and M. E. Hay. 2000. Geographic variation in camouflage specialization by a decorator crab. *American Naturalist* 156:59–71.

Suchanek, T. H. 1986. Mussels and their role in structuring rocky shore communities. In *The ecology of rocky coasts,* ed. P. G. Moore and R. Seed, 70–96. New York: Columbia University Press.

Sutherland, J. P. 1974. Multiple stable points in communities. *American Naturalist* 108:859–73.

Taylor, M. S., and M. E. Hellberg. 2003. Genetic evidence for local retention of pelagic larvae in a Caribbean reef fish. *Science* 299:107–09.

Thompson, J. N. 1998. Rapid evolution as an ecological process. *Trends in Ecology and Evolution* 13:329–32.

———. 1999. The evolution of species interactions. *Science* 284:2116–8.

Thompson, J. N., and B. M. Cunningham. 2002. Geographic structure and dynamics of coevolutionary selection. *Nature* 417:735–38.

Travis, J. 1996. The significance of geographic variation in species interactions. *American Naturalist* 148:S1–S8.

Underwood, A. J. and E. J. Denley. 1984. Paradigms, explanations, and generalizations in models for the structure of intertidal communities on rocky shores. In *Ecological communities: Conceptual issues and the evidence,* ed. D. R. Strong, D. Simberloff, L. G. Abele, and A. B. Thistle, 151–80. Princeton, NJ: Princeton University Press.

Vermeij, G. J. 1974. Marine faunal dominance and molluscan shell form. *Evolution* 28:656–64.

———. 1977. Patterns in crab claw size: The geography of crushing. *Systematic Zoology* 26:138–51.

———. 1978. *Biogeography and adaptation: Patterns of marine life.* Cambridge, MA: Harvard University Press.

———. 1980. Geographic variation in the strength of Thaidid snail shells. *Biological Bulletin* 158:383–89.

Vitousek, P. M. 1994. Beyond global warming: Ecology and global change. *Ecology* 75:1861–76.

Wares, J. P., and C. W. Cunningham. 2001. Phylogeography and historical ecology of the North Atlantic intertidal. *Evolution* 55:2455–69.

Weiher, E. and P. Keddy. 1999. Introduction: The scope and goals of research on assembly rules. In *Ecological assembly rules: Perspectives, advances, retreats,* ed. E. Weiher, and P. Keddy, 1–22. New York: Cambridge University Press.

Wethey, D. S. 1983. Geographic limits and local zonation: The barnacles *Semibalanus* (*Balanus*) and *Chthamalus* in New England. *Biological Bulletin* 165:330–41.

———. 1984. Sun and shade mediate competition in the barnacles *Chthamalus* and *Semibalanus*: A field experiment. *Biological Bulletin* 167:176–85.

———. 2002. Biogeography, competition, and microclimate: The barnacle *Chthamalus fragilis* in New England. *Integrative and Comparative Biology* 42:872–80.

Wiens, J. A., J. F. Addicott, T. J. Case, and J. Diamond. 1986. Overview: The importance of spatial and temporal scale in ecological investigations. In *Community ecology,* ed. J. Diamond and T. J. Case, 145–53. New York: Harper and Row.

Witman, J. D., and K. P. Sebens. 1992. Regional variation in fish predation intensity: A historical perspective in the Gulf of Maine. *Oecologia* 90:305–15.

Witman J. D., and F. Smith. 2003. Rapid community change at a tropical upwelling site in the Galapagos Marine Reserve. *Biodiversity and Conservation* 12:25–45.

CHAPTER FIFTEEN

THE SUBTIDAL ECOLOGY OF ROCKY COASTS: LOCAL-REGIONAL-BIOGEOGRAPHIC PATTERNS AND THEIR EXPERIMENTAL ANALYSIS

SEAN D. CONNELL AND ANDREW D. IRVING

Introduction

Marine ecologists are becoming increasingly aware that they are working at scales where complexity is often greatest (i.e., local or small scale: Anderson et al. 2005; Fowler-Walker, Connell, and Gillanders 2005). At local scales, patterns are likely to represent the outcome of special and unique events that incorporate variation from local to broad scales. Unsurprisingly, we tend to be captivated by the description and explanation of local variation while being pessimistic about the existence of broader patterns that may be understood through observation and experimentation.

Our success with experimental ecology may well have hampered tests of ecological generality (Brown 1995). While the phenomenal rise of experimental ecology transformed us from a "descriptive" to a "rigorous" science, it has also held us to the intense study of local phenomena at select localities. Hence, ecological understanding is often fragmentary, incomplete (tests of subsets of processes), but rigorous to the extent that uncertainty is narrowed to the point of acceptance by peers (unambiguous and precise). It is clear that lack of assessments of the extent to which local patterns are representative of larger areas contributes to a culture of interest in the discovery of new details (dissimilarity in patterns) and publication of idiosyncratic patterns and processes (Underwood, Chapman, and Connell 2000).

The recent recognition of meaningful patterns at broad scales has fostered discussion about which avenues of enquiry may further identify general patterns and responses in nature (Gaston and Blackburn 2000). The accumulation of detail in the description and explanation of local scale variation may increase confidence in our predictive understanding of local phenomena, but it contributes little toward knowledge of the broad spatial generality of such patterns (Keddy 2001). Similarly, macroecology cannot afford to become an isolated subdiscipline that is disconnected from the scales at which most ecological phenomena are researched and understood. We have long recognized that local processes may be at least as important as broad-scale processes, and local variance cannot be ignored as simple statistical nuisance (Horne and Schneider 1995).

The problem confronting contemporary ecologists is not whether one should test for the existence of general phenomena (macroecology) or specific phenomena, but what balance should be sought between the two. One approach focuses on tractable components of a system amenable to experimentation with clear and unambiguous results at local scales that become less certain as the results are scaled up (Thrush et al. 1997). The other is an integrative and holistic approach, which searches for simple structures at broad scales that may appear overly simplistic at local scales where complexity is large (Peters 1991). Integration of both approaches (fig. 15.1) offers solutions to "the problem of relating phenomena across scales (which) is the central problem in biology and all of science" (Levin 1992, p.1961; also see Witman and Roy, this volume).

A promising area of research centers on local-regional-biogeographic patterns in the subtidal ecology of exposed coastal Australasia. The rocky coastline of Southern Australia represents the world's longest east-west temperate coastline, providing a unique opportunity to understand the effects of scale-dependency unfettered by tropical-temperate gradients. This chapter demonstrates how structured sampling and replicated experiments over broad scales have reconciled conflicting studies and provided fresh opportunities to understand local patterns within their regional and biogeographic contexts.

Identifying Local-Regional Patterns

Processes operating from local (i.e., m–km) to regional scales (i.e., thousands of km) may produce regional scale patterns. Three general models may be applicable. First, a regional pattern occurs because of some local scale process that occurs repeatedly across large distances, but differs in intensity among regions. Second, some truly regional scale process occurs in some regions,

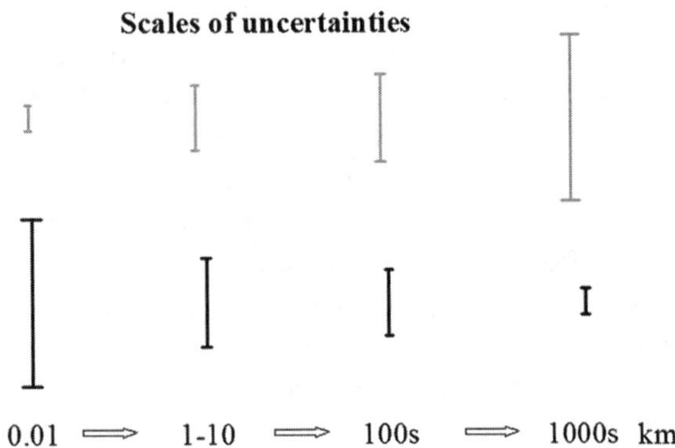

FIGURE 15.1 Conceptual diagram of the magnitude of uncertainty (size of bar) associated with spatial scale and methodology (small-scale experiments versus broad-scale observation). Broad-scale descriptions may overly simplify local patterns (but provide clear regional contexts for their interpretation) and local experiments may provide incomplete and uncertain outcomes for understanding generality (but provide unambiguous and precise information at local scales). A challenge is to integrate both approaches (after Connell 2007a). Reproduced with permission from Connell (2007) "Subtidal temperate rocky habitats: Habitat heterogeneity at local to continental scales."

but not in others. Third, interactions exist between local and regional scale processes where a regional process (e.g., upwelling) negates the effect of a local process (e.g., grazing) in one region, but such local effects are detectable in other regions where the regional scale process is absent. These models have consequences for how we assess macroecological patterns.

Assessments of Pattern from Independent Studies

Comparisons of local studies can provide a difficult basis to judge the generality or otherwise of pattern and process. Most reviews of the ecological literature represent an attempt to compare, synthesize, and identify generalities in pattern and process. Of their nature, such efforts also involve an assessment of the spatial generality of independent studies typically done at local scales across disparate parts of the globe. Perhaps because complexity tends to be greatest at local scales, such comparisons often highlight inconsistencies from place to place at all scales. Inconsistencies among local studies can lead to confusion about how representative a particular study is of other localities, and cultivate pessimism about the existence of generality.

Comparisons of local observations of canopy-understorey associations across Australia have, until recently, suggested massive variation across all scales, reinforcing the idea that few general patterns exist at any scale (fig 15.2. cf. Kendrick, Lavery, and Phillips 1999 of Western Australia v. Melville and Connell 2001 of South Australia v. Daume, Brand-Gardner, and Woelkerling 1999 of South and Southeast Australia v. Kennelly 1987 of East-

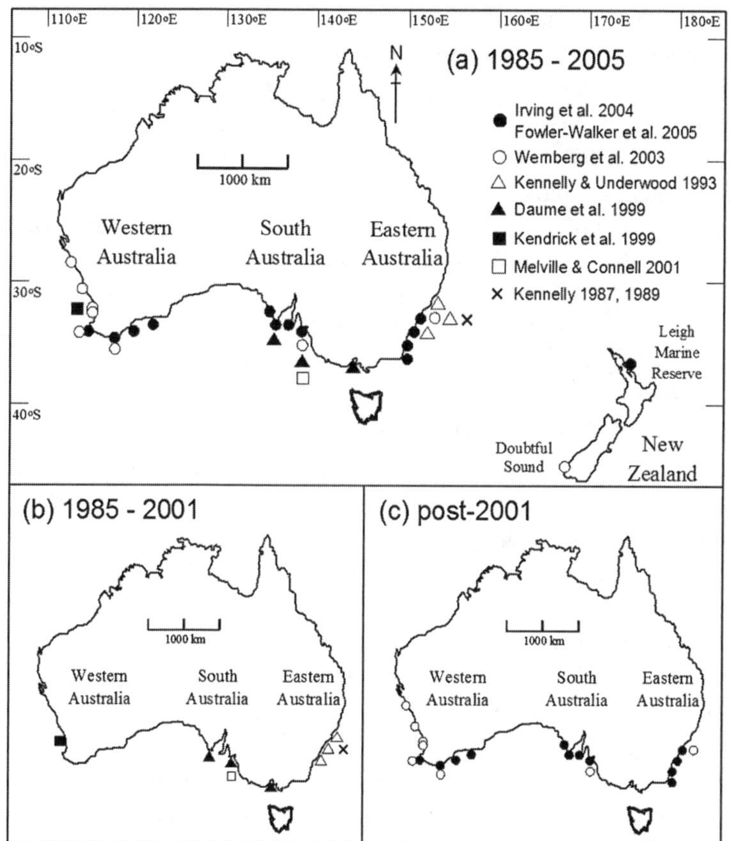

FIGURE 15.2 Map of Australasia showing the spatial arrangement of locations used to assess kelp canopy-understorey associations. These include locations of (A) studies done between 1985 and 2005 across multiple regions (Wernberg et al. 2003; Irving, Connell, and Gillanders 2004; Fowler-Walker, Connell, and Gillanders 2005), at multiple sites within regions (Kennelly and Underwood 1993; Daume, Brand-Gardner, and Woelkerling 1999), or at a single site within a region (Kennelly 1987; Kennelly 1989; Kendrick, Lavery, and Phillips 1999; Melville and Connell 2001). (B) studies done between 1985 and 2001 (primarily local scale). (C) studies done post-2001 (broad scale).

ern Australia). Subtidal ecologists (e.g., Daume, Brand-Gardner, and Woelkerling 1999) emphasized local complexity and inadequacy of simple models, citing work from similar subtidal systems across Australasia (i.e., Schiel 1990; Kennelly and Underwood 1992; Schiel, Andrew, and Foster 1995). In frustration with the way such differences were used to highlight inconsistent effects at all scales, Fowler-Walker and Connell (2002) tested whether local variation in canopy-understorey associations transcended local through regional scales. By testing scale dependency of such patterns across Australia, they established that patterns can emerge at broader scales (hundreds and thousands of km) from substantial complexity at local scales (i.e., among site patterns, 1 to 10 km apart).

These similarities/dissimilarities also highlight that comparisons of studies done at small scales, even if done at several sites in a locality, can provide a difficult basis to understand generality due to large variation at local scales. The realization of broad-scale patterns in canopy-understorey associations has reconciled apparently conflicting studies, provided context for the interpretation of within-region consistencies (e.g., Eastern Australia: Kennelly and Underwood 1993) and inspired confidence in subsequent assessments that incorporated local pattern in the context of regional scales (encompassing distances of >3000 km; Irving, Connell, and Gillanders 2004). It is clear that inconsistencies among local studies can be poor evidence for models that suggest variation at local scales is an overwhelming feature of natural systems, and such comparisons may erroneously reinforce the idea that few spatial generalizations are possible.

In situ Assessments of Scale-Dependent Patterns

The appreciation of scale-dependent patterns is not novel; what is pioneering is the recent application of concepts in sampling and experimental design to tests about scale dependency from local to broad scales. A review of scale-dependent assessments of marine systems revealed thirteen such studies at scales of ~100 km and only six at scales > thousands of kilometers (Fraschetti, Terlizzi, and Benedetti-Cecchi 2005). The lack of such assessments may, in part, stem from their costs, highlighting the need for broad-scale assessments to be well planned. The need for careful assessments of local through regional scale patterns is of particular concern because a relatively small number of assessments result in stronger emphasis being placed on fewer papers.

Possibly one of the best-understood approaches to assessments of scale dependency centers on the explicit incorporation of spatial scale into sampling and experimental design such that a hierarchy of spatial scales are included (Green 1979; Underwood 1997). The application of nesting observational or experimental units within successively larger scales that span the region(s) of interest is widely recognized. However, the decision to not use hierarchical sampling, or intensely focus on a few replicate locations, can result in the failure to detect real patterns. An example centers on two studies testing for regional patterns of kelp morphology across Australia. One study emphasized local variation (relatively unstructured approach with few replicate observations: Wernberg et al. 2003) and the other emphasized local variation nested within regional patterns (hierarchical approach within many replicate observations: Fowler-Walker, Connell, and Gillanders 2005). Critically, these differences in interpretation cannot be easily explained by differences in the variables measured or the localities and times of sampling. Both studies are remarkably similar in all ways except sampling design (fig. 15.2, panel C). It appears that the inability to adequately estimate the within-region variation by replicate samples prevents interpretable comparison among regions and leads to greater emphasis on local variation. An advantage of hierarchical sampling across scales of interest is that it provides an estimate of the contribution of each scale to the total variation across regions. By understanding the proportion of total variation that is attributable to each scale, we are in a stronger position to identify the scales at which general patterns emerge.

Greater attention to replication within regions is often required if we are to enable regional comparisons. Where variation within regions (among replicate sites) is as great as variation among regions, local processes appear at least as important as larger-scale processes. Alternatively, greater variation among regions suggests pattern at regional scales, notwithstanding substantial variation that may occur within regions. Such outcomes can lead to the discovery of new patterns (e.g., scaling up; Irving, Connell, and Gillanders 2004), providing fresh perspectives on seemingly well-understood local phenomena (Ricklefs 2004). Furthermore, understanding variation from local to regional scales can identify the spatial extent of similarities, identifying with greater confidence the limits from which local knowledge can be scaled up to broader areas of coast (Thrush et al. 1997).

Future Challenges in the Application of Structured Assessments

There are costs involved in the search for generality. These costs often sacrifice specific information for breadth and ignore some special feature of the environment which, when taken into account, could improve predictive power (Wiens 1989). Local processes may be at least as important as regional processes in generating patterns, and ignorance of local patterns may hamper the detection of regional patterns. For example, while canopy-understorey associations vary enormously at local scales across temperate Australasia, such variation may be a consequence of specific local drivers (e.g., type of stand: mixed-species versus monospecific canopies) as well as regional drivers. If this local pattern (e.g., canopy-understorey association) is strongly associated with an unrecognized feature of the environment (e.g., type of stand), then tests of broad-scale patterns may be compromised. Indeed, for every square meter of Western and South Australian rocky coast that supports the laminarian canopy-forming alga *E. radiata* (0 to 18 meters depth), it is estimated that more than half occur as mixed-species stands (*E. radiata* and fucalean canopy-formers) and the rest as monospecific stands (Goodsell et al. 2004). Importantly, the effect of stand as a driver of benthic patterns is sufficiently strong (Irving and Connell 2006) that its effects appear to scale up to generate regional patterns in benthos (Irving, Connell, and Gillanders 2004), highlighting the value of detailed local knowledge to regional assessments.

To date, tests of generality often compare replicate sites (sometimes ordered into a hierarchy) between widely separated regions (e.g., New Zealand in South Pacific versus Oregon in North Pacific: Menge et al. 2002) with the expectation that similarity between distant regions provides powerful inferences for generality (i.e., across regions not studied). While these approaches may create the opportunity for rapid progress, interpretation of spatial generalities are hampered because of a lack of insight into the scales and places where similarity ends (e.g., spatial extent of generalities at "middle" scales). While the use and advantages of the hierarchical approach are widely accepted, there remains much-needed discussion on its use in extending ecological knowledge beyond high context dependency of local phenomena and low predictive value to new sites (Noda 2004).

We recognize the need for ecologists to continually balance interpretations of dissimilarity and similarity in patterns and response. Dissimilarity (natural variation) is an ecological phenomenon that ecologists first ignored, then recognized (McIntosh 1980) and now embrace (Underwood 1997; Benedetti-Cecchi 2003). Indeed, the word *variation* is common to the titles of many papers intent on identifying and explaining *differences*. This tunnel vision for dissimilarity promotes a culture that is intent on the discovery of new details (dissimilarity in patterns) and publication of idiosyncratic patterns and processes (Peters 1991), biasing ecology against the development of ecological generality. Restoring balance would require explicit interpretations of similarities and identification of common components in patterns and experimental responses.

Understanding Benthic Heterogeneity: Observation and Experimentation

Observation of canopy-morphology and understorey associations across Australia has revealed two types of kelp forest: one characteristic of Western Australia and South Australia, and the other of Eastern Australia. Individual plants in Eastern Australia are smaller and connected to longer and thicker stipes, whereas Western and South Australian plants are generally larger and connected to shorter and slender stipes (fig. 15.3, panel A; Fowler-Walker, Connell, and Gillanders 2005). The shorter and more flexible forms are associated with more extensive covers of prostrate understorey taxa (en-

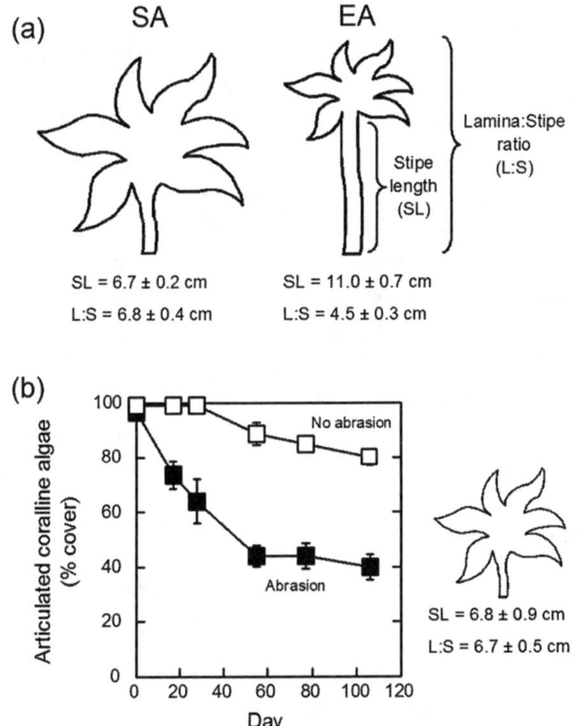

FIGURE 15.3 (A) Comparison of the mean (± SE) values of morphological variables (stipe length and lamina:stipe length ratio) of *Ecklonia radiata* between South and Eastern Australia (after Fowler-Walker, Connell, and Gillanders 2005). The South Australian morphology of *E. radiata* matches closely with morphologies known to produce strong negative effects of abrasion on erect taxa (B), such as articulated coralline algae. Panel (B) partially reproduced with permission from Irving and Connell (2006).

crusting coralline algae in Western and South Australia), whereas the taller and more rigid plants are associated with more extensive covers of erect taxa (invertebrates, foliose and turf-forming algae in Eastern Australia; Fowler-Walker et al. 2005).

Regional differences in the structure of understorey assemblages may be partially explained by regional differences in kelp morphology and its local effects on physical variables like shade and abrasion of the substratum. There are regional differences in the intensity with which kelp cause direct physical disturbance to the substratum. In Eastern Australia, abrasion has negligible effects on several species of understorey algae and sessile invertebrates (Kennelly 1989), whereas in South Australia, it has large negative effects on sessile invertebrates (Connell 2003b) and erect algae (Irving and Connell 2006). The intensity of abrasion decreases with increasing stipe length (Kennelly 1989), and kelp in Eastern Australia generally have longer stipes and are more rigid than plants in South Australia (Fowler-Walker, Connell, and Gilllanders 2005). The averages of morphological dimensions of kelp across Western and South Australia match closely with those known to produce strong effects of abrasion on erect algae (fig. 15.3). Experimental tests demonstrate that this regional variation in the morphology of *E. radiata* contributes substantially to regional variation in the intensity of abrasion and its local effects on understorey taxa, linking a local scale process (abrasion) with regional scale patterns (morphology and understorey; B. D. Russell and S. D. Connell, unpublished data).

Integration of local-regional-biogeographic knowledge (fig. 15.1) will not only rely on linking local scale processes with regional patterns (i.e., spatially replicated manipulations of key factors that drive local patterns that are repeated across space), but also link these to regional and biogeographic phenomena (e.g., oceanographic drivers of regional differences in canopy productivity and morphology).

Regional Processes: Top-Down and Bottom-Up

The paradigm of fishing's impact on coastal kelp habitats cascading down to much-simplified urchin-barrens has proven to be very general (Steneck et al. 2002). Yet, there is tremendous need to recognize local through to regional and global patterns in the strength of top-down processes (e.g., Witman and Sebens 1992) and their interaction with bottom-up processes (Connell 2007b).

Top-Down Forcing

Overfishing of vertebrate predators at the top of the food chain can trigger increases in herbivore populations, leading to widespread deforestation (also see Tittensor, Worm, and Myers, this volume). Australasia's best-understood example of this top-down-driven process occurs in northern New Zealand where the closure of fishing (Marine Protected Areas) resulted in widespread expansion of kelp forests (*E. radiata*), reflecting a concomitant decline in sea urchin populations with increasing densities of predatory fish (fig. 15.4; Shears and Babcock 2002). Such effects appear possible in Eastern Australia (fig. 15.2; latitudes >33° to 37° S), but not at equivalent latitudes in Western and South Australia.

A key distinction between these regions centers on the density of, and efficacy with which benthic grazers maintain surfaces free from overgrowth by erect algae (fig. 15.5). Grazers that scrape the substratum (i.e., remove algae at their base) form dense populations in Eastern Australia, but are sparsely distributed in Western and South Australia (Fowler-Walker and Connell

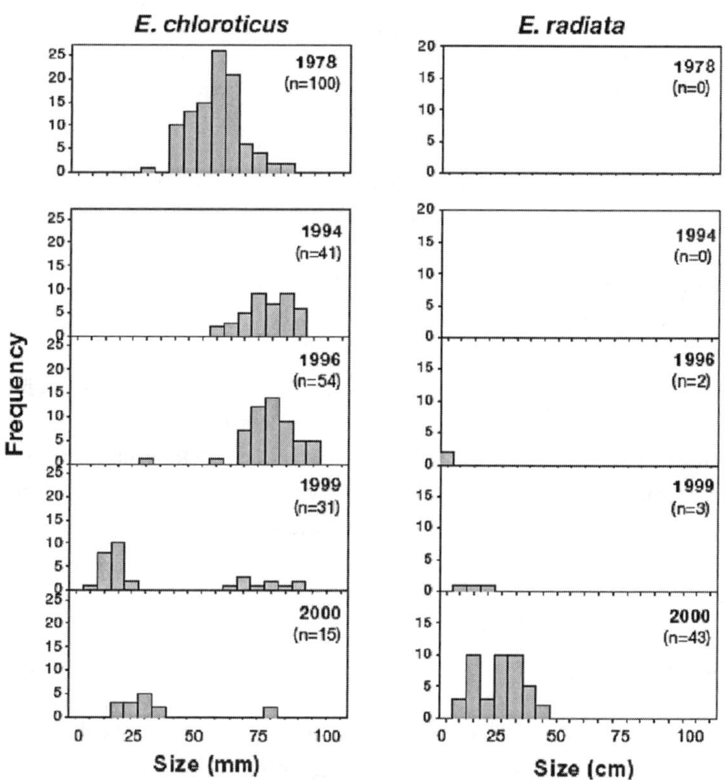

FIGURE 15.4 Changes in urchin (*Evechinus chloroticus*) and kelp (*Ecklonia radiata*) size frequency distributions at Waterfall Reef (Leigh Marine Reserve, New Zealand) between 1978 and 2001. Reproduced with permission from Shears and Babcock (2003).

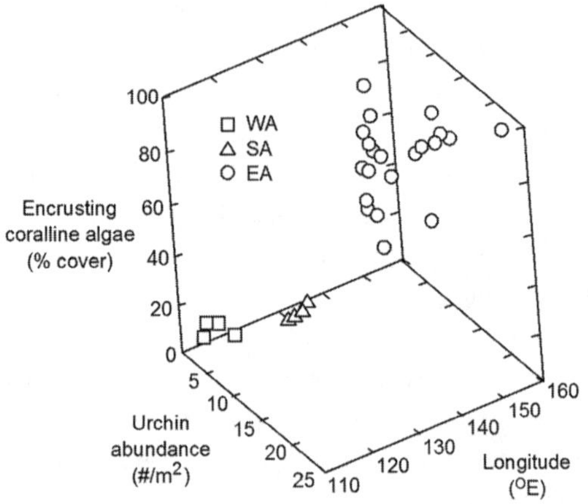

FIGURE 15.5 Variation in the benthic covers of encrusting coralline algae and the abundance of urchins (grazers) with longitude across temperate Australia (WA = Western Australia, SA = South Australia, EA = Eastern Australia). Data were compiled from the following papers: Andrew and Underwood 1989; Underwood, Kingsford, and Andrew 1991; Andrew 1993; Andrew and Underwood 1993; Andrew and O'Neill 2000; Fowler-Walker and Connell 2002; Hill et al. 2003; Irving, Connell, and Gillanders 2004; Vanderklift and Kendrick 2004.

2002; Vanderklift and Kendrick 2004). Moreover, the ability of these grazers to maintain surfaces free from erect algae is barely detectable in South Australia (fig. 15.6), but in Eastern Australia are a dominant force that creates and maintains spatial heterogeneity of subtidal habitats (i.e., barrens versus kelp forest; Andrew 1993).

It is entirely possible that the extensive barrens of Eastern Australia, which are notably absent in the other regions (fig. 15.7), are indirectly maintained by overfishing of urchin predators (e.g., labrids *Achoerodus viridus* and sparids *Pagrus auratus*, in particular). Loss of predators may have enabled urchin populations to reach extraordinarily high densities (notably *Centrostephanus rodgersii*; Fowler-Walker and Connell 2002, and sometimes *Heliocidaris erythrogramma*; Wright et al. 2005, particularly in localities where kelp forests are surprisingly difficult to find (fig. 15.7). Ecosystem-wide effects of reduced fishing are more likely in systems of top-down control with intense grazing pressure. Comparison of marine protected areas (large units of reduced fishing pressure) provides some insights into the strength of top-down influences, albeit they are seldom replicated in ways that provide unequivocal interpretations. A good example is the contrasting outcomes of marine protected areas in New Zealand (Leigh, established in 1978) and South Australia (West Island, established in 1975). Cessation of fish-

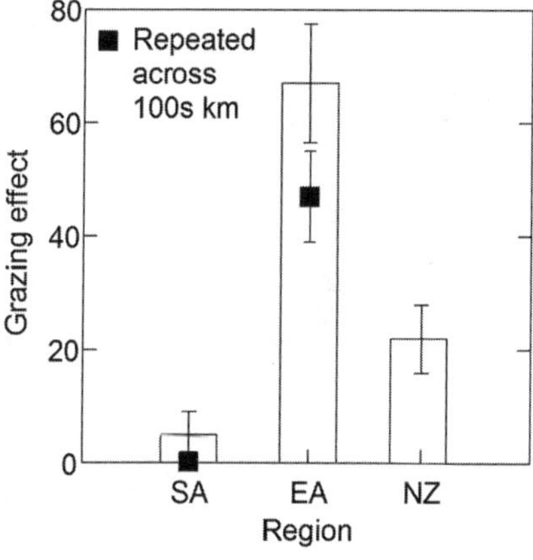

FIGURE 15.6 The efficacy at which herbivores (benthic scrapers) maintain space free from overgrowth by erect algae in SA (South Australia), EA (Eastern Australia) and NZ (New Zealand). These data represent the percentage difference in covers of erect algae between treatments that exclude grazers or allow them to forage freely. The black square represents the single study that repeated these experiments at several sites within each of several locations across hundreds of kilometers within each region (Gorman, Hart, and Connell, unpublished data). Otherwise, Eastern Australia is represented by six studies within the Sydney locality, South Australia by two studies at West Island, and New Zealand by one study in the North Island and one in the South Island.

ing at Leigh and the concomitant increase in predators, decline in grazers, and switch from barrens to kelp forest has been observed in other marine protected areas in northern New Zealand (Shears and Babcock 2002). Yet, cessation of fishing at West Island, while having positive effects on grazers directly targeted by fishing, has not caused changes to habitat that distinguishes this site from comparable non-MPA sites of South Australia. This lack of change occurs because the benthic grazers (abalone, urchins) primarily feed on large amounts of drifting algae instead of attached algae. Experimentally maintained populations of grazers at some of the highest densities observed across South Australia have barely detectable effects in South Australia (fig. 15.6). Hence, it is not surprising that it is not possible to detect top-down effects (removing grazers) in South Australia while observing large and consistent top-down effects among identical manipulations in Eastern Australia (fig. 15.6).

Bottom-Up Forcing

Ideas about bottom-up forcing in marine systems typically consider variation in the nutrient concentration of the surrounding water column as a

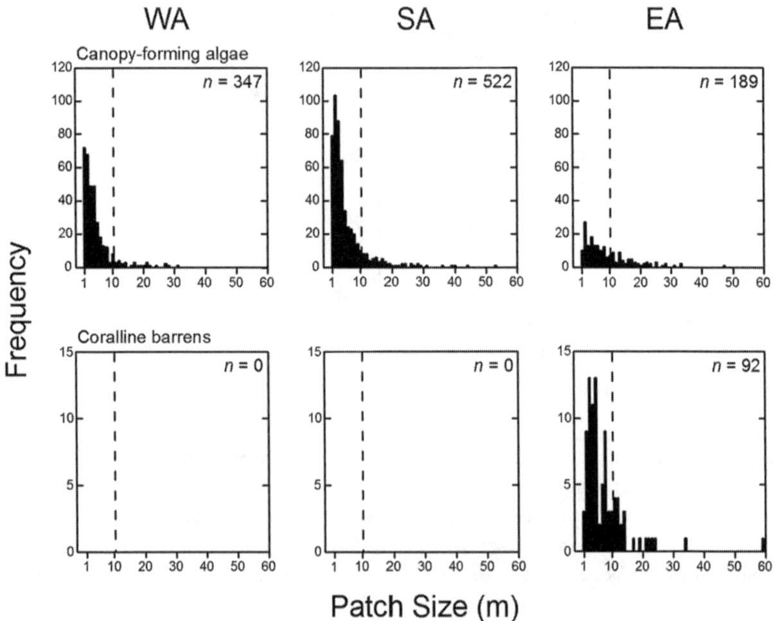

FIGURE 15.7 Frequency distributions of the size of individual patches of habitat created by canopy-forming algae and coralline barrens within each region of temperate Australia (WA = Western Australia, SA = South Australia, EA = Eastern Australia). Broken lines indicate the 10 m size class. Reproduced with permission from Connell and Irving (2008).

key factor shaping patterns in the distribution and abundance of benthic life (Menge 2000; Underwood 2000). Importantly, sources of nutrients can be local (e.g., sewage outfall: Littler and Murray 1975), and may also occur over much larger scales (e.g., coastal upwelling: Gibbs 2000). Extensive covers of large and slow-growing macroalgae such as kelp (and their associated biodiversity: Steneck et al. 2002; Graham 2004) occur across a range of ambient nutrient concentrations at temperate latitudes, yet their replacement by small, opportunistic, and fast-growing species of "turf-forming" algae appears associated with coasts of poor quality (e.g., urbanized localities; Worm et al. 1999; Benedetti-Cecchi et al. 2001; Eriksson, Johansson, and Snoeijs 2002). The top-down influence of grazers can control blooms of fast-growing algae, but localities of weak grazing pressure appear to be primarily nutrient controlled (Worm et al. 2000). Critically, local scale experiments that enrich nutrient concentrations have highlighted strong effects consistent with the expansion of turf-forming algae on human dominated coasts (Worm et al. 1999; Gorgula and Connell 2004).

The abundance of turf-forming algae can vary enormously at local scales (e.g., tens of m to 1 to 2 km: Lavery and Vanderklift 2002), yet broad-scale patterns do exist. Across temperate Australia, the cover of filamentous turfs growing as epiphytes on the laterals of *E. radiata* is greater in Eastern Australia than Western or South Australia, but also varies considerably among locations separated by hundreds of km within Eastern Australia (fig. 15.8, panel A; Russell et al. 2005). Such variation in the cover of epiphytic turfs correlates positively with variation in the concentration of chlorophyll a (a proxy for nutrient concentration; fig. 15.8, panel A), suggesting a causative link between the ambient concentration of nutrients and the cover of epiphytic turfs over scales of thousands of kilometers. A recent experimental test of this model demonstrated that nutrient enrichment increased the cover of epiphytic turfs in both Eastern and South Australia, but that the magnitude of such increases was disproportionately greater in South Australia (considered to be an oligotrophic coast) relative to Eastern Australia

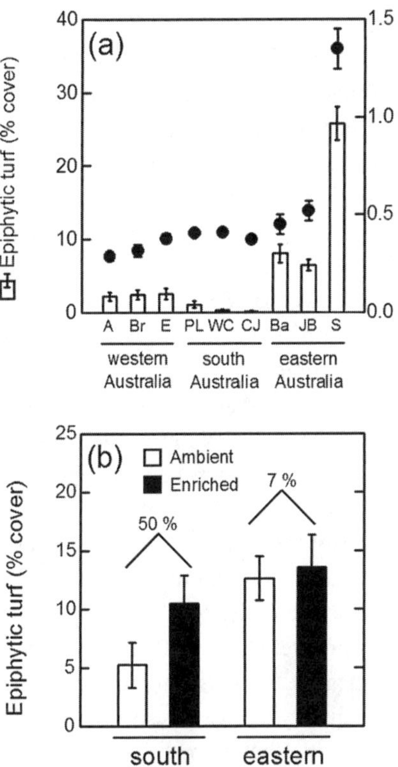

FIGURE 15.8 (A) Bar graph of the percentage cover (mean ± SE) of epiphytic turf-forming algae growing on laterals of *Ecklonia radiata* across temperate Australia plotted in relation to concentrations of chlorophyll *a* (solid circle). All data are plotted by locations within regions (A = Albany, Br = Bremer Bay, E = Esperance, PL = Port Lincoln, WC = West Cape, CJ = Cape Jervis, Ba = Batemans Bay, JB = Jervis Bay, S = Sydney; after Russell et al. 2005). (B) Percentage increase (mean ± SE) of epiphytic turf-forming algae growing on laterals of *Ecklonia radiata* after experimental nutrient enrichment across three sites, nested within each of three locations in each region (Eastern and South Australia). The figures above bars represent the magnitude of difference caused by nutrients in each region (after Connell and Elsdon, unpublished data).

(fig. 15.8, panel B; Connell and Elsdon unpublished data). Disproportionate responses of algae to nutrient enrichment are well known (Duarte 1995), and appear useful to developing theory about the efficiency of nutrient use among nutrient-rich versus nutrient-poor environments (Chapin 1980; Vitousek 1982).

Bottom-up effects caused by factors other than nutrients, such as prey abundance, have also been demonstrated over large distances (e.g., Menge et al. 2002, Witman et al. 2003). Here, variation in the strength of top-down effects (predation) that is associated with variation in bottom-up processes (prey supply) raises the interesting prospect of bottom-up regulation of top-down processes (Witman et al. 2003).

Macroecology and the Management of Coastal Resources

The identification of regional differences in strength of top-down and bottom-up control has profoundly affected the way that scientists and managers view Australia's temperate coast. The regional context of research and policy is starting to be understood, particularly by managers of coasts for which traditional models of trophically structured habitats (Steneck et al. 2002) appear to have poor applicability and alternate models have been overlooked (Connell 2007b). For example, recognition that oligotrophic coasts maybe more susceptible to land-based activities (e.g., release of nutrients from urban catchments), may explain the puzzling anomaly of why previous benthic studies on relatively less oligotrophic coasts de-emphasize the role of urban discharge around Australia's largest city (Sydney: Chapman et al. 1995) relative to those identified around a substantially smaller city (Adelaide: Gorgula and Connell 2004; associated with up to 70 percent loss of kelp forests spanning >30 years, Connell et al. 2008). It appears that such subsidies can bring disproportionately greater ecological change where the disparity in resource availability between donor and recipient systems is greatest (Marczak, Thompson, and Richardson 2007), hence the relationship between human populations and ecological impacts may be dependent on regional oceanography. Indeed, integration of MPAs with other spatial arrangements for conservation is a priority for Southern Australia (Environment Australia 2003) because MPAs placed in proximity to disturbances (e.g., urban catchments) may compromise the utility of a reserve as a biodiversity repository (Connell 2007b; Connell et al. 2008). Without knowledge of regional contexts for different benthic responses to stressors (e.g., changes to trophic structure and water quality), coastal resources may be unwittingly mismanaged because of disparities between ecology and management strategies.

Interactions between Taxonomic and Spatial Scale

Of its nature, biogeographic scale research necessarily crosses the boundaries of species distributions, and we are often confronted with the issue of how to classify organisms. One approach has been to establish morphological and functionally equivalent identities across breaks. Broader taxonomic identities transcend biogeographic breaks, whereas species identities typically do not.

A preoccupation with species identities may restrict the ability to detect patterns that form the basis of useful generalities. Perhaps our efforts to predict changes in abundance or species composition is as difficult as Brown's (1995) analogy of a physicist attempting to understand the path of individual molecules, compared with the general process of diffusion as an emergent property of "any large collection of gas molecules." The highly predictable pattern of diffusion of molecules (random diffusion) may be equivalent to patterns observed between spatial and taxonomic scales (fig. 15.9). That is, across broad taxonomic scales, broad patterns can be observed with encouraging consistency. A focus on broader taxonomic groups may provide a more tractable basis to understand ecological behaviors from local through regional scales in the way theories of diffusion as a general process (random diffusion) has general properties that explain highly predictable patterns in physics. Hence, understanding the relationship between taxonomic scales and spatial scales is a promising area of research.

Phylogenetic Relationships

Initially, researchers tested whether identifying taxa into broad phylogenetic groups provided similar patterns using less resources (Chapman 1998), but perhaps now could be used to assesses scale dependency in taxonomy. Multivariate analyses of fauna of kelp holdfasts within a biogeographic region of New Zealand (i.e., not crossing well-understood biogeographic boundaries) indicated that variability at larger scales (localities separated by hundreds of kilometers) becomes less important (and local scales more important) as taxonomic scales are broadened (i.e., from species to phyla; fig. 15.9). Variation in the composition of species behaves in the opposite way. Together, these observations suggest that variation in species composition at a particular location is driven by large-scale processes, while variation in relative abundance at local scales is driven by smaller-scale processes (Anderson et al. 2005). Lack of variation in proportional abundances of broad taxonomic groups at broad scales suggests that some consistency in pattern may emerge at larger scales, even in the presence of considerable small-scale variation.

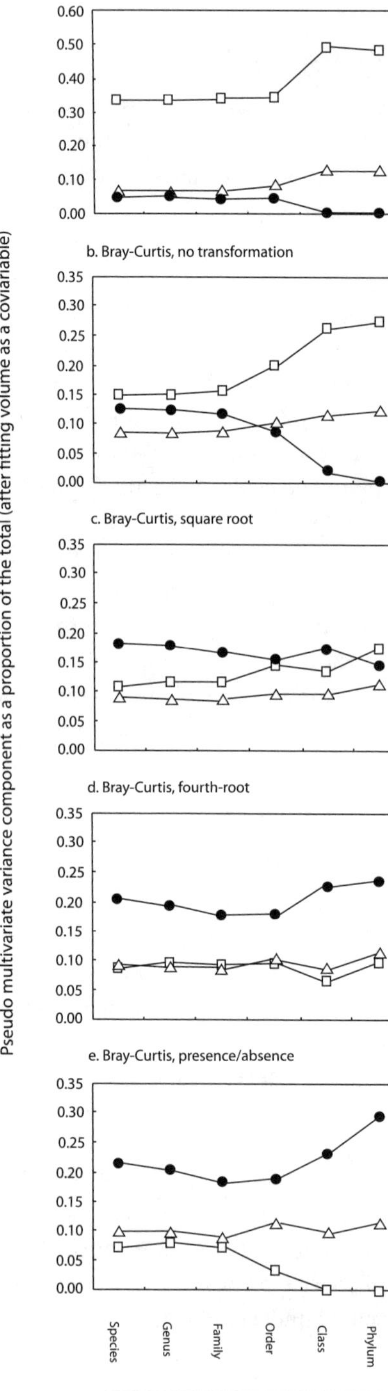

FIGURE 15.9 Proportion of variability accounted by multivariate variance components at scales of meters (area), tens of kilometers (sites) and hundreds of kilometers (location) within a region. Reproduced with permission from Anderson et al. (2005).

Variation in species composition (and abundance) is often greatest at local scales and decreases with increasing spatial scale (Anderson et al. 2005) until our observations cross "biogeographic breaks," which, by definition, means that a disproportionately large number of species identities and abundances change abruptly. Hence, we can anticipate variation in abundances of species to be greatest among regions, but decreases as taxonomic scales increase (species to order). The relative consistency of patterns in biodiversity across broad spatial and taxonomic scales is encouraging. It suggests that regional-scale patterns may emerge from local complexity and provide predictive capacity to new situations and localities with different taxonomic composition.

Ecological Relationships

Our understanding of generalities may have been hindered by our preoccupation with species and phylogenetic relationships rather than the use of ecological attributes, such as trophic levels, life history, and disturbance resistance (i.e., functional form and group hypotheses; Padilla and Allen 2000). In many cases, analyses of coarser phylogenetic units (e.g., order) may reveal less about patterns than ecological traits. Species interactions have high context dependency, because knowledge of their outcomes is specific to particular places and species composition. When knowledge is based on species names, and its local species pool, scaling-up of this knowledge becomes an exercise in increasing uncertainty.

Ecological groups based on the way herbivorous fish feed (scraping herbivores) are key to predictions about the resilience of coral reefs (Bellwood et al. 2004) and reconciling local-regional similarities/dissimilarities in habitat heterogeneity across temperate Australia (see top-down section, discussed earlier). There is certainly a need to assess the extent to which ecological groupings represent their component species (Phillips, Kendrick, and Lavery 1997), but such efforts maybe usefully extended to test hypotheses about which subsets of species share particular traits and the circumstances in which they have generality and predictive value. The plastic morphology of many algae (i.e., a single species can adopt forms of different morphological groups) signifies that phylogeny and morphology may seldom match, but such mismatches may be more concerning to those interested in species-species interactions and population biology. For those interested in broader-level issues (ecosystems and macroecology), understanding which traits (and which subset of species) occur within a particular set of environmental conditions would be invaluable. Such information would not only enable more insightful tests of general patterns, but also promises a predictive capacity to new situations and localities with different phylogenetic composition (Keddy 1990). Hence, increasing the breadth of our taxonomic identities, while understanding the complexities this causes, is a fertile area of research that is needed to assist us increase the scale of our observations and test underlying ecological concepts.

Looking Toward the Future

Some of the most celebrated scientific achievements have come about by the development of a whole theory to solve a problem, rather than by a piecemeal process of studying bits and pieces that might be assembled into a coherent theory (Kuhn 1996). Indeed, it is becoming increasingly clear that the accumulation of local details provides a difficult basis for the interpretation of general or broad-scale patterns. No science has succeeded in understanding the structure and dynamics of a complex system by reducing the problem to the study of its parts alone (Brown 1995) and relating phenomena across scales is central to the science of ecology (Levin 1992).

Nature, being variable, allows endless possibilities for an infinite study of singular observations. Over thirty years ago ecology was considered to suffer "from a surfeit of fascinating but unrelated observations, superimposed upon an acute shortage of general theories" (Lawton 1974). Today, ecologists remain concerned about the accumulating observations that lack theoretical basis and with little resolve for advances in prediction or conceptual understanding (Underwood, Chapman, and Connell 2000). Indeed, criticism of contemporary ecology centers on the lack of predictive power of ecological theories, resulting largely from the lack of pursuit of spatial generality of locally observed patterns and responses (Peters 1991; Keddy 2001). Macroecological research provides the opportunity to progress our understanding beyond high context dependency of local scales and have a quantitative understanding of the magnitude of predictability of local phenomena across spatial scales. Such tests have the capacity to highlight new patterns and provide fresh perspectives on well-understood local ones (Connell and Irving 2008). If progressive avenues of ecological inquiry are to be judged by their predictive capacity (Peters 1991), then studies of macroecological patterns may be one of the more useful tools in the provision of general and predictive understanding of ecological phenomena.

Incorporation of Intensively Studied Sites in Macroecology

That broad-scale patterns can emerge from local variation is an encouraging feature of macroecology (e.g., Chesson 1996; Fowler-Walker, Connell, and Gillanders 2005). There can be little doubt that working at more localities produces a better understanding of generalities, rather than intensively focusing on a few localities. Disproportionately large variation at local scales not only enables the endless opportunity to discover new details about processes and their unique combinations and interactions, but also permits the development of theory at a single locality that lacks relevance to broader patterns within the same system. Hence, the intense study of a few localities may well have contributed to uncertainty about the generality of ecological theory. This uncertainty may be redressed by explicitly incorporating some of the world's best-understood localities into broader-scale studies that link patterns and processes. In this way, the importance and generality of theories developed at single localities can be judged within their own biological system.

The successful use of manipulative experiments to link pattern and process at local scales is also possible in macroecology (i.e., matching natural and experimental effect sizes; Weldon and Slauson 1986). Preliminary attempts to link broad patterns (across 5,000 km) began with an experiment at a single location (Russell et al. 2005) before providing the confidence needed to repeat these tests within a successive hierarchy of scales spanning the regions of interest (fig. 15.8, panel B, Connell and Elsdon unpublished data). While the description of pattern is fundamental to ecology (Underwood, Chapman, and Connell 2000), there remains a great need for ecologists to link quantitative assessments of patterns with experimental tests of their cause.

We advocate greater discussion about the methods needed for tests of scale dependency. While alternate approaches exist, and provide for spirited discussion in general ecology (e.g., Cottingham, Lennon, and Brown 2005 and replies), hierarchical approaches in macroecology have demonstrable advantages in their capacity to compare variation in patterns and responses across local to regional scales. Manipulative experiments on macroecological scales present a daunting task, least of which because many patterns are beyond experimentation or cost. Nevertheless, repeated local scale manipulations over the scales appropriate to the original pattern may be possible, and facilitate considerable progress in understanding the regional context of local studies and the global context of regional studies.

Replication of Experiments

Over the last thirty years, ecology has swung from a descriptive science to embrace experimentation. Some adherents to experimental approaches chastise descriptive ecology for being poor on rigor and have found rigor by adapting practices that embrace certainty—but this has come at the cost of understanding generality. For example, an experiment done at one particular place and time has a greater capacity of delivering an unambiguous outcome, than replicating it at another location(s) in which the result is not repeated. Indeed, the extra effort required to repeat experiments at multiple sites, associated with uncertain outcomes, is often not an attractive proposition (see Witman and Roy, this volume). Arguably, however, these outcomes are not rigorous because the high-context dependency of local studies reduces their repeatability.

A knee-jerk reaction would be to demand experiments to be replicated among sites. We do not advocate experiments to be mindlessly replicated for the sake of replication. Tests of generality require strategic thought in their design (e.g., choice of appropriate factors, placement and level of replication) judged within a specific ecological context and desire to test generality. It is often difficult to interpret generality from experiments that are repeated at a select few sites, regardless of whether this is done at local or regional scales. Future challenges, therefore, may be less concerned about whether we should repeat experiments, and maybe more concerned with the integration of knowledge from an intensively studied locality with broader-scale phenomenon.

Conclusions

The problem confronting ecologists is not whether one should test for the existence of general or specific phenomena, but what balance should be sought between the two and the magnitudes of uncertainty associated with favoring one aspect over the other. If we recognize that broad-scale patterns may be overly simplistic in the understanding of local patterns (but provide a regional context for their interpretation) and that local experiments provide uncertain outcomes for understanding generality (but provide unambiguous information at local scales), then we are in a position to use both techniques to our advantage (fig. 15.1; also see fig. 13.2 in Witman and Roy, this volume). Application of both modes across multiple scales provides a way in which this uncertainty is winnowed. Such knowledge is not only key to generality, but also has relevance to society, particularly given the range of scales over which humans modify marine systems (e.g., fishing, pollution, climate), and the coarse measures and spatial scales that governments and their agencies can effectively manage natural systems.

In conclusion, we are becoming increasingly aware that ecologists work on local patterns that are likely to represent the outcome of special and unique events that incorporate variation from broad to local scales. For those interested in the discovery of detail, local patterns appear to have an infinite supply. For those interested in the existence of generalities, it is encouraging to observe that general patterns and responses can emerge from complexity at local scales. This realization, together with the need for a renewed effort for carefully planned sampling and experimentation across broad scales, suggests that there are opportunities to test some of the more interesting questions about the relative importance of processes across vast parts of the world's coast. This future may parallel the pioneering efforts of early nineteenth-century biogeographers who discovered general patterns across our planet, and in doing so left an indelible mark on scientific thinking (e.g., Darwin 1859). More than 150 years later, but with considerable experimental, statistical, and technological advantages, ecologists appear set to make pioneering discoveries at scales that can shape and refine our empirical and theoretical understanding of nature and her future.

ACKNOWLEDGMENTS

This book demonstrates that we have colleagues in marine macroecology, a benefit that was not available when we started our continental scale research. In a positive way, we acknowledge the dictum that ecology is best based on species-level classifications, and is best progressed by the intense study of a few choice localities. While this has an element of truth, our macroecological research came as a reaction to those who enforced this view on our work. We thank those who shared our rebellion and traveled the vast fringe of the great southern land; spending days under sea and nights under stars. For now—we break free. This work was supported by Australian Research Council grants to SDC and we thank those who must have also shared this vision by making this funding opportunity possible.

REFERENCES

Anderson, M. J., S. D. Connell, B. M. Gillanders, C. E. Diebel, W. M. Blom, J. E. Saunders, and T. J. Landers. 2005. Relationships between taxonomic resolution and spatial scales of multivariate variation. *Journal of Animal Ecology* 74:636–46.

Andrew, N. L. 1993. Spatial heterogeneity, sea urchin grazing, and habitat structure on reefs in temperate Australia. *Ecology* 74:292–302.

Andrew, N. L., and A. L. O'Neill. 2000. Large-scale patterns in habitat structure on subtidal rocky reefs in New South Wales. *Marine and Freshwater Research* 51:255–63.

Andrew, N. L., and A. J. Underwood. 1989. Patterns of abundance of the sea urchin *Centrostephanus rodgersii* (Agassiv) on the central coast of New South Wales, Australia. *Journal of Experimental Marine Biology & Ecology* 131:61–80.

———. 1993. Density-dependent foraging in the sea urchin *Centrostephanus rodgersii* on shallow subtidal reefs in New South Wales, Australia. *Marine Ecology Progress Series* 99:89–98.

Bellwood, D. R., T. P. Hughes, C. Folke, and M. Nystrom. 2004. Confronting the coral reef crisis. *Nature* 429:827–33.

Benedetti-Cecchi, L. 2003. The importance of the variance around the mean effect size of ecological processes. *Ecology* 84:2335–46.

Benedetti-Cecchi, L., F. Pannacciulli, F. Bulleri, P. S. Moschella, L. Airoldi, G. Relini, and F. Cinelli. 2001. Predicting the consequences of anthropogenic disturbance: Large-scale effects of loss of canopy algae on rocky shores. *Marine Ecology Progress Series* 214:137–50.

Brown, J. H. 1995. *Macroecology*. Chicago: The University of Chicago Press.

Chapin, F. S. I. 1980. The mineral nutrition of wild plants. *Annual Review of Ecology and Systematics* 11:233–60.

Chapman, M. G. 1998. Relationships between spatial patterns of benthic assemblages in a mangrove forest using different levels of taxonomic resolution. *Marine Ecology Progress Series* 162:71–78.

Chapman, M. G., A. J. Underwood, and G. A. Skilleter. 1995. Variability at different spatial scales between a subtidal assemblage exposed to the discharge of sewage and 2 control assemblages. *Journal of Experimental Marine Biology and Ecology* 189:103–22.

Chesson, P. 1996. Matters of scale in the dynamics of populations and communities. In *Frontiers of population ecology*, ed. R. B. Floyd, A. W. Sheppard, and P. J. De Barro, 353–68. Melbourne: CSIRO.

Connell, S. D. 2003. Negative effects overpower the positive of kelp to exclude invertebrates from the understorey community. *Oecologia* 137:97–103.

———. 2007a. Subtidal temperate rocky habitats: Habitat heterogeneity at local to continental scales. In *Marine ecology*, ed. S. D. Connell and B. M. Gillanders, 378–401. Melbourne: Oxford University Press.

———. 2007b. Water quality and the loss of coral reefs and kelp forests: Alternative states and the influence of fishing. In *Marine ecology*, ed. S. D. Connell and B. M. Gillanders, 556–68. Melbourne: Oxford University Press.

Connell, S. D., and A. D. Irving. 2008. Integrating ecology with biogeography using landscape characteristics: A case study of subtidal habitat across continental Australia. *Journal of Biogeography* 35:1608–21.

Connell, S. D., B. C. Russell, D. J. Turner, S. A. Shepherd, T. Kildae, D. J. Miller, L. Airoldi, and A. Cheshire. 2008. Recovering a lost baseline: Missing kelp forests from a metropolitan coast. *Marine Ecology Progress Series* 360:63–72.

Cottingham, K. L., J. T. Lennon, and B. L. Brown. 2005. Regression versus ANOVA. *Frontiers in Ecology and the Environment* 3:356–58.

Darwin, C. 1859. *On the origin of species*. London: Murray.

Daume, S., S. Brand-Gardner, and W. J. Woelkerling. 1999. Community structure of nongeniculate coralline red algae (Corallinales, Rhodophyta) in three boulder habitats in southern Australia. *Phycologia* 38:138–48.

Duarte, C. M. 1995. Submerged aquatic vegetation in relation to different nutrient regimes. *Ophelia* 41:87–112.

Environment Australia. 2003. Australia's south-east marine region: A user's guide to identifying candidate areas for a regional representative system of marine protected areas. Commonwealth of Australia.

Eriksson, B. K., G. Johansson, and P. Snoeijs. 2002. Long-term changes in the macroalgal vegetation of the inner Gullmar fjord, Swedish Skagerrak coast. *Journal of Phycology* 38:284–96.

Fowler-Walker, M. J., and S. D. Connell. 2002. Opposing states of subtidal habitat across temperate Australia: Consistency and predictability in kelp canopy-understorey associations. *Marine Ecology Progress Series* 240:49–56.

Fowler-Walker, M. J., S. D. Connell, and B. M. Gillanders. 2005. Variation at local scales need not impede tests for broader scale patterns. *Marine Biology* 147:823–31.

Fowler-Walker, M. J., B. M. Gillanders, S. D. Connell, and A. D. Irving. 2005. Patterns of association between canopy-morphology and understorey assemblages across temperate Australia. *Estuarine Coastal and Shelf Science* 63:133–41.

Fraschetti, S., A. Terlizzi, and L. Benedetti-Cecchi. 2005. Patterns of distribution of marine assemblages from rocky shores: Evidence of relevant scales of variation. *Marine Ecology Progress Series* 296:13–29.

Gaston, K. G., and T. M. Blackburn. 2000. *Pattern and process in macroecology*. Cambridge: Blackwell Science.

Gibbs, M. T. 2000. Elevated chlorophyll a concentrations associated with a transient shelfbreak front in a western boundary current at Sydney, south-eastern Australia. *Marine and Freshwater Research* 51:733–37.

Goodsell, P. J., M. J. Fowler-Walker, B. M. Gillanders, and S. D. Connell. 2004. Variations in the configuration of algae in subtidal forests: Implications for invertebrate assemblages. *Australian Ecology* 29:350–57.

Gorgula, S. K., and S. D. Connell. 2004. Expansive covers of turf-forming algae on human-dominated coast: The relative effects of increasing nutrient and sediment loads. *Marine Biology* 145:613–19.

Graham, M. H. 2004. Effects of local deforestation on the diversity and structure of southern California giant kelp forest food webs. *Ecosystems* 7:341–57.

Green, R. H. 1979. *Sampling design and statistical methods for environmental biologists*. New York: Wiley.

Hill, N. A., C. Blount, A. G. B. Poore, D. Worthington, and P. D. Steinberg. 2003. Grazing effects of the sea urchin *Centrostephanus rodgersii* in two contrasting rocky reef habitats: Effects of urchin density and its implications for the fishery. *Marine and Freshwater Research* 54:691–700.

Horne, J. K., and D. C. Schneider. 1995. Spatial variance in ecology. *Oikos* 74:18–26.

Irving, A. D., and S. D. Connell. 2006. Physical disturbance by kelp abrades erect algae from the understorey. Marine Ecology Progress Series 324:127–37.

Irving, A. D., S. D. Connell, and B. M. Gillanders. 2004. Local complexity in patterns of canopy-benthos associations produce regional patterns across temperate Australasia. *Marine Biology* 144:361–68.

Keddy, P. A. 1990. The use of functional as opposed to phylogenetic systematics: A first step in predictive community ecology. In *Biological approaches and evolutionary trends in plant*, ed. S. Kawano, 387–406. London: Academic Press.

———. 2001. Extending the generality of field experiments. In *Competition*, 2nd ed., vol. 26, 317–32. Dordrecht: Kluwer.

Kendrick, G. A., P. S. Lavery, and J. C. Phillips. 1999. Influence of *Ecklonia radiata* kelp canopy on structure of macro-algal assemblages in Marmion Lagoon, Western Australia. *Hydrobiologia* 398/399:275–83.

Kennelly, S. J. 1987. Physical disturbances in an Australian kelp community. II. Effects on understorey species due to differences in kelp cover. *Marine Ecology Progress Series* 40:155–65.

———. 1989. Effects of kelp canopies on understorey species due to shade and scour. *Marine Ecology Progress Series* 50:215–24.

Kennelly, S. J., and A. J. Underwood. 1992. Fluctuations in the distributions and abundances of species in sublittoral kelp forests in New South Wales. *Australian Journal of Ecology* 17:367–82.

———. 1993. Geographic consistencies of effects of experimental physical disturbance on understorey species in sublittoral kelp forests in central New South Wales. *Journal of Experimental Marine Biology & Ecology* 168:35–58.

Kuhn, T. S. 1996. *The structure of scientific revolutions*, 3rd ed. Chicago: The University of Chicago Press.

Lavery, P. S., and M. A. Vanderklift. 2002. A comparison of spatial and temporal patterns in epiphytic macroalgal assemblages of the seagrasses *Amphibolis griffithii* and *Posidonia coriacea*. *Marine Ecology Progress Series* 236:99–112.

Lawton, J. H. 1974. Review of J. M. Smith, Models in ecology. *Nature* 248:537.

Lenihan, H. S., C. H. Peterson, J. E. Byers, H. Grabowski, G. W. Thayer, and D. R. Colby. 2001. Cascading of habitat degradation: Oyster reefs invaded by refugee fishes escaping stress. *Ecological Applications* 11:764–82.

Levin, S. A. 1992. The problem of pattern and scale in ecology. *Ecology* 73:1943–67.

Littler, M. M., and S. N. Murray. 1975. Impact of sewage on the distribution, abundance and community structure of rocky intertidal macro-organisms. *Marine Biology* 30:277–91.

Marczak, L. B., R. M. Thompson, and J. S. Richardson. 2007. Meta-analysis: Trophic level, habitat, and productivity shape the food web effects of resource subsidies. *Ecology* 88:140–48.

McIntosh, R. P. 1980. The background and some current problems of theoretical ecology. In *Conceptual issues in ecology*, ed. E. Saarinen, 1–61. Holland: D. Reidel.

Melville, A. J., and S. D. Connell. 2001. Experimental effects of kelp canopies on subtidal coralline algae. *Australian Journal of Ecology* 26:102–8.

Menge, B. A. 2000. Top-down and bottom-up community regulation in marine rocky intertidal habitats. *Journal of Experimental Marine Biology & Ecology* 250:257–89.

Menge, B. A., E. Sanford, B. A. Daley, T. L. Freidenburg, G. Hudson, and J. Lubchenco. 2002. Inter-hemispheric comparison of bottom-up effects on community structure: Insights revealed using the comparative-experimental approach. *Ecological Research* 17:1–16.

Noda, T. 2004. Spatial hierarchical approach in community ecology: A way beyond high context-dependency and low predictability in local phenomena. *Population Ecology* 46:105–17.

Padilla, D. K., and B. J. Allen. 2000. Paradigm lost: Reconsidering functional form and group hypotheses in marine ecology. *Journal of Experimental Marine Biology & Ecology* 250:207–21.

Peters, R. H. 1991. *A critique for ecology*. New York: Cambridge University Press.

Phillips, J. C., G. A. Kendrick, and P. S. Lavery. 1997. A test of a functional group approach to detecting shifts in macroalgal communities along a disturbance gradient. *Marine Ecology Progress Series* 153:125–38.

Ricklefs, R. E. 2004. A comprehensive framework for global patterns in biodiversity. *Ecology Letters* 7:1–15.

Russell, B. D., T. S. Elsdon, B. M. Gillanders, and S. D. Connell. 2005. Nutrients increase epiphyte loads: Broad-scale observations and an experimental assessment. *Marine Biology* 147:551–58.

Schiel, D. R. 1990. Macroalgal assemblages in New Zealand: Structure, interactions and demography. *Hydrobiologia* 192:59–76.

Schiel, D. R., N. L. Andrew, and M. S. Foster. 1995. The structure of subtidal algae and invertebrate assemblages at the Chatham Islands, New Zealand. *Marine Biology* 123:355–467.

Shears, N. T., and R. C. Babcock. 2002. Marine reserves demonstrate top-down control of community structure on temperate reefs. *Oecologia* 132:131–42.

———. 2003. Continuing trophic cascade effects after 25 years of no-take marine reserve protection. *Marine Ecology Progress Series* 246:1–16.

Steneck, R. S., M. H. Graham, B. J. Bourget, D. Corbett, J. M. Erlandson, J. A. Estes, and M. J. Tegner. 2002. Kelp forest ecosystems: Biodiversity, stability, resilience and future. *Environmental Conservation* 29:436–59.

Thrush, S. F., D. C. Schneider, P. Legendre, R. B. Whitlach, P. K. Dayton, J. E. Hewitt, A. H. Hines, et al. 1997. Scaling-up from experiments to complex ecological systems: Where to next. *Journal of Experimental Marine Biology & Ecology* 216:243–54.

Underwood, A. J. 1997. *Experiments in ecology: Their logical design and interpretation using analysis of variance.* Cambridge: Cambridge University Press.

———. (2000) Experimental ecology of rocky intertidal habitats: What are we learning? *Journal of Experimental Marine Biology & Ecology* 250:51–76.

Underwood, A. J., M. G. Chapman, and S. D. Connell. 2000. Observations in ecology: You can't make progress on processes without understanding the patterns. *Journal of Experimental Marine Biology & Ecology* 250:97–115.

Underwood, A. J., M. J. Kingsford, and N. L. Andrew. 1991. Patterns in shallow subtidal marine assemblages along the coast of New South Wales. *Australian Journal of Ecology* 16:231–49.

Vanderklift, M. A., and G. A. Kendrick. 2004. Variation in abundances of herbivorous invertebrates in temperate subtidal rocky reef habitats. *Marine & Freshwater Research* 55:93–103.

Vitousek, P. 1982. Nutrient cycling and nutrient use efficiency. *The American Naturalist* 119:553–72.

Weldon, C. W., and W. L. Slauson. 1986. The intensity of competition versus its importance: An overlooked distinction and some implications. *Quarterly Review of Biology* 61:23–43.

Wernberg, T., M. Coleman, A. Fairhead, S. Miller, and M. Thomsen. 2003. Morphology of *Ecklonia radiata* (Phaeophyta: Laminarales) along its geographic distribution in south-western Australia and Australasia. *Marine Biology* 143:47–55.

Wiens, J. A. 1989. Spatial scaling in ecology. *Functional Ecology* 3:385–97.

Witman, J. D., S. J. Genovese, J. F. Bruno, J. W. McLaughlin, and B. I. Pavlin. 2003. Massive prey recruitment and the control of rocky subtidal communities on large spatial scales. *Ecological Monographs* 73:441–62.

Witman, J. D., and K. P. Sebens. 1992. Regional variation in fish predation intensity: A historical perspective in the Gulf of Maine. *Oecologia* 90:305–15.

Worm, B., H. K. Lotze, C. Bostrom, R. Engkvist, V. Labanauskas, and U. Sommer. 1999. Marine diversity shift linked to interactions among grazers nutrients and propagule banks. *Marine Ecology Progress Series* 185:309–14.

Worm, B., H. K. Lotze, and U. Sommer. 2000. Coastal food web structure, carbon storage, and nitrogen retention regulated by consumer pressure and nutrient loading. *Limnology & Oceanography* 45:339–49.

Wright, J. T., S. A. Dworjanyn, C. N. Rogers, P. D. Steinberg, J. E. Williamson, and A. G. B. Poore. 2005. Density-dependent sea urchin grazing: differential removal of species, changes in community composition and alternative community states. *Marine Ecology Progress Series* 298:143–56.

CONTRIBUTORS

MARK D. BERTNESS
Department of Ecology and Evolutionary Biology
Brown University
Providence, RI 02912

JOHN J. BOLTON
Botany Department
University of Cape Town
South Africa

ANDREW CLARKE
British Antarctic Survey
Cambridge CB3 0ET
United Kingdom

SEAN D. CONNELL
Southern Seas Ecology Laboratories, School of Earth & Environmental Sciences
University of Adelaide
South Australia 5005
Australia

SEAN R. CONNOLLY
School of Marine and Tropical Biology
James Cook University
Townsville, Queensland 4811
Australia

GINNY ECKERT
Department of Biology
University of Alaska
Juneau, AK 99801

RON J. ETTER
Biology Department
University of Massachusetts
Boston, MA 02125

STEVEN D. GAINES
Marine Science Institute
University of California Santa Barbara
Santa Barbara, CA 93106

BRIAN GAYLORD
Bodega Marine Lab and Section of Evolution & Ecology
University of California, Davis
Bodega Bay, CA 94923

PHILIP A. HASTINGS
Scripps Institution of Oceanography
University of California San Diego
La Jolla, CA 92093-0208

ANDREW D. IRVING
SARDI Aquatic Sciences
PO Box 120
Henley Beach, SA 5022
Australia

BRIAN P. KINLAN
Marine Science Institute
University of California Santa Barbara
Santa Barbara, CA 93106

JAMES J. LEICHTER
Scripps Institution of Oceanography
University of California San Diego
La Jolla, CA 92093-0208

SARAH E. LESTER
Department of Ecology, Evolution, and Marine Biology
University of California Santa Barbara
Santa Barbara, CA 93106

WILLIAM LI
Department of Fisheries and Oceans
Ecosystem Research Divison
Bedford Institute of Oceanography
Dartmouth, Nova Scotia
Canada B2Y 4A2

ENRIQUE MACPHERSON
Centro de Estudios Avanzados de Blanes (CSIC)
C. acc. Cala Sant Francesco
Girona, Spain

CRAIG R. MCCLAIN
National Evolutionary Synthesis Center
2024 W. Main St., Suite A200
Box 104403
Durham, NC 27705

ISABEL MENESES
CONICYT
Bernarda Morín 551, 3er piso, Providencia
Santiago, Chile

RANSOM A. MYERS (DECEASED)
Department of Biology
Dalhousie University
Halifax, Nova Scotia B3H 4J1
Canada

SEAN NEE
Institute of Evolutionary Biology
School of Biological Sciences
Edinburgh EH9 3JT
UK

MICHAEL A. REX
Department of Biology
University of Massachusetts-Boston
Boston, MA 02125

D. ROSS ROBERTSON
Smithsonian Tropical Research Institute
Balboa, Republic of Panamá

KAUSTUV ROY
Section of Ecology, Behavior & Evolution
Division of Biological Sciences
University of California, San Diego
La Jolla, CA 92093-0116

RAPHAEL SAGARIN
Ocean and Coastal Policy
Nicholas Institute for Environmental Policy Solutions
Duke University
Durham, NC 27708

ERIC SANFORD
Bodega Marine Lab and Section of Evolution & Ecology
University of California, Davis
Bodega Bay, CA 94923

BERNABÉ SANTELICES
Departamento de Ecología
Facultad de Ciencias Biológicas
Pontificia Universidad Católica de Chile
Santiago, Chile

GRAHAM STONE
Institute of Cell, Animal and Population Biology
Ashworth Laboratories
University of Edinburgh
Edinburgh EH9 3JT
UK

DEREK P. TITTENSOR
Department of Biology
Dalhousie University
Halifax, Nova Scotia B3H 4J1
Canada

JAMES W. VALENTINE
Museum of Paleontology
University of California Berkeley
Berkeley, CA 94720

JON D. WITMAN
Department of Ecology and Evolutionary Biology
Box G-W, Brown University
Providence, RI 02912

BORIS WORM
Department of Biology
Dalhousie University
Halifax, Nova Scotia B3H 4J1
Canada

INDEX